CONTAMINATED SURFACE SOILS IN-PLACE TREATMENT TECHNIQUES

CONTAMINATED SURFACE SOILS IN-PLACE TREATMENT TECHNIQUES

by

Ronald Sims, Darwin Sorensen, Judith Sims
Joan McLean, Ramzi Mahmood, Ryan Dupont
Jerry Jurinak

Utah State University
Logan, Utah

Kathleen Wagner

JRB Associates
McLean, Virginia

np

NOYES PUBLICATIONS

Park Ridge, New Jersey, U.S.A.

Library of Congress Catalog Card Number 86-5162
ISBN: 0-8155-1085-3
ISSN: 0090-516X
Printed in the United States

Published in the United States of America by
Noyes Publications
Mill Road, Park Ridge, New Jersey 07656

10 9 8 7 6 5 4 3 2 1

Library of Congress Cataloging-in-Publication Data

Contaminated surface soils in-place treatment
techniques.

(Pollution technology review ; no. 132)
Bibliography: p.
Includes index.
1. Soil pollution. 2. Land treatment of wastewater.
I. Sims, Ronald. II. Series.
TD878.C67 1986 628.5'5 86-5162
ISBN 0-8155-1085-3

Foreword

In-place treatment technologies applicable to contaminated surface soils—less than two feet in depth—are described in this book. The technologies covered are potentially applicable for immediate removal, planned removal, and remedial action as defined in the National Contingency Plan. They could be extremely cost-effective in situations where contamination from hazardous materials has not yet reached groundwater, but might if no action were taken. Examples of problem situations include leakage from unpaved drum storage areas, spills, during truck and rail transport, spills in transfer of chemicals, and leakage from shallow lagoons or burial pits.

The techniques discussed have the potential for attenuating, extracting, immobilizing, degrading, or reducing the volatilization of organic and inorganic soil contaminants.

The book is divided into two parts. Part I discusses the selection of appropriate in-place treatment technology for a particular site and provides specific information on each technology. Data provided include descriptions, wastes amenable to treatment, status of technology, ease of application, potentially achievable level of treatment, reliability of method, secondary impacts, equipment and exogenous reagents, information requirements, and sources of information. Selection of any specific technology would involve assessment of wastes, soil, and site-specific variables. Part II provides monitoring information and data on the behavior and fate of hazardous constituents in soil systems. This additional detailed information provides appropriate background material for those who must make complex decisions concerning site, soil and waste interactions. This book is a vital communications link between the researcher and the user community.

The information in the book is from *Review of In-Place Treatment Techniques for Contaminated Surface Soils—Volume 1: Technical Evaluation. Volume 2: Background Information for In Situ Treatment,* prepared by Ronald Sims, Darwin Sorensen, Judith Sims, Joan McLean, Ramzi Mahmood, Ryan Dupont, and Jerry Jurinak of Utah State University and Kathleen Wagner of JRB Associates for the U.S. Environmental Protection Agency, September 1984 (Volume 1) and November 1984 (Volume 2).

The table of contents is organized in such a way as to serve as a subject index and provides easy access to the information contained in the book.

Advanced composition and production methods developed by Noyes Publications are employed to bring this durably bound book to you in a minimum of time. Special techniques are used to close the gap between "manuscript" and "completed book." In order to keep the price of the book to a reasonable level, it has been partially reproduced by photo-offset directly from the original reports and the cost saving passed on to the reader. Due to this method of publishing, certain portions of the book may be less legible than desired.

NOTICE

ACKNOWLEDGEMENTS

The Utah Water Research Laboratory (UWRL) prepared the initial draft of this report for EPA's Office of Research and Development under contract to JRB Associates. Ms. Naomi Barkley was the Project Officer. Dr. Ronald Sims was the principal investigator. Dr. Darwin Sorensen, Ms. Judith Sims, Ms. Joan McLean, Mr. Ramzi Mahmood, Dr. Ryan Dupont, and Dr. Jerry Jurinak contributed to the project. Special thanks are extended to Ms. Kathy Wagner of JRB Associates and Mr. Donald Sanning of the Municipal Environmental Research Laboratory, Cincinnati, OH for their direction and valuable contributions.

Arthur D. Little, Inc. prepared the final report, based on information contained in UWRL's draft. Ms. Norma Lewis was Project Officer. Mr. Jeffrey Bass, Arthur D. Little's Project Manager, Ms. Joo Hooi Ong, Dr. Joan Berkowitz, and Dr. John Ehrenfeld were principal contributors. Mr. Bruce Goodwin and Dr. Warren Lyman also contributed.

Contents and Subject Index

I. TECHNICAL EVALUATION

II. BACKGROUND INFORMATION

I. Technical Evaluation

Executive Summary

PURPOSE AND SCOPE

The purpose of this report is to provide state-of-the-art information on in-place treatment technologies for soils contaminated to a shallow depth. Such situations are commonly encountered when the source of contamination is at or near the soil surface. Examples include leakage from unpaved drum storage areas, spills in truck and rail transport, spills in transfer of chemicals, and leakage from shallow lagoons or burial pits.

The in-place technologies discussed have the potential for attenuating, extracting, immobilizing, degrading, or reducing the volatilization of both inorganic and organic soil contaminants. The particular application methods emphasized in this report are generally limited to contaminated soils that lie above the water table. With different application methods, not discussed in detail in this report, the basic technologies can be adapted to treatment of contaminants in the saturated zone.

This report has been divided into two volumes. Section 2 of Volume 1 provides a guide for selection of in-place treatment technologies. Section 3 provides a discussion of each in-place treatment technology, including process description, information requirements for application of the technology, wastes amenable to treatment, current status of the technology, ease of application, potentially achievable levels of treatment, reliability of the technology over the long term, secondary impacts, and equipment materials required to implement the technology. Section 4 discusses engineering methods for modifying the oxygen content, moisture content, nutrient content, pH and temperature of the soil to optimize the effectiveness of in-place treatment. In addition, data for estimating the costs of implementing in-place treatment are provided in Section 4 and in an appendix on cost information.

Background information for characterization and evaluation of immobilization, degradation, and volatilization processes in waste/soil systems is provided in Volume 2 of the report, available through NTIS.

SELECTION PROCESS FOR IN-PLACE TREATMENT TECHNOLOGY

The technologies discussed in this report are potentially applicable for immediate removal, planned removal, and remedial action as defined in the National Contingency Plan (NCP). They could be particularly cost-effective in situations in which contamination has not yet reached groundwater, but would be expected to do so if no action were taken. They also have the potential for controlling or preventing further migration of contaminants to groundwater. In the latter case, in-place treatment of near-surface soils might be sufficient in and of itself to achieve remedial objectives, or it might be used in conjunction with other source control and offsite remedial actions directed at containment or cleanup of already contaminated groundwater.

The procedure for selecting appropriate remedial action technologies (including in-place treatment technologies) is specified in the National Contingency Plan. The prescribed steps are as follows:

- Preliminary assessment (scoping) — Determination on the basis of available data of the possible need for remedial response.

- Remedial investigation — Gathering of sufficient information, generally through field sampling and monitoring, to determine the nature and extent of both the problem and the remedial action that may be necessary.

- Setting of objectives — Defining the goal(s) or desired end-result(s) of the remedial program. In general, objectives will include an environmental goal (e.g., reduce exposure via the direct contact, air, groundwater, and surface water pathways to some level of acceptability) and a cost effectiveness component (e.g., to achieve the environmental goal(s) using the least cost alternative that is technically feasible and reliable).

- Identification of alternatives with potential for meeting the environmental goal(s) — the identification should be as comprehensive as possible so that the most cost-effective alternative is not overlooked.

- Screening alternatives — Evaluation of the technical effectiveness, reliability, engineering practicality, costs (capital, operating, and maintenance), and environmental impacts of implementing each alternatives.

- Refinement and detailed specification of alternatives that pass screening —Detailed cost estimation, including distribution of costs over time; specification and evaluation of an engineering implementation plan; detailed assessment of how well each alternative meets the remedial objectives; and detailed analysis of any adverse environmental impacts, mitigation measures, and associated costs.

- Selection of the most cost-effective alternative — This will generally be the lowest cost alternative of those analyzed in detail that meets the objectives of the remedial program.

When the remedial investigation shows contamination of near-surface soils as contributing to risk, Table 2-1 should be consulted to identify the in-place technologies that might be appropriate. In the identification step, the primary consideration is the chemical nature of the contaminants of concern. In particular, some technologies are designed primarily to control the release of heavy metals, while others are designed primarily to control the release of organics. When both types of contaminants are of concern, the possibility of combining two or more in-place technologies to meet remedial objectives should be considered.

SPECIFIC IN-PLACE TREATMENT TECHNIQUES

There are five major categories of in-place treatment techniques. These are:

- Extraction,
- Immobilization,
- Degradation,
- Attenuation, and
- Reduction of volatilization.

Specific technologies within each category are discussed briefly below.

EXTRACTION (SOIL FLUSHING)

Extraction is the washing of contaminants from the soil with a suitable solvent such as water, or other aqueous or non-aqueous solutions. The method is potentially applicable to all types of soil contaminants, provided that a suitable non-polluting solvent can be found. Effects of the solvent on the physical, chemical, and biological properties of the soil need to be evaluated. Furthermore, a system for profusing the contaminated area with solvent and for capturing the elutriate needs to be designed. This might involve flooding the site and collecting the elutriate in a series of shallow well points or subsurface drains, or appropriate placement of recharge and discharge wells. After treatment, the elutriate might be discharged to a receiving water body or sewer, or it might be recycled through the contaminated area. This technology derives from the mining industry, where it has been used for in-place extraction of metal values from ores.

IMMOBILIZATION

Immobilization includes a wide range of in-place treatment technologies designed to reduce the rate of release of contaminants from the soil so that resultant concentrations along pathways of exposure are held within acceptable limits. The primary immobilization mechanisms are sorption, ion exchange, and precipitation. Sorption is potentially applicable to both organic and inorganic contaminants. Ion exchange is applicable to organic and inorganic cation and anion species. Precipitation is limited, for most practical purposes, to heavy metal contaminants.

Sorption of Heavy Metals

Heavy metal contaminants in soil may be immobilized by mixing the contaminated soil with a good adsorbent, or mixture of adsorbents, for the metals present, or with a metal complexing agent, such as tetren, that binds the metal into a complex that strongly adsorbs to soil. The adsorbent, or complexing agent, may be mixed into the soil with conventional power implements or tillers. Suitable adsorbents for heavy metals include various agricultural products and by-products, such as straw, sawdust, peanut hulls, bark, and activated carbon. For maximum adsorption efficiency, the pH of the soil generally needs to be above 6.5. To maintain a high pH, lime can be mixed into the soil with the adsorbent. The amount of adsorbent added must be sufficient to tie up the releasable heavy metals, and periodic reliming may be necessary to maintain the treated system at a pH above 6.5.

Tetren-metal complexes are strongly adsorbed to soil clays. For contaminated soils rich in clays, mixing of tetren (and lime, if the soil is acidic) into the contaminated area may be sufficient to achieve immobilization. If the contaminated soils are of low clay content, a combination of tetren, clay materials, and lime (if necessary) may be used. Tetren generally forms 1:1 complexes with metals and, hence, should be added in amounts slightly in excess of the number of moles of heavy metal contaminants present.

Sorption of Organics

The distribution of an organic chemical between soil and infiltrating precipitation at equilibrium is directly proportional to the fraction of organic carbon in the soil matrix. Therefore, addition of organic matter can increase the sorption capacity of the soil, and can be particularly effective for soils with low organic carbon content, such as sand and gravel. Possible sorbents include sewage sludge, agricultural products and by-products, organic soil materials, such as soils of the Histosol soil order, and activated carbon. The sorbents may be mixed into contaminated surface soils using conventional power implements and tillers.

Ion Exchange

Certain clays, synthetic resins, and zeolites have the capacity to release ions of one type and to preferentially adsorb ions of another type at the vacated lattice position (i.e., to exchange one type of ion for another). Cation exchangers have replaceable cations and anion exchangers have replaceable anions.

Increasing the clay content of a soil can increase its capacity to immobilize cationic compounds. The cationic exchange capacity (CEC) of a soil is defined as the number of milli-equivalents of exchangeable ions per hundred grams dry weight of soil. Clays tend to have the highest CEC values among natural soils, and typically exchange calcium ions for cations of heavy metals or organics (see Table 3-7).

Synthetic resins have been developed that carry either exchangeable cations (e.g., Chelex 100) or exchangeable anions (e.g., Dowex 1-X8).

Zeolites are a family of crystalline-hydrated alumino silicates with high selectivity and capacity for adsorption of heavy metal cations. Zeolites are stable over the pH range of 6-12, but begin to degrade around pH 4-5 and below. Therefore, zeolites should be used only in alkaline or limed soils where the pH is maintained above 6.

Precipitation of Heavy Metals

The addition of sulfides, carbonates, phosphates, and hydroxides to metal-contaminated soils can immobilize metals by precipitation of a highly insoluble compound. Precipitating agents that may be mixed into the soil by standard agricultural techniques include lime for hydroxide precipitation, limestone for carbonate precipitation, trebel superphosphate for phosphate precipitation, and calcium sulfide or sodium sulfide for sulfide precipitation. The use of phosphate should be avoided if arsenic is present in the soil matrix, because of the potential for the formation and release of arsenate into infiltrating precipitation. Under some conditions, heavy metals may form soluble phosphate, carbonate, and hydroxide complexes that are more mobile than the free metal ions. By controlling the rate of addition of precipitating reagents and the pH of the contaminated soil system, it should be possible to optimize precipitation over a complex formation.

In-place precipitation of sulfides presents special complications in soil systems. The heavy metal sulfides are generally the least soluble of the heavy metal compounds. While the solubilities of metal sulfides decrease with increasing pH (except for arsenic, which precipitates only at pH 5), metal sulfides have very low solubilities, even at a pH as low as 4. Some heavy metals form soluble sulfide complexes that may be more mobile in the environment than the free metal ion. Controlled addition of sulfide reagent, controlled pH, and thorough mixing of the contaminated soil with the precipitating reagent can help to favor precipitation over soluble complex formation. More serious over the long term is the potential for oxidation of the precipitated sulfides to form soluble metal sulfates under aerobic soil conditions. Maintenance of a high soil pH, is particularly important, since acid conditions will not only dissolve the metals, but also lead to the release of hydrogen sulfide. The latter is a highly toxic gas, and the odor, even if concentrations are below toxic thresholds, can result in a major community relations problem.

CHEMICAL DEGRADATION

Oxidation, reduction, and polymerization reactions may be carried out in-place to transform soil contaminants into less toxic or less mobile products.

Oxidation

Chemical oxidation of organic contaminants with half-wave potentials less than the redox potential (0.8 volt) of a well-aerated soil can be oxidized in-place in the presence of clay catalysts. This method is conceptual, and the oxidation products are not necessarily less toxic or less mobile than the original contaminants. As previously discussed, increasing the clay content of surface soils can aid in immobilization of contaminants. Clay-catalyzed oxidation of organics may be an additional benefit, or an undesired side effect, depending upon the products of oxidation. These should be investigated in laboratory or field treatability studies undertaken to develop design parameters for in-place treatment with clay.

Introduction of chemical oxidants into the soil system is another method for promoting the oxidation of organics. Possible oxidizing agents include ozone or hydrogen peroxide. The agents may be applied in water solutions directly onto the soil surface, injected into the subsurface, or introduced through injection wells. Since oxidants are relatively non-selective and may act on natural organics in the soil as well as organic contaminants, laboratory and field treatability studies would be required to assess the reactions that occur and to develop data for design of a full-scale treatment system if results are favorable.

Reduction

In-place reduction reactions may be brought about by the addition of reducing agents to the contaminated soil. Possible reagents include catalyzed metal powders of iron, zinc or aluminum, for example, or sodium borohydride. These reagents have been shown to degrade toxic organic constituents, although there have been very few demonstrations of in-place treatment of contaminated soils. Catalyzed metal powders can be applied to the soil surface and mixed into the contaminated soil with conventional agricultural equipment. Sodium borohydride can be applied in an alkaline water solution via irrigation, subsurface injection, or injection wells.

Hexavalent chromium is highly toxic and mobile in soils, generally in the form of a chromate ion. Trivalent chromium, the reduced form, is less toxic and readily precipitated by hydroxides over a wide pH range. Hexavalent chromium is readily reduced in natural soils. Mixing of ferrous iron, leaf litter, or acid compost into the surface soils can assist in the conversion. Liming of the soil once reduction is complete will precipitate trivalent chromium hydroxide compounds.

Hexavalent selenium, as selenate, is also highly soluble and mobile in soils. Selenates are readily reduced in natural soils to elemental selenium, which is virtually immobile, and also to selenites. The solubility and potential leachability of selenites increases with increasing pH. Hence the high pH that is generally required to immobilize other heavy metal contaminants could result in the release of selenium if any is present.

Polymerization

If a soil is contaminated with polymerizable organics such as styrene, vinyl chloride, isoprene, acrylonitrile, or methyl methacrylate, mixing of iron and sulfates into the contaminated area may catalyze an in-place polymerization reaction. The polymers are commonly less toxic and less mobile than the monomers from which they are formed. Generally, a 2:1 ratio of volume of catalysts and activator to volume of contaminant is required. The catalysts and activator are generally applied separately, and a wetting agent added to promote rapid and uniform dispersion of solutions through the contaminated area. The technical feasibility of the method and parameters for design of a full-scale system would need to be established by a combination of laboratory bench-scale and pilot field demonstrations.

BIODEGRADATION

Biodegradation, as the term is used here, refers to the breakdown of organic compounds in soils by the action of micro-organisms such as bacteria, actinomycetes, and fungi. Treatment generally consists of optimizing conditions of pH, temperature, soil moisture content, soil oxygen content, and nutrient concentration to stimulate the growth of micro-organisms that will feed on the particular contaminants present. Alternatively, genetically engineered organisms may be added to the soil system and conditions established within the soil to optimize their growth. Optimum conditions of application generally need to be established in laboratory bench-scale studies and small field pilot test spots. Some of the hazardous constituents present in a contaminated soil may be most readily biodegraded under aerobic conditions, while others are more readily degraded under anaerobic conditions. Treatment might therefore consist of alternate aerobic and anaerobic cycles.

While generally biodegradation is used as a detoxification mechanism for organic contaminants in soil, some micro-organisms will also interact with metallic species. For example, aerobic heterotrophic bacteria oxidize arsenite to arsenate. Further treatment with ferrous sulfate would then immobilize the arsenic in the form of highly insoluble ferric arsenate.

PHOTOLYSIS

Photochemical reactions require the absorption of light energy, generally from sunlight in natural systems. Since light does not penetrate very far into soils, photodegradation of soil contaminants is limited to soil surfaces. The addition of proton donors in the form of polar solvents, such as methanol, can enhance surface photodegradation of soil contaminants. For example, photolysis of dioxin (TCDD) on soil surfaces has been reported in the presence of methanol. Photodecomposition of PCBs at soil surfaces has been reported in the presence of triethylamine as a proton donor.

ATTENUATION

Attenuation is the mixing of contaminated surface soil with clean soil, using conventional power implements or tillers, so that the concentrations of hazardous components in the mixture are reduced to acceptable levels. Although potentially applicable to any type of waste, guidelines for acceptable concentration levels in surface soils are available only for heavy metals. The amount of clean soil required to assure that the mixture lies within an acceptable range is a practical limitation that needs to be evaluated.

REDUCTION OF VOLATILIZATION

If soil contaminants are volatile, and the air transport pathway could lead to adverse exposure, it may be important to suppress volatilization. This could have the added benefit of retaining compounds within the soil system for a long enough time to allow for in-place treatment by one of the mechanisms described above. The most practical method of suppressing volatilization is to reduce the soil vapor pore volume through which the transport of vapors occurs. This can be done by compaction or addition of water to reduce the air-filled pore spaces within the soil relative to the water-filled pore spaces. Another technique is to decrease the temperature of the soil, since vapor pressure of volatile constituents generally decreases with decreasing temperature.

METHODS FOR IMPLEMENTATION OF IN-PLACE TREATMENT

The in-place methods discussed in this report have many elements in common. Most involve application of absorbents or reagents and thorough mixing with the contaminated soil. Many involve liming for pH control and adjustment of the moisture content of the soil to optimize conditions of treatment.

Solid or semi-solid agents are typically applied to the soil surface by means of spreaders that may be truck-mounted, hitch-mounted, or tractor-drawn. Liquids can be applied with hydraulic sprayers or subsurface injectors. Mixing of treatment agents with the contaminated soil can be accomplished with plows, rotary tillers, subsoilers, disc harrows, spike harrows, or spring-tooth harrows. Moisture can be controlled with sprinklers and perforated pipe drains. Tilling, which breaks, mixes, and aerates the soil, tends to increase oxygen content; compaction and flooding tend to decrease oxygen content. Irrigation increases surface moisture; drainage and well points remove excess water from the surface. Various additives are also available for moisture control. For example, synthetic substances that store water can enhance the water retention capacity of the soil; water-repelling agents can accelerate soil drainage and improve water infiltration and percolation.

RESEARCH NEEDS

There has been little operating experience with the use of in-place treatment of contaminated soils. Since the National Contingency Plan requires that any alternative selected for implementation as part of a remedial program must be technically feasible and reliable, research on in-place treatment technologies is critically needed in the following areas to assure that in-place methods can be properly evaluated as potential remedial alternatives.

1) Information on soil processes interfacing with atmospheric processes (i.e., controlling volatilization and photodegradation), including:

   - investigation of photochemical reactions that enhance biodegradation of refractory compounds and those that can produce toxic or undesired breakdown products;

   - laboratory and field-scale analysis of the feasibility of using volatilization/photodegradation as a viable treatment method for volatile/photoreactive hazardous chemicals;

- evaluation of the importance of soil-photochemical reactions to the fate and behavior of photoreactive compounds within the environment;

- refinement of containment/transport models that address volatilization, vapor adsorption, biodegradation, and leaching through the soil matrix; and

- continued development of rapid partition coefficient estimation methods for predicting compound partitioning among air/water/soil systems.

2) Information on chemical reactions relating to the transformation and immobilization of constituents in soil systems, including:

- effects of chemical addition (oxidants, reductants, and polymerizing agents) on the soil properties affecting treatment;

- constituent sorption and precipitation studies at concentrations in the soil matrix characteristic of those found at remedial action sites;

- Eh-pH information for metal species in soils at concentrations representative of remedial action site concentrations; and

- behavior of specific metals, viz., arsenic, beryllium, silver, selenium, mercury, and chromium in soil systems, including the investigation of the reaction of chromium (VI) to chromium (III).

3) Information on biological reactions to stimulate the biodegradation of constituents in soil systems, including:

- potential for combinations of chemical and biological treatment methods for accelerating soil treatment;

- potential for composting of hazardous waste contaminated soil for accelerating biological reactions and for detoxifying and degrading recalcitrant organic constituents;

- evaluation of the biodegradation kinetics and extent of hazardous constituents and metabolites of hazardous constituents sorbed to clay and soil organic matter;

- degradation rates and pathways for chemical classes, including alkyl halides and highly chlorinated organics (PCBs) in the soil treatment zone (upper 5 feet of soil); and

- information concerning the effectiveness of micro-organism seeding compared with no seeding, specifically in terrestrial systems.

4) Information on immobilization reactions in the soil/waste system, including:

- investigation of enhancement of sorption by adding different sorbents, such as activated carbon, straw, synthetic resins. (The specific information needed includes application rates of sorbent in relation to concentrations of contaminant in soil);

- continued development of estimation methods for predicting soil adsorption constants by chemical and physical properties of compounds;

- long-term effectiveness of immobilization of sorbed contaminants on soil; and

- effects of solvents on mobility of organic contaminants in soil systems.

5) Information on leaching in the soil/waste system, including:

- laboratory experiments to estimate parameters for transport models that can be used to predict behavior under field conditions;

- field instrumentation to validate transport models under imposed environmental conditions, incorporating the variability under field conditions by designing appropriate sampling stations; and

- effect of different environmental conditions, including soil moisture, redox potential, and organic content on transport parameters.

1. Introduction

1.1 PURPOSE AND FUNCTION OF THE REPORT

Unplanned and uncontrolled disposal of hazardous wastes frequently results in the contamination of the upper 2 feet of soil. This may severely affect public health, damage terrestrial systems, and destroy or diminish opportunities for land use. Solutions to problems associated with contaminated sites generally follow the procedure specified in the National Contingency Plan (NCP). The NCP process, which is presented in Section 2, is required for sites which receive Federal Superfund money. It also serves as the model for clean-up of non-Superfund sites.

Three types of remedial responses are outlined in the NCP:

- Immediate removal;
- Planned removal; and
- Remedial action.

Immediate removal is employed when there is an "immediate and significant risk of harm to human life or health or to the environment . . ." (40CFR 300.55). Planned removal involves the continuation of immediate response actions, or the implementation of remedial measures at sites not on the National Priorities List. Remedial actions are defined as "responses to realeases on the National Priorities List that are consistent with permanent remedy to prevent or mitigate the migration of a release of hazardous substances into the environment" (40CFR 300.68). The in-place technologies discussed in this report are potentially applicable to each type of response. Selection of in-place treatment as a remedial action response, however, is considered in detail.

The NCP differentiates among initial, source control, and off-site remedial actions. Initial remedial actions are those measures which prevent immediate exposure while long-term solutions are being considered. A security fence and temporary cover are examples of initial remedial actions. Source control remedial actions address realized and potential on-site contamination including, for example, drummed wastes, leaky storage tanks, or contaminated soil. Source control remedial actions contain the hazardous substances where they are located, or eliminate potential contamination by treatment or removal. Off-site remedial actions apply to hazardous substances that have migrated away from the original source of contamination. Off-site measures include provisions for alternative drinking water supplies or treatment of contaminated aquifers. At a given hazardous waste site, any or all of the three types of remedial actions may be applicable.

One of the most common problems found at uncontrolled sites is the contamination of soil and underlying strata. Such contamination arises from leaking drums and tanks, leaking impoundments, direct discharge of wastes in pits, and leaching from solid residues. When large volumes of soil are contaminated and pose a continuing hazard, the development of a remedial action plan will almost certainly require consideration of source control options. In-place treatment is one of a family of source control measures described in the NCP. Others include: no action, excavation and off-site disposal, on-site treatment (e.g., excavation, treatment, and replacement), and physical containment (e.g., cap, hydraulic barrier). In-place treatment can be used to contain the source of contamination (e.g., by immobilization), or to remove it through treatment (e.g., by degradation).

Selection of the appropriate remedial alternative is made through the process outlined in the NCP and is based on site-specific considerations. In some cases, in-place treatment might be considered the most cost-effective option, as indicated by the following scenarios:

- An industrial facility that has produced heavy metal-based chemicals for the past 50 years is closed and the owner wishes to sell the land. Soil analyses, however, indicate that several acres of soil (to a depth of up to 10 feet) at the facility are contaminated with heavy metals on the order of 100 ppm. Excavation of such large quantities of soil would be prohibitively expensive and the benefit relatively low, since the facility is in a highly industrialized area. In addition, it is likely that the new owner would not accept a no-action alternative, and physical containment would interfere with land use. After further investigation, immobilization through in-place treatment is considered the most suitable alternative.

- A farmer decided to supplement his income by disposing of drummed chemical waste in trenches at the edge of his property. A fire in the trenches, however, alerted authorities to his activities. Since the farmer was not able to pay for clean-up, the site was put on the National Priority List making it eligible for Superfund money. Based on the remedial investigation and feasibility study, the U.S. Environmental Protection Agency decided to excavate and remove the drummed wastes and treat the residual contaminated soil in place. An impervious cap was also constructed over the trenches and the site monitored annually.

- A train carrying several tankers of various chemicals derailed in a remote area. During the accident, a piece of track pierced a tank car carrying benzene, creating two large holes from which the chemical drained. Much of the benzene volatilized immediately. The rest soaked into surface soils. An emergency crew called to the scene estimated that 25,000 gallons of benzene had spilled from the tanker, 10,000 gallons of which remained in the surface soils. Because of the remote location, enhanced microbial degradation was used to remove the benzene from the soil.

The purpose of this report is to provide state-of-the-art information on in-place treatment technologies for contaminated soil. While many of the technologies discussed are applicable for in-place treatment regardless of the depth of contamination, treatment of contamination in the upper 2 feet of soil is emphasized. The report is intended to be used by on-scene coordinators, engineers, regulatory agencies and researchers in evaluating in-place treatment of contaminated surface soils as a remedial alternative.

The report provides a discussion of the major in-place treatment technologies which can be used to treat contaminated surface soil. A methodology for selection of the appropriate technology for a particular site and techniques to modify soil properties are also discussed. The report, however, is not intended to be a guidance manual. Rather, it provides the technical basis necessary to understand the applicability of in-place treatment technologies as a source control remedial measure for contaminated surface soil.

1.2 OVERVIEW

This manual has been organized into two volumes. Volume 1 discusses in-place treatment technologies applicable to contaminated surface soil. Volume 2, available through NTIS, provides background information on the characterization and evaluation of the fundamental processes applicable to site/soil/waste systems.

Volume 1 contains four sections. Section 1 presents the purpose and function of the report and discusses information gaps for in-place treatment. Section 2 gives a general discussion of the various in-place treatment technologies as well as the methodology for selecting the appropriate technology for a particular site. Section 3 presents state-of-the-art information on the technologies. Each technology discussion includes:

- description,
- wastes amenable to treatment,
- status of technology,
- ease of application,
- potential achievable level of treatment,
- reliability of method,
- secondary impacts,
- equipment and exogenous reagents,
- information requirements, and
- sources of information.

A summary matrix with a brief evaluation of each technology is also provided. Section 4 briefly discusses soil modification techniques applicable to several technologies. Soil properties discussed include:

- oxygen content,
- moisture content,
- nutrient content,
- pH, and
- temperature.

The section also contains a table linking soil modification techniques to in-place treatment technologies.

Volume 2 contains three sections and an appendix of chemical data. Section 1 is an introduction. Section 2 provides information on monitoring soil/waste systems. Section 3 addresses fundamental processes in soil/waste systems, including site conditions physical soil properties, soil sorption, soil microbiology, volatilization and degradation as related to hazardous waste treatment in soil systems. This information is included to provide the user with additional information for making more complex decisions in the analysis of site/soil/waste systems. It is also useful for evaluating the application of additional treatment techniques not specifically discussed in this report. The Appendix provides compound properties and adsorption, degradation and volatilization parameters for various chemicals. These data are important in assessing soil/waste interactions. A glossary is also provided.

2. Selection of In-Place Treatment Technology

2.1 INTRODUCTION

Selection of appropriate remedial action technologies follows the procedure specified in the National Contingency Plan (NCP). This procedure is outlined in Figure 2-1. First, a preliminary assessment, called scoping, is conducted to determine the type of response expected (initial, source control, or off-site remedial action). Scoping relies on available information and provides the basis for funding requests. This is followed by a remedial investigation to determine "the nature and extent of the problem presented by the releases. This includes sampling and monitoring, as necessary, and includes the gathering of sufficient information to determine the necessity for and the proposed extent of remedial action." (40 CFR 300.68 (f)). The remedial investigation also re-evaluates the conclusions of the scoping phase. Alternative remedial technologies are then suggested and screened according to the factors given in the figure. Finally, the most promising technologies are analyzed in detail and the most cost-effective alternative selected.

Evaluation of in-place treatment technologies is a part of this overall process. The specific methodology for selecting an in-place treatment technology (beginning with the remedial investigation) is depicted in Figure 2-2. It is assumed here that the results of scoping indicate source control remedial action to be appropriate. It is also assumed that contaminated surface soils are involved, so that the technologies in this report are potentially applicable.

This Section begins with a general discussion of in-place treatment technologies, followed by four sections that parallel the procedure outlined in Figure 2-2. The development of alternatives in selecting in-place treatment type and the detailed analysis to select appropriate in-place treatment technologies are emphasized. Finally, consideration is given to information gaps and the need for further research pertaining to in-place treatment as a remedial action alternative.

2.2 DISCUSSION OF IN-PLACE TREATMENT TECHNOLOGIES

There are five major categories of in-place treatment technologies that can be defined in terms of their primary action on the contaminants contained within the soil:

1. extraction,
2. immobilization,
3. degradation,
4. attenuation,
5. reduction of volatilization.

2.2.1 Extraction Techniques

Extraction techniques actually remove the undesired contaminant from the soil by dissolution in a fluid which is subsequently recovered and treated either on site or at another location. This technique offers a more or less permanent solution to the problems pre-existing at the remedial action site. The problem of ultimate disposal of the contaminants

Figure 2-1. Detailed sequence – phase VI – remedial action (40 CFR, part 300.68)

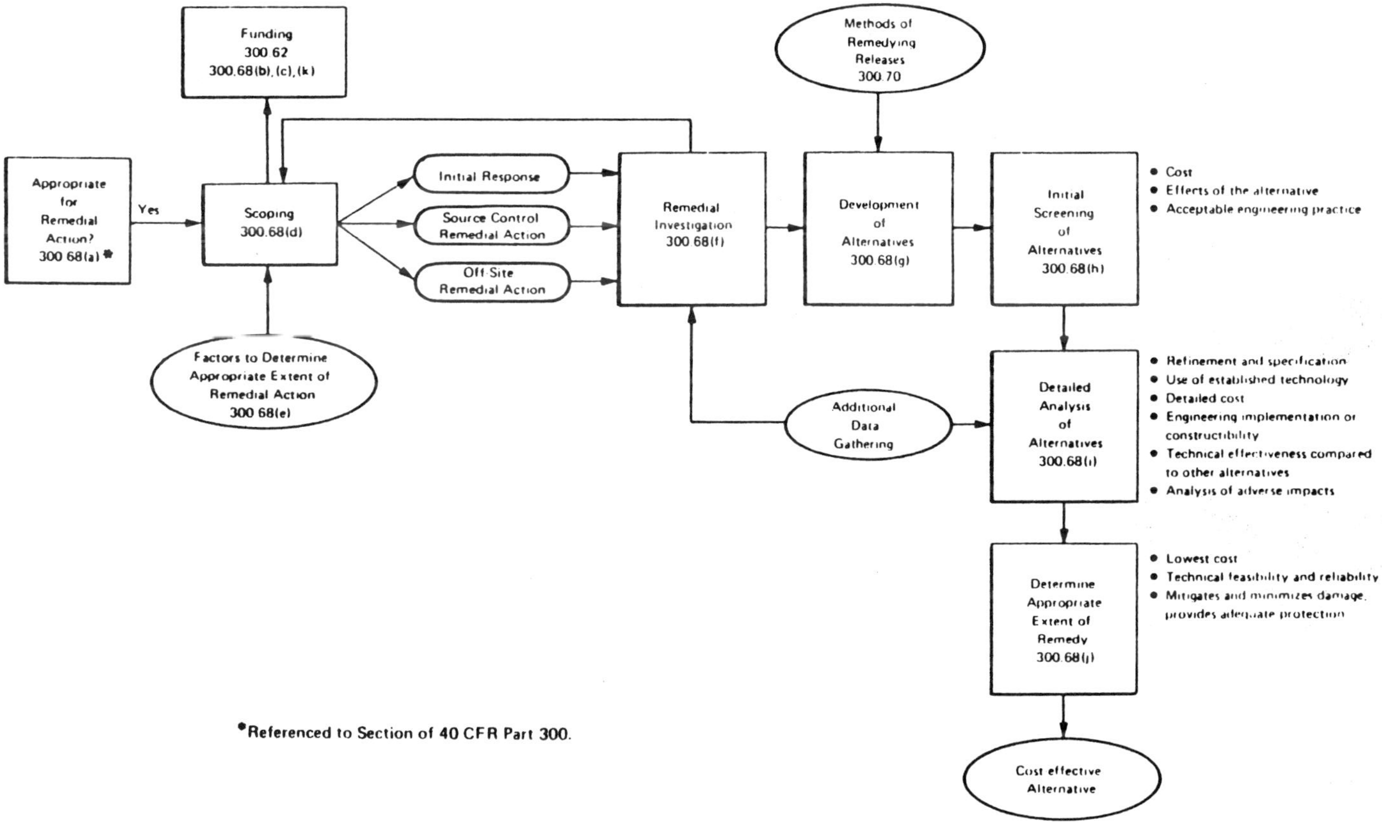

*Referenced to Section of 40 CFR Part 300.

Source: Ehrenfeld and Bass, 1983.

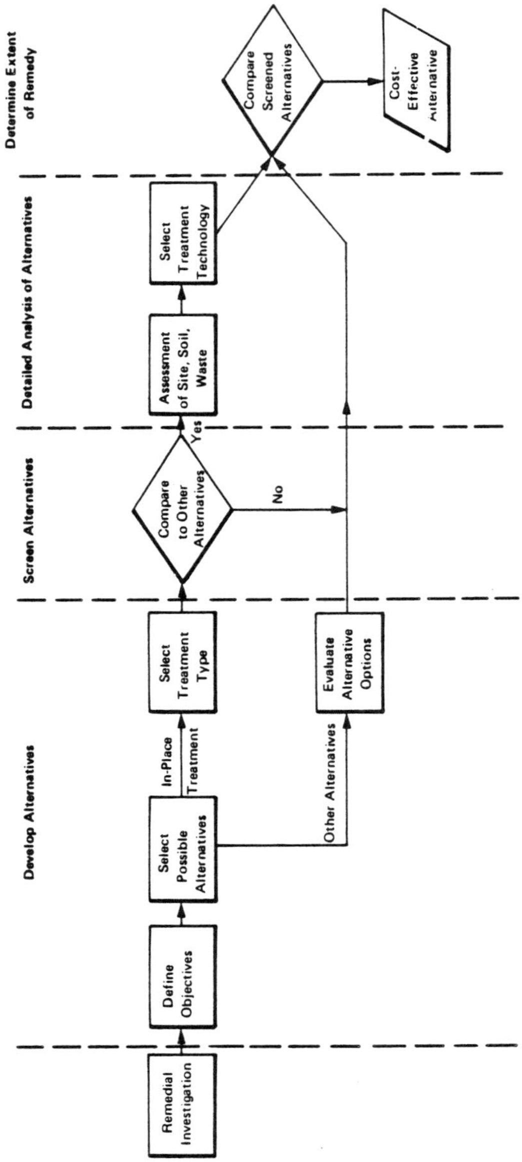

Figure 2-2. Methodology for selecting in-place treatment technology.

is, however, moved to another location and another set of processes. The ultimate treatment and disposal of the contaminated extraction fluid can often be carried out under more favorable technical conditions, at significantly lower risk, and at reduced costs compared to other in-place treatment options.

2.2.2 Immobilization Techniques

Immobilization techniques are designed to capture the species within the contaminated soil mass. Immobilization reduces the tendency of the contaminant to enter the groundwater, surface water, or atmospheric exposure pathways. The immobilized contaminants, however, remain in the soil, leaving open the possibility for exposure via direct contact or contaminant migration under changed conditions in the future.

The three major classes of immobilization techniques are: adsorption (sorption), ion exchange, and precipitation.

Adsorption includes techniques designed to capture the contaminants on the soil particles by adjusting the properties within the soil system. In addition, adsorptive materials, such as activated carbon can be admixed into the soil to enhance its inherent adsorptive properties. Adsorption can be applied to both inorganic and organic contaminants.

Ion exchange is a means for immobilizing inorganic species similar in nature to physical adsorption. Initially mobile metal ions exchange positions with innocuous cations and become bound to clay particles in the soil system. Many heavy metal contaminants that pose significant risks are quite tightly bound by common clay minerals. The activity of natural ion exchange in soils can be enhanced by admixing synthetic ion exchange resins. Such resins can be tailored to be highly selective for particular metals. Ion exchange techniques are applicable only to inorganic contaminants.

The final immobilization technique is precipitation. Inorganic contaminants that are initially soluble are caused to react and form compounds of very low solubility. The insoluble products are thereby retained by the soil system within the initially contaminated region instead of being removed by infiltration and groundwater flow. The precipitating agent can be added directly, for example, in the form of sodium sulfide (Na_2S), or it can be generated in-place by a chemical or biochemical reaction.

2.2.3 Degradation Techniques

The third category of treatment technologies is degradation. This is a family of methods which converts the contaminant species into an innocuous or less toxic compound or compounds. In general, degradation is applicable to organic compounds. There are a number of limited cases in which degradation is applicable to inorganic species. Highly toxic hexavalent chromium, for example, can be reduced to the less toxic trivalent form by the addition of reducing agents. Degradation is often used in combination with immobilization for inorganic contaminants. Following the above example, trivalent chromium can be readily precipitated in the hydroxide form. By and large, however, the degradation techniques are of broader applicability to organic contaminants than to inorganic.

Chemical degradation techniques convert contaminant species by promoting the natural capacity of the soil to support oxidation or reduction reactions or by adding suitable reagents. Reactive organic compounds, such as vinyl monomers, may be polymerized in-place by the addition of catalysts to form immobile species.

Biological techniques utilize the action of micro-organisms to break down organic compounds into innocuous or less toxic metabolic products. The naturally occurring micro-organisms found in nearly every soil system can often be used to degrade organic contaminants. The soil system may have to be modified to promote the activity of naturally found organisms. Promotion includes addition of nutrients and aeration of the soil. If the intrinsic micro-organism flora do not work on some set of contaminants, selectively adapted innoculants or even perhaps genetically engineered species, can be added to the soil mass.

The last degradation technique involves photodegradation (photolysis). The action of the ultraviolet portion of sunlight can result in the breakdown of many complex and toxic organic compounds. Organic species brought to the surface by volatilization or mechanical mixing may be broken down by sunlight incident on the surface. The compounds must move toward the surface in order for this technique to be practical.

2.2.4 Attenuation

Attenuation, the fourth category of in-place treatment techniques, involves admixing clean soils or other bulking agents to reduce contamination to an acceptable level. This approach has been used to reduce the hazards due to ingestion or direct contact.

2.2.5 Reduction of Volatilization

Finally, reduction of volatilization, the last major category of treatment techniques, is applicable to exposure via the air pathway. Volatile organic contaminants pose a significant risk via this pathway. Techniques for reducing volatility include those which impede the diffusion of volatile contaminants and others which reduce the inherent tendency for volatilization to take place. Flow toward the surface can be inhibited by reducing the soil vapor pore volume. Alternatively, the volatility of the contaminant materials can be reduced by techniques such as cooling the soil to lower the vapor pressure.

2.3 DEVELOPMENT OF ALTERNATIVES

The first phase in selecting the appropriate in-place treatment technology, as shown in Figure 2-2, is the development of alternatives. This takes place after scoping and the remedial investigation have been completed. This means that the type of remedial response has been determined. For this discussion, it is assumed that source control remedial actions are appropriate, and that information from the remedial investigation is available to determine the nature and extent of the contamination problem. Such information would include:

- type of waste present (organic, inorganic, chemical components),
- estimate of waste quantity,
- waste form (free liquid, solid, contaminated soil),
- waste location (areal extent, depth, relationship to water table),
- exposure pathways (groundwater, surface water, air, direct contact),
- population threatened (human, environmental).

The remedial investigation provides the basis for the selection processes through the comparison and screening of initial remedial alternatives.

2.3.1 Definition of Objectives

The first step in developing alternatives is to define the objective of the remedial program. This step, which is not explicitly specified in the NCP, is iterative and occurs throughout the selection process. The objectives are refined and modified as new information is obtained. It is necessary, however, to have a clear understanding of the remedial objectives before considering alternative remedial actions, since selection depends on the desired result.

An objective can be defined as an aim or goal of the remedial program. Based on current legislation, the primary objective for remedial action at Superfund sites is to reduce risk to an acceptable level, while minimizing cost. The purpose of the selection process, therefore, is to determine which alternative is most "cost-effective" (defined in 40 CFR 300.68(j) as "the lowest cost alternative that is technologically feasible and reliable and which effectively mitigates and minimizes damage to and provides adequate protection of public health, welfare, or the environment"). Secondary objectives can be used to further define the primary objective. They may consider a particular exposure pathway or address other technical requirements, local regulations, or political concerns. Together, the objectives specify the end-result or outcome desired to be achieved through implementing the selected remedial action technologies.

Figure 2-3. Exposure pathways.

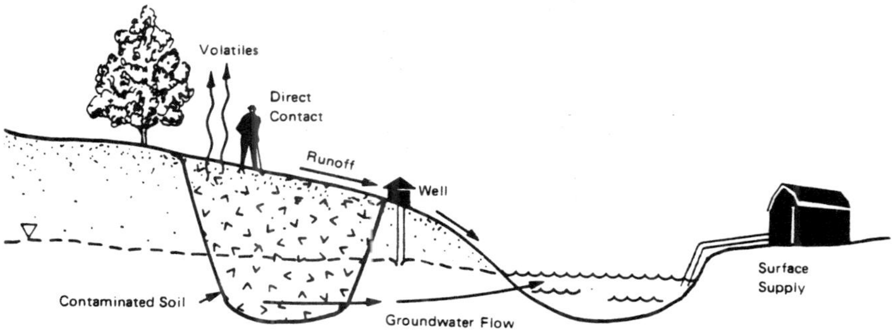

Source: Arthur D. Little, Inc.

Figure 2-3 depicts a generalized site containing a large quantity of contaminated soil. The remedial objective at the site might be to reduce risk (a function of the contamination, the pathways and the receptors) by minimizing exposure along each of the pathways depicted (direct contact, air, groundwater, surface water). Alternatively, if there was no drinking water use downgradient or downstream of the site, the objectives might be to minimize exposure along the air and direct contact pathways.

2.3.2 Selection of Possible Alternatives

Once objectives have been determined, it is necessary to select alternatives that would meet the objectives from the many available source control remedial action technologies. For the example in Figure 2-3, a security fence, cap and slurry wall combination, removal and off-site disposal, on-site treatment, and in-place treatment would all potentially reduce exposure along each of the pathways. They, therefore, can be considered in the initial development of alternatives, along with any other technology or combination of technologies which have a similar effect. The purpose of the initial selection of possible alternatives is to consider as many technologies as possible so that the cost-effective alternative is not overlooked. Subsequent screening and detailed analysis of alternatives serve to single out those technologies that are most cost-effective in meeting the objectives.

The remainder of this Section emphasizes the considerations particular to selecting in-place treatment technologies. The evaluation of other technologies follows a similar path which is not discussed, except where the two paths converge.

2.3.3 Selection of an In-place Treatment Type

The selection process described herein fits into the two-step procedure commonly used in remedial investigation/feasibility studies (RI/FS). It begins with a preliminary evaluation of all the potential in-place treatment methods available and the identification of all those that are potentially applicable to the site, and that are consistent with the objectives. A very rough screen can be made on the basis of the exposure pathways and the chemical nature of the contaminants. Table 2-1 shows a general applicability matrix. Reference to this table permits selection of the family of techniques applicable to any combination of pathway and compound. For example, if the objectives contain a requirement to reduce the risk from organic compounds via direct contact, then extraction or any of the degradation techniques would be potentially applicable. None of the immobilization techniques would be useful. On the other hand, if the objective were to reduce the risk from inorganic species via groundwater, then removal and immobilization via adsorption, ion exchange, or precipitation would be potentially applicable. Degradation would have limited applicability, depending on the specific inorganic species.

TABLE 2-1. GENERAL APPLICABILITY MATRIX

Technology	Air	Groundwater	Surface Water	Direct Contact	Inorganic	Organic
EXTRACTION	X	X	X	X	X	X
IMMOBILIZATION						
Adsorption	–	X	X	–	X	X
Ion exchange	–	X	X	–	X	L
Precipitation	–	X	X	–	X	–
DEGRADATION						
Chemical	X	X	X	X	L	X
Biological	X	X	X	X	L	X
Photolysis	X	X	X	X	–	X
ATTENUATION	L	–	–	X	X	X
VOLATILE REDUCTION	X	–	–	–	L	X

X = Potentially Applicable
– = Not Applicable
L = Limited Applications

If multiple pathways are involved, as is very often the case, technologies potentially applicable to all pathways should be considered. For example, dioxin-contaminated soils may pose a risk through both ingestion of surface soils and groundwater contamination. Although immobilization techniques via adsorption would be potentially applicable for groundwater, as shown in Table 2-1, the inapplicability of this approach when direct contact is included would rule it out, or it would at least require adsorption to be used in combination with another technology.

The preliminary screening procedure should be broadly construed. If there is any question, a general class of in-place treatment should be left in the portfolio of options to be evaluated in the detailed steps further along the process. Combinations of techniques should also be included in the first round.

The list of potentially applicable alternatives can also be narrowed down by consideration of the time available for treatment, or by the rough reduction levels if available at this stage of the analysis. Some of the processes take a very long time (e.g., photodegradation) and may be inappropriate at a site requiring clean-up over a limited time period.

Using only information available from the initial remedial investigation, one may find it possible to eliminate some of the technologies, or at least note that the likelihood of their feasibility is lower than that of other options. Immobilization techniques depend perhaps more than any other option on the intrinsic properties of the soil system. Adsorption can be expected to be fairly high in soil systems containing high clay and organic content. Waste sites where soils are predominantly sandy in nature would probably be poor candidates for techniques based on adsorption, although additives could be used to enhance the poor intrinsic soil properties.

The effectiveness of any one technique or combination of techniques depends on many site-specific and waste-specific factors. The procedure for introducing these factors into the evaluation and final selection of the single or small family of cost-effective techniques is developed in the following sections of this document.

2.4 SCREENING OF ALTERNATIVES

Once a list of alternative technologies and technology combinations has been developed, it then becomes necessary to narrow the list for further detailed analysis. The three criteria used for this screening are:

- costs,
- effects of the alternative, and
- acceptable engineering practice.

Costs include capital, operating, and maintenance costs. Technologies that are significantly more expensive than other alternatives without providing a greater degree of benefit or reliability can be excluded from further consideration. For example, if removal by extraction costs X dollars and removal by excavation costs 10X, excavation would no longer be considered, assuming that extraction would meet the second and third criteria.

Evaluating the effects of the alternative involves two considerations:

(1) whether implementing the alternative causes adverse environmental effects. Some biodegradation techniques, for example, produce toxic metabolic products which bio accumulate in the course of detoxifying the target contaminant(s); and

(2) whether the technology has sufficient capabilities relative to the objectives and associated performance requirements. In other words, can the technology achieve the required level of source control?

If either consideration produces an unsatisfactory finding, the technology option would not be considered further.

Acceptable engineering practice means that the alternative is feasible, given site-specific conditions, and is considered an applicable and reliable means of addressing the problem.

These screening criteria are evaluated based on available information according to the judgment of the lead agency or other decision-makers. The screening should result in a shorter list of technologies to be evaluated in detail.

2.5 DETAILED ANALYSIS OF ALTERNATIVES

The detailed analysis of alternatives involves further investigation in five areas (40 CFR 300.68(i)(2)):

1) refinement and specification of alternatives in detail;

2) detailed cost estimation, including distribution of costs over time;

3) evaluation in terms of engineering implementation or constructability;

4) an assessment of how well the alternative meets the remedial objectives; and

5) an analysis of any adverse environmental impacts, methods for mitigating these impacts, and costs of mitigation.

For in-place treatment technologies these steps involve (1) assessment of waste, soil, and site-specific variables, and (2) selection (including detailed specification) of an in-place treatment technology within the treatment category.

2.5.1 Assessment of Waste, Soil and Site-specific Variables

Many waste, soil, and site-specific variables affect the technical feasibility and effectiveness of in-place treatment. Some of these variables can be conveniently established as part of the remedial investigation. Others may require independent bench-scale or pilot studies directly related to treatability. Each of the important variables is discussed below under the following sub-headings:

- waste characteristics,
- site characteristics, and
- waste/soil system characteristics.

These discussions are followed by a discussion of laboratory and pilot-scale testing.

2.5.1.1 Waste Characteristics

Soil core and soil pore water samples should be obtained and analyzed to determine the chemical contaminants present, the areal extent and depth of contamination, and the range of contaminant concentrations at presumed source locations and along expected pathways of migration.

Soil and groundwater samples from contaminated areas and control (background) areas should be analyzed for:

- Elemental and Inorganic Constituents:
  - — Metals and metalloids (As, B, Ca, Cd, Cr, Cu, K, Mg, Mo, Na, Ni, P, Pb, Se, V, Zn),
  - — Total organic carbon (TOC),
  - — Total Kjeldahl nitrogen (TKN),
  - — Sulfate
  - — Chloride.
- Organic Constituents:
  - — Priority pollutants (volatiles, acid extractables, and base/neutrals),
  - — Others known or expected to be present, based on past activities at the site.

Based on the analytical results, a list of contaminants within each group that are present at levels in excess of background should be compiled. Maps should be prepared showing contaminant locations and concentrations.

To assist in evaluating pathways of migration, and in designing in-place treatment processes, the following physical/chemical properties should be derived from the literature or estimated for each of the identified waste constituents:

- solubility in water (and in other solvents, if flushing is an alternative of interest);
- vapor pressure;
- soil/water partition coefficient, K_{oc}, for organic waste constituents (K_{oc} is defined as the Freundlich adsorption coefficient (K) divided by the weight fraction organic carbon (f_{oc}) in the soil, i.e., $K_{oc} = K/f_{oc}$);

- Henry's Law Constant (used in evaluating the partitioning of a contaminant between soil pore water and the vapor phase); this may be derived from the vapor pressure/water solubility ratio for chemicals of limited water solubility (<1m);

- standard electrode potential (used in evaluating oxidation/reduction alternatives).

2.5.1.2 Site Characteristics

Significant site variables are ones that affect pathways of migration of waste constituents, and/or design and implementation of in-place treatment processes. Site characteristics that are important in the selection of in-place treatment methodology include:

- Location — While the significant characteristics of a given site are usually unique, useful hypotheses about pathways of migration and estimates of parameters needed to calculate migration rate can often be developed from available regional data, and keyed to location, topography, surface drainage patterns, flood potential, subsurface stratigraphy, groundwater flow patterns, and climate.

- Topography — Topographic variables are important in evaluating surface drainage, runon and runoff potential to and from contaminated areas of the site, and logistics of moving and/or placing equipment for in-place treatment.

- Stratigraphy — The nature of subsurface soils, determined by examination of soil core borings to bedrock, is an important input to evaluation of pathways of migration in both the unsaturated and saturated zones.

- Groundwater Levels (Equipotential Surfaces) — Seasonal maps of water table contours and piezometric surfaces, developed by analysis of groundwater monitoring well data, are important in predicting groundwater flow directions and hydraulic gradients.

- Groundwater Flow — Information on permeability and porosity of subsurface strata, combined with data on hydraulic gradients, is important in predicting groundwater flow velocities and in estimating contaminant transport times.

- Meteorology and Climate

  — Wind velocity and direction is important in determining the potential for migration of airborne particulate matter and volatile waste constituents.

  — Temperature, including seasonal or monthly means and duration of the frost-free period, is important in estimating rates of biological and chemical reactions in-place, and in evaluating the technical feasibility of in-place methods. If possible, data on soil temperatures as well as ambient temperatures should be obtained, since they often differ.

  — Precipitation, including annual, seasonal or monthly rain and snowfall, is an important parameter in determining a water balance for the site, and in evaluating leachate potential.

  — Evapotranspiration is also important in developing a water balance for the site. It is often estimated from temperature and the nature of vegetative growth at the site.

- Soil Properties

  — pH is an important variable in evaluating mobility of many metal contaminants, as well as organic acids and bases, and in designing several of the in-place processes discussed in this report.

  — Cation exchange capacity (CEC) is an important determinant of the mobility of metallic species in soils; if the CEC is sufficiently high to adequately immobilize the heavy metals present in the soil, no further action may be necessary; if one remedial objective is to reduce heavy metal migration from expected "no action" levels, a potential in-place treatment method might involve increasing the intrinsic CEC of the soil.

  — Redox potential (Eh) is important in determining the stability of various metallic and organic species in the subsurface environment of the site, and it also might be modified by in-place treatment.

  — Organic carbon content is a major variable affecting adsorption, and hence mobility, of organic species in the subsurface environment.

  — Microbial content as a function of depth is an important variable, if potentially biodegradable constituents are present in the waste.

  — Soil type (e.g., clay, till, sand, fractured bedrock) is a major variable affecting rates and routes of groundwater migration and contaminant transport.

  — Hydraulic conductivity is important in determining feasibility and spacing of drains and wells.

- Trafficability — Factors that affect the mobility and/or placement of equipment on the site, and the ability to perform tillage and other in-place treatment operations need to be considered in the remedial action design. Significant variables include: bearing capacity, traction capacity, soil strength, slipperiness, stickiness, moisture content, clay content, presence of debris, structures and/or vegetation, and slope of the terrain (topography), as mentioned earlier.

- Potentially Exposed Human Populations and Sensitive Ecological Environments

  — Groundwater and surface water usage, especially downgradient of the site, are important in evaluating risks and environmental benefits of remedial alternatives.

  — Size of population and nature of ecological resources downgradient and downwind of the site are also important variables for risk assessment.

2.5.1.3 Waste/Soil System Characteristics

Design and implementation of an in-place treatment process requires information on characteristics of the waste/soil system as a whole. The important variables are:

- Depth of Contamination — If contamination is limited to the upper 6-8 inches of the soil and it is well above the water table, in-place treatment techniques may be much more easily applied than if the contamination extends well below the plow layer and into the seasonally high water table.

- Contaminant Concentrations and Quantities — The efficiency and effectiveness of many in-place processes depend upon both contaminant concentration levels and the total quantity of each contaminant present in a given area.

- Treatability — Based on the various waste, soil, and system characteristics discussed above, an analysis can be made of routes and rates of contaminant migration and the potential for damage to human health and the environment as a function of time under conditions of no action. The in-place methods discussed in this report are intended for consideration in situations in which naturally occurring immobilization, degradation, attentuation, and volatilization processes are insufficient in and of themselves to meet defined objectives within an acceptable time frame. For practical purposes, therefore, in-place treatment will require physical modification of the waste/soil system and/or addition of exogenous agents to accelerate the rate of naturally occurring remediation processes or to induce processes in-place to meet the remedial objectives.

The in-place methods discussed in this report are basically intended to alter the waste, soil, and/or waste/soil system characteristics determined to exist at the site prior to treatment. Any one of the generic methods under consideration, however, needs to be specifically adapted to each particular site contamination situation in order to achieve desired system characteristics after treatment. Table 2-2 summarizes the treatment variables that need to be determined to assess the capability of in-place treatment alternatives to achieve remedial objectives.

Waste and soil characteristics are useful in prescreening in-place alternatives for potential applicability in meeting remedial objectives. However, the present state of the art of in-place treatment is not sufficiently advanced to have a comprehensive data base, or to allow calculation of optimum conditions for degradation, detoxification, and/or immobilization of waste constituents. For those alternatives identified as potentially applicable, bench scale treatability studies will generally be necessary to assess technical feasibility further and to establish design parameters for implementation and costing. The combined use of waste, soil and system characteristics in the assessment of several generic in-place alternatives is illustrated below.

- No Action Alternative

In situations where the intrinsic properties of the soil are adequate to block migration of waste constituents along pathways of concern, the no action alternative is likely to be the most cost-effective. In all cases, the inherent assimilative capacity of the soil for the waste constituents of concern should be evaluated (see Overcash and Pal, 1979 and USEPA, 1983).

Even if naturally occurring degradation, immobilization, and attentuation processes are deemed inadequate to meet remedial objectives, assessment of these processes provides a useful baseline for designing additional in-place remedial measures. For example, soil systems containing high clay and organic content may adequately immobilize both heavy metal and organic chemical constituents of waste by natural adsorption processes, provided that the assimilation capacity of adsorption sites is not exceeded. Waste sites where soils are predominantly sandy would have little potential for immobilizing waste constituents. No action might be appropriate for the clay soils, but not for the sandy soils. For two equally contaminated sites of the same size, less additional adsorbent might be required to immobilize constituents in the clay soil where its natural assimilative capacity has been exceeded than in the sandy soil.

- Extraction Alternative

Flushing or "solution mining" of contaminants requires identification and bench-scale testing of potentially suitable solvents and the design of a recharge/discharge system to effect flushing in the field.

- Immobilization Alternative

Immobilization processes are designed essentially to reduce leaching (or volatilization) of hazardous waste constituents to acceptable levels. Depending upon the particular waste/soil system, immobilization may be accomplished through physical or chemical adsorption, ion exchange, or precipitation. In general, immobilization of contaminants at uncontrolled hazardous waste sites requires the incorporation of physical or chemical adsorbents into

TABLE 2-2. TREATMENT VARIABLES FOR IN-PLACE TREATMENT TECHNOLOGIES

Technology	Treatment Variables	Comments
EXTRACTION	• Solubility of waste constituents in solvent • Concentrations of waste constituents • Rate of dissolution of adsorbed species	Solubilities of constituents vary with different solvents.
IMMOBILIZATION		
Sorption	• Adsorptive capacity of sorbent for specific waste constituents • Concentrations of sorbates, including naturally occurring substances • Soil pH • Soil moisture • Soil/water partition coefficient (K_{oc})	All sorbates compete for adsorptive capacity of sorbent.
Ion Exchange	• Cation (or anion) exchange capacity of clay, resins, or zeolites • Concentration of ions in soil water • Soil pH	Clays and zeolites are used to treat cations (metal ions), resins for both cations and anions.
Precipitation	• Concentration of soluble metals • Stability of precipitate relative to dissolved species • Soil pH • Oxidation state of metal ions • Soil oxygen content • Reactivity of precipitate	Sulfides may transform to more soluble sulfates under long periods of aerobic soil conditions
DEGRADATION		
Chemical	• Redox potential of waste/soil system • Soil oxygen content • Presence of catalyst(s) • Polymerization potential of wastes • Soil pH • Soil moisture • Soil temperature	Oxidation/reduction products may be more mobile and/or more toxic than the parent compound; chemical oxidants are non-selective and may preferentially oxidize organic matter in soils.

(continued)

TABLE 2-2 (Continued)

Technology	Treatment Variables	Comments
Biological	• Biodegradability of waste constituents (half-life) • Soil pH • Micro-organisms present (type; population) • Soil oxygen content • Soil moisture • Soil nutrient content (C:N:P ratio) • Soil temperature • Rates of biodegradation, or rates of leaching	Caution is needed to ensure that contaminants are not toxic to micro-organisms. Chemical-specific micro-organisms may be encouraged by manipulating treatment variables.
Photolysis	• Absorption spectra of waste constituents (> 290-nm wavelength range) • Half-life of photolysis; photolysis products • Volatility of waste constituents (vapor pressure, Henry's Law Constant)	Photolysis products may be more toxic than parent compound.
ATTENUATION	• Site-soil assimilative capacity • Feasibility of mixing uncontaminated material with contaminated soils	Assimilative capacities are available for many heavy metals, but not for organics.
REDUCTION OF VOLATILIZATION	• Volatility of waste constituents and dependence on temperature (vapor pressure, Henry's Law Constant) • Soil moisture • Soil temperature	

Source: Arthur D. Little, Inc.

the contaminated soil, or the addition of ion exchange or precipitating agents. Optimum additive concentrations are best determined using laboratory treatability studies with the waste/soil system. Optimum conditions for in-place immobilization can be selected by varying the treatment parameters and determining the minimum rate of leaching of the constituents of concern. The minimum achievable leaching rate should then be compared with the desired rate to meet remedial objectives.

- Degradation Alternative

Management of the soil/site system to increase the rate of biodegradation is a potential option to be considered for sites contaminated with organic chemical wastes. As discussed above, part of the site characterization will involve a determination of the presence or absence of micro-organisms (e.g., aerobic bacteria, anaerobic bacteria, actinomycetes, fungi, and algae) for representative areas of contamination. Part of the waste characterization will involve analyses to determine the carbon:nitrogen:phosphorus ratio. If the C:N:P ratio is not optimal for micro-organism growth, bench-scale experiments should be conducted to determine whether adjustment of this ratio will increase the concentration of micro-organisms in the soil and the rate of biodegradation. These initial experiments should be done with samples of contaminated soils from the site at dissolved oxygen concentration levels characteristic of the respective contaminated areas if possible.

Experiments should also be done to determine the effect of various oxygen levels on soil micro-organism count and rate of biodegradation. If nutrient and oxygen adjustments are insufficient to stimulate the growth of micro-organisms or to increase the biodegradation rate adequately, experiments might be performed to innoculate the soil/waste mixture with activated sludge, commercially developed micro-organisms likely to feed on the particular contaminants in the soil, or soil from the site or control area that has a high micro-organism count. Rates of biodegradation should be measured systematically as a function of nutrient concentrations, oxygen, and micro-organisms to establish optimum conditions for degradation.

Once these conditions have been determined, an in-place method capable of achieving these conditions in the field would have to be designed. There are a number of possibilities. For contaminated areas within 2 feet of the surface, simple tilling of the surface soils may introduce sufficient oxygen into the system to accelerate the rate of biodegradation. Depending upon the laboratory results, it may also be necessary to innoculate the surface with additional micro-organisms, adjust the pH and nutrients, and/or adjust the moisture content. For contaminated areas at greater depths, oxygen nutrient and possibly micro-organism additions will almost certainly be required to achieve satisfactory degradation rates. Conceivably, deeper constituents could be flushed out of the soil and reapplied in a controlled manner to the soil surface for biological treatment as described above for surface contaminants.

In areas where contaminant concentrations are so high as to be toxic to micro-organisms, the contaminated soil might be mixed with uncontaminated soil to reduce overall concentrations to levels that could then be successfully biodegraded in-place.

In the soil/waste system, biodegradation under the optimum conditions established in the laboratory experiments will generally compete with leaching, or volatilization of the waste constituents or their intermediate degradation products. Laboratory experiments should be performed to compare degradation rates with leaching and volatilization rates. If leaching or volatilization is rapid compared to biodegradation, it may be necessary to modify soil properties for some distance along the pathways of migration to immobilize the biodegradable constituents.

The laboratory experiments performed to establish the optimum achievable rate of degradation should determine not only the rate of disappearance of the parent compound, but also the chemical identity and rate of disappearance of the degradation products over time. The toxicity of partially degraded fractions should be determined either from the literature or via a bioassay. If the biodegradation products are toxic and do not degrade, then biodegradation would generally have to be considered infeasible.

Depending upon the nature of the contaminants present and the treatment objectives, chemical reagents may also have to be incorporated into contaminated surface soils, or injected into deeper contaminated layers, to bring about detoxification or degradation reactions. Such reactions might involve oxidation, reduction, or polymerization, for example. Appropriate reagents, reaction rates, and the leachability and toxicity of reaction products will generally have to be determined by laboratory treatability studies.

- Attenuation Alternative

Federal and State guidelines for land application of sewage sludges specify acceptable levels of heavy metal accumulation in soils (mg/kg). If these or other specified levels are exceeded at a site, they can be attenuated to acceptable levels by mixing with clean soil. The amount of soil required is determined by the metal that must be attenuated to the greatest extent to meet standards. Attenuation is also applicable, in principle, to organic contaminants. In practice, assimilative capacities of soils, which must be known to design attenuation systems, have only been established for heavy metals.

- Reduction of Volatiles

Suppression of vapor transport from contaminated soils may be desirable if contaminants or their degradation products can escape from the untreated soils at levels and rates potentially damaging to target organisms. Simple models are available to estimate the loss and transport of volatiles from soil systems. These models generally require data or estimates of the vapor pressure and solubility of the contaminants, as well as knowledge of the meteorological conditions at the site. Effectiveness of methods to reduce volatiles should be tested in the laboratory to establish design parameters for a field system (Farmer et al., 1980).

2.5.1.4 Laboratory and Pilot-scale Testing

Laboratory (bench-scale) and pilot-scale testing may be required to evaluate the technical and practical feasibility of in-place treatment methods for contaminated soil prior to full-scale implementation. Testing may be used to establish the following at in-place treatment sites:

1) The critical soil level for the waste at which treatment (degradation, detoxification, and/or immobilization) is ineffective due to toxicity or mass flow conditions.

2) The rate of degradation or detoxification of organic constituents, i.e., the half-life, and the extent of immobilization of inorganic constituents.

3) The mobility, toxicity, and biodegradability of partially degraded waste constituents or waste fractions.

4) Criteria for management of the soil and site to enhance the natural ability of the soil to attenuate constituents by determining optimum conditions for degradation, detoxification, and/or immobilization.

5) Parameters and constituents that should be monitored to indicate contaminant migration to receiver systems including groundwater, surface water, and atmosphere.

6) Technical feasibility and potential costs associated with using techniques based on management of the site/soil assimulative capacity for accomplishing the treatment required for a particular site/soil/waste system.

In principle, the direct costs for labor and equipment for laboratory or pilot-scale testing, as well as the indirect costs of delaying implementation until test results are obtained, will be determined by the complexity of the experimental design. That will depend, in turn, on the number of experimental variables to be controlled and/or monitored, the types of equipment required, the length of time and number of samples required to obtain the desired information with the requisite precision, or reliability, and prior operating experience with the process. Each of these factors will be specific to the process under study and thus cannot be specified *a priori*.

Certain elements of the testing program, however, are likely to be common to most other laboratory and pilot-scale testing programs. Among those elements are statistical sampling requirements, sampling equipment, sample collection, preservation, shipping and storage, and chemical analysis. Basic considerations and reference sources for each of these areas are described briefly under the respective headings below.

2.5.1.4.1 Statistical Sampling Requirements

The purpose of sampling is to permit estimation of the properties of interest in a cost- and time-effective manner. The sampling program chosen to achieve this objective will depend on the precision required of the estimation. Sampling plans commonly used for soil testing programs include simple random sampling, stratified random sampling, and systematic sampling. Statistical considerations associated with these and other types of sampling plans are discussed in detail in Black, 1965; Walsh and Beaton, 1973; and "Guidelines for Data Acquisition and Data Quality Evaluation in Environmental Chemistry," 1980.

2.5.1.4.2 Sampling Equipment

The sampling equipment used to obtain samples for subsequent testing depends on the type of sample required and on the soil type and depth from which the sample is to be obtained. The materials from which the sampling equipment is constructed must be chosen so that those surfaces in direct contact with the soil will not contaminate the sample with any species which may interfere in the chemical analysis. General descriptions of the types of equipment which may be used for soil sampling with directions for their use are in Soil Conservation Service, 1972; and U.S. EPA, 1982.

2.5.1.4.3 Sample Collection, Preservation, Shipping and Storage

Procedures used for sample-handling and storage will depend on the characteristics to be determined. In the simplest case, the sample may be air-dried under ambient conditions and subsequently stored in clean containers until analyses are performed. However, special precautions may be required in cases where the characteristic to be determined may be affected by the conditions of handling, shipping, and storage. Those precautions may include addition of chemicals to fix or retard changes in the concentration of the species of interest, freezing the sample, or performing the analysis at the time of collection. Guidelines for sampling handling and storage are provided in U.S. EPA, 1982; and American Public Health Association, 1975.

2.5.1.4.4 Chemical Analysis

Methods applicable to the determination of a wide variety of organic and inorganic compounds, elements, anions, and physical and chemical properties are readily available in the technical literature. Representative sources include: Walsh and Beaton, 1973; U.S. EPA, 1982; American Public Health Association, 1975; David, 1978; and Purdue University, 1977.

2.5.1.4.5 Monitoring

In general, laboratory and pilot-scale testing do not simulate field conditions exactly. Therefore, the effectiveness of the treatment process selected must be monitored after implementation. Soil core and soil pore water in the treatment zone and along pathways of migration must be sampled and analyzed to determine whether the treatment process is functioning according to design. Appropriate parameters to monitor would generally be selected from among the waste constituents expected to be treated, and the degradation products of treatment identified in treatability studies or in the literature as indicators of the success of treatment. If the treatment process is not as effective in the field as had been predicted from the laboratory or pilot studies, improvement may be possible by adjusting operating conditions. A more complete discussion on monitoring is provided in Volume 2.

2.5.2 Selection of In-Place Treatment Technology

The assessment of waste, soil, and site-specific variables described above will result in identification of technically feasible in-place treatment technology. Further analysis of each of the technically feasible alternatives is required to determine which (if any) meet the cost-effectiveness criteria of the National Contingency Plan.

The data assembled for assessing technical feasibility should be sufficient to form the basis of initial detailed engineering designs and operating specifications for each in-place treatment system. The designs and specifications in turn form the basis for developing detailed capital, operating and maintenance cost estimates over the lifetime of the remedial program.

The initial implementation plan should then be examined in some detail, and refined as appropriate. Several iterations may be necessary. First, the plan should be examined for completeness and logical consistency with respect to engineering implementation and constructability. Have any steps for successful implementation, operation, and monitoring been overlooked? Are all steps in logical order? Is the schedule realistic? Have all steps been properly costed? Next, an assessment needs to be made of how well each alternative meets the remedial objectives. This may require mathematical modeling of contaminant migration patterns, based on the modified waste/soil system characteristics that would be anticipated if the remedial alternatives were implemented. In addition, any adverse environmental impacts should be identified and assessed. Methods and costs for mitigating these impacts should be incorporated into a revised implementation plan. Finally, each in-place treatment process design that meets the remedial objectives should be rank ordered in terms of annualized cost, and the most cost-effective in-place alternative(s) should be selected for comparison with other alternatives.

2.6 DETERMINATION OF EXTENT OF REMEDY

The process described above, along with a similar process for other technologies, should provide enough information to enable selection of the most cost-effective alternative. The results of the analysis for the in-place treatment alternatives that survive screening are compared with other alternatives that have also been analyzed in detail. The lowest cost alternative that reduces risk to an acceptable level and meets the other objectives of the remedial program is then selected.

2.7 IN-PLACE INFORMATION GAPS

The evaluation of the in-place techniques discussed briefly in this section and in more detail in the next section is hampered by the general lack of information pertaining to remedial applications. The missing information is of two types. The first is general information about the technologies in any setting, remedial action, land treatment, agriculture, and the like. A number of the techniques have been developed for and applied in the latter two areas, but even these are often in the early stages of development.

The second type of information gap is related to specific information on individual technologies. Data on the natural, intrinsic (assimilative) capacity of soil systems to support these in-place technologies are needed, as are data on means to enhance the natural capacity by the addition of reagents, or the modifications of the soil/waste system.

To begin to address some of the information gaps described below, the Utah Water Research Laboratory has initiated studies — in cooperation with the U.S. Environmental Protection Agency, Robert S. Kerr Environmental Research Laboratory — to investigate in-place treatment of complex hazardous wastes for application to in-place remedial action as well as to land treatment systems. The experimental studies focus on high soil concentrations and assimilation capacities for complex hazardous wastes, and on volatilization of hazardous constituents from soil systems. The goal of the work is to be able to relate waste composition on a constituent basis to soil-treatment capacities. Assimilation capacities with respect to degradation, detoxification, and immobilization will be determined to generate a data base for establishing the capabilities and limitations for soil treatment of hazardous waste constituents.

2.7.1 General Information Gaps

The ability to utilize in-place treatment will be improved by developing a broader data base containing the following:

1) Information obtained from the controlled application of *complex* hazardous wastes (including solid, liquid, and semi-solid wastes) to soil systems in laboratory or field experiments. Information must specifically include mass balance for degradation, mobilization, volatilization, photodegradation, and rates of reaction.

2) Information obtained from *field*-scale applications of hazardous wastes, either controlled or uncontrolled, in which environmental sampling includes waste and soil parameters through the soil profile, soil core and soil pore-liquid, volatile fugitive emissions, and groundwater.

3) Information obtained from controlled experiments on a field scale, comparing effectiveness of adding treatment agents (e.g., chemicals, micro-organisms, adsorbents) compared to a non-treatment agent addition for soil treatment (degradation, transformation, immobilization) of constituents in complex hazardous wastes.

2.7.2 Technology-specific Information Gaps

Several areas in which specific information concerning in-place treatment is scarce or completely lacking have also been identified. These areas relate to the treatment processes in a soil system, viz., degradation, transformation, and immobilization. Such information needed to determine the assimilative capacity, or quantitative ability of the soil system to prevent contamination of air, groundwater, and surface water is described below.

1) Information on soil processes interfacing with atmospheric processes (i.e., controlling volatilization and photodegradation), including:

- investigation of photochemical reactions that enhance biodegradation of refractory compounds and those that can produce toxic or undesired breakdown products;

- laboratory and field-scale analysis of the feasibility of using volatilization/photodegradation as a viable treatment method for volatile/photoreactive hazardous chemicals;

- evaluation of the importance of soil-photochemical reactions to the fate and behavior of photoreactive compounds within the environment;

- refinement of containment/transport models that address volatilization, vapor adsorption, biodegradation; and leaching through the soil matrix; and

- continued development of rapid partition coefficient estimation methods for predicting compound partitioning among air/water/soil systems.

2) Information on chemical reactions relating to the transformation and immobilization of constituents in soil systems, including:

- effects of chemical addition (oxidants, reductants, and polymerizing agents) on the soil properties affecting treatment;

- constituent sorption and precipitation studies at concentrations in the soil matrix characteristic of those found at remedial action sites,

- Eh-pH information for metal species in soils at concentrations representative of remedial action site concentrations; and

- behavior of specific metals, viz., arsenic, beryllium, silver, selenium, mercury, and chromium in soil systems, including the investigation of the reaction of chromium (VI) to chromium (III).

3) Information on biological reactions to stimulate the biodegradation of constituents in soil systems, including:

- potential for combinations of chemical and biological treatment methods for accelerating soil treatment;

- potential for composting of hazardous waste contaminated soil for accelerating biological reactions and for detoxifying and degrading recalcitrant organic constituents;

- evaluation of the biodegradation kinetics and extent of hazardous constituents and metabolites of hazardous constituents sorbed to clay and soil organic matter;

- degradation rates and pathways for chemical classes, including alkyl halides and highly chlorinated organics (PCBs) in the soil treatment zone (upper 5 feet of soil); and

- information concerning the effectiveness of micro-organism seeding compared with no seeding, specifically in terrestrial systems.

4) Information on immobilization reactions in the soil/waste system, including:

- investigation of enhancement of sorption by adding different sorbents, such as activated carbon, straw, synthetic resins. (The specific information needed includes application rates of sorbent in relation to concentrations of contaminant in soil);

- continued development of estimation methods for predicting soil adsorption constants by chemical and physical properties of compounds;

- long-term effectiveness of immobilization of sorbed contaminants on soil; and

- effects of solvents on mobility of organic contaminants in soil systems.

5) Information on leaching in the soil/waste system, including:

- laboratory experiments to estimate parameters for transport models that can be used to predict behavior under field conditions;

- field instrumentation to validate transport models under imposed environmental conditions, incorporating the variability under field conditions by designing appropriate sampling stations; and

- effect of different environmental conditions, including soil moisture, redox potential, and organic content on transport parameters.

3. Technologies for In-Place Treatment

3.1 INTRODUCTION

This section presents detailed information on specific in-place treatment technologies. Each technology was selected on the basis of its potential or demonstrated ability to augment natural soil processes to accomplish in-place treatment.

The technology discussions are divided into the five treatment categories presented in Section 2; viz., extraction, immobilization, degradation, attenuation, and reduction of volatiles. Each discussion is presented in the following format:

- Description — a qualitative discussion of the technology and principles on which it is based.
- Wastes amenable to treatment — a discussion of the kinds of wastes suited to the treatment in terms of waste type (organic/inorganic), chemical class, or certain required properties of the class.
- Status of technology — a measure of the availability of the technology and degree to which it has been demonstrated for soil treatment. "Field" means that it has been used for soil treatment in practice, either large-scale or in small pilot studies. "Laboratory" means that the technology has been demonstrated for soil treatment in small experiments in laboratories. "Conceptual" means either that it is still purely theoretical, or that it has been used for treatment other than soil treatment but no laboratory or field work exists for soil treatment.
- Ease of application — a qualitative description of the relative difficulty of implementing the technology in the field. "Easy" means that application problems are not expected under most conditions, while "difficult" means that problems are usually expected. "Easy to difficult" means that ease of application varies considerably, depending on site-specific conditions.
- Potential achievable level of treatment — a description (high, low, variable) of the conceptual or theoretical level of treatment which could be obtained independent of site-specific considerations. A "high" potential achievable level of treatment for reduction, for example, means that wastes applicable to treatment by reducing agents may be effectively degraded by reduction. Site conditions, however, may interfere with technology performance.
- Reliability of method — a brief discussion of the long-term effectiveness of the technology, including reversibility of treatment and retreatment needs. If there is a lack of information on long-term effectiveness, the reliability is designated as "unknown."
- Secondary impacts — a discussion of the effects of technology implementation in addition to its intended primary effect. The impacts discussed include effects on soil properties, site conditions, and the enhancement or retardation of other in-place treatment processes.

- Equipment and exogenous reagents — a brief description of the equipment and materials that may be required to implement the technology. Additional information on equipment and reagents is presented in the appendix.

- Information requirements — a list of information, including waste, soil, and site factors discussed in Section 2, that is needed for application of the technology.

- Sources of information — a listing of the publications used in preparing the technology discussions, including some that are not specifically referenced.

A matrix of important summary information for each of the technologies is presented in Table 3-1.

3.2 EXTRACTION (SOIL FLUSHING)

Description

Removal of soil contaminants can be accomplished through extraction (soil flushing or solution mining). This involves the elutriation of organic and/or inorganic constituents from soil for recovery and treatment. The site is flooded with the appropriate flushing solution and the elutriate is collected in a series of shallow well points or subsurface drains. The elutriate is collected, treated, and/or recycled back into the site. Collection of elutriate is required to prevent uncontrolled contaminant migration through uncontaminated soil and into receiver systems including groundwater and surface water. An example of a soil flushing system with elutriate recycle is given in Figure 3-1.

Flushing solutions may include water, acidic aqueous solutions (sulfuric, hydrochloric, nitric, phosphoric, and carbonic acid), basic solutions (e.g., sodium hydroxide), and surfactants (e.g., alkylbenzene sulfonate). Water can be used to extract water-soluble or water-mobile constituents. Acidic solutions are used for metals recovery and for basic organic constituents, including amines, ethers, and anilines, and basic solutions for metals, including zinc, tin, and lead and for some phenols, complexing and chelating agents, and surfactants (U.S. EPA, 1982).

Addition of any flushing solution to the system requires careful management and knowledge of reactions that may adversely affect the soil system. For example, a sodium addition as sodium hydroxide to soil systems may adversely affect soil permeability by affecting the soil sodium absorption ratio. It is not only important to understand the chemical reaction(s) between solvent and solute, but also between solvent and site/soil system.

For a site contaminated by organic constituents, it may be possible to recycle the elutriate back through the soil for treatment by biodegradation. Proper control of the application rate, based on hazardous waste land treatment principles (U.S. EPA, 1983), would provide for effective in-place treatment at soil concentrations that would allow controlled biodegradation of the waste constituents. Using this approach may eliminate the need for separate processes for treatment and disposal of the collected waste solution, or at least provide for a combination of pretreatment/land application that may be considerably more economical than using unit operations alone for treatment of elutriate.

For soils contaminated with inorganic and organic constituents, a combination of pretreatment land applications where the metal constituent(s) is reduced or eliminated in the elutriate by precipitation, followed by land application of the elutriate, may be a feasible, cost-effective method of treatment.

Soil flushing and elutriate recovery may also be appropriate for situations where chemical oxidizing or reducing agents are used to chemically degrade waste constituents resulting in the production of large amounts of oxygenated, mobile, degradation products. The most conservative and safest approach may be to flush the soil after treatment to recover and possibly reapply the elutriate in a controlled manner to the soil surface.

TABLE 3-1. SUMMARY MATRIX OF TREATMENT TECHNOLOGIES

Technology	Wastes Amenable to Treatment	Status	Ease of Application	Potential Level of Treatment	Reliability
EXTRACTION	Soluble organics and inorganics	Laboratory	Easy – Difficult	Variable	Good
IMMOBILIZATION					
Sorption (heavy metals)					
Agri. products	Heavy metals	Field	Easy – Difficult	High	Retreatment required
Activated carbon	Heavy metals	Conceptual	Easy – Difficult	Unknown	Unknown
Tetren	Heavy metals	Laboratory	Easy – Difficult	High	Unknown
Sorption (Organics)					
Soil moisture	Organics, non-volatile, $K_d < 10$	Conceptual	Easy – Difficult	High	Retreatment required
Agri. products	Organics	Laboratory	Easy – Difficult	High	Retreatment required
Activated carbon	Organics, low water solubility	Field	Easy – Difficult	Low – High	Unknown
Ion Exchange					
Clay	Cationic components	Laboratory	Easy – Difficult	High	Good
Synthetic resins	Certain cationic and anionic compounds	Laboratory	Easy – Difficult	Variable	Unknown
Zeolites	Heavy metals	Conceptual	Easy – Difficult	Unknown	Unknown
Precipitation					
Sulfides	Heavy metals	Conceptual	Difficult	High	Fair
Carbonates, phosphates and hydroxides	Heavy metals	Laboratory	Easy – Difficult	Unknown	Retreatment required
DEGRADATION					
Oxidation					
Soil catalyzed reactions	Aliphatic organics, other organics	Limited field	Easy – Difficult	Variable	Good
Oxidizing agents	Various organics	Limited field	Moderate – Difficult	High	Good
Reduction					
Organics	Chlorinated organics, unsaturated aromatics, aliphatics	Limited field	Easy – Difficult	High	Retreatment required
Chromium	Hexavalent chromium	Limited field	Easy – Difficult	High	Retreatment required
Selenium	Hexavalent selenium	Limited field	Easy – Difficult	High	Retreatment required

(continued)

TABLE 3.1 (Continued)

Technology	Wastes Amenable to Treatment	Status	Ease of Application	Potential Level of Treatment	Reliability
Sodium reduction	PCB, dioxin, halogenated compounds	Conceptual	Moderate	High	Good
Polymerization	Aliphatics, aromatics, oxygenated organic compounds	Expt. field	Moderate – Difficult	Variable	Unknown
Modification of soil Properties (biodegradation)					
Soil moisture	Organics	Field (Agri.)	Easy – Difficult	Low – High	Retreatment required
Soil oxygen – aerobic	Organics	Field	Easy – Difficult	Low – High	Retreatment required
Soil oxygen – anaerobic	Halogenated organics	Conceptual	Moderate – Difficult	Low – High	Retreatment required
Soil pH	Organics	Field	Easy – Difficult	High	Retreatment required
Nutrients	Organics	Field	Easy – Difficult	High	Retreatment required
Nonspecific organic amendments	Organics, arsenite wastes	Laboratory	Easy – Difficult	Low – High	Retreatment required
Analog enrichment for cometabolism	Some organics with analogs	Laboratory	Easy – Difficult	Low – High	Unknown
Exogenous acclimated or mutant micro-organisms	Various organics	Field	Easy – Difficult	High	Retreatment required
Cell-free enzymes	Organics	Laboratory	Difficult	High	Unknown
Photolysis					
Proton donors	Some organics, including TCDD, Kepone, PCB	Field	Easy – Difficult	High	Unknown
Enhance volatilization	Specific organics	Conceptual	Easy – Difficult	High	Good
ATTENUATION					
Metals	Metals	Field	Easy – Difficult	High	Good
Organics	Organics	Limited field	Easy – Difficult	High	Good
REDUCTION OF VOLATILIZATION					
Soil vapor volume	Volatile organics and inorganics	Laboratory	Easy – Difficult	Low – Medium	Retreatment required
Soil cooling	Volatile organics	Expt., limited field	Difficult	Low – Medium	Retreatment required

Source: Arthur D. Little, Inc.

Figure 3-1. Schematic of an elutriate recycle system.

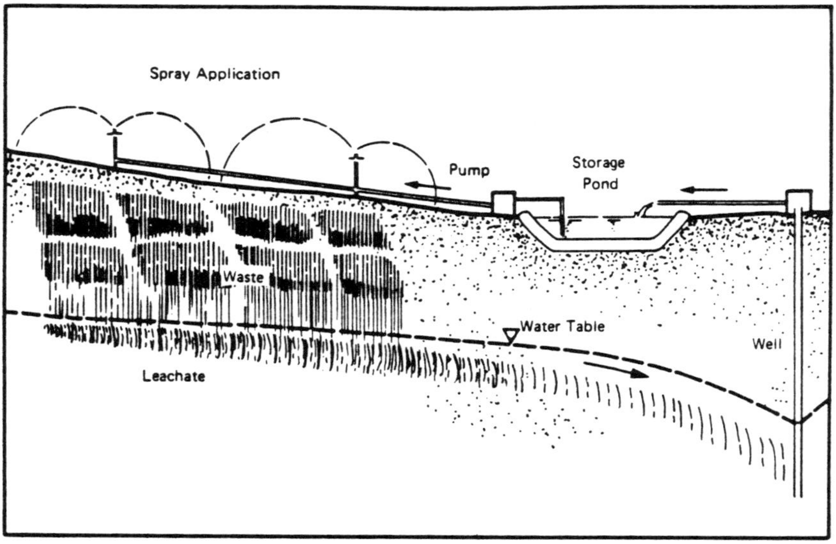

Source: Ehrenfeld, J.R., and Bass, J.M., 1983.

Wastes Amenable to Treatment

Both inorganics and organics are suitable for soil flushing treatment if they are sufficiently soluble in an inexpensive solvent that is obtainable in a large enough volume. Surfactants can be used for hydrophobic organics.

Status of Technology

This technology is currently at the laboratory stage. Studies have been conducted to determine appropriate solvents for mobilizing various classes and types of waste constituents.

Ease of Application

This technology may be easy or difficult to apply depending on the ability to flood the soil with the flushing solution and the installation of collection wells or subsurface drains to recover all the applied liquids. Provisions must also be made for ultimate disposal of the elutriate.

Potential Achievable Level of Treatment

The achievable level of treatment is variable, depending on the contact of flushing solution with waste constituents, the appropriateness of solutions for the wastes, and the hydraulic conductivity of the soil. This technology is more applicable to highly permeable soils because the level of treatment will probably be higher.

Reliability of Method

Although the level of treatment may be variable, once the waste components are removed from the soil, the treatment is not reversible and no additional retreatment is necessary.

Secondary Impacts

The solutions used for the flushing may themselves be potential pollutants. They may have toxic and other environmental impacts on the soil system and water receiver systems. The soil system after treatment is altered from its original state. Its physical, chemical, and biological properties may be altered adversely; e.g., the pH may be lowered by the use of an acidic solvent, or the soil may be compacted from being flooded. These soil properties may have to be restored to assure that other treatment processes can occur (e.g., biodegradation).

Equipment and Exogenous Reagents

Equipment used for soil flushing include drains and an elutriate collection and distribution system. The solvents for flushing are required.

Reapplication of collected elutriate may require construction of a holding tank for the elutriate.

Information Requirements

- characterization and concentration of waste constituents;
- depth, profile, and areal distribution of contamination;
- partitioning of waste constituents between solvent(s) and soil;
- effects of flushing agent (solvent) on soil physical, chemical, and biological properties;
- suitability of site to flooding and installation of wells and/or subsurface drains;
- trafficability of soil and site.

Sources of Information

Botre, C., et al., 1978; U.S. EPA, 1982, 1983.

3.3 IMMOBILIZATION

3.3.1 Sorption

The sorption of a pollutant refers to processes which result in a higher concentration of the chemical at the surface or within the solid phase than is present in the bulk solution of soils. Actual sorption mechanisms are often not known. Sorption is the major general retention mechanism for many organic compounds and metals. Adsorbed compounds or ions are in equilibrium with the soil solution and are capable of desorption (Bonazountas and Wagner, 1981).

Several processes are involved in sorption including:

- ion exchange.
- physical sorption through weak atomic and molecular interaction forces (van der Waal forces).

- specific adsorption exhibited by anions involving the exchange of the ion with surface ligands to form partly covalent bonds, and
- chemisorption involving a chemical reaction between the compound and the surface of the sorbent.

We will discuss ion-exchange separately as another immobilization technology in Section 3.3.2.

Soil sorption is perhaps the most important soil-waste process affecting immobilization of toxic and recalcitrant fractions of hazardous wastes. Leaching potential and the residence time in soil for constituents which undergo degradation are directly affected by the extent of immobilization.

The sorption process is usually described by an adsorption isotherm, which expresses the relationship between the amount of constituent adsorbed onto a solid (soil, activated carbon, zeolite, organic matter, etc.), and the concentration of solute in solution at equilibrium. One frequently used relationship is the Freundlich isotherm, which is expressed as:

$$S = KC^N \qquad (3\text{-}1)$$

where

S = amount of constituent adsorbed per unit dry weight of soil,
K and N are constants, and
C = solution phase equilibrium concentration.

The percentage sorbed under natural moisture conditions can be estimated by:

$$\% \text{ sorbed} = \frac{100}{1 + \left(\frac{S}{K}\right)^{1/N} \frac{\Theta}{S}} \qquad (3\text{-}2)$$

where θ = fraction soil moisture content (weight basis). When the sorption is linear, N = 1, and the percent sorbed is not a function of the amount sorbed per unit weight of soil (S). However, when the sorption is not linear, the percent sorbed becomes a function of S. Figure 3-2 shows the percent of chemical sorbed as a function of K values and for different S values.

Desorption of constituents is also important in affecting treatment effectiveness through the extent of release of chemicals from soil into percolating water moving through the soil. Generally the extent of desorption follows the Freundlich isotherm, but with constants different from the ones used for adsorption. This is due to hysteresis of a soil system exhibited during the sorption-desorption cycle. Factors directly associated with desorption include the amount of leachate (soil/water ratio) and the amount of constituent contaminating the soil (soil/constituent ratio). The extent of desorption will decrease with an increase of these ratios.

Treatment techniques to enhance immobilization of constituents by controlling or augmenting the sorption process in soils have been developed based on fundamental principles and applied land treatment techniques. These treatment techniques also provide engineering management tools for affecting treatment of constituents through degradation, especially biological degradation.

Because the mechanisms involved in the sorption of organics are significantly different from that of heavy metals, we will discuss these two groups separately.

Figure 3-2. Extent of sorption as a function of amount sorbed and K for θ = 0.1 and N = 0.9.

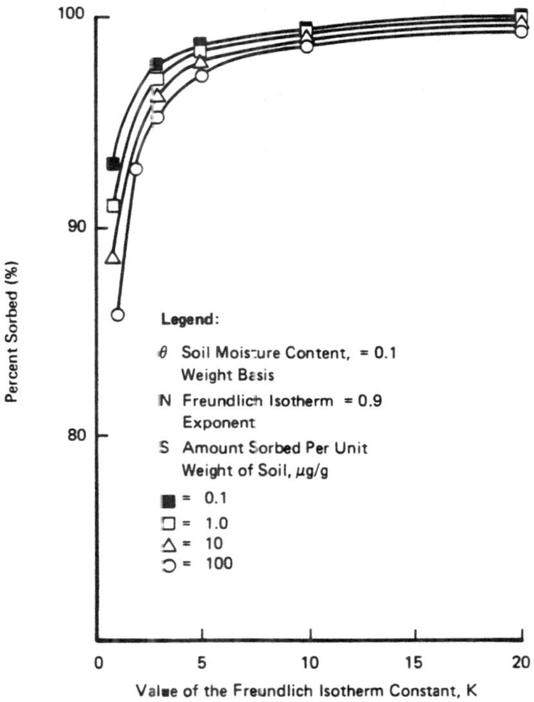

Source: Mahmood and Sims, 1984.

3.3.1.1 Heavy Metals — Addition of Sorbents

Many heavy metals have a strong affinity for organic matter. The retention of added metals is often well correlated with soil organic matter. Metals are readily chelated and/or complexed by functional groups in organic matter. These include -COOH, phenolic, alcoholic and enolic-OH and carbonyl (C = O) structures of various types. The stability of these metal-organic complexes increases with pH due to the increased ionization of the functional groups. Sorption processes are affected by 1) pH, 2) competing cations, 3) solvent present in a complexed waste, 4) presence of chelating agents, 5) solution ionic strength, and 6) types of anions present in the soil solution.

Theoretically, the addition of organic matter to a contaminated soil should remove metals from the soil solution, thus preventing their leaching in groundwater. Organic materials most conducive for use with soils include: agricultural products and by-products, and activated carbon.

3.3.1.1.1 Addition of Agricultural Products and By-products for Heavy Metals

Description

The use of various agricultural products and by-products, such as straw, sawdust, peanut hulls, and bark, for the removal of heavy metals from wastewater solutions has been recently suggested by Larsen and Schierup (1981) and Henderson et al. (1977 a,b). Larsen and Schierup (1981) found that 1 gram of barley straw was able to adsorb amounts of Zn, Cu, Pb, Ni, and Cd ranging from 4.3 to 15.2 mg, while pine sawdust removed 1.3 to 5.0 mg. The selective order of metal sorption for straw was Pb > Cu > Cd ≅ Zn ≅ Ni. The concentration of metals and the pH value of the metal solutions also affected the amount of metals sorbed. The obvious advantage of the use of waste organic materials compared to the use of other sorbing materials, such as activated carbon, is the lower cost involved when organic waste materials are used.

Agricultural products and by-products (e.g., animal manures, plant residues, and food processing wastes) have been used extensively as soil conditioners. Sewage sludge and animal manures have been used as fertilizer sources and as soil conditioners. The use of such waste materials for the removal of metals in soils has not been extensively studied.

Sewage sludges from municipal areas often contain high concentrations of heavy metals themselves. The use of such sludges should be avoided. Waste materials may also contain soluble organic matter that chelates metals and increases their mobility. More research is needed to identify the extent of increased mobility. Other factors also affect the sorption of metals, including the presence of competing cations and the ionic strength of the soil solution.

To obtain maximum sorption of metals by organic matter, soil pH must be adjusted and/or maintained at greater than 6.5. The addition of organic materials may result in a decreased pH, requiring continued pH adjustment.

Wastes Amenable to Treatment

Heavy metal wastes and organic wastes are amenable to treatment by the addition of organic materials. This discussion is devoted to heavy metals.

Status of Technology

Addition of agricultural materials has been used in the field for soil conditioning. In the laboratory, it has been conceptually evaluated for use in the removal of metals from wastewater.

Ease of Application

The quantity of organic material required cannot be predicted from stoichiometric relationships. Laboratory sorption studies are required. Liming may be required to maintain the pH of the soil system at greater than 6.5. The application process may be easy to difficult depending on the trafficability of the site and the depth of contamination. Thorough mixing is required for maximum waste and organic material contact. Because tilling is used, erosion control techniques may have to be implemented.

Potential Achievable Level of Treatment

This is an effective method for removal of metals from wastes. However, agricultural products and by-products are highly susceptible to microbial activity. Degradation of the materials may result in the release of metals. The long-term level of treatment achievable is probably much smaller than in the short-term unless treatments are repeated.

Reliability of Method

As mentioned above, the mineralization of the organic materials may result in the release of the sorbed metals. Repeated treatments may be necessary periodically. Reliming is likely to be a necessity as well, since mineralization of the organics would tend to reduce the pH of the soil system. Another important factor is the potential competition of the metals with organics which are also sorbable on organic materials.

Secondary Impacts

Organic materials may alter the properties of the soil system including:

- degree of structure;
- water-holding capacity;
- bulk density;
- immobilization of nutrients, hindering degradation of organic wastes;
- reduction in soil erosion; and
- soil temperature.

Organic materials may also result in excessive nitrate levels in receiving waters, depending on the nitrogen content and degree of mineralization of the material. Since tilling is used to incorporate the organic material into the soil, wind and water erosion may effect the soil surface adversely.

Equipment and Exogenous Reagents

The equipment used are power implements, tillers, and applicators. Organic material and liming materials are the exogenous reagents required.

Information Requirements

- characterization and concentration of metals present in waste;
- depth, profile, and areal distribution of contamination;
- soil pH;
- adsorptive selectivity and capacity of organic material for metals at the site;
- soil biological activity (mineralization potential of organic material); and
- trafficability of soil and site.

Sources of Information

Bonazountas, M., and Wagner, J., 1981; Brown, L.S., 1972; Coffey, D.L., and Warren, G.R., 1969; Coun. for Agri. Sci. & Tech., 1975; Damanakis, M., et al., 1970; Jacobs, L.W., 1977; Ko, W.H., and Lockwood, J.L., 1968; Shin, Y., et al., 1970; Tames, R.S., and Hance, R.J., 1969; Taylor, J.M., et al., 1980; U.S. EPA, 1976, 1977, 1978, 1980 1981; Voerman, S., and Tammes, P.M. 1969; Walker A., and Crawford, D.V., 1968.

3.3.1.1.2 Addition of Activated Carbon for Heavy Metals

Description

In water and wastewater treatment, most of the research and development effort related to the use of activated carbon is oriented toward organics, with relatively little attention being given to inorganics, specifically heavy metals. The soil literature lacks information on the use of activated carbon as a treatment method. The following discussion is based on the use of activated carbon in water treatment.

Several factors can contribute to heavy metal removal capability of activated carbon: 1) specific surface area, 2) pore size and distribution, and 3) surface chemical properties. Huang and Ostoric (1978) suggested that the chemical reactivity of the surfaces is of major importance. They found that, in general, powdered activated carbon (e.g., Nuchar S-N and Nuchar S-A) has a low pH at the zero point of charge (zpc) (Table 3-2) and excellent sorption capacity for heavy metals. Granular activated carbon (e.g., Darco HD 3000 and Filtrasorb 400), having a high pH_{zpc} is rather poor for metal ion adsorption. The pH_{zpc} value reflects the acidic nature of surface functional groups.

The effect of solution pH on Cd sorption by Nuchar-SA is illustrated in Figure 3-3. The data were obtained from a batch equilibrium study using a synthetic Cd-fluoroborate-plating wastewater and activated carbon (Huang and Smith, 1981). The initial linear segments illustrate that sorption of Cd by Nuchar-SA increases with pH. Above pH 6 to 7, precipitation reactions, $CdCO_3$ and $Cd(OH)_2$, become the major mechanism of removal of Cd from solution. Precipitation reactions have long been used as the method of removing metals from wastewater. Precipitation is often, however, a kinetically slower process than sorption.

Larsen and Schierup (1981) found that granular activated carbon removed from solution 6.2 to 19.5 mg of Zn, Cu, Pb, Ni, and Cd per g of carbon.

Figure 3-4 illustrates the extent of Cd sorption as affected by the carbon to cadmium ratio. Huang and Smith (1981), using these data, recommended a carbon-to-metal ratio of 0.25 g/mg at pH 7 for maximum efficiency of removal of Cd from wastewater. Work by Pillie et al. (1975) indicates that carbon-to-metal ratios of 0.1 g/mg are needed for maximum efficiencies in treating wastewater. Soils may require ratios in excess of this amount because sorption is affected by numerous factors, such as the presence of chelating agents, competing cations, and ionic strength of the soil solution. Sorption of metals by activated carbon occurs over a wide pH range, with maximum efficiency at a pH greater than 6.5. Adjustment of the pH of acid soils with liming materials may be necessary.

TABLE 3-2. TYPICAL SURFACE PROPERTIES OF ACTIVATED CARBONS

Carbon Type	Specific Surface Area (m^2/g)	Particle Size (mm)	pH_{zpc}
Nuchar S-N	1400-1800	0.04-0.10	5.84
Nuchar S-A	1400-1800	0.03-0.10	3.84
Darco HD 3000	550-650	0.84-4.75	a
Filtrasorb 400	941	0.90-1.00	7.10

a. Value not determined, but estimated to be 5-7 based on comparison with values of other Darco brand granular carbons and observations of pH drift in various solution.

Source: Huang and Smith, 1981. (See Copyright Notice)

Figure 3-3. Adsorption characteristics of Cd(II) from synthetic Cd(II)-plating wastewaters (Cd-BF_4 solution) as affected by pH.

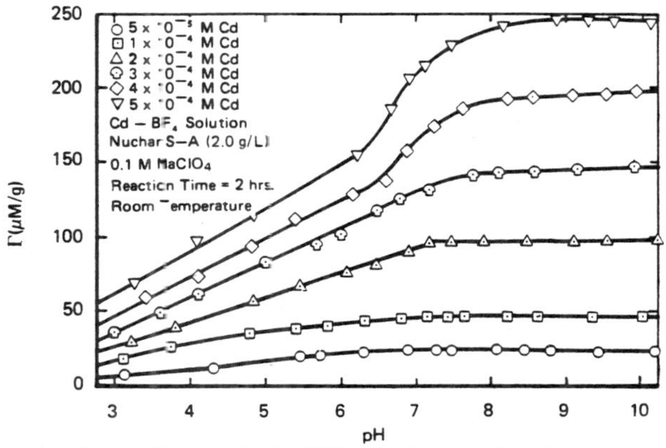

Source: Huang and Smith, 1981. (See Copyright Notice)

Figure 3-4. Extent of Cd(II) adsorption as affected by carbon/cadmium ratio at various pH.

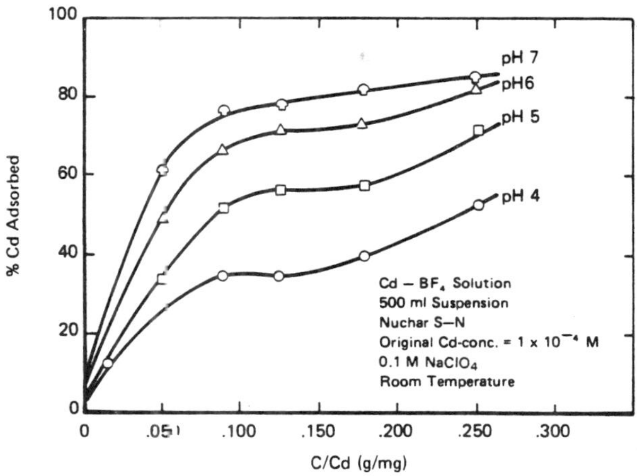

Source: Huang and Smith, 1981. (See Copyright Notice)

Activated carbon should be relatively resistant to degradation in the soil environment.

Wastes Amenable to Treatment

Heavy metal wastes (and organic wastes) are amenable to treatment with activated carbon.

Status of Technology

This technology is used conventionally in wastewater treatment, but its use in soil treatment is still conceptual in nature.

Ease of Application

Liming may be necessary to keep the soil in the pH range of optimum sorption. The application of this technology is easy to difficult, depending on the trafficability of the soil and site and the depth of contamination. Thorough mixing is required for maximum contact between the waste and the activated carbon. Because of the tilling that is required for incorporation of the activated carbon into the soil, controls to prevent erosion from runoff and controls for runon may have to be installed, depending on the site.

Potential Achievable Level of Treatment

This technology has been effective in wastewater treatment. However, removal efficiency is metal-specific. The potential achievable level of treatment for metals in soil is unknown because of competition with organics and because of the number of complicating variables in the soil system.

Reliability of Method

The reliability of this method is unknown. Desorption may be a problem in the long term because of competition for the activated carbon by organics in the soil. Also, changes in pH in the soil will affect the sorption. Reliming may be necessary to keep the metals adsorbed on the activated carbon.

Secondary Impacts

Tilled soils are usually more susceptible to water and wind erosion.

Equipment and Exogenous Reagents

The equipment needed includes power implements, tillers, and applicators. Activated carbon and liming materials are the exogenous reagents used.

Information Requirements

- characterization and concentration of metals present in waste;
- depth, profile, and areal distribution of contamination;
- soil pH;
- adsorptive selectivity and capacity of activated carbon for metals at site;
- zero point of charge of activated carbon;
- trafficability of soil and site.

Sources of Information

Huang and Astoric, 1978; Huang and Smith, 1981; Pillie et al., 1975.

3.3.1.2 Heavy Metals — Chelation with Tetren

Description

Metals will react with tetraethylenepentamine (tetren) to form stable metal chelates. Most chelating agents, such as EDTA, will increase metal mobility in soils and their use is to be avoided. Metal complexes with tetren are, however, strongly sorbed by soil clays, thus immobilizing the metal by enhancement of the natural soil sorption system. Smeulders et al. (1983) showed that the use of tetren increases the extent of metal sorption by soil clays. Their study indicated that the addition of tetren to a clay soil ensures the nearly complete sorption of rather large doses of Cu, Zn, Ni, and Cd, with efficiencies comparable to those for specific adsorption (which is restricted to a very small fraction of the exchange capacity)

To be effective, tetren must be applied to a soil relatively high in clays. Tetren-metal complexes are not sorbed by soil organic matter. If tetren is applied to an organic soil low in clays, the tetren-metal complex will remain in the soil solution and will be susceptible to leaching.

It is possible that the utility of tetren in adsorption to clay-like soils is due to the presence of metallic ions, such as iron, in the clay. To obtain a better understanding of the long-term stability of the tetren-metal complexes, it will be necessary to determine the prevalent form(s) of the complex for each metallic species.

It is possible, for example, to form a complex with a metal ion in which several of the amine groups in one tetren molecule are bonded to the same metal ion. Alternatively, only the terminal amine group (or possibly only the secondary amine groups) may be available for coordination to the metal of interest. The question of single versus multiple coordination of the amines to the metal of interest is extremely important, since it directly affects the effective stability of the resultant complex. For example, if multiple amines are coordinated to a single metal ion, exchange processes that remove only one amino group will not result in mobility of the metal ion. Indeed, if it is held by the remaining amino group coordination, subsequent exchange to permit recomplexation of the displaced amine group is still possible.

In addition, the question of complex type is also important with respect to adsorption to the clay materials. The higher the number of amine groups that are complexed to the clay substrate, the fewer there are available for the metal ion of interest. Tetren is a molecule which maximizes the number of amine groups available for metal complexation in a small stereochemical area, and deserves further study as a fixation method for metals known to form strong amine complexes.

The stability of tetren-metal complex against decomposition or degradation is not yet known. More research is also needed on soil factors affecting metal complexation with tetren and resulting sorption by clays, including soil moisture, pH and the ionic strength of the soil solution.

Wastes Amenable to Treatment

Heavy metal wastes are amenable to treatment with tetren. Tetren will likely form strong complexes with those metallic species that interact strongly with amines.

Status of Technology

There have been limited studies in the laboratory and greenhouse. It is still at the conceptual stage in the field. The stability of the tetren-metal complex against decomposition and degradation is not yet known. The soil factors affecting metal complexation with tetren and the resulting sorption of the complex by clays are not well understood. These soil factors probably include soil moisture, pH, and the ionic strength of the soil solution.

Ease of Application

Tetren should be added in sufficient amounts for the formation of a 1:1 complex with the metals present. To ensure the formation of such a complex, a ratio of at least 1:1.15 metal/tetren is advisable. Addition of clay materials may be necessary to increase sorption of tetren-metal complex. Liming may also be necessary if the soil is too acidic. The application may be easy to difficult, depending on the trafficability of the soil and the site and the depth of contamination. Thorough mixing is required for maximum contact between the waste and the tetren. Erosion and runon controls may be necessary because the tilling of the site may increase its susceptibility to erosion.

Potential Achievable Level of Treatment

The achievable level of treatment is potentially high, depending on the characteristics of the soil at the site and the complexation capacity of tetren for the metals present at the site.

Reliability of Method

The stability of the tetren-metal complex is not yet known. Also, its interaction with soil factors like soil moisture, pH, and soil ionic strength is not known. These uncertainties affect the predictability of the tetren-metal complexation effectiveness in soil. If tetren is applied to a soil low in clays, the tetren-metal complex will remain in solution and be susceptible to leaching. In general, this technology cannot be considered reliable at the present time.

Secondary Impacts

Tilling of the soil to incorporate the tetren may increase the susceptibility of the site to erosion by wind and water.

Equipment and Exogenous Reagents

Equipment needed for the application of this technology includes power implements, tillers, and applicators. Tetren and liming materials are required. If the soil at the site is low in clay, it might be necessary to incorporate clay as well.

Information Requirements

- characterization and concentration of metals present in waste;
- depth, profile, and areal distribution of contamination;
- soil pH;
- complexation capacity of tetren with metals at site;
- clay content of soil at site;
- organic matter content of soil at site;
- degree of sorption of tetren-complex with soils at site;
- trafficability of soil and site.

Sources of Information

Smeulders et al., 1983.

3.3.1.3 Organics — Reduction of Soil Moisture

Description

Controlling the moisture content of a soil can be utilized to immobilize constituents in contaminated soils, as well as allowing for additional time to accomplish biological degradation. Where immobilization of constituents using this technique is to be followed by anaerobic decomposition in a treatment train, anaerobiosis must be achieved with techniques other than flooding, such as soil compaction or organic matter addition.

Using the Freundlich isotherm, discussed previously, the percent sorbed under natural moisture conditions can be estimated by using Equation 3-2 (see subsection 3.3.1). When N is equal to 1 (i.e., linear isotherm), Equation 3-2 becomes:

$$\% \text{ sorbed} = \frac{100}{1 + (\Theta/K_d)} \qquad (3\text{-}3)$$

where K_d is the distribution coefficient of the chemical between soil and soil water and θ is the soil moisture content (weight basis). Figure 3-5 shows percent sorbed as a function of K_d for different values of θ. From Figure 3-5, it is seen that with lower moisture content, the percent of adsorbed chemical is higher. This is especially important for constituents with relatively small K_d values, i.e., constituents not strongly sorbed to soil. When the adsorption isotherm is nonlinear ($N < 1$) for a given amount sorbed per unit weight soil (S), controlling soil moisture will also follow the same trend as in the case of the linear isotherm, as shown in Figure 3-6.

Figure 3-5. Extent of sorption as a function of soil moisture θ and K_d.

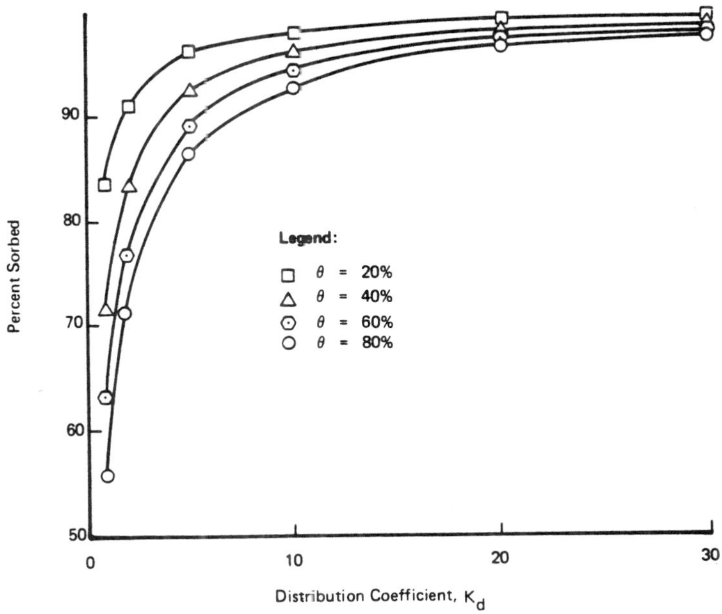

Source: Mahmood and Sims, 1984.

Figure 3-6. **Extent of sorption as a function of amount sorbed and K for a range of moisture contents and N = 0.7.**

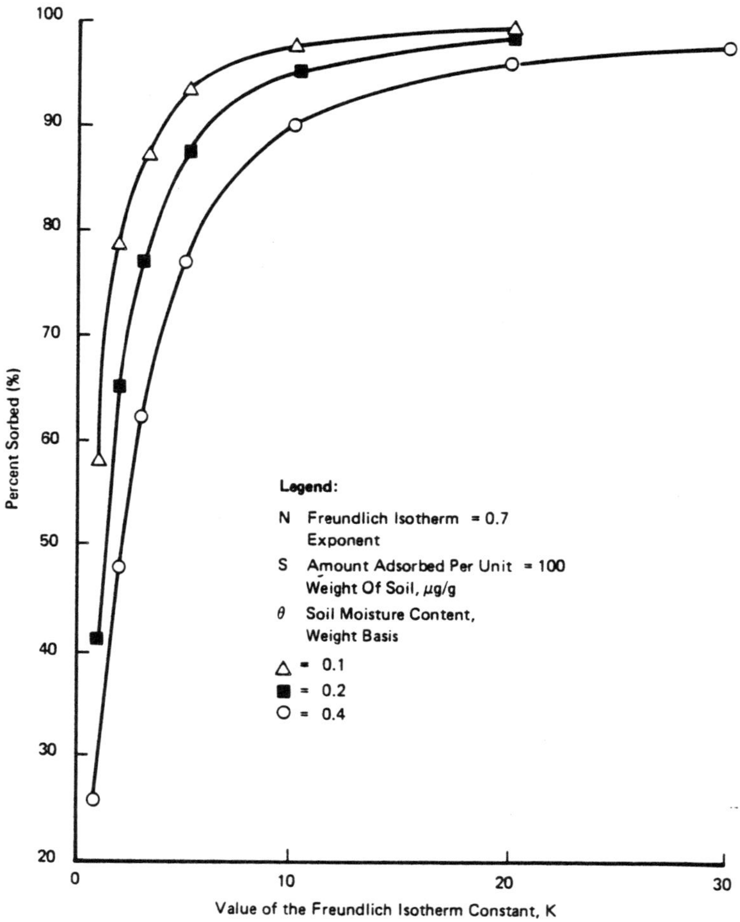

Source: Mahmood and Sims, 1984.

Wastes Amenable to Treatment

This technology is more effective for those organics that are not strongly sorbed (K_d less than 10) than for those that are strongly sorbed. However, the wastes must not be volatile, as drying of the soil will increase volatilization.

Status of Technology

This technology is conceptual at this point in time.

Ease of Application

The site may require tillage to incorporate wastes and to enhance soil drying. Installation of a drainage system may be necessary to reduce soil moisture. Measures to increase soil temperature may enhance soil drying (see Section 4). Theoretically, this is a simple technology that would be easy to apply. However, difficulties that could make it difficult to maintain low soil moisture may arise at the site.

Potential Achievable Level of Treatment

With suitable characteristics of the waste, soil, and the site, the achievable level of treatment may be potentially high.

Reliability of Method

This method is totally reversible To maintain an effective low level of soil moisture, constant maintenance of the drainage system and tilling is necessary.

Secondary Impacts

Decreased soil moisture may result in possible retardation of microbial activity and increased volatilization of volatile waste components. Clay-catalyzed degradation may be enhanced, and a reduction of toxic concentrations due to volatilization may enhance microbial activity if the concentrations were toxic initially. Tilling will increase the susceptibility of the site to erosion.

Equipment and Exogenous Reagents

No exogenous reagents are required. Equipment for drainage, tillage, and power implements are necessary.

Information Requirements

- characterization and concentration of wastes, particularly organics at site;
- depth, profile, and areal distribution of contamination;
- sorption based on organic carbon content (K_{oc});
- octanol/water partition coefficient (K_{ow});
- distribution coefficient (K_d);
- soil moisture (θ);
- soil pH;
- trafficability of soil and site.

Sources of Information

Mahmood and Sims, 1984.

3.3.1.4 Organics — Addition of Sorbents

Organic matter can be used to sorb organics in soil. Sorbents may immobilize the organic constituents, and may also allow additional time for further treatment by biodegradation. Addition of an adsorbent is also useful for relatively "immobile" compounds if cracks and fractures exist in the soil which may increase the potential for mobilization and groundwater contamination. The sorbents which can be used in soils include agricultural products and by-products, sewage sludges, other organic matter and activated carbon.

3.3.1.4.1 Addition of Agricultural Products and By-Products, Sewage Sludges, and Other Organic Matter

Description

Addition of organic matter including sewage sludge, agricultural products and by-products, and organic soil materials to a site/soil/waste system increases the organic content of the soil. This is useful both for waste attenuation (see Section 3.5) and for immobilizing organics in wastes by sorption. Addition of organic matter for increasing the sorption capacity for toxic constituents is especially important and effective in soils with low organic matter content, such as sandy and strip-mined soils.

Sludge may be defined as a semi-liquid waste having a suspended solid content ranging between 0-25 percent dry solids. Sewage sludges are generated by different processes during wastewater treatment. They may be classified either according to stage of treatment (e.g., primary, secondary, tertiary), or according to the process by which the sludges were produced (e.g., activated, digested, etc.). Municipal sludges contain organic and inorganic fractions with 90 percent by weight or more water. Average sludge characteristics of primary and digested sludge are shown in Table 3-3. However, it should be realized that sewage sludge characteristics are quite variable, depending on the source and type of industries associated with the makeup of wastewater (Table 3-4).

Additional advantages of using municipal sludges in organic waste treatment are:

1. Sewage sludges contain active indigenous populations of micro-organisms with degradative potential. This enhances the possibility of degrading toxic constituents.

2. Sewage sludges provide necessary nutrients for biodegradation.

A major disadvantage of using sewage sludges is that the heavy metal content of the sludge may increase potential for groundwater pollution. Therefore, careful management practices are required in terms of application rates of sludges that have high levels of heavy metal. Table 3-5 shows the accumulation of metals allowed in agricultural land (assuming it is safe with respect to leaching), based on the cation exchange capacity (CEC) of the soil.

Organic materials have been added to soil systems to increase sorptive capacity for pesticides. Walker and Crawford (1968) added straw to different soils and noticed an increase in adsorption of four s-triazine herbicides (atrazine, propazine, prometone, and prometryne). The application rate to achieve immobilization was the amount necessary to raise the organic matter content of soil to 2.5 percent. If an organic content of soil of 0.5 percent is assumed, the amount of straw needed is 2 percent or 20,000 ppm, and if a 6-inch depth of incorporation is assumed, the application rate is 45 tons/hectare. The adsorption capacity of soil mixed with the straw remained constant for 81 weeks.

Other materials that have been investigated include plant roots, muck soil, fungal mycelium, and baker's yeast. Tames and Hance (1969) have shown that the addition of freshly killed roots of five plants (oats, beans, peas, cucumber, and radish) to soil increased its adsorptive capacity. The study was conducted with the herbicides atrazine, diuron, linuron, and monolinuron. The adsorptive capacity of the plant roots, however, was less than the capacity of soil organic matter. Coffey and Warren (1969) used a muck soil in the adsorption of several herbicides. With low

TABLE 3.3. AVERAGE CONCENTRATION OF CONSTITUENTS IN PRIMARY AND DIGESTED SLUDGES FROM 33 U.S. TREATMENT PLANTS.

Constituent	Raw or Primary (geometric mean[a])	Digested or Stabilized (geometric mean[a])
	mg/kg except where noted	
Nitrogen	80,000	37,100
Phosphorus	9,070	16,700
Sulfur	3,100	6,010
Boron	775	380
Cadmium	27	4.3
Cobalt	410	290
Copper	740	1,270
Mercury	8.2	6.5
Manganese	460	475
Nickel	420	530
Lead	1,150	2,210
Silver	355	190
Strontium	175	290
Zinc	1,740	2,900
% Volatile Solids	74.4	51.9
Btu/lb (heat of combustion)	7,910	5,850

a. The *nth* root of the product of *n* observed values.

Source: Jacobs, L.W., 1977.

TABLE 3-4. ELEMENTAL ANALYSES OF TREATMENT PLANT SLUDGE SAMPLES EXPRESSED AS RANGES ON A DRY WEIGHT BASIS (in mg/kg)[a]

Chemical Element	Domestic Wastewaters	Industrial and Domestic Wastewaters	Industrial Wastewaters
Phosphorus	2,900-19,600	11,000-22,800	12,700-38,300
Calcium	4,100-120,000	13,200-40,000	32,000-128,000
Magnesium	500-5,400	2,700-4,300	3,000-7,600
Potassium	400-6,000	700-1,700	1,600-4,000
Sodium	200-7,000	500-2,300	800-5,400
Iron	2,300-12,600	15,320-47,600	64,500-225,000
Aluminum	3,800-13,400	3,850-12,000	10,800-70,000
Zinc	1,000-1,800	800-4,600	3,200-14,000
Copper	95-700	960-2,300	1,640-4,700
Nickel	110-400	200-900	440-2,800
Cadmium	<10-400[b]	90-240	<40-200
Manganese	100-300	500-6,100	640-6,100
Chromium	50-200	260-2,650	1,240-2,700
Cobalt	20-<400	400-500	<40-500
Lead	200-<500	760-2,790	1,280-8,300
Silver	7-100	20-300	200-1,680
Strontium	100-200	100-1,600	80-2,100
Beryllium	<10-<100	<10-<100	<40-<100
Vanadium	<500-1,000	<200-500	100-2,000
Barium	600-1,000	700-1,350	2,600-6,400
Boron	50-400	c	c
Mercury	1.0-11.2	2.6-5.0	0.6-3.0

a. EPA analysis data – sludges collected from four plants for each type of wastewater received, 12 treatment plants in all.

b. An unusually high value for Cd obtained from only one sludge source.

c. Analyses not included.

Source: Jacobs, L.W., 1977.

TABLE 3-5. TOTAL AMOUNT OF SLUDGE METALS ALLOWED ON AGRICULTURAL LAND

Metal	Soil Cation Exchange Capacity (meq/100g)[a]		
	<5	5-15	>15
	Maximum amount of metal (lb/acre)		
Lead (Pb)	500	1,000	2,000
Zinc (Zn)	250	500	1,000
Copper (Cu)	125	250	500
Nickel (Ni)	50	100	200
Cadmium (Cd)	5	10	20

[a]Determined by the pH 7 ammonium acetate procedure.

Source: Jacobs, L.W., 1977.

TABLE 3-6. EFFECT OF INCREASING THE CONCENTRATION OF MUCK SOIL AND BENTONITE CLAY ON THE ADSORPTION OF HERBICIDES

Herbicide	Concentration of Herbicide in ppm to Give 50% Root Inhibition of the Test Plant				
	Control	0.01%		1.0%	
	(No Adsorbent)	Muck	Bentonite	Muck	Bentonite
2,4-D	0.12	0.13	0.12	0.43	0.12
DNBP	30	37	34	83	36
Amiben	6.8	6.8	8.3	11	8.3
DCPA	10.5	10.5	10.5	38	10.5
Dipheramid	1.4	1.4	1.4	4.3	5.1

Source: Coffey, D.L., and Warren, G.F., 1969. (See Copyright Notice)

adsorbent concentrations added to the growth medium (0.01 percent), negligible adsorption was observed for muck soil. However, when the concentration was increased to 1.0 percent, adsorption was increased for muck soil (Table 3-6). Ko and Lockwood (1968) have shown that addition of living mycelium *(R. Solani)* to oxidized soil restored its ability to immobilize PCNB and dieldrin. Shin et al. (1970) also demonstrated the beneficial effect of mycelia *(R. Solani)* on soil sorption of DDT. Voerman and Tammes (1969) demonstrated effective sorption of lindane and dieldrin by baker's yeast. With yeast and mycelia cells, nonliving cells exhibited greater sorption capacity than living cells.

Wastes Amenable to Treatment

Organics are amenable to sorption by organic materials.

Status of Technology

Laboratory studies have been conducted demonstrating the sorptive capacity of various organic materials.

Ease of Application

Organic material may be ground, if necessary, and applied to the soil surface or injected below the surface and thoroughly mixed with the contaminated soil. Controls to prevent runoff and runon of precipitation may be necessary. This technology is therefore easy to difficult to apply, depending on the trafficability of the soil and site and the depth of contamination.

Potential Achievable Level of Treatment

The level of treatment achievable is potentially high.

Reliability of Method

Periodic treatment may be required if wastes are recalcitrant. Mineralization of organic materials may release waste constituents if the additives biodegrade more quickly than the waste.

Secondary Impacts

Organic materials may affect soil properties, including:

- degree of structure;
- water-holding capacity:
- bulk density:
- immobilization of nutrients, hindering degradation of organic wastes;
- reduction in soil erosion potential;
- soil temperature.

Organic materials may also result in excessive nitrate levels in receiving waters, depending on the nitrogen content and degree of mineralization of the material. Tillage may make the site more susceptible to water and wind erosion.

Equipment and Exogenous Reagents

Power implements, tillers, and applicators are required. Organic materials are the only exogenous reagents required.

Information Requirements

- characterization and concentration of wastes, particularly organics at site;
- depth, profile, and areal distribution of contamination;
- adsorption isotherm constants for specific added organic material (K, N);
- soil pH;
- soil biological activity (soil C:N:P ratio, soil oyxgen content, soil moisture, etc.);
- trafficability of soil and site.

Sources of Information

Jacobs, 1977; Walker and Crawford, 1968; Tawes and Hance, 1969; Coffey and Warren, 1969; Ko and Lockwood, 1968; Shin et al., 1970; Voerman and Tammes, 1969.

3.3.1.4.2 Addition of Activated Carbon for Organics

Description

Activated carbon is a very strong sorbent for certain classes of chemicals and has been used extensively in industrial and advanced domestic wastewater treatment processes. The aggregate structure of activated carbon (i.e., the pores distributed as macropores (channels) and micropores (capillaries)) enhances its ability for adsorption. Powdered activated carbon is the preferred form because of its high surface area to mass ratio ($\sim$1000 m^2/g). Per unit mass of activated carbon, the powder form offers the advantages of maximum rate of sorption, ease of mixing with the soil, and maximum extent of contact of waste constituents with carbon.

One assumption used to determine an application rate for activated carbon is that the amount of activated carbon needed to make constituents unavailable for plant uptake will be enough to immobilize the constituent. Use of this assumption allows a starting point based on several examples from the literature for short-term treatability studies with site/soil/waste systems. However, it should be recognized that the amount of activated carbon required varies with type and amount of constituent and plant involved. Much more research is required with hazardous waste constituents in different site/soil/waste conditions to develop a useful data base for designing efficient treatability studies and for determining effective activated carbon application rates.

Activated carbon has been used to reduce phytotoxicity and uptake of pesticides from soil by crops (Ahrens and Kring 1968; Andersen 1968; Lichtenstein et al. 1968; Coffey and Warren, 1969; Gupta, 1976; Weber and Mrozek, 1979; Strek et al., 1981). Application rates for the compounds used in these studies including PCB's, aldrin, dieldrin, heptachlor, heptachlor epoxide, atrazine, monuron, bromacil, DPCA, DMPA, CIPC, and trifuralin, varied from 1 to 7 metric ton/hectare. Cost calculations were based on high application rates. For pesticide chemicals, it was demonstrated that activated carbon was more effective on non-ionic compounds, and that adsorption was sustained for long periods of time, with negligible desorption. However, for ionic compounds, desorption may be significant.

Requirements for activated carbon for other constituents (kg activated carbon per kg constituent) are 10-20 kg/kg for toluene and 10-100 kg/kg for sodium dodecyl benzene sulfonate (Jensen, 1982). Activated carbon sorption for a large number of energy processes related pollutants, based on the Freundlich isotherm model, is included in the U.S. EPA publication, "Carbon Adsorption Isotherms for Toxic Organics" (Dobbs and Cohen, 1980).

Wastes Amenable to Treatment

Organic wastes, particularly high molecular weight compounds with low water solubility, low polarity, and low degree of ionization, are amenable to treatment with activated carbon. This treatment has been used on pesticides. Activated carbon is more effective with non-ionic than ionic compounds. Highly water soluble organics, which often contain two or more hydrophilic groups, are difficult to remove using activated carbon.

Status of Technology

In the field, activated carbon has been used to immobilize pesticides and herbicides in soils. The long-term physical and chemical stability of activated carbon in soil systems is unknown.

Ease of Application

Activated carbon may be applied to the soil surface and thoroughly mixed. Wind conditions must be assessed during application to avoid blowing of carbon. The carbon should be applied during periods of low wind activity or wetted before application. Alternatively, the carbon may be injected below the soil surface and then thoroughly mixed with the soil. Activated carbon has been used to immobilize pesticides at rates of 1-7 metric ton/hectare. The ease of application depends on the trafficability of the soil and the site and the depth of contamination. Controls to prevent runoff and runon may be necessary because erosion could be a problem.

Potential Achievable Level of Treatment

The level of treatment achievable is dependent on the sorptive capacity of activated carbon for the specific waste constituents present at the site and the extent of contamination. The treatment achieved may range from low to high.

Reliability of Method

As mentioned earlier, the long-term physical and chemical stability of activated carbon in soil systems is unknown. Because it has been used in the field successfully for pesticides, it may be considered reliable at least in the short run. Desorption may be a problem for ionic compounds because of competition with non-ionic compounds which may be present in the soil.

Secondary Impacts

Immobilization of organics by activated carbon may adversely affect the rate of biodegradation in soils. Tillage will increase the susceptibility of the site to water and wind erosion.

Equipment and Exogenous Reagents

Power implements, tillers, and applicators are needed to apply the activated carbon.

Information Requirements

- characterization and concentration of wastes, particularly organic at site;
- depth, profile, and areal distribution of contamination;
- soil pH;
- adsorptive selectivity and capacity of activated carbon for metals at site;
- adsorption isotherm constants (K, N) for specific waste constituents;
- trafficability of soil and site.

Sources of Information

Ahrens, J.F., and Kring, J.B., 1968; Andersen, A.H., 1968; Lichenstein, E.P., et al., 1968; Coffey D.L., and Warren, G.F., 1969; Gupta, O.P., 1976; Weber, J.B., and Mrozek, Jr., E., 1979; Strek, H.J., et al., 1981; Jensen, R.A., 1982; Dobbs, R.A., and Cohen, J.M., 1980.

3.3.2 Ion Exchange

Ion exchange is a process in which certain minerals and resins in contact with a solution, particularly an aqueous one, release ions in preference for ions of another type present in the solution. Clay minerals in soil can remove cations in exchange for an equivalent amount of calcium. Ion exchangers can be classified into two principal types: cationic and anionic. Cation exchangers have replaceable cations which are exchanged for other cations in the waste.

TABLE 3-7. CATION EXCHANGE CAPACITY (CEC) FOR VARIOUS MATERIALS

Material	CEC Value (m.e./100g)
Humus	200
Montmorillonite	100
Vermiculite	150
Hydrous Mica and chlorites	30
Kaolinite	8
Hydrous oxides	4

Source: From Brady, N.C., 1974.

Correspondingly, anion exchangers have replaceable anions which are exchanged for other anions in the waste. Three types of ion exchangers will be discussed below as technologies usable in soils for immobilization of pollutants in soil. They are: clays, synthetic resins, and zeolites.

3.3.2.1 Addition of Clays

Description

Clays have high surface areas and are especially effective in immobilizing cationic compounds. A mixture of clay and organics could be used for wastes containing both ionic and hydrophobic compounds. The cationic exchange capacity (CEC) of a soil is usually defined as the number of milliequivalents of the ion that can be exchanged per 100g dry weight of soil. With clays, the exchanged ion is often calcium:

$$M^{++} + [clay] \cdot Ca \rightleftharpoons Ca^{++} + [clay] \cdot M \qquad (3\text{-}4)$$

Among soils, humic material and clays tend to have the highest CEC values. Typical CEC values for various materials are shown in Table 3-7. Brady (1974) also showed a range of CEC values between 2-60 m.e./100g for various soils. The cation exchange capacity is variable in a particular soil; in most soils the exchange capacity increases with pH (Brady, 1974).

Bentonite clay (primarily montmorillonite) has been used for inactivating pesticide phytotoxicity (Coffey and Warren, 1969). Table 3-8 illustrates the sorption of herbicides by bentonite clay at a clay concentration of 25 mg/25 ml solution. As expected for pesticides in anionic form (Weber et al., 1965), neither amiben nor 2,4-D were sorbed. Paraquat, a strongly cationic and water-soluble chemical, was strongly sorbed by bentonite. Paraquat cations penetrate the basal plane of the bentonite, and sorption is relatively irreversible. Hydrogen saturated montmorillonite exhibited greater adsorption than sodium saturated montmorillonite for pesticides including S-triazines, substituted ureas, phenyl carbamates, anilines, anilides, and picolinic acids (Bailey et al., 1968).

Wastes Amenable to Treatment

Cationic compounds, both organic and inorganic are amenable to this treatment. This technology is not effective if the cations to be immobilized are present in the wastes in relatively low concentrations when compared to certain other cations (e.g., Ca^{++}, Na^{+}, Fe^{++}, and K^{+}). The latter cations may overwhelm the exchange capacity of the clay and low concentrations of the former cations, including those with higher affinities, will not be significantly exchanged. Landfill leachates and industrial wastes are often concentrated solutions with high total dissolved solids content (including Na^{+}, Ca^{++}, etc.). Trace cations may not be effectively exchanged in these wastes. For heavy metals, a high pH (greater than 7) would limit the role of ion exchange in preference for precipitation (Bonazountas and Wagner, 1981).

TABLE 3-8. CONCENTRATIONS IN PPM OF HERBICIDE SOLUTIONS ADDED TO THE ADSORBENTS TO GIVE 50 PERCENT INHIBITION OF THE TEST PLANT

	Herbicide							
Adsorbent	CIPC	Tri-fluralin	2,4-D	Di-phenamid	DCPA	DNBP	Amiben	Paraquat
Muck soil	0.78	0.64	0.13	1.4	10.5	37	6.8	31.0
Bentonite Clay	0.45	0.58	0.12	1.4	10.5	34	8.3	81.0
Control (no adsorbent)	0.26	0.14	0.12	1.4	10.5	30	6.8	21.0

Source: Coffey, D.L., and Warren, G.F., 1969. (See Copyright Notice)

Status of Technology

Studies have been conducted at the laboratory stage showing immobilization of pesticides by clays.

Ease of Application

The rate of application (mass per volume of soil) is determined in short-term treatability studies. Clay is applied to the soil surface and thoroughly incorporated through the depth of contamination. Runoff and runon controls may be appropriate if erosion or drainage problems occur. This technology is easy to apply using traditional equipment and methods. However, if the site is not trafficable, it might be difficult to apply.

Potential Achievable Level of Treatment

For certain cationic compounds, the level of treatment achievable is high. As mentioned earlier, other ions present in the waste mixture could effectively reduce the immobilization expected.

Reliability of Method

The interactions of ions in the soil system are very complex. This technology is reliable in that the addition of clay to a soil is likely to cause immobilization of some cations. However, the effectiveness may not be as high as anticipated from theory or tests in the laboratory.

Secondary Impacts

Clay addition may affect the chemical, physical, and biological properties of natural soil. The extent and type of effects are dependent on the characteristics of both native soil and added clay. Tillage will increase susceptibility of the site to erosion.

Equipment and Exogenous Reagents

Power implements, tillers, and applicators are necessary to incorporate the clay into the soil.

Information Requirements

- characterization and concentration of wastes, primarily ionic forms present;
- depth, profile, and areal distribution of contamination;
- cationic exchange capacity of clay;
- soil pH;
- selectivity of clay for cations in waste;
- competing, naturally occurring ions at site;
- trafficability of soil and site.

Sources of Information

Brady, N.C., 1974; Coffey, D.L., and Warren, G.F., 1969; Weber, J.B., et al., 1965; Bailey, G.W., et al., 1968; Bonazountas, M., and Wagner, J., 1981.

3.3.2.2 Addition of Synthetic Resins

Description

Synthetic resins consist of a network of hydrocarbon radicals to which ionic functional groups are attached. These hydrocarbon molecules are cross-linked in a three dimensional arrangement, rendering the resin insoluble. The cross-linking of these molecules affects the pore structure of the resin. Ions penetrate these pores in order to be exchanged. Hence, the appropriate resin allows the exclusion of a given size of ions. Synthetic resins can be: 1) those carrying exchangeable cations (acidic), and 2) those carrying exchangeable anions (basic). The advantage of synthetic resins over activated carbon is that they are effective in immobilizing ionic compounds. Despite the high cost and lower sorption capacity of synthetic resins compared with activated carbon, use of synthetic resins is attractive because they can be made selective for specific compounds. The availability and cost of resins do limit their use in large-scale situations like those in hazardous waste sites.

Competition from ionic species found naturally in soil is a problem, decreasing the effectiveness of synthetic resins in the real world.

Wastes Amenable to Treatment

Both organic and inorganics are amenable to treatment with synthetic resins. Cations and anions can be immobilized on synthetic resins with the right choice of resins.

Status of Technology

Synthetic resins (cationic exchange resin, e.g., Chelex 100, and anionic exchange resin, e.g., Dowex 1-X8) have been shown to reduce phytotoxicity of pesticides in the laboratory. Synthetic resins have been effective in removing ionic compounds such as paraquat, DNBP, 2,4-D and amiben from solution.

Ease of Application

The rate of application (mass per volume of soil) should be determined in short-term treatability studies. The resin is applied to the soil surface and thoroughly incorporated through the depth of contamination. Runoff and runon controls may have to be installed to prevent erosion and drainage into the site. This technology should be easy to apply if the depth of contamination is relatively shallow and the site is trafficable.

Potential Achievable Level of Treatment

The achievable treatment level is variably dependent on resin characteristics and competing, naturally occurring ions in the soil.

Reliability of Method

The long-term desorption and stability of resins in soil systems is unknown at the present time. This increases the risk of this technology because of the lack of certainty and the unpredictability of the ion-exchange capacity for a particular ion in natural soils.

Secondary Impacts

Immobilization of organics by synthetic resins may adversely affect rates of biodegradation. Tillage increases the susceptibility of a site to water and wind erosion.

Equipment and Exogenous Reagents

Power implements, tillers, and applicators are necessary to apply resins into the soil.

Information Requirements

- characterization and concentration of waste, primarily ionic forms present;
- depth, profile, and areal distribution of contamination;
- ion exchange capacity of synthetic resin;
- soil pH;
- selective characteristics for specific ions present at site;
- trafficability of soil and site.

3.3.2.3. Addition of Zeolites

Description

Molecular sieve zeolites are crystalline, hydrated aluminosilicates with chemical formulae such as $Na_2Al_2Si_4O_{12}$. Ion exchange selectivities in zeolite do not follow the typical rules and patterns exhibited by other inorganic or organic ion exchangers. Zeolites provide certain combinations of selectivity, capacity, and stability characteristics superior to the more common organic and inorganic cation exchangers. Breck (1974) presents detailed information on the properties of zeolites.

Zeolites are highly selective for particular metals. Clinoptilolite and mordenite both show selective sorption of heavy metals (Sherman, 1978):

- Clinoptilolite $Cu \leq Zn \leq Cd << Pb$
- Mordenite $Ni < Zn$; and $Co < Cu < Mn$

The maximum theoretical cation exchange capacity for these two zeolites is 2.6 m.e./g. Zeolites have been used to soften water.

Zeolites are stable over a wide alkaline pH range (6-12) and do not biologically degrade. The chief restriction in the use of zeolite ion exchangers is their limited acid resistance. Most zeolite ion exchangers should not be employed below about pH 4-5. At lower pH's, the zeolite will itself degrade. Because of this, it would be recommended that zeolites be used only in alkaline soils or limed soils (pH greater than 6).

As with all cation exchangers, sorption of metals by zeolite is affected by 1) pH, 2) competing cations, 3) choice of solvent, 4) presence of complexing agents, 5) solution ionic strength, and 6) type of anions present. The effect of these variables upon the overall ion exchange performance of zeolites is generally less complex and more predictable than with resin exchangers.

The lack of data, both from laboratory and field studies, on the use of zeolites for metal clean-up in soils makes it difficult to comment on possible undesirable factors that may be involved. Research is needed on potential long-term immobilization.

Wastes Amenable to Treatment

Zeolites can be used to immobilize wastes containing heavy metals.

Status of Technology

This technology is conceptual for immobilization of heavy metals in soil systems. It has been used in agricultural applications for retention of ammonium and potassium. Research is needed to investigate the long-term potential for metal immobilization, although it is effective for removing heavy metals (especially cadmium, copper, lead, and zinc) from wastewater.

Ease of Application

Liming might be necessary to increase the pH to greater than 6. Thorough mixing is required for maximum waste/zeolite contact. This technology may be easy or difficult to apply depending on the trafficability of the site and the depth of contamination.

Practical Achievable Level of Treatment

Zeolites have been found to be effective for removing heavy metals (especially Cd, Cu, Pb, and Zn) from wastewater. They look promising for use in soils, but research is required to investigate the long-term potential for metal immobilization.

Reliability of Method

There are a lot of uncertainties associated with this method. At present, its reliability is unknown.

Secondary Impacts

Tillage would increase the susceptibility of the site to wind and water erosion.

Equipment and Exogenous Reagents

The equipment needed are power implements, tillers, and applicators. Zeolites and liming material are also needed.

Information Requirements

- characterization and concentration of waste, primarily ionic forms present;
- depth, profile, and areal distribution of contamination;
- ion exchange capacity and selectivity for metals at site;
- competing naturally occurring ions at site;
- soil pH;
- trafficability of soil and site.

Sources of Information

Breck, 1974; Sherman, 1978; Coffey, D.L., and Warren, G.F., 1969; Weber, J.B., et al., 1965.

TABLE 3-9. SOLUBILITY PRODUCT CONSTANTS FOR METAL SULFIDES

Substance	Equilibrium	Solubility Product Constants
Sulfides		
Cadmium Sulfide	$CdS = Cd^{++} + S$	6×10^{-27}
Cobalt Sulfide	$CoS = Co^{++} + S$	5×10^{-22}
Cupric Sulfide	$CuS = Cu^{++} + S$	4×10^{-36}
Ferrous Sulfide	$FeS = Fe^{++} + S$	4×10^{-17}
Lead Sulfide	$PbS = Pb^{++} + S$	4×10^{-26}
Manganous Sulfide	$MnS = Mn^{++} + S$	8×10^{-14}
Mercuric Sulfide	$HgS = Hg^{++} + S$	1×10^{-50}
Nickelous Sulfide	$NiS = Ni^{++} + S$	1×10^{-22}
Silver Sulfide	$Ag_2S = 2Ag^{+} + S$	1×10^{-50}
Zinc Sulfide	$ZnS = Zn^{++} + S$	1×10^{-20}

Source: Overcash, M.R., and Pal, D., 1979.
(See Copyright Notice)

3.3.3 Precipitation

Another immobilization technology for metals in soils is precipitation. Theoretically, precipitation occurs when the solubility product of the ions forming the precipitate is exceeded in the solution. Metals may be precipitated as sulfides, carbonates, phosphates, and hydroxides. We will discuss sulfides separately from the rest because the method for precipitating sulfides is different from that for precipitating the others.

3.3.3.1 Precipitation as Sulfides

Description

Heavy metals will react with sulfide ions to form insoluble metal sulfides (Table 3-9). These metal sulfides have very low solubilities, even at acidic pH values. The extent of metal sulfide precipitation is a function of 1) pH, 2) type of metal, 3) sulfide content, and 4) interfering ions. A high salt content of the waste will reduce the theoretical extent of precipitation.

With divalent heavy metals the important metal-sulfide reactions are:

$$H_2S \rightleftharpoons H^+ + HS^- \qquad (3\text{-}5)$$

$$HS^- \rightleftharpoons H^+ + S^{2-} \qquad (3\text{-}6)$$

$$Me^{2+} + S^{2-} \rightleftharpoons MeS \qquad (3\text{-}7)$$

The solubilities of metal sulfides decrease with increasing pH (arsenic is an exception, precipitating only at pH <5). With sulfide precipitation, a residual concentration less than 0.1 mg/l metal (for Cd, Cu, Pb, and Zn) can be achieved at $4 \leq pH \leq 12$. Therefore, a single pH level may be used to remove several heavy metals.

Metal sulfides are the least soluble of the metal compounds likely to form in the soil system. Competition from other anion species present would be negligible. The high stability of metal sulfides makes it possible to precipitate metals even in the presence of organic liquids such as chelating agents.

The kinetic aspects of dissolution and precipitation reactions involving heavy metals in the soil matrix have not been studied in detail. Therefore, the time required for maximum precipitation is not known.

Some heavy metals will form soluble sulfide complexes: ZnS_2^{2-}, HgS_2^{2-}, HgS_2H^-, AgS^-, and $As_2S_2^{2-}$. The complexed metal may be more mobile than the free-metal ion. Controlled additions of sulfide by accurate control of pH or sulfide concentration may help prevent the formation of these complexes.

Wastewater treatments have employed several effective sources of sulfide. The sodium salts of sulfide (Na_2S or NaHS) are highly soluble, so that concentrated solutions of sulfide can be prepared. However, addition of Na may have adverse effects on soil physical properties. Calcium sulfide (CaS) has been used, but must be prepared as a slurry because of its low solubility. Iron sulfide (FeS) can additionally reduce Cr^{6+} to Cr^{3+} but, because of its very low solubility, it would not be of practical use in soils as a source of sulfide.

Diking the site, adding gypsum and organic matter, and then flooding the site will provide an anaerobic environment for the reduction of SO_4^{2-} to S^{2-}. This process could be used to provide sulfide for precipitation in soils having sufficiently high sulfate concentrations. However, the possibility of leaching other hazardous constituents must be investigated.

Sulfide precipitates in a natural soil system may be an important mechanism for regulating the solution concentration of heavy metals under reduced conditions only. Sulfides will be oxidized to form soluble metal sulfates under aerobic soil conditions. However, there is no published literature discussing the use of sulfides for reclaiming soils contaminated with metals.

Wastes Amenable to Treatment

Inorganic wastes containing heavy metals, particularly metals with highly insoluble sulfides like copper, cadmium, lead, mercury, and zinc

Status of Technology

This technology has been used in the field to treat river water and wastewaters. On the laboratory scale, there have been additional studies with wastewater. The treatment of soils by precipitation of sulfides is purely conceptual at the present time.

Ease of Application

Theoretically, one mole of sulfide reacts with one mole of divalent metal. At a waste site, metals are encountered as compounds with various anions. Excess sulfide must be added to ensure that the precipitation is as complete as possible. Calcium sulfide may be applied as a slurry and incorporated. Sodium sulfide may be applied in an irrigation system or with sprayers. The soil must be maintained at reduced conditions. Otherwise, the oxidation of precipitated sulfides to more soluble sulfate compounds may occur under aerobic soil conditions. Runon and runoff controls may have to be installed depending on whether the site has been tilled.

Potential Achievable Level of Treatment

Excellent treatment of wastewaters containing copper, cadmium, lead, mercury, and zinc has been obtained for $4 \leq pH \leq 12$. This technology may prove promising for soil systems, with appropriate soil conditions and careful management.

Reliability of Method

If the precipitated sulfides are subjected to aerobic conditions for long periods of time, they may be oxidized to the more soluble sulfates and be leached from the soil into groundwater. Also, if the pH is lowered, the metals will dissolve, releasing hydrogen sulfide. The soil conditions have to be maintained carefully if the metals are to remain as immobile precipitates.

Secondary Impacts

The use of sodium sulfide salts may affect soil permeability. However, soils low in clays and native sodium may be conducive to sodium sulfide treatments without adverse effects. Incomplete sulfide precipitation under reducing conditions may result in leaching of nonprecipitated metals and other soluble constituents, and the formation of volatile toxic metal compounds.

Hydrogen sulfide, a poisonous flammable gas with an offensive odor, may be formed in the precipitation treatment. However, Bhattacharyya et al. (1981) found no formation of H_2S gas during sulfide precipitation reactions. They attributed this to the high reactivity of sulfides and metals. Pillie et al. (1975) noted that the use of a small amount of NaOH prevented the evolution of H_2S gas. The rate of formation of metal-sulfide precipitates in soils may be slow in comparison with the rate in water. The likelihood of the evolution of H_2S will increase as the reactivity of sulfides and metals decrease. Hydrogen sulfide is neutralized in alkaline soils.

A favorable secondary impact may be a reduction of selenium and chromium as a result of precipitating under reducing conditions.

Equipment and Exogenous Reagents

If calcium sulfide is used as the source of sulfide, power implements, tillers, and an applicator for applying slurries are required. Otherwise, if sodium sulfide is used in solution, sprayers or sprinklers may be used.

Information Requirements

- characterization and concentration of metals in waste, primarily heavy metals;
- depth, profile, and areal distribution of contamination;
- soil pH;
- soil oxygen content;
- solubility of metal sulfide(s);
- oxidation/reduction (redox) potential of waste constituents at site;
- trafficability of soil and site;
- susceptibility of soil permeability to change by addition of sodium.
- hydraulic conductivity of soil

Sources of Information

Bhattacharyya, D., et al., 1981; Pillie, R.J., et al., 1975; Weast, R.C., 1983; King, 1981; Overcash, M.R., and Pal, D., 1979.

3.3.3.2 Precipitation as Carbonates, Phosphates, and Hydroxides

Description

Theoretically, many metals will form insoluble compounds with carbonates, phosphates, and hydroxides. Table 3-10 lists the solubility product constants for these compounds.

TABLE 3-10. SOLUBILITY PRODUCT CONSTANTS FOR METAL CARBONATES, PHOSPHATES, AND HYDROXIDES

Substance	Equilibrium	Solubility Product Constants
Carbonates		
Barium carbonate	$BaCO_3 = Ba^{++} + CO_3^{--}$	1.6×10^{-9}
Cadmium carbonate	$CdCO_3 = Cd^{++} + CO_3^{--}$	5.2×10^{-12}
Calcium carbonate	$CaCO_3 = Ca^{++} + CO_3^{--}$	6.9×10^{-9}
Cobalt carbonate	$CoCO_3 = Co^{++} + CO_3^{--}$	8×10^{-13}
Cupric carbonate	$CuCO_3 = Cu^{++} + CO_3^{--}$	2.5×10^{-10}
Lead carbonate	$PbCO_3 = Pb^{++} + CO_3^{--}$	1.5×10^{-13}
Magnesium carbonate	$MgCO_3 = Mg^{++} + CO_3^{--}$	4×10^{-5}
Manganous carbonate	$MnCO_3 = Mn^{++} + CO_3^{--}$	9×10^{-11}
Nickelous carbonate	$NiCO_3 = Ni^{++} + CO_3^{--}$	1.4×10^{-7}
Silver carbonate	$Ag_2CO_3 = 2Ag^{+} + CO_3^{--}$	8.2×10^{-12}
Strontium carbonate	$SrCO_3 = Sr^{++} + CO_3^{--}$	7×10^{-10}
Zinc carbonate	$ZnCO_3 = Zn^{++} + CO_3^{--}$	2×10^{-10}
Phosphates		
Variscite	$Al(H_2PO_4)(OH)_2 = Al^{3+} + H_2PO_4^{-} + 2OH^{-}$	3×10^{-31}
Strengite	$Fe(H_4PO_4)(OH)_2 = Fe^{3+} + H_2PO_4^{-} + 2OH^{-}$	1×10^{-35}
Octocalcium phosphate	$Ca_4H(PO_4)_3 = 4Ca^{2+} + H^{+} + 3PO_4^{3-}$	1×10^{-47}
Fluorapatite	$Ca_{10}(PO_4)_6(FH)_2 = 10Ca^{2+} + 6PO_4^{3-} + 2F^{-}$	4×10^{-119}
Hydroxyapatite	$Ca_{10}(PO_4)_6(OH)_2 = 10Ca^{2+} + 6PO_4^{3-} + 2OH^{-}$	2×10^{-114}
Hydroxides		
Aluminum hydroxide	$Al(OH)_3 = Al^{+++} + 3(OH)^{-}$	5×10^{-33}
Cadmium hydroxide	$Cd(OH)_2 = Cd^{++} + 2(OH)^{-}$	2.0×10^{-14}
Chromic hydroxide	$Cr(OH)_3 = Cr^{+++} + 3(OH)^{-}$	7×10^{-31}
Cobaltous hydroxide	$Co(OH)_2 = Co^{++} + 2(OH)^{-}$	2.5×10^{-16}
Cupric hydroxide	$Cu(OH)_2 = Cu^{++} + 2(OH)^{-}$	1.6×10^{-19}
Ferric hydroxide	$Fe(OH)_3 = Fe^{+++} + 3(OH)^{-}$	6×10^{-38}
Ferrous hydroxide	$Fe(OH)_2 = Fe^{++} + 2(OH)^{-}$	2×10^{-15}
Lead hydroxide	$Pb(OH)_2 = Pb^{++} + 2(OH)^{-}$	4×10^{-15}
Magnesium hydroxide	$Mg(OH)_2 = Mg^{++} + 2(OH)^{-}$	8.9×10^{-12}
Manganese hydroxide	$Mn(OH)_2 = Mn^{++} + 2(OH)^{-}$	2×10^{-13}
Mercuric hydroxide	$HgO + H_2O = Hg^{++} + 2(OH)^{-}$	3×10^{-26}
Nickel hydroxide	$Ni(OH)_2 = Ni^{++} + 2(OH)^{-}$	0.6×10^{-16}
Zinc hydroxide	$Zn(OH)_2 = Zn^{++} + 2(OH)^{-}$	5×10^{-17}

Source: Overcash, M.R., and Pal, D., 1979. (See Copyright Notice)

Santillan-Medrano and Jurinak (1975) obtained experimental data from soil column studies using both Pb and Cd. Their data showed that for the calcareous Nibley soil: 1) the solubility of Pb decreases with increasing soil pH, which is the usual trend for most heavy metals, 2) lead phosphate compounds could be regulating the activity of Pb^{2+} ion in solution, and 3) mixed compound precipitation cannot be precluded between pH 7.5-8.0 because of the convergence of the solubility isotherms for $PbCO_3$, $Pb(OH)_2$, $Pb_3(PO_4)_2$ and $Pb_5(PO_4)_3Cl$. Similar results were seen for the solubility of Cd in Nibley soil. However, the solubility of Cd is considerably higher than Pb at any pH, and at high pH values the soil solution is undersaturated with respect to the compounds considered. This research showed that precipitation reactions involving native soil carbonates, hydroxides, and phosphates are important in regulating solution concentrations of metals added to the soil.

McBride (1980) concluded that calcite serves as a site for chemisorption of Cd^{2+} at low levels of Cd additions, while $CdCO_3$ precipitation occurs at higher Cd concentrations. Carbonate surfaces are known to chemisorb other heavy metals such as Zn^{2+} (Jurinak and Bauer, 1956) and Mn^{2+} (McBride, 1979), providing a mechanism of metal retention that can lower the solution activity below that predicted by the solubility product of the least soluble pure mineral phase. McBride (1980) found that the initial chemisorption of Cd^{2+} on calcite was very rapid, while $CdCO_3$ precipitation of higher Cd^{2+} concentrations was slow.

Artiole and Fuller (1979) tested the effectiveness of using a limestone barrier as a landfill liner to retard the migration rate of metals. The test was conducted in a laboratory column study using a leachate generated from municipal solid waste. The use of a limestone barrier increased the retention of the metals studied (Be, Cd, Fe, Ni, Zn and Cr) as is shown in Table 3-11.

TABLE 3-11. RELATIVE EFFECT OF CRUSHED LIMESTONE LINER PLACED OVER SOIL ON THE PREVENTION OF HEAVY METALS IN MUNICIPAL SOLID WASTE LANDFILL LEACHATE FROM MIGRATING

	Pore Volume Displacement to Achieve the Same C/C_o[a]											
	With Limestone Layer With Soil						No Limestone Layer, Soil Only					
Element	1[b]	2	3	4	5	6	1	2	3	4	5	6
Be	30	29	35	30	30	25	10	16	15	15	18	10
Cd	30	35	40	38	35	35	16	18	28	15	18	6
Cr (pH 2.5)	35	52	85	88	86	35	10	15	5	10	10	6
Cr (pH 4.0)	28	49	125	144	111	109	2	21	39	25	17	15
Fe	31	31	35	36	30	25	21	26	10	25	10	10
Ni	40	35	30	35	28	24	20	15	11	15	7	10
Zn	30	25	30	32	25	30	10	11	15	18	12	16

a. Determination used in column studies, where C = concentration of effluent, C_o = concentration of influent. The higher the pore volume displacement, the greater the volume that was passed through the column to achieve the same C/C_o, and therefore the greater the compound was retained in the column.

b.

Soil Characteristics	pH	Texture
1. Davidson	6.2	clay
2. Ava	4.5	silty-clay-loam
3. Anthony	7.7	sandy-loam
4. Mohave	7.3	sandy-loam
5. Kalkaska	4.7	sand
6. Wagram	4.2	loamy-sand

Source: Artiole, J., and Fuller, W.H., 1979. (See Copyright Notice)

TABLE 3-12. SOME PROBABLE BIVALENT METAL COMPLEXES WITH INORGANIC LIGANDS IN SOIL SOLUTIONS

OH	Cl	SO_4	CO_3	PO_4
MOH^+	MCl^+	$MHSO_4^+$	$MHCO_3^+$	$MH_2PO_4^+$
$M(OH)_2^0$	MCl_2^0	MSO_4^0	$M(HCO_3)_2^0$	$MHPO_4^0$
$M(OH)_3^-$	MCl_3^-	$M(SO_4)_2^{2-}$	MCO_3^0	MPO_4^-
$M(OH)_4^{2-}$	MCl_4^{2-}		$M(CO_3)_2^{2-}$	

Source: Mattigod, S.W., et al., 1981. (See Copyright Notice)

For maximum treatment effectiveness, soil pH must be maintained over time. In a calcareous soil this should not pose any problems. In an acid soil, however, the pH of the soil must be adjusted to a pH greater than 6 and maintained with continued liming of the soil.

In determining the amount of $CaCO_3$ to be added to increase the pH and thus precipitate the metals present, consideration must be given to the lime requirement of the soil. The lime requirement calculation indicates the amount of lime required to reach a predetermined pH. Lime in excess of this amount will also provide an adsorbing surface for a number of heavy metals.

When applying relatively soluble treble superphosphate fertilizer $[Ca(H_2PO_4)_2]$ to the soil to precipitate metals as phosphates, consideration must be given to the competitive interaction for phosphate of the soil constituents. In acid soils this would be iron and aluminum compounds, whereas excess calcium existing in alkaline soils is competitive for the phosphate anion.

Maximum insolubility of metal-hydroxides occurs at different pH values for each heavy metal. Selection of an optimum pH for all metals that may be involved in a treatment process is not possible. Because of this, Kim (1980) and Bhattacharyya et al. (1983) recommended the use of sulfide in treating wastewater for metals. This same pH-dependency on optimum precipitation is also true for carbonates and phosphates.

Heavy metals may also form soluble phosphate, carbonate, and hydroxide complexes (Table 3-12). These complexed metals may be more mobile than the free metal ion. Controlled additions of the anions and optimizing pH for precipitation versus complex-formation may help prevent the formation of these complexes.

The kinetic aspects of dissolution and precipitation reactions involving heavy metals in the soil matrix have not been studied in any detail. Also, high salt content of any accompanying waste as well as chelating agents and other competing reactions may reduce performance of these treatments.

Wastes Amenable to Treatment

Inorganic wastes containing metals, particularly heavy metals, are amenable to this treatment.

Status of Technology

Hydroxide precipitation has been used effectively in wastewater treatment of metals. The use of limestone as a barrier to retard the migration of heavy metals from landfill leachate has been successfully demonstrated in the laboratory.

Ease of Application

Hydroxide salts are highly corrosive and their direct application to soil is not recommended. Hydroxide ions are formed when a soil is treated with liming materials, and they are available for hydroxide-metal precipitation reactions. After adjusting soil pH to the optimum pH for precipitation, theoretically, for divalent metals, one mole of carbonate (two moles of hydroxide) reacts with one mole of metal. However, at a hazardous waste site, the complexity of the soil matrix and wastes make simple estimates impossible and excess carbonate or phosphate should be used. Treble superphosphate fertilizer, which is relatively soluble, may be used to provide phosphate ions.

Limestone and treble superphosphate are both easy to handle agricultural chemicals. No special equipment or safety considerations are necessary. Soil pH must be maintained at a value conducive to maximum insolubility of the specific metal precipitate. Thorough mixing is required for maximum waste and reagent contact. Controls to prevent runon and runoff may be necessary because tillage will increase the susceptibility of the site to erosion.

Potential Achievable Level of Treatment

Insoluble metal compounds of phosphate, carbonate, and hydroxide are known to participate in regulating solution concentrations of metals in natural soil systems. Little, however, is known about using precipitation as a means of treating metals in soils. The kinetics of metals precipitation in soils may limit the effectiveness of this treatment. Also, high salt content of waste, as well as any chelating agents and other competing reactions, may reduce the performance of this treatment.

Reliability of Method

Because precipitation reactions are pH-dependent, it is necessary to maintain the pH of the soil system at high levels to prevent dissolution of the precipitates and subsequent leaching. Reliming is necessary at intervals.

Secondary Impacts

If arsenic is present in the system, the use of phosphate may cause the release of arsenate to the soil solution. Heavy metals may form soluble phosphate, carbonate, and hydroxide complexes which are more mobile than the free metal ion. Tillage will increase the susceptibility of the site to erosion.

Equipment and Exogenous Reagents

Power implements, tillers and applicators are necessary. Calcium carbonate and treble superphosphate fertilizer are required as the precipitating agents.

Information Requirements

- characterization and concentration of metals in waste, primarily heavy metals;
- depth, profile, and areal distribution of contamination;
- soil pH;
- oxidation/reduction (redox) potential of waste constituents at site;
- iron and aluminum compounds, calcium (phosphate fixation capacity);
- soil arsenic (oxidation state and concentration);
- solubility of precipitates (carbonates, phosphates, and hydroxides);
- trafficability of soil and site.

TABLE 3-13. OXIDATION REACTIVITY FOR ORGANIC CHEMICAL CLASSES

Reactivity		
High	Moderate	Low
Phenols	Alcohols	Halogenated hydrocarbons
Aldehydes	Alkyl-substituted aromatics	Saturated aliphatic compounds
Aromatic amines	Nitro-substituted aromatics	Benzene
Certain organic sulfur compounds	Unsaturated alkyl groups	Chlorinated insecticides
	Aliphatic ketones	
	Aliphatic acids	
	Aliphatic esters	
	Aliphatic amines	

Source: JRB, 1982.

Sources of Information

Santillan-Medrano, J., and Jurinak, J.J., 1975; Overcash, M.R., and Pal, D., 1979; McBride, M.B., 1979, 1980; Jurinak, J.J., and Bauer, N., 1956; Antiole, J., and Fuller, W.H., 1979; Kim, B.E., 1981; Bhattacharyya, D., et al., 1983; Mattigod et al., 1981 Dept. of Army, 1982.

3.4 DEGRADATION

3.4.1 Chemical

Chemicals naturally undergo reactions in soil that may transform them into more or less toxic products, or which may increase or decrease their mobility within the soil system. Chemical treatment of contaminated soils entails the reaction of pollutants with reagents, resulting in products which are less toxic, or which become immobilized in the soil column. These reactions may be classified as oxidation reactions, reduction reactions, and polymerization reactions.

3.4.1.1 Oxidation

Chemical oxidation is a process in which the oxidation state of an atom is increased. This is accomplished by removal of electrons or addition of oxygen to the atom. Chemical oxidation represents a significant treatment process in soil systems. As a result of oxidation, a substance may be transformed, degraded, and/or immobilized in soil. Oxidation reactions within the soil matrix may occur through management of the natural processes in a soil, or through addition of an oxidizing agent to the soil-waste complex. Certain compounds are more oxidizable in soils than others. General oxidation reactivity for organic chemical classes is summarized in Table 3-13.

The following discussion is primarily concerned with the oxidation of organics. Oxidation of heavy metals is not usually effective as a treatment method because the higher the oxidation state, the more mobile the heavy metal tends to be. Arsenic is an exception. It is discussed here rather than separately because there is very little information on its oxidation.

Arsenate (As(V)) is less toxic and forms less soluble compounds than arsenite (As(III)). Treatment, therefore, consists of oxidizing As(III) to As(V) followed by the addition of 1:1 ferrous sulfate to precipitate ferric arsenate

($FeAsO_4$), a highly insoluble compound. Aluminum sulfate, zinc sulfate, organic matter, or lime can also be used as a means of fixing As(V). Volatile arsines are formed under anaerobic conditions which, therefore, must be avoided. There is a lack of research dealing with methods for oxidation of arsenic. Aeration may be adequate. Also, there is presently no analytical technique for distinguishing between As(III) and As(V).

3.4.1.1.1 Soil Catalyzed Reactions

Description

Iron, aluminum, trace metals within layer silicates, and adsorbed oxygen have been identified as catalysts promoting free-radical oxidation of constituents in soil systems (Page, 1941; Solomon, 1968; Theng, 1974; Furukawa and Brindley, 1973; and Hirschler, 1966). General characteristics of the organic chemicals likely to undergo oxidation include: 1) aromaticity, 2) fused ring structures, 3) extensive conjugation, and 4) ring substituent fragments.

For oxidation to occur in soil systems, the redox potential of the solid phase must be greater than that of the organic chemical contaminant. Therefore, the half-cell potentials, $E_{1/2}$, of chemical contaminants need to be below the redox potential, 0.8V of a well oxidized soil (Dragun and Baker, 1979).

Another characteristic that is significant with respect to soil-catalyzed oxidation is the solubility of the organic contaminant. The oxidation reaction site is the hydrophilic clay mineral surface, and sorption to the surface precedes soil-catalyzed oxidation. Therefore, more water-soluble compounds should be more readily oxidized in clay-catalyzed systems.

Soil water content (degree of saturation) may also play a very important role in controlling, and therefore managing, soil-catalyzed oxidation. Greater oxidation of chemical contaminants is expected in less saturated soils (Dragun and Helling, 1982). Therefore the technique for immobilization by control of soil moisture is completely compatible with this treatment technique.

Techniques for immobilization by control of soil moisture or by incorporation or addition of uncontaminated soil should not only augment sorption, but they should also augment clay catalyzed reactions in the soil B-horizon where the clay fraction of soil is predominant. Thus a physical-chemical and biological treatment system may be achieved in a layered system through the soil where sorption of hydrophobic constituents occurs in the upper soil layers, or where organic matter content is high, and chemical reactions for hydrophilic constituents occur in lower soil layers where the clay fraction predominates. Biological activity may be expected to increase the extent of degradation of constituents as the retention time of constituents is increased through sorption.

Wastes Amenable to Treatment

Organic wastes which are water-soluble and have half-cell potentials below the redox potential of a well oxidized soil are amenable to this treatment. Table 3-14 lists chemicals which do not undergo free-radical oxidation at soil and clay surfaces. Included in this group of chemicals is the aliphatic class of compounds.

Status of Technology

Soil-catalyzed oxidation reactions have been verified in the field for several chemical classes, including S-triazines and organo-phosphate compounds. Some other compounds have been verified in the laboratory. Otherwise, this technology is at the conceptual stage.

Ease of Application

If required, clay may be applied to the soil surface and thoroughly incorporated through the depth of contamination. The site may require installation of drainage systems to reduce soil moisture. Tillage may be used to dry and

TABLE 3-14. SOME CHEMICALS THAT DO NOT OXIDIZE AT SOIL AND CLAY SURFACES

Chemical Name	
Acetamide	β-Carotene
Acetone, anisilidene-	Cyclohexylamine
-, dianisilidene-	Monoethanolamine
-, dicinnamylidene-	Triethylamine
-, dibenzylidene-	

Source: Dragun, J., and Helling, C.S., 1982.

aerate the soil. Increasing soil temperatures may enhance soil drying and increase the rate of reaction. This treatment technology requires aerobic soil conditions to be maintained which may be easy or difficult, depending on the site and the depth of contamination as it affects the ability to incorporate clays.

Potential Achievable Level of Treatment

The level of treatment achievable is variable, depending on the oxidation potential of the waste constituents and the aeration of the soil.

Reliability of Method

The chemical degradation of a compound does not guarantee less mobile or less toxic products. Care must be taken that oxidation of the waste will not produce substances more problematic than the parent compounds in the waste.

Secondary Impacts

Decreased soil moisture may result in possible retardation in microbial activity or increased volatilization of volatile waste constituents. Volatilization may present a public health hazard, but also may reduce toxic concentrations to soil micro-organisms. Toxic concentrations may also be reduced by attenuation with the added soil. Other clay-catalyzed degradation may be enhanced by this treatment. Tillage will increase the susceptibility of the soil to erosion.

Equipment and Exogenous Reagents

Equipment to set up a drainage system may be necessary, together with applicators and tillers to incorporate the clay.

Information Requirements

- characterization and concentration of wastes, particularly organics at site;
- potential for oxidation of waste constituents (half-potentials, $E_{1/2}$);
- oxidation products (particularly hazardous products);
- solubility of organics;
- depth, profile, and areal distribution of contamination;
- soil moisture;
- soil type and profile;
- catalysts for oxidation present in soil;
- trafficability of soil and site.

TABLE 3-15. RELATIVE OXIDATION POWER OF OXIDIZING SPECIES

Species	Oxidation Potential Volts	Relative Oxidation Power
Fluorine	3.06	2.25
Hydroxyl radical	2.80	2.05
Atomic oxygen	2.42	1.78
Ozone	2.07	1.52
Hydrogen peroxide	1.77	1.30
Perhydroxyl radicals	1.70	1.25
Hypochlorous acid	1.49	1.10
Chlorine	1.36	1.00

Source: Rice, 1981. (See Copyright Notice)

Sources of Information

Page, 1941; Solomon, 1968; Theng, B.K.G., Furakawa, T., and Brindley, 1973; Hirschler, 1966; Dragun and Baker, 1979; Dragun and Helling, 1982.

3.4.1.1.2 Addition of Oxidizing Agents

Description

Oxidizing agents may be utilized to degrade organic constituents in soil systems. Oxidation reactions are usually limited in application due to their substrate specificity and pH dependence. Two powerful oxidizing agents considered for in-place treatment include ozone and hydrogen peroxide. The relative oxidizing ability of these chemicals, compared with other well known oxidants is indicated in Table 3-15. A serious potential limitation to the use of oxidizing agents for soil treatment is the additional consumption of the oxidizing agent(s) by nontarget constituents comprising the soil organic matter.

Ozone is an oxidizing agent that may be used to degrade recalcitrant compounds directly, to create an oxygenated compound without chemical degradation, and/or to increase the dissolved oxygen level in the water for enhancing biological activity. Ozone is a colorless gas characterized by a pungent odor and very high oxidation potential.

The rate of decomposition of ozone is also strongly influenced by pH. Ozone reactions are believed to be of two fundamental types: 1) direct reaction of ozone with the organic compounds, and 2) free radical reaction of ozone, which involves a hydroxyl free-radical intermediate. Direct reaction of ozone with solute achieves the most rapid decomposition of solute. At high pH, the hydroxyl free-radical reactions have been observed to dominate over the direct ozone reactions. Thus the relative rate of ozone reaction can be controlled by adjusting the pH of the medium.

The most efficient and cost-effective uses of ozone in soil system decontamination appear to be in the treatment of contaminated water extracted from contaminated soil systems through recovery wells, and in the stimulation of biological activity in saturated soil (Nagel et al., 1982). If the specific organic constituents present in contaminated soil are relatively biodegradable, ozone treatment may be very effective as an enhancement of biological activity. However, if a large fraction of the matrix is relatively biorefractory, the amount of ozone that will be required to directly treat the waste by chemical destruction will be a direct function of the organic matter present in the solution and in the soil, and will represent a greatly increased cost of treatment. The presence of natural soil organic matter will greatly increase ozone dosage and consumption necessary to treat the target constituents.

Groundwater contaminated with oil products were treated with ozone to reduce dissolved organic carbon (DOC) concentration (Nagel et al., 1982). Dosages of 1 gram ozone per gram dissolved organic carbon resulted in residual water ozone concentration of 0.1 to 0.2 ppm. The treated water was then infiltrated into the aquifer through injection wells. An increase in dissolved oxygen (DO) in the contaminated water was demonstrated. The increase in DO increased microbial activity in the saturated soil zone which stimulated microbial degradation of the organic contaminants. Thus, ozone can be effectively used to treat contaminated water extracted from the soil system and to stimulate biological activity in deeper saturated zones of contaminated soils.

Hydrogen peroxide is an oxidant that has been successfully used in wastewater treatment to degrade compounds that are resistant to biological treatment (recalcitrant). It has also been used to modify the mobility of some metals.

Hydrogen peroxide can react in three major ways:

1) Direct reaction with substrate as shown in Equation 3-8, where the peroxide reacts with silver nitrate to form elemental silver and nitric acid:

$$2AgNO_3 + H_2O_2 \longrightarrow 2Ag + O_2 + 2HNO_3 \qquad (3\text{-}8)$$

2) Hydrogen peroxide can be degraded by UV light to form hydroxyl free radicals as shown in Equation 3-9:

$$H_2O_2 + UV \rightarrow 2OH^- \qquad (3\text{-}9)$$

2,537A

3) It can undergo auto-decomposition in the presence of a metal catalyst as in Equation 3-10:

$$6\ H_2O_2 \xrightarrow{\text{Metal}} 6\ H_2O + 3\ O_2 \qquad (3\text{-}10)$$

Hydrogen peroxide has also been used in conjunction with ozone to degrade compounds which are refractory to either material individually (Nakayama et al., 1979).

Peroxide, as demonstrated in Equations (3-8) and (3-10), can be used to increase oxygen levels in the soil. A previous study (Nagel et al., 1982) has shown an increase in microbial activity and microbial degradation of organic contaminants with increasing oxygen content in soil/groundwater systems.

Hydrogen peroxide is a strong oxidant and, as a result, it is nonselective. If this material is added to the soil, it will react with any oxidizable material present in the soil. This will be a major concern because the concentration of natural organic material in the soils may be lowered, resulting in decrease sorption capacity for some organics.

Thus, the effectiveness of peroxide may be inhibited because it simultaneously increases mobility and decreases possible sorption sites.

Wastes Amenable to Treatment

Organic wastes are amenable to treatment by the addition of oxidizing agents, subject to considerations of the production of more toxic or more mobile oxidation products.

TABLE 3-16. HAZARDOUS PRODUCTS OF OZONE REACTIONS

Parent Compound	Reaction Product	Oxidation of Product with Ozone
Aldrin	Dieldrin	Very slow
Heptachlor	Heptachlor epoxide	Stable to further oxidation
DDT	DDE	—
Parathion	Paraoxon	Nitrophenols, phosphoric acid
Malathion	Malaoxon	—

Source: Utah Water Research Laboratory.

Some hazardous compounds are known to be non-reactive with ozone. Unreactive chemical species are usually characterized by inorganic compounds in which cations and anions are in their highest oxidation state, or organic compounds which are highly halogenated. There are many hazardous chemicals for which no information currently exists concerning their susceptibility to ozone oxidation.

Some general rules concerning chemical destruction of organic constituents include the following:

1) Saturated aliphatic compounds which do not contain easily oxidized functional groups are not readily reactive with ozone. Examples include saturated aliphatic hydrocarbons, aldehydes, and alcohols.

2) For aromatic compounds, reactivity with ozone is a function of the number and type of substituent(s). Generally, substituents which withdraw electrons from the ring deactivate the ring toward ozone. Examples include halogens, nitro, sulfonic acid, carbonyl, and carboxyl groups. Substituents which release electrons activate the ring toward ozone. Examples include alkyl, methoxyl, and hydroxyl.

The following general patterns concerning reactivity with ozone have been identified:

1) phenol, xylene > toluene > benzene
2) pentachlorophenol < dichloro-, trichloro-, tetrachlorophenol

Ozonation of hazardous pesticides may actually be detrimental in many instances. Table 3-16 presents specific examples in which reactions of ozone with parent compounds result in the production of hazardous products, which are often degraded very slowly with ozone.

Hydrogen peroxide has been demonstrated to be effective for oxidizing cyanide, aldehydes, dialkyl sulfides, dithionate, nitrogen compounds, phenols, and sulfur compounds (FMC Corp., 1979).

Peroxide also reacts with many chemical classes with a resultant increase in mobility for the products (JRB, 1982). Table 3-17 shows chemical groups having incompatible reactions with peroxides (i.e., reaction products are more mobile).

Status of Technology

Oxidizing agents are used in wastewater treatment, but there is little experience with their use in terrestrial systems. Hydrogen peroxide is used in septic tank drainfields experiencing failure due to biological clogging.

TABLE 3-17. CHEMICAL GROUPS THAT REACT WITH PEROXIDES TO FORM MORE MOBILE PRODUCTS

Acid chlorides and anhydrides	Cyanides
Acids, mineral, non-oxidizing	Dithio carbamates
Acids, mineral oxidizing	Aldehydes
Acids, organics	Metals and metal compounds
Alcohols and glycols	Phenols and cresols
Alkyl halides	Sulfides, inorganic
Azo, diazo compounds, hydrazine	Chlorinated aromatics/alicycles

Source: Utah Water Research Laboratory.

Ease of Application

The soil should not be disturbed to avoid dilution of contaminants in the soil. The oxidizing agents may be applied in water solutions directly onto the soil surface, injected into the subsurface, or applied through injection wells, depending on the depth and location of contamination. Loading rates can be determined in short-term treatability studies.

Application may be moderate to difficult because oxidizing agents are dangerous to handle and require special treatment. Controls to manage runon and runoff may also be necessary.

Potential Achievable Level of Treatment

The achievable level of treatment is potentially high for wastes susceptible to oxidation, in soils without large quantities of competing oxidizable substances, and for limited areas of contamination.

Reliability of Method

Because ozone and hydrogen peroxide are very strong oxidizers, they are not particularly discriminating in the substances which they will oxidize in the soil. As a result, much of the oxidant will be wasted on oxidizing non-target compounds. Treatment may have to be repeated should initial applications be insufficient, or if the non-target compounds were more susceptible to oxidation than the oxidizable compounds which were problematic at the site.

Secondary Impacts

Oxidizing agents may result in violent reactions with certain classes of compounds (e.g. metals), and may be corrosive to application equipment. Their use may also affect soil hydraulic properties (e.g., infiltration rate), especially in structured soils. Oxidation of soil organic matter may decrease sorption sites for nonoxidizable waste constituents. Oxygenated degradation products are expected to be more polar than the parent compounds and, therefore, potentially more mobile. The chemical reaction may produce a large quantity of potentially mobile constituents in a relatively short period of time, necessitating the installation of recovery wells. The oxygenated products may also be toxic to soil systems, human health, and the environment. Some products may be more refractory than the parent compounds. The use of oxidizing agents may also increase the mobility of some metals. Oxidizing agents may have beneficial effects on microbial degradation processes by adding O_2 to the soil water solution.

Equipment and Exogenous Reagents

Power implements are required. If ozone is used, an ozone generator is necessary. Depending on the application method, an irrigation system, applicators, and/or injection wells may be needed. Exogenous reagents needed are oxidizing agents, ozone or hydrogen peroxide.

Information Requirements

- characterization and concentration of wastes, particularly organics at site;
- potential for oxidation of waste constituents (half-cell potentials, $E_{1/2}$);
- oxidation products (particularly hazardous products);
- depth, profile, and areal distribution of contamination;
- soil and waste pH;
- other naturally occurring oxidizable substances in soil at site;
- selectivity of oxidizing agent(s) for specific wastes present at site;
- trafficability of soil and site.

Sources of Information

Nagel, G., et al., 1982; Nakayama, S., et al., 1979; FMC Corp., 1979; JRB Inc., 1982; Pringle, Jr., H.W., 1977; Rice, R.G., 1981; Griffin, R.A., and Shimp, N.F., 1978; Metsy, A.A., 1980; Overcash, M.R., and Pal, D., 1979; Woolson, E.A., 1977.

3.4.1.2 Reduction

Chemical reduction is a process in which the oxidation state of an atom is decreased. Reducing agents are electron donors, with reduction accomplished by the addition of electrons to the atom. Reduction of chemicals may occur naturally within the soil system. Certain compounds are more susceptible to reduction than others because they will accept electrons. Addition of reducing agents to soil to degrade reducible compounds can be used as an in-place treatment technology.

Reducing agents and conditions of reduction vary with organics and with metals. The following discussion is divided into organics, chromium, and selenium for ease of organization of the information.

3.4.1.2.1 Organics — Addition of Reducing Agents

Description

Chemical reduction using catalyzed metal powders and sodium borohydride has been shown to degrade toxic organic constituents. Reduction with catalyzed iron, zinc, or aluminum affect treatment through mechanisms including hydrogenolysis, hydroxylation, saturation of aromatic structures, condensation, ring opening, and rearrangements to transform toxic organics to innocuous forms.

The use of catalyzed metal powders has been used successfully for aqueous solutions passed through beds of reactant diluted with an inert solid (Sweeney, 1981). The process may be adaptable to terrestrial application, although this has not been directly demonstrated at this time. The process has been used successfully on the following specific constituents:

hexachlorocyclopentadiene	PCBs
p-nitrophenol	chlordane
trichloroethylene	chlorinated phenoxyacetic acid
chlorobenzene	di- and tri-nitrophenols
kepone	atrazine

Iron powders are preferred for soil systems and are also the most cost-effective and available. Reactions of iron with some organic constituents are as follows:

1) Removal of halogen atom and replacement by hydrogen in halogenated organic species:

$$Fe + H_2O + RCl \rightleftharpoons RH + Fe^{++} + OH^- + Cl^- \quad (3\text{-}11)$$

An example is the transformation of DDT to DDA.

2. Replacement of a halogen by a hydroxyl group:

$$Fe + 2H_2O + 2RCl \rightleftharpoons 2ROH + Fe^{++} + 2Cl^- + H_2 \quad (3\text{-}12)$$

3. Saturation of an aromatic structure:

$$Fe + 2H_2O + RCH = CHR \rightleftharpoons RCH2CH2R + Fe^{++} + 2OH^- \quad (3\text{-}13)$$

An example is the transformation of chlorobenzene to cyclohexanol.

4. Condensation of species:

$$Fe + 2RCl \rightleftharpoons R\cdot R + 2Cl^- + Fe^{++} \quad (3\text{-}14)$$

An example is the condensation of DDT to TTTB.

Consumption of metal occurs through the reactions discussed above and also through reactions of the active metal with water:

$$Fe + 2H_2O \rightleftharpoons Fe^{++} + 2OH^- + H_2 \quad (3\text{-}15)$$

The total consumption of metal from these reactions in aqueous solutions of industrial wastewaters produces 1 to 5 mg/l of metal (F_e^{++}) in the solution when low toxicant levels are treated.

Organic chemical constituents in soil may also be chemically reduced through the use of sodium borohydride and zinc. These chemicals have been successful for in-place, small-scale field experiments with soils (Staiff et al., 1981). Results of reductive treatment for degradation of paraquat in soil is summarized in Table 3-18. Results indicate that sodium borohydride and powdered Zn/acetic acid combinations accomplished very effective degradation of paraquat in soil and sand media. Toxic products that may be produced as a result of reductive treatment and other by-products were not investigated to any significant extent in this study.

Wastes Amenable to Treatment

Chlorinated organics, unsaturated aromatics and aliphatics, and other organics that are susceptible to reduction will be amenable to treatment by reducing agents.

Status of Technology

Reduction with catalyzed metal powders has been utilized in wastewater treatment systems. Reduction of paraquat with sodium borohydride and powdered zinc has been demonstrated in small-scale field plots.

TABLE 3-18. CHEMICAL REDUCTIVE TREATMENT FOR DEGRADATION OF PARAQUAT IN SOIL

Chemical Treatment	Paraquat in Soil (ppm) Initial (1 day)	4 Months	Comment
None	9,590	6,300	--
$NaBH_4$-soil	None detected	None detected	Violent foaming
$NaBH_4$-sand	None detected	None detected	No foaming
Powdered Zn acetic acid	60	69	Some bubbling

Source: Staiff, D.C., et al., 1981. (See Copyright Notice)

Ease of Application

The soil should not be disturbed prior to treatment to avoid dilution of contaminants in the soil. For metal-catalyzed powders, stoichiometric excess of reductant powder should be applied to the soil surface and mixed with the contaminated soil to achieve maximum contact. For chemical reducing agents, a sodium borohydride-stabilized water solution should be applied to the soil at 50 percent stoichiometric excess. A solution used in a small-scale study contained:

Sodium borohydride	12 ± 0.5%
Sodium hydroxide	42 ± 2%
Water	balance

Iron may be more desirable than zinc or aluminum, since iron is naturally present in most soils. Aluminum is toxic to biological systems and contributes to soil acidity. Soil pH must be maintained at pH 6-8 for maximum treatment effectiveness. Soil water should be controlled at less than saturated conditions (60-80% of field capacity) to provide an aqueous environment for reductive reactions to occur while preventing leaching.

Controls may be necessary for runon and runoff management. Depending on the trafficability of the site and the depth and the extent of contamination, application of this technology may be easy or it may be difficult.

Potential Achievable Level of Treatment

The achievable level of treatment is potentially high for wastes susceptible to reduction, and for limited areas of contamination. The soil must be without large quantities of competing constituents susceptible to reduction, or the level of treatment may be greatly decreased.

Reliability of Treatment

Treatment may have to be repeated if the reducing agents are not sufficient because of high levels of naturally occurring reducible compounds in the soil.

Secondary Impacts

The use of reducing agents may also degrade soil organic matter. The extent of impact on soils is not known at the present time. The products of reduction may present problems with respect to toxicity, mobility, and degradation. Little information is available at the present time. The addition of metals to soil adds to the metal contaminant load.

Iron appears to be the least damaging to the soil system, though iron has a secondary drinking water standard and is of concern with respect to aesthetics. Addition of metals with acetic acid may possibly increase metal mobility by decreasing soil pH. Addition of sodium borohydride may adversely impact soil permeability, depending on the type and content of clay and ionic constituents in the soil solution.

Equipment and Exogenous Reagents

Power implements, tillers, and applicators are needed. An irrigation and drainage system is needed to apply the sodium borohydride and to maintain the soil moisture at an optimum level. Reducing agents are necessary.

Information Requirements

- characterization and concentration of wastes, particularly organics at site;
- potential for reduction of waste constituents;
- reduction products;
- depth, profile, and areal distribution of contamination;
- soil and waste pH;
- soil moisture;
- selectivity of reducing agent(s) for specific wastes present at site;
- trafficability of soil and site.

Sources of Information

Staiff, D.C., 1981; Sweeney, K., 1981.

3.4.1.2.2 Chromium — Addition of Reducing Agents

Description

Hexavalent chromium is highly toxic and highly mobile in soils. Treatment consists of reducing Cr (VI) to Cr (III), which is less toxic and is readily precipitated by hydroxides over a wide pH range. In a study of the relative mobility of metals in soils at pH 5, Cr (III) was found to be the least mobile (Griffin and Shimp, 1978).

Acidification agents (such as sulfur) and reducing agents (such as leaf litter, acid compost or ferrous iron) may serve in the conversion of Cr (VI) to Cr (III) (Grove and Ellis, 1980). Hexavalent chromium itself is a strong oxidizing agent under acidic conditions and, as such, Cr (VI) will be readily reduced to Cr (III), even without the addition of strong reducing agents. After reduction, liming of the soil will precipitate Cr (III) compounds. Precipitation of Cr (III) occurs at pH 4.5-5.5, so little soil pH adjustment is necessary. Caution is required, however, since trivalent chromium can be oxidized to Cr (VI) under conditions prevalent in many soils, e.g., under alkaline and aerobic conditions in the presence of manganese.

Wastes Amenable to Treatment

Wastes containing hexavalent chromium are amenable to this treatment technology.

Status of Technology

Laboratory data support the theory of this treatment method. Treatment of soils has been performed under field conditions.

Ease of Application

Acidification requirements for the particular soil need to be determined. Three moles of ferrous sulfate will reduce one mole of Cr (VI). Excess of this amount will probably be needed to account for other reduction reactions occurring in the soil. The quantity of organic material cannot be predicted from stoichiometric considerations. Liming materials, after the acidification and reducing steps, are applied to raise the pH to greater than 5.

Leaf litter and compost are easily applied to soils by standard agricultural practices, if site/soil trafficability is suitable. Ferrous sulfate may be applied directly to soil or through an irrigation system. Reduction of Cr (VI) must occur under acidic conditions, followed by liming to precipitate Cr (III). Acidification and liming are standard agricultural practices.

Runon and runoff controls may have to be installed to prevent erosion and drainage problems.

Potential Achievable Level of Treatment

The potential achievable level of treatment is high.

Reliability of Method

The pH of the system has to be maintained at greater than 5. Reliming may be necessary at intervals to ensure that the chromium is immobilized in the soil.

Secondary Impacts

Tillage increases the susceptibility of the site to water and wind erosion, and organic materials may have many effects on soil properties, including:

- degree of structure;
- water holding capacity;
- bulk density;
- immobilization of nutrients, hindering degradation of organic wastes;
- reduction in soil erosion potential;
- soil temperature.

Organic materials may also result in excessive nitrate levels in receiving waters, depending on the nitrogen content and degree of mineralization of the material.

Equipment and Exogenous Reagents

Power implements, tillers, applicators, and an irrigation system are necessary to apply the reducing, acidifying, and liming materials.

Information Requirements

- characterization and concentration of metals, particularly Cr (VI), arsenic, mercury, and other constituents whose treatment requirements may be incompatible;
- depth, profile, and areal distribution of contamination;
- soil pH;
- soil organic matter;
- acidification, reduction, and liming reaction rates;

- trafficability of soil and site.

Sources of Information

Griffin and Shimp, 1978; Grove and Ellis, 1980; Metsy, 1980.

3.4.1.2.3 Selenium — Addition of Reducing Agents

Description

Hexavalent selenium (as selenate (SeO_4^{2-})) is highly mobile in soils. It is the dominant form of selenium in calcareous soils. Elemental selenium and selenite (Se(IV)) are less mobile in soils. Hexavalent selenium can be reduced to Se (IV) or Se° under acid conditions. Reduction of selenium occurs naturally in soils. Elemental selenium is virtually immobile in soils. Se (IV) will participate in sorption and precipitation reactions, but unlike the metals discussed previously, selenite is an anion (SeO_3^{2-}) and its potential leachability will increase with increasing pH. Therefore, at a site that contains selenium as well as other metals, selenium could not be treated if increased pH were required as part of the treatment for the other metals.

Wastes Amenable to Treatment

Wastes containing hexavalent selenium (SeO_4^{2-}) that do not contain significant amounts of other metallic constituents are the most amenable to treatment by this method.

Status of Technology

Studies have been limited to those involving basic chemistry of selenium in soils.

Ease of Application

The soil requires acidification with sulfur or another agricultural acidifying agent to pH 2-3. Acidification requirements for the particular soil need to be determined experimentally. Leaf litter or compost are easily applied to soils by standard agricultural practices, if site trafficability is suitable. Ferrous sulfate may be applied directly to soil or through an irrigation system. Two moles of ferrous sulfate will reduce one mole of Se(VI). Excess of this amount should be added to account for other reduction reactions that may occur in the soil. The quantity of organic material cannot be predicted from stoichiometric considerations.

Runoff and runon controls may be necessary to prevent erosion and drainage problems.

Potential Achievable Level of Treatment

The addition of reducing agents speeds up the natural process of selenium reduction in soils. The potential level of treatment should be high.

Reliability of Method

Once reduction has occurred, the soil must be kept acidic. Reapplication of an acidifying agent may be necessary as required to maintain the pH to between 2 and 3.

Secondary Impacts

Low pH will adversely affect microbial activity and degradation of organic waste constituents. Tillage increases the susceptibility of the site to water and wind erosion. Organic materials may have many effects on soil properties, including:

- degree of structure;
- water holding capacity;
- bulk density;
- immobilization of nutrients, hindering degradation of organic wastes;
- reduction of soil erosion potential;
- soil temperature.

Organic matter may also result in excessive nitrate levels in receiving waters, depending on the nitrogen content and the degree of mineralization of the material.

Equipment and Exogenous Reagents

Power implements, tillers, and applicators are needed to prepare the site and apply the acidifying agent and reducing agent.

Information Requirements

- characterization and concentration of metals, particularly selenium (Treatment of selenium is incompatible with treatment of all other metals; oxidation state of metallic ions);
- depth, profile, and areal distribution of contamination;
- soil pH;
- soil organic matter;
- clay content of soil;
- acidification and reduction reaction rates;
- trafficability of soil and site.

Sources of Information

Griffin, R.A., and Shimp, N.A., 1978; Sharma, S., and Singh, R., 1983.

3.4.1.2.4 Sodium Reduction/Dehalogenation for PCBs and Dioxins

Description

Several processes to detoxify PCB's and, potentially dioxins have been developed in the past several years. All employ a sodium-based chemical reagent to remove chlorine from the very stable PCB and dioxin molecules. The residue structures are generally non-toxic or of lower toxicity than the original compound. These processes were originally developed for the treatment of PCB-containing oils; several have been suggested for application to contaminated soils. The sodium strips off the chlorine to form sodium chloride (common salt).

One process, Acurex, follows a two-step procedure. First, dioxins (or PCB's) are extracted from the soil with a special blend of solvents. The solvents are then treated with a proprietary sodium-based reagent to destroy the contaminants (Mille, G.J., 1982). A second process, developed by the Franklin Research Institute, applies a reagent directly to the soil. The reagent is a chemically-modified sodium polyethylene glycol (NaPEG) complex. The reagent is quite stable, does not contain metallic sodium, and is not sensitive to small quantities of water (Franklin Research Institute, 1981).

Wastes Amenable to Treatment

PCB's and dioxins are detoxified. Other halogenated compounds will also be attached by the reagents.

Status of Technology

The technology has been applied to the treatment of PCB-containing oils, but is still in the developmental stage with respect to soil decontamination.

Ease of Application

The reagents are liquids and should be applied relatively easily. They will react with excess moisture and thus must be applied under carefully controlled conditions.

Potential Achievable Level of Treatment

Results with PCB oils have shown high levels of detoxification. The same levels are potentially realizable in soils.

Reliability

Once the compounds are dechlorinated, there is no mechanism to reverse the reaction. If the application is successful in the first place, no further retreatment should be needed.

Secondary Impacts

The reagents are powerful reducing agents and may react with soil organic matter. The consequences of such reactions are unknown at present.

Equipment and Exogenous Reagents

Power implements, tillers, and applicators are needed. Some sort of temporary cover might be needed to keep rain off of the treatment area until the reactions are complete. Soil may have to be drained or tilled and allowed to dry. Reducing agents are needed in sufficient quantity to react with the target contaminants and with other waste and natural constituents.

Information Requirements

- depth, profile, and areal distribution of contamination;
- soil moisture content.
- soil temperature

Source of Information

Mille, G.J., 1982; Franklin Research Institute, 1981.

3.4.1.3 Polymerization

Description

A polymer is a large molecule built up by the repetition of small, simple chemical units. A polymerization reaction is the conversion of a particular compound to a larger chemical multiple of itself (Kirk Othmer Encyclopedia of Chemical Technology, 1982). The resulting polymer often has different physical and chemical properties from the initial unit and could be less mobile in the soil system.

It has been demonstrated that naturally occurring iron and sulfates in contaminated soil may catalyze initial polymerization of contaminants. Treatment solutions containing sulfate-related constituents have been successfully used in polymerization reactions in the soil (Williams, 1982). Acrylate monomer (4200 gallons) contaminating glacial sand and gravel layers was polymerized in-place by injecting a catalyst, activator, and wetting agent.

Mercer, B.W., et al. (1970), working with grouting materials and polymeric agents, found the process of in-place immobilization with these agents expensive and complicated by the logistics of obtaining widespread coverage without an excessive number of injection wells.

Wastes Amenable to Treatment

Chemical polymerization is most effective for immobilization of organic constituents, preferably those with more than one double bond. General categories of constituents, applicable to polymerization include aliphatic, aromatic, and oxygenated monomers, such as styrene, vinyl chloride, isoprene, acrylonitrile, etc. For multi-organic contamination, catalysts and activators necessary to achieve polymerization may interact with one another.

Status of Technology

Experimental studies have been conducted in the field.

Ease of Application

A 2:1 ratio of volume of catalyst and activator to volume of contaminant is used. The catalyst and activator should be applied in two applications and applied separately to prevent reactions before contact with wastes. A wetting agent is added to promote rapid and uniform dispersion of solutions through the contaminated area. If ground temperature falls below 50°F, it may be necessary to warm the treatment solution to 50°F before use. Because of the acidic nature of treatment reagents, corrosion-resistant application equipment is required.

If the surficial zone is too shallow to tolerate sufficient injection pressure for dispersing catalyst and activator solutions, installation of exfiltration galleries (e.g., a 2-inch diameter perforated PVC casing, buried in trenches below ground surface across the contaminated zone) is required. A riser pipe and manifold header connect each gallery to solution tanks which contain catalyst and activator.

This technology is moderate to difficult to apply. In a field study, it was found that obtaining widespread coverage was difficult without an excessive number of injection wells.

Potential Achievable Level of Treatment

The level of treatment achievable is variable, depending on the waste and soil conditions. The potential for long-term immobilization is unknown at this time.

Reliability of Method

The reliability of the treatment is unknown since there is no information on its long-term effectiveness.

Secondary Impacts

The polymerized area may exhibit decreased infiltration and permeability.

Equipment and Exogenous Reagents

Catalysts and activators are needed. Vendors should be consulted as to the equipment to be used.

Information Requirements

- characterization and concentration of wastes, particularly organics at site;
- potential for polymerization of waste constituents;
- polymerization products;
- depth, profile, and areal distribution of constituents;
- iron and sulfate content in soil;
- catalysts and activators present in soil;
- trafficability of soil and site.

Sources of Information

Kirk Othmer Encyclopedia of Chemical Technology, 1982; Mercer, B.W., et al., 1970; Williams, E.B., 1982.

3.4.2 Biological

Biodegradation is an important environmental process causing the breakdown of organic compounds. It is a significant loss mechanism in soil in the mineralization process by which organics are converted to inorganics.

Micro-organisms, principally bacteria, actinomycetes, and fungi, are the most significant group of organisms involved in biodegradation, and soil environments contain a diverse microbial population. The parameters influencing the rate of biodegradation are of two types:

1) those that determine the availability and concentration of the compound to be degraded or that affect the microbial population site and activity; and

2) those that control the reaction rate.

Important parameters affecting biodegradation include pH, temperature, soil moisture content, soil oxygen content, and nutrient concentration, among many others (Bonazountas and Wagner, 1981).

Table 3-19 shows rate constants of organic compounds in soil. Table 3-20 shows rates in anaerobic systems.

Biological treatment methods are directed toward enhancing biochemical mechanisms for detoxifying or decomposing hazardous waste materials in contaminated soils. Soil micro-organisms, principally the bacteria, actinomycetes, and fungi, are important in decomposition or detoxification processes. Therefore, treatments applied to the soil to enhance biological processes must not alter the physical environment in such a way that it would severely restrict microbial growth and/or biochemical activity. In general, this means that soil temperatures should be between 50°C and 60°C (Atlas and Bartha, 1981); soil water potential should be greater than − 15 bars (Sommers et al., 1981), pH should be between 5 and 9 (Alexander, 1977; Atlas and Bartha, 1981), and oxidation-reduction (redox) potential should be between pe + pH of 17.5 to 2.7 (Baas and Becking et al., 1960). Soil pH and redox boundaries should be carefully monitored when chemical and biological treatments are combined. With these restrictions in mind several treatments are considered, both tried and theoretical, which may enhance microbial activity in hazardous waste contaminated soil.

TABLE 3-19. BIODEGRADATION RATE CONSTANTS FOR ORGANIC COMPOUNDS IN SOIL (day^{-1})

Compound	Test Method Die-Away	 $^{14}CO_2$ Evolution
Aldrin, Dieldrin	0.013	
Atrazine	0.019	0.0001
Bromacil	0.0077	0.0024
Carbaryl	0.037	0.0063
Carbofuran	0.047	0.0013
Dalapon	0.047	
DDT	0.00013	
Diazinon	0.023	0.022
Dicamba	0.022	0.0022
Fonofos	0.012	
Glyphosate	0.1	0.0086
Heptachlor	0.011	
Lindane	0.0026	
Linuron	0.0096	
Malathion	1.4	
Methyl parathion	0.16	
Paraquat	0.0016	
Parathion	0.029	
Phorate	0.0084	
Picloram	0.0073	0.0008
Simazine	0.014	
TCA	0.059	
Terbacil	0.015	0.0045
Trifluralin	0.008	0.0013
2,4-D	0.066	0.051
2,4,5-T	0.035	0.029

Source: Lyman, W.J., et al., 1982.

TABLE 3-20. BIODEGRADATION RATE CONSTANTS FOR ORGANIC COMPOUNDS IN ANAEROBIC SYSTEMS (day^{-1})

Compound	In Soil Die-Away	In Soil $^{14}CO_2$ Evolution	In Sewage Sludge
Carbofuran	0.026		
DDT	0.0035		
Endrin	0.03		
Lindane		0.0046	
PCP		0.07	
Trifluralin	0.025		
Mirex			0.0192
Methoxyclor			9.6
2,3,5,6-Tetrachlorobenzene			12.72
Bifenox			6.27

Source: Lyman, W.J., et al., 1982.

3.4.2.1 Modification of Soil Properties

Since the activity of micro-organisms is so dependent on soil conditions, modification of soil properties is a viable method to enhance the microbial activity in the soil. These soil properties are: soil moisture, soil oxygen content, soil pH, and nutrient content. Details of how to apply the soil modification techniques are given in Section 4.

3.4.2.1.1 Soil Moisture

Description

When natural precipitation is insufficient to maintain soil moisture within a range that is near optimal for microbial activity, irrigation may be necessary. Although many microbial functions continue in soils at −15 bars or drier, optimum biochemical activity is usually observed at soil water potentials of −0.1 to 1.0 bar (Sommers et al., 1981).

In a review of soil water potential on decomposition processes in soils, including pesticide degradation, Sommers et al. (1981) suggested that the effect of soil water potential on pesticide degradation is to alter general microbial activity and to affect the kinds of micro-organisms which are metabolically active in the soil. Ou et al. (1983) observed rapid mineralization of methyl parathion in soils at −0.1 and −0.33 bar soil moisture tension, along with the formation of bound residues. The ratio of the degradation products (p-nitro-phenol to p-aminophenol) increased as the soils became drier. In dry soil (−15 bar), mineralization of methyl parathion and bound residue formation were slower. Limited experimentation with the effects of soil moisture on degradation indicates that degradation rates are highest at soil water potential between 0 and −1 bar.

The degradation of hazardous organic compounds may be accelerated by soil moisture optimization, and this approach may be sufficient to bring about required degradation, especially for constituents relatively easy to degrade. However, more rapid treatment of the contaminated soil may be achieved when moisture augmentation is used in combination with other techniques.

Wastes Amenable to Treatment

Biodegradable organic compounds are amenable to this treatment.

Status of Technology

Moisture control is widely practiced in agriculture. However, little information is available on the use of moisture control to stimulate biological degradation of hazardous materials in soil. Most experimental work in the laboratory on pesticides and other xenobiotic compound degradation has been conducted at or near optimal soil moisture.

Ease of Application

An irrigation and drainage system is required. Irrigation water is applied using standard irrigation practices. Irrigation should be applied frequently in relatively small amounts, without exceeding field capacity, to minimize leaching. Depending on the ease of controlling water at the site and on the availability of a suitable water source (e.g., transport distance, drilling of new wells, availability and cost of energy for pumping), it may be easy or difficult to apply this technology. Controls for erosion and proper drainage due to runoff are necessary.

Potential Achievable Level of Treatment

The achievable level of treatment may be low to high, depending on the biodegradability of waste constituents and suitability of the site and soil for effective moisture control. The effectiveness of this technology may be enhanced by the use of other treatment techniques to increase biological activity.

Reliability of Method

This technology is reliable in that it has been used in agriculture, but retreatment is necessary.

Secondary Impacts

Leaching of soluble hazardous compounds may occur. Erosion may also be a problem.

Equipment and Exogenous Reagents

An irrigation system and water are required.

Information Requirements

- characterization and concentration of wastes, particularly organics at site;
- micro-organisms present at site;
- biodegradability of waste constituents (half-life, rate constant);
- biodegradation products (particularly hazardous products);
- depth, profile, and areal distribution of constituents;
- soil moisture;
- other soil properties for biological activity (pH, oxygen content, nutrient content, organic matter, temperature, etc.);
- trafficability of soil and site.

Sources of Information

Bonazountas, M., and Wagner, J., 1981; Atlas and Bartha, 1981; Sommers, L.E., et al., 1981; Alexander, M., 1977; Bass et al., 1960; Lyman, W.J., et al., 1982; Ou et al., 1983.

3.4.2.1.2 Soil Oxygen Content for Aerobic Biodegradation

Description

One reason for the common practices of tilling and/or draining the soil in agriculture is to stimulate organic matter decomposition in an aerobic environment so that nutrients will be mineralized and made available for plant assimilation. Aerobic metabolism is more energy-efficient, and microbial decomposition processes are, in general, more rapid under aerobic conditions. Although the decomposition of some xenobiotic organic compounds appears to require anaerobic metabolism, the majority of organisms in soils shown to be active in pesticide and other xenobiotic compound decomposition are aerobic (Alexander 1977; Pal et al., 1980, Baker and Mayfield, 1980; Brunner and Focht, 1983; Sims and Overcash, 1981). In many instances, therefore, assuring the aerobiosis of the soil will enhance the rate of biological decomposition.

Tilling the soil for aeration is common practice in agriculture and has been recommended for hazardous waste-contaminated soil reclamation by practitioners and researchers (Arthur D. Little, Inc., 1976; Thibault and Elliott, 1979). Soils with high water tables that restrict aeration may also be drained using common agricultural techniques.

Wastes Amenable to Treatment

Organic wastes that are acted upon by micro-organisms under aerobic conditions are amenable to treatment. Most organics fall into this category of compounds.

Methods of Technology

In the field, tilling the soil for aeration is common practice in agriculture to enhance crop-residue degradation, and it has been recommended for hazardous waste-contaminated soil reclamation.

Ease of Application

If the site is too wet, a drainage system should be installed. The soil is tilled at periodic intervals to achieve aeration. Controls to prevent runon and runoff of precipitation are necessary. This technology is easy to difficult, depending on the characteristics of the soil and the site and the trafficability of the site.

Potential Achievable Level of Treatment

The level of treatment achievable is from low to high, depending on the biodegradability of the waste constituents and the suitability of the site and soil for maintenance of aerobic conditions.

Reliability of Method

Retreatment at periodic intervals is necessary to assure that the soil oxygen is at a sufficiently high level.

Secondary Impacts

Tillage will increase the susceptibility of the site to erosion.

Equipment and Exogenous Reagents

Power implements and tillers are necessary.

Information Requirements

- characterization and concentration of wastes, particularly organics at site;
- micro-organisms present at site;
- biodegradability of waste constituents (half-life, rate constant);
- biodegradation products (particularly hazardous products);
- depth, profile, and areal distribution of constituents;
- soil oxygen content;
- other soil properties for biological activity (soil moisture, pH, nutrient content, organic matter, temperature, etc.);
- soil texture;
- trafficability of soil and site.

Sources of Information

Pal, et al., 1980; Baker, M.D., and Mayfield, 1980; Brunner, W., and Focht, D.D., 1980; Sims, R.C., and Overcash, M.R., 1981; Arthur D. Little, Inc., 1976; Thibault, G.T., and Elliott, N.W., 1979; Lyman, W., et al., 1982; Atlas and Bartha, 1981; Bass, J., et al., 1960.

3.4.2.1.3 Soil Oxygen Content for Anaerobic Biodegration

Description

There is increasing evidence that some halogenated xenobiotic compounds may be dehalogenated or completely degraded under anaerobic conditions (Sulflita, J.M., et al., 1982; Sulflita, J.M., and Tiedje, J.M., 1983; Horowitz, A., et al., 1983; Sulflita, J.M., et al., 1983; Kobayashi and Rittman, 1982; Pfaender and Alexander, 1972). Therefore, manipulation of contaminated soil to create an anaerobic, reducing environment to enhance the decomposition of certain hazardous waste constituents should be considered. Apparently, the redox potential (Eh) of the environment must be below 0.35V for significant reductive dechlorination to take place, but exact requirements depend upon the individual compounds being reduced (Kobayashi and Rittman, 1982). Reductive reactions may be catalyzed by both abiotic and biochemical means in anaerobic environments.

Once a recalcitrant compound has been altered by reductive reactions under anaerobic conditions, it may be more amenable to decomposition under aerobic conditions. For example, the reductive dechlorination of 2,4,5-trichlorophenoxyacetic acid (2,4,5-T) under anaerobic conditions to 2,4-dichlorophenoxyacetic acid (2,4-D) which is readily degraded in the soil under aerobic conditions has been described (Munnecke et al., 1982). Laboratory experimentation may show that anaerobic soil conditions followed by aeration may enhance biological decomposition of some hazardous waste constituents. However, Marinucci and Bartha (1979) evaluated weekly alterations between anaerobic and aerobic conditions for enhancement of biodegradation of trichlorobenzenes in soil and found no improvement in mineralization rates. The trichlorobenzenes were mineralized most rapidly under continuous aerobic conditions. Other classes of compounds may not follow this pattern, and more research is needed to further evaluate the potential for treatment by using alternating anaerobic and aerobic conditions. Longer periods between alternation may be appropriate.

Arthur D. Little, Inc. (1976), also reported that the only proven method of creating anaerobic conditions is to dike and flood the soil in a fashion similar to that used to grow rice. They cite unpublished work by W. Farmer at the University of California at Riverside in which a 1.5-acre DDT-contaminated field was amended with organic matter, flooded, and the soil temperature increased. Complete transformation of the DDT to DDD was observed in 18 days. They suggest that, without this treatment, the transformation to DDD would have taken more than 2 years. Flooding the soil presents opportunities for leaching of hazardous materials from contaminated soil and is probably not advisable in most hazardous waste-contaminated soils. However, it should be possible to lower the redox potential of the soil by adding excessive amounts of readily biodegradable organic matter, compacting the soil to reduce oxygen diffusion through large soil pore spaces, keeping the soil wet without exceeding the gravitational water potential (field capacity), and perhaps deep mulching to impair oxygen diffusion to the soil surface.

Applications of this method of inducing anaerobiosis have not been identified in the literature as used either in the field or under laboratory conditions. Reductive dehalogenations or other reductive reactions that lead to decomposition or detoxification of specific hazardous waste constituents should be verified from the literature or from experimentation before this treatment is used.

Wastes Amenable to Treatment

Organic wastes that are biodegradable under anaerobic conditions, e.g., halogenated compounds, are amenable to this treatment.

Status of Technology

In the field, this technology is only conceptual. Neither are there any reported laboratory studies using this technique.

Ease of Application

Organic materials are applied and incorporated into the soil. Irrigation water is applied using standard irrigation practices. Irrigation water should be applied in amounts large enough to create anaerobic sites in the soil, but not to cause leaching. Mulches may also be applied to act as a barrier to oxygen diffusion into the soil. The soil may be compacted to reduce porosity. Runon and runoff controls are necessary. This technology involves a complex combination of soil manipulations (i.e., moisture control, organic amendment, compaction, and mulching) and may range from moderate to difficult to apply.

Potential Achievable Level of Treatment

Depending on the degradative pathway of the constituents, the achievable level of treatment may range from low to high. This technology may result only in partial degradation, requiring the establishment of aerobic conditions to complete treatment.

Reliability of Method

Retreatment is necessary as frequently as required to maintain anaerobic conditions.

Secondary Impacts

This technology may result in the formation of toxic volatile forms of metals (e.g., methylated mercury and arsines), hydrogen sulfide, and other nuisance odor compounds. Leaching of hazardous constituents may also occur if water addition is not carefully controlled. Organic materials may have many effects on soil properties, including:

- degree of structure;
- water-holding capacity;
- bulk density;
- immobilization of nutrients, hindering degradation of organic wastes;
- reduction in soil erosion potential;
- soil temperature.

Organic materials may also result in excessive nitrate levels in receiving waters, depending on the nitrogen content and degree of mineralization of the material.

Equipment and Exogenous Reagents

Power implements, compactors and an irrigation system are necessary. The exogenous materials required are irrigation water and organic materials, e.g., mulches.

Information Requirements

- characterization and concentration of wastes, particularly organics at site;
- micro-organisms present at site;
- biodegradability of waste constituents (half-life, rate constant);
- biodegradation products (particularly hazardous products);
- depth, profile, and areal distribution of constituents;
- soil oxygen content;
- other soil properties for biological activity (soil moisture, pH, nutrient content, organic matter, temperature, etc.);
- soil water holding capacity;
- suitability of site to flooding and drainage;
- trafficability of soil and site.

Sources of Information

Sulflita, J.B., et al., 1982; Sulflita, J.M., et al., 1983; Sulflita, J.M., and Tiedje, J.M., 1983; Horowitz, A., et al., 1983; Sulflita, J.M., et al., 1983; Kobayashi H., and Rittman, B.E., 1982; Pfaender, F.K., and Alexander, M., 1972; Munnecke, D.M., et al., 1982; Marinucci, A.C., and Bartha, R., 1979; Arthur D. Little, Inc., 1976; Alexander, 1977; Atlas and Bartha, 1981; Bass et al., 1960; Lyman et al., 1982; Bonazountas, M., and Wagner, J., 1981; Munnecke, D.M., 1980.

3.4.2.1.4 Soil pH

Description

Depending on the nature of the hazardous waste components contaminating the soil, it may be advantageous to optimize the soil pH for a particular segment of the microbial population, since microbial community structure and activity are affected by soil pH (Gray, 1978; Alexander, 1977). Some fungi have a competitive advantage at slightly acidic pH, while actinomycetes flourish at slightly alkaline pH (Alexander, 1977). Soil pH has also been shown to be an important factor in determining the effect various pesticides have on soil micro-organisms (Anderson, 1978). Near neutral pH values are probably most conducive to microbial functioning in general. The recent discovery that fungal metabolism of polynuclear aromatics (PNAs) is qualitatively similar to mammalian metabolism in that mutagenic arene oxides (epoxides) are produced as initial oxidation products (Cerniglia and Gibson, 1979; Cerniglia et al., 1979) suggests that fungal degradation of PNAs in the environment should be discouraged. Although the effect of soil pH on the formation of epoxides from PNAs has not been demonstrated, it may be advantageous to maintain pH near neutrality to encourage a relatively higher bacterial activity in soils contaminated with these compounds.

It may be necessary to treat contaminated soil with crushed limestone or lime products to raise the pH to the desired range, or with acid-producing materials or sulfur to lower the pH. Methods for determining the lime requirement of agricultural soils that take into account the buffering capacity of the soil (McLean et al., 1966; McLean, 1982) have been developed but guidelines for reducing soil pH are not readily available, and the addition of acidifying agents must be determined experimentally in the laboratory.

Wastes Amenable to Treatment

Organic wastes that are biodegradable are amenable to this treatment.

Status of Technology

Liming is a common agricultural practice. Acidification is much less commonly required.

Ease of Application

A lime requirement test may be performed to determine loading rate for increasing soil pH. Acidification requirements for a particular soil have to be determined experimentally. Buffering capacity of the waste must also be considered.

Thorough mixing is required in the zone of contamination to effect pH change. Because the soil is tilled, runoff and minor controls are necessary to control drainages and erosion. This technology ranges from easy to difficult to apply, depending on the trafficability of the soil and the depth of contamination.

Potential Achievable Level of Treatment

The achievable level of treatment is high, depending on the biodegradability of the wastes and the suitability of the site and the soil.

Reliability of Method

Reliming or reacidification is necessary as treatment progresses.

Secondary Impacts

Dissolution or precipitation of materials within the soil are affected by changes in soil pH. Care must be taken to assure that hazardous materials do not increase in mobility due to the raising or lowering of soil pH. Tillage increases the susceptibility of the site to erosion.

Equipment and Exogenous Reagents

Applicators, tillers, and power implements are required. Depending on the wastes and the soil characteristics, liming or acidifying material is required.

Information Requirements

- characterization and concentration of wastes, particularly organics at site;
- micro-organisms present at site;
- biodegradability of waste constituents (half-life, rate constant);
- biodegradation products (particularly hazardous products);
- depth, profile, and areal distribution of constituents;
- soil pH;
- other soil properties for biological activity (oxygen content, soil moisture, nutrient content, organic matter, temperature, etc.);
- trafficability of soil and site.

Sources of Information

Gray, 1978; Alexander, 1977; Anderson, 1978; Cerniglia, and Gibson, 1979; Cerniglia, et al., 1979; McLean, 1982; McLean, et al., 1966.

3.4.2.1.5 Soil Nutrients

Description

Although most micro-organisms are capable of very efficient extraction of inorganic nutrients from their environment, their activity may be limited by nutrient availability. This is especially true if available carbon is in large excess relative to the nitrogen and/or phosphorus required by the micro-organisms that degrade it. Determination of soil organic carbon, organic nitrogen, and organic phosphorus allows the determination of its C:N:P ratio and an evaluation of nutrient availability. If the ratio of organic C:N:P is wider than about 300:15:1 and available (extractable) inorganic forms of N and P do not narrow the ratio to within these limits, supplemental nitrogen and/or phosphorus should be added (Alexander, 1977; Kowalenko, 1978). Excesses or deficits of nitrogen or phosphorus brought about by addition of any organic amendments should be taken into account, and commercial fertilizers can be used to make up any deficit.

Adding fertilizer to hasten the decomposition of crop residues is used in agriculture (Alexander, 1977), and this procedure has been used in the treatment of hazardous waste (oil spill)-contaminated soils (Thibault, and Elliott, 1980). Skujins, et al. (1983) studied the biodegradation of waste oils at a disposal site whose soils were amended with calcium hydroxide, phosphate, and urea. Within 4 years, 90 percent of the applied oil (added 7.5 percent by weight to the surface 10 cm) was degraded.

Wastes Amenable to Treatment

Biodegradable organics are amenable to this treatment.

Status of Technology

In the field, this technology is commonly used in agriculture. It has also been used in the treatment of oil wastes.

Ease of Application

Fertilizers are added to the soil using standard agricultural methods. Depending on the nutrient type, physical state, solubility of the fertilizer, and depth of contamination, the fertilizer is incorporated into the soil as necessary. Sufficient N and P is applied to ensure that these nutrients do not limit the microbial and metabolic activity. Controls to manage the runon and runoff from the site are necessary to prevent drainage and erosion problems. This technology ranges from easy to difficult to apply, depending on trafficability considerations.

Potential Achievable Level of Treatment

If nutrient availability is limiting or retarding microbial degradation or detoxification of organic hazardous waste constituents, the achievable level of treatment would be high if the site and soil characteristics are suitable.

Reliability of Method

Retreatment may be necessary at intervals as nutrients are used up in the process. Liming and reliming may be necessary to maintain optimal pH for biological activity.

Secondary Impacts

Caution must be used in the application of nitrogen to the soil to avoid excessive application. Nitrate or other forms of nitrogen in the soil oxidized to nitrate may be leached to groundwater. Also, some nitrogen fertilizers tend to lower the soil pH, necessitating a liming program to maintain pH optimal for biological activity. Tillage will increase the susceptibility of the site to erosion.

Equipment and Exogenous Reagents

Power implements, tillers, and applicators are required to apply the fertilizer to the soil.

Information Requirements

- characterization and concentration of wastes, particularly organics at site;
- micro-organisms present at site;
- biodegradability of waste constituents (half-life, rate constant);
- biodegradation products (particularly hazardous products);
- depth, profile, and areal distribution of constituents;
- soil nutrient (concentration of all essential nutrients, particularly limiting nutrients);
- other soil properties (pH, soil moisture, oxygen content, organic matter, temperature, etc.);
- trafficability of soil and site.

Sources of Information

Alexander, 1977; Kowalenko, 1978; Thibault and Elliott, 1980; Skujins, et al., 1983; Bass et al., 1960; Bonazountas and Wagner, 1981; Lyman, et al., 1982; Atlas and Bartha, 1981.

3.4.2.2 Addition of Non-specific Organic Amendments

Description

Stimulating general soil microbial activity and population size through organic matter addition improves the opportunities for selection of organisms that can degrade hazardous waste components. High microbial activity provides opportunity for cometabolic processes to act on recalcitrant hazardous waste components. Addition of manures, plant materials, and/or wastewater treatment digester sludge at levels characteristic of composting may prove valuable to biological treatment of hazardous waste-contaminated soils (Kaplan, and Kaplan, 1982a; Doyle and Isbister, 1982).

Extensive laboratory research has shown that supplemental carbon and energy sources can stimulate the metabolism of xenobiotic, often recalcitrant compounds. The breakdown or transformation of these compounds can be through cometabolism (Alexander, 1981), or metabolism of the compound may simply be stimulated by the supplemental carbon and energy source (Yagi and Sudo, 1980). Composting of contaminated soil has been shown to degrade hexahydro-1,3,5-trinitro-1,3,5-triazine (RDX), while the ring structure of 2,4,6-trinitrotoluene (TNT) was not mineralized. The TNT residues were apparently strongly sorbed to the compost (Doyle and Isbister, 1982). Camoni, et al., (1982) demonstrated that addition of organic compost to soil had no significant effect on the half-life (1 year) of 2,3,7,8-TCCD in soil. The degradation of pentachlorophenol and pentachloronitrobenzene in a laboratory composting system has also been studied (Sikora, et al., 1982).

Laboratory experimentation may be needed to determine the biochemical fate of given hazardous compounds in organically enriched soil or compost, and the environmental hazards associated with any residues evaluated (Kaplan and Kaplan, 1982b). Residues may be more or less toxic than the parent compounds. Residues of hazardous compounds may not be extractable from organically enriched soil using ordinary solvents, suggesting strong binding to organic matter or other soil constituents (Doyle and Isbister, 1982; Khan, 1982; Wallnöfer, et al., 1981; Bartha, 1980; Stevenson, 1972). Enzymatic activities of soil micro-organisms can be responsible for coupling xenobiotic compounds and their breakdown products to soil humic materials (Bollag, 1983; Bollag, et al., 1983; Sjobland and Bollag, 1981; Liu, et al., 1981; Sulflita and Bollag, 1980; Bollag, et al. 1980; Bollag, et al. 1978). Careful monitoring for bound hazardous organic compounds, including toxic metabolites of hazardous parent compounds should be performed. Humus-bound xenobiotic compounds may be slow to mineralize or be transformed to innocuous forms (Khan, 1982; Chowdhury, et al., 1981). In such cases, increasing the humic content of the soil may not be the method of choice.

Microbial decomposition of humic matter that contains bound hazardous organic compounds can release these compounds to the soil solution, where they are subject to leaching, volatilization, or reattachment to soil organic matter. This potential mobility of bound hazardous compounds suggests that treatment is not complete until their absence or safe level in the soil can be demonstrated (Bartha, 1980; Saxena and Bartha, 1983b; Khan and Iverson, 1982).

No examples of field trials of this treatment technique are available. Doyle and Isbister (1982) observed 55 percent degradation of RDX in compost incubated in a greenhouse in 6 weeks. In the same study, TNT levels were reduced by more than 99 percent within 3 weeks, but very little decomposition (mineralization) was observed.

Aerobic heterotrophic bacteria oxidize arsenite (As^{+3}) to arsenate (As^{+5}) with the consumption of 0_2. An available reserve of organic matter must be present in the soil for the oxidation to occur. Therefore, when arsenite contaminates a soil in concentrations that are below toxic levels for soil heterotrophs, it will be possible to oxidize the arsenite to arsenate by amending the soil with readily available organic matter and maintaining aerobic conditions in the soil. Oxidation of arsenite in laboratory soil perfusion systems has been described by Quastel and Scholefield (1953). The microbial biochemistry of arsenic has been reviewed by Alexander (1977) and Konetzka (1977). Further treatment with ferrous sulfate will form highly insoluble $FeASO_4$.

Wastes Amenable to Treatment

Biodegradable organic wastes and arsenite wastes are amenable to this treatment.

Status of Technology

There are no examples of field trials. Chemical precipitation is usually used to treat arsenite wastes at landfill sites. Extensive laboratory research has shown that supplemental carbon and energy sources can stimulate metabolism of even recalcitrant compounds. Experimental soil systems have demonstrated the microbial oxidation of arsenite to arsenate.

Ease of Application

The quantity of organic material required must be determined in treatability studies. Nonspecific, readily biodegradable organic matter should be added and mixed into the soil as dry materials or as slurries. The soil moisture level should be optimized, and frequent mixing is required to maintain aerobic conditions. Runon and runoff controls are required. This technology is easy or difficult, depending on the trafficability of the soil and site, and the depth of contamination.

Potential Achievable Level of Treatment

The potential achievable level of treatment ranges from low to high, depending on the solubility, sorption, and biodegradability of the organic constituents in the waste. Some arsenite may be bound to the soil and will not be available for oxidation. Available (extractable) arsenite should be quickly and completely oxidized.

Reliability of Method

This technolgy may require reapplications to complete treatment.

Secondary Impacts

Hazardous constituents may be initially bound to organic materials, but later released as organic materials decompose.

Under anaerobic conditions, the added organic matter may result in the reduction and methylation of arsenic to volatile forms. Anaerobic conditions must be avoided. However, anaerobic microsites are known to exist even in well-aerated soil, and some volatile metal compounds may be produced even in carefully managed soils. Tillage will increase the susceptibility of the site to erosion.

Organic materials may have many effects on soil properties, including:

- degree of structure;
- water-holding capacity;
- bulk density;
- immobilization of nutrients, hindering degradation of organic wastes;
- reduction in soil erosion potential;
- soil temperature.

Organic materials may also result in excessive nitrate levels in receiving waters, depending on the nitrogen content and degree of mineralization of the material.

Equipment and Exogenous Reagents

Power implements, tillers, applicators, and proper drainage are required. The exogenous reagent required is organic material.

Information Requirements

- characterization and concentration of wastes, particularly organics at site;
- micro-organisms present at site;
- biodegradability of waste constituents (half-life, rate constant);
- biodegradation products (particularly hazardous products);
- depth, profile, and areal distribution of constituents;
- soil organic matter;
- other soil properties (pH, soil moisture, oxygen content, nutrient content, temperature, etc.);
- trafficability of soil and site.

Sources of Information

Kaplan and Kaplan, 1982a and 1982b; Doyle, Isbister, 1982; Alexander, 1981; Yagi, and Sudo, 1980; Camoni, et al., 1982; Sikora, et al., 1982; Khan, 1982; Wallnöfer, et al., 1981; Bartha, 1980; Stevenson, 1982; Bollag, 1983; Bollag, et al., 1983; Sjobland and Bollag, 1981; Liu, et al., 1981; Sulflita and Bollag, 1980; Saxena and Bartha, 1983a and 1983b; Khan and Iverson, 1982; Quastel and Scholefield, 1953; Konetzka, 1977; Lyman, et al., 1983; Bonazountas and Wagner, 1981; Sikora L.J., et al., 1982.

3.4.2.3 Analog Enrichment for Cometabolism

Description

Adding a chemical analog of a hazardous compound to a contaminated soil or to culture media can accomplish cometabolism of the hazardous compound (Sims and Overcash, 1981; Pal et al., 1980; Furukawa, 1982; Focht and Alexander, 1970). Apparently, enzymes proliferated by micro-organisms to metabolize an energy-yielding substrate with structural similarity to a recalcitrant xenobiotic compound can, in certain cases, transform the recalcitrant molecule cometabolically (Alexander, 1981). For compounds where the transformation product of the cometabolic process is not hazardous or is degradable by other organisms in the soil microbial community, analog enrichment may be an effective treatment for contaminated soil.

Sims and Overcash (1981) used analog enrichment with phenanthrene to increase the rate of degradation of benz(a)pyrene, resulting in a decrease of 35 percent in the half-life. Biphenyl has been used to stimulate cometabolic degradation of PCBs (Furukawa, 1982).

Close chemical analogs to hazardous compounds or their degradation products may be hazardous. Therefore, care must be used in selecting and using analog exogenous agents for treatment.

Wastes Amenable to Treatment

Organic waste containing constituent(s) having analogs with high rates of degradation by organisms without producing toxic products are amenable to treatment.

Status of Technology

No information is available on the field level. In the laboratory, studies that are limited in number and range of compounds have shown that analog enrichment can accomplish cometabolism.

Ease of Application

Analog compounds are added in amounts large enough to stimulate microbial activity, but not enough to be toxic to microbial functions or to adversely affect public health and the environment. Treatability studies are required to determine the feasibility, loading rate, and effectiveness of the analog(s). The analogs may be applied as solids, liquids, or slurries and mixed thoroughly with the contaminated soil. Fertilization may be required to maintain microbial activity. Controls may be necessary to prevent drainage and erosion problems. This technology may range from easy to difficult to apply, depending on the trafficability and the depth of contamination.

Potential Achievable Level of Treatment

The level of treatment may range from low to high, depending on the susceptibility of the hazardous constituent to cometabolism.

Reliability of Method

The reliability of this technology is unknown.

Secondary Impacts

Tillage will increase the susceptibility of the site to erosion.

Equipment and Exogenous Reagents

Power implements, tillers, and applicators are required to apply the analog compounds.

Information Requirements

- characterization and concentration of wastes, particularly organics at site;
- micro-oganisms present at site;
- biodegradability of waste constituents (half-life, rate constant);
- availability of structural analog(s) to waste constituent(s);
- degradation pathway for analog;
- biodegradation products (particularly hazardous products);
- depth, profile, and areal distribution of constituents;
- other soil properties (pH, soil moisture, soil nutrient content, oxygen content, organic matter, temperature, etc.);
- trafficability of soil and site.

Sources of Information

Sims and Overcash, 1981; Pal, et al., Furukawa, 1982; Focht, and Alexander, 1970; Alexander, 1981; Bonazountas and Wagner, 1981; Lyman, et al., 1982; Alexander, 1977.

3.4.2.4 Augmentation with Exogenous Acclimated or Mutant Micro-organisms

Description

Biological treatment methods described thus far have relied on the stimulation of microbial activity in the soil or on the natural selection of populations of micro-organisms, which can degrade toxic waste constituents. These approaches show considerable promise for treating many kinds of organic hazardous waste constituents. However, the metabolic range of the natural soil microbiota may not include the capability to degrade certain compounds or classes of compounds. In addition, microbial metabolic specialists may not develop large enough populations under limited substrate conditions to degrade xenobiotic compounds rapidly enough to meet treatment criteria. In situations such as these, it may be advisable to add exogenously grown micro-organisms to the soil. Microbial inoculants are available commercially with the broad range of metabolic capabilities, and experience in their use in both soil and aquatic systems contaminated with waste chemicals is expanding (Anonymous, 1981, 1982; Thibault and Elliot, 1980; Walton and Dobbs, 1980; Thibault and Elliot, 1979). Table 3-21 lists suppliers of biological products that have been or may be used to treat hazardous waste-contaminated soils. Frequently, the application of microbial amendments to the soil is combined with other treatment techniques such as soil moisture management, aeration, and fertilizer addition.

Laboratory trials have recently demonstrated the potential of exogenously grown, xenobiotic compound degrading bacteria to quickly degrade target compounds. Edgehill and Finn (1983) inoculated pentacholorophenol (PCP) degrading *Arthrobacter* and observed rapid degradation of PCP ($t_{1/2} < 1$ day) when the soil was incubated at 30°C. In soil treated under a roof where temperatures ranged from 8 to 16°C, PCP degradation was much slower, but mixed inoculated soil lost PCP faster than the control.

Kilbane et al. (1983) used repeated applications of a "genetically engineered" *Pseudomonas cepacia* with the ability to mineralize 2,4,5-trichlorophenoxyacetic acid (2,4,5-T) obtaining 90 percent to essentially complete removal

TABLE 3-21. COMMERCIAL MICROBIAL AUGMENTATION PRODUCTS OR PROCESSES USED TO TREAT HAZARDOUS WASTE CONTAMINATED SOILS

Vendor	Address	Product Name(s)	Product Description	Treatment	Price $/Unit	Price $/Acre
Flow Laboratories Environmental Cultures Division	Ingelwood, CA	DBC Plus; Types A, A-2,B,F, and H-1.	Formulated from specifically cultured bacteria preserved by freeze drying and air drying techniques.	25 lb/acre	10.50-15.80/lb	$263-395.00
General Environmental Science	Beachwood, OH	LLMO	Mixture of 7 bacterial strains (*Bacillus, Pseudomonas, Nitrosomonas, Nitrobacter, Cellulomonas, Acrobacter, Rhodopseudomonas*) in liquid suspension.	(Site dependent)	$16.00/gallon	
Groundwater Decontamination Systems, Inc.	Waldwick, NJ	GDS process	Technique involves circulating water from the soil into an environmentally controlled tank. Nutrients are added and the water is aerated. Treated water is returned to the soil. Air may be injected into the soil to stimulate further biodegradation.	(Site dependent)	$0.02/gal treated	
Polybac/Cytox Corporation	Allentown, PA; San Francisco, CA; Gonzales, FL	Polysoil process	Mutant bacteria formulation, nitrogen and phosphorus fertilizer, and biodegradable emulsifier.	100 lb. organisms + 400 lb. fertilizer and emulsifier if needed.		$3227-8067.00 per application[a]
		Chemico-biological augmentation process	Uses chemical treatment ahead of biological treatment to shorten treatment time (currently in experimental and demonstration stages).			$40,300-161,300 for total treatment
Sybron Biochemical	Birmingham, NJ; Salem, VA	Detoxsol	Formulation of mutant bacteria, buffer nutrients, growth stimulator, and detoxifying agents.	363 lb/acre	$27/lb	$9,801[b]

a. Includes labor, equipment, and products. Usually 2 to 6 applications required, depending on degradability with the Polysoil process.

b. Prices for treatment of areas larger than 2000 ft^2 are negotiable.

Source: Utah Water Research Laboratory

of 2,4,5-T from contaminated soil within 6 weeks. When the 2,4,5-T was exhausted, the population of *P. cepacia* became undetectable, but when 2,4,5-T was added to the soil 12 weeks after initial treatment, regrowth occurred.

This method may be most effective against one compound or closely related compounds. Toxicity or the inability of the micro-organisms to metabolize a wide range of substrates may limit their effectiveness.

Recent research advancements in genetic engineering, particularly in interstrain and interspecies genetic transfer, hold out hope for development of organisms with extraordinary abilities to degrade xenobiotic compounds (Chakrabarty, 1982; Johnston and Robinson, 1982; Chakrabarty, 1980). More information is required, however, on the ability of genetically engineered organisms to survive, grow, and function in the soil environment (Stotsky and Krasovsky, 1981, Liang et al., 1982).

Wastes Amenable to Treatment

Compounds or classes of compounds which may be degraded by mutant or selected bacterial cultures which are available commercially are shown in Table 3-22.

State of Technology

This technology has been demonstrated in the laboratory and has been used in several full-scale soil decontamination operations. Case histories of treatment of chemical spill sites (oil spill, orthochlorophenol spill, and acrylonitrile spill) reported by Thibault and Elliott (1979, 1980) and Walton and Dobbs (1980) were deemed successful. However, some hazardous waste clean-up practitioners are skeptical about the use of this technology, since the soil environment is so important in determining microbial activity and hence the success of applying exogenous organisms (Anonymous, 1982). More information is required on the ability of exogenous organisms to service, grow, and function in the soil environment.

Ease of Application

Methods for application are determined in consultation with the vendor of the micro-organisms. The micro-organisms may be applied in liquid suspension or with a solid carrier. Depending on the method of application, runon and runoff controls may be necessary. The ease of application depends on the trafficability and the depth of contamination.

Potential Achievable Level of Treatment

With waste constituents that are susceptible to degradation by the added micro-organisms and when the site and soil are conducive to microbial activity, the potential level of treatment is high.

Reliability of Method

Relatively long periods of time may be required to complete treatment. Excessive precipitation may "wash out" the inoculum, necessitating retreatment.

Secondary Treatment

Tillage, if used, will increase the susceptibility of the site to erosion.

Equipment and Exogenous Reagents

These vary according to the micro-organisms used, as recommended by the vendor. Micro-organisms are usually applied by the vendors.

TABLE 3-22. COMPOUNDS OR CLASSES OF COMPOUNDS THAT HAVE BEEN (OR COULD BE) DEGRADED BY COMMERCIALLY AVAILABLE MICROBIAL AUGMENTATION PRODUCTIONS

Alcohols	Esters
n-Butyl alcohol	Methacrylates
Dimethylaminoethanol	
Alkyl Halides	**Ketones**
Ethylene dichloride (1,2-Dichloroethane)	Acetone
Methylene chloride (Dichloromethane)	**Nitriles**
Propylene dichloride (1,2-Dichloropropane)	Acrylonitrile
Amines	**Phenols**
Dimethylaniline	Phenol
Trimethylamine	Metachlorophenol
Aromatic Hydrocarbons	Orthochlorophenol
Divinyl Benzene	Pentachlorophenol
Polynuclear Aromatic Hydrocarbons (PNA's)	Resorcinol (1,3-Benzenediol)
Styrene (Vinyl Benzene)	t- Butylcatechol
Chlorinated Aromatics	Crude and refined oils
	Emulsifiers
Polychlorinated biphenyls (PCB's)	Detergents

Source: Utah Water Research Laboratory

Information Requirements

- characterization and concentration of wastes, particularly organics at site;
- micro-organisms present at site;
- metabolic capability of exogenously grown micro-organisms;
- pathogenicity to susceptible populations;
- biodegradability of waste constituents (half-life, rate constant);
- biodegradation products (particularly hazardous products);
- depth, profile, and areal distribution of constituents;
- other soil properties (pH, soil moisture, nutrients, oxygen content, organic matter, temperature, etc.);
- trafficability of soil and site;
- climate, particularly precipitation.

Sources of Information

Anon., 1982; Thibault and Elliott, 1979, 1980; Walton and Dobbs, 1980; Edgehill and Finn, 1983; Kilbane, et al., 1983; Chakrabarty, 1980, 1982; Johnson and Robinson, 1982; Stotsky and Krasovsky, 1981; Bonazountas and Wagner, 1981; Lyman, et al., 1982; Alexander, 1977.

3.4.2.5 Application of Cell-free Enzymes

Description

Enzymes, produced by micro-organisms, which can transform hazardous compounds to nonhazardous or more labile products, could possibly be harvested from cells grown in mass culture and applied to contaminated soils. Crude or purified enzyme extracts are commonly used in industry either in solution or immobilized on glass beads, resins, or fibers to catalyze a variety of reactions, including the breakdown or transformation of carbohydrates and proteins. Munnecke, et al. (1982) discussed the enzymology of selected pesticide degradation, and suggested that extracted (cell-free) enzymes might be used to quickly transform pesticides in soils. They pointed out that a bacterial enzyme preparation has been used to detoxify organophosphate pesticide waste from containers (Munnecke, 1980), and that the enzyme parathion hydrolase hydrolyzed 1 percent parathion or diazinon within 24 hours in contaminated soil.

Enzyme activity can often be preserved in environments which are not hospitable to micro-organisms. Enzymes could possibly be used in soils with extremes of pH and temperature, high salinity, or high solvent concentrations, i.e., in soils where microbial growth may be restricted. In milder soil environments, enzymatic hydrolysis or oxidation of a compound may make it more susceptible to decomposition by the soil micribiota (Munnecke, et al., 1982).

To function outside the cell in the soil environment an enzyme must not require co-factors or co-enzymes — a requirement that will limit the application of many enzymes. Enzymes may also be chemically and/or biologically degraded. They may be leached out of the treatment zone, and they may be inactive or have lower activity if they are bound to clay or humus in the soil. Outside of biochemical and environmental constraints, logistics and costs for producing enzymes in large enough quantities may limit current use of this concept.

Wastes Amenable to Treatment

Organic wastes primarily are amenable to this treatment.

Status of Technology

Theoretically, enzymes would quickly transform hazardous compounds if they remained active in soil. There have been laboratory experiments with parathion hydrolase.

Little information is available on the use of this technique in soil. The only example at the present time is the work by Munnecke, et al. (1982). There is no information available from the field.

Ease of Application

For treatment of pesticides, application rates of 1 mg crude protein per 100 mg of pesticide in soil have been suggested (Munnecke, personal communication, July 1983). Thorough mixing with the soil is necessary. Enzymes are sprayed on soil in solution or suspension, or spread with solid carrier, using sprayers or fertilizer spreaders. The poor availability of appropriate enzymes for hazardous waste constituents makes application difficult. Depending on application method, controls to prevent runon and runoff may be necessary.

Potential Achievable Level of Treatment

Given the appropriate enzyme and if the enzyme remains active in the soil, the potential achievable level of treatment is high.

Reliability of Treatment

The reliability of this technology is unknown.

Secondary Impacts

Enzymatic degradation products may not be less hazardous than parent compound(s). Products may be more water soluble and/or mobile in the soil. Tillage, if used, will increase the susceptibility of the site to erosion.

Equipment and Exogenous Reagents

Power implements, tillers, and applicators are necessary for application of this treatment. Enzymes for treating hazardous waste constituents are not currently in bulk production. Although increases in the use of industrial enzymes have been projected (O'Sullivan, 1981), only eight companies accounted for 90 percent of worldwide production in 1981. Five of these companies are located in western Europe. Only 16 enzymes (primarily amylases, proteases, oxidases, and isomerases) accounted for 99 percent of the 1981 market. This information suggests that specialized enzyme production, even on a large scale, may be quite expensive. Current prices for bulk enzyme materials range in price from $1.45 to $164 per pound. If the enzyme can be produced through chemical synthesis, it will be much less expensive than if it is produced by micro-organsims in fermenters (Miles Laboratories, personal communication, July 1983).

Information Requirements

- characterization and concentration of wastes, particularly organics at site;
- micro-organisms present at site;
- biodegradability of waste constituents (half-life, rate constant);
- biodegradation products (particularly hazardous products);
- cell-free enzymetic activity for transformation of the compound(s) of interest;
- co-factor requirements of enzyme;
- stability of enzyme under soil environmental conditions;
- depth, profile, and areal distribution of constituents;
- other soil properties (pH, salinity, soil moisture, nutrients oxygen content, organic matter, temperature, etc.);
- trafficability of soil and site.

Sources of Information

Munnecke, et al., 1980, 1982; O'Sullivan, 1981.

3.4.3 Photolysis

Utilization of the lower atmosphere as a treatment medium requires an analysis of both the photoreaction potential and the volatility of the compounds of interest. Volatilization and dilution alone are not considered as acceptable treatment methods. An adequate assessment of the potential for the use of photodegradation requires information regarding the compound's atmospheric reaction rate (log $K_{OH^{\circ}}$) and anticipated reaction products. While this information is available for a selected number of compounds (Cupitt 1980; Lemaire et al., 1980), much more data are required if photodegradation is to become a viable treatment option.

If a compound is determined to be poorly photoreactive, e.g., a $t_{1/2}$ in the atmosphere greater than 1 day, volatilization suppression may be required to maintain safe ambient air concentrations at the site. Volatilization suppression techniques are described in Section 3.6.

Photodegradation is the use of incident solar radiation to carry out photoreaction processes. Both direct photolysis (photoreactions due to direct light absorption by the substrate molecule) and sensitized photo-oxidation (photoreactions mitigated by an energy-transferring sensitizer molecule) are possible under environmental conditions. Sensitized photoreactions are characteristically ones of photo-oxidation resulting in substrate molecule oxidation rather than substrate isomerism, dehalogenation, or dissociation characteristic of direct photolysis reactions.

The rate of photoreaction is influenced by the nature of light reaching the reaction medium, the absorption spectrum of the reacting species of sensitizer, the concentration of reacting species, the energy yield produced upon light energy absorption, the nature of the media in which the reaction is taking place, and the interactions that occur between the contaminant and its surroundings. Overall, reactions are a complicated function of the characteristics listed above. Photolysis reaction rates and breakdown products are only crudely understood.

Although the occurrence of the soil photoreaction of adsorbed chemicals has been identified, the importance of this reaction as compared to aqueous or vapor photoreactivity has not been identified. Soil photodecomposition will be of concern if the compound or compounds remain relatively stationary within the contaminated soil, e.g., high values of K_D (soil:water partition coefficient) and high values for K_W (air:water partition coefficient). Soil characteristics including soil organic content (Spencer, et al., 1980), transition metal content (Nilles and Zabik, 1975), and soil pigment content (Hautala, 1978) have been indicated as affecting photochemical reactions within soil systems. Moisture content and its effect upon chemical partitioning within the air/water/soil matrix within a soil system will also potentially have a great impact upon soil photoreactions (Burkhard and Guth, 1979; Hautala, 1978).

Information regarding the photolysis of pesticides in air is generally available due to the relatively high volatility of pesticides and the concern for their transport via the air medium. The major photoreaction taking place with pesticides in the atmosphere is oxidation (Crosby, 1971; Plimmer, 1971) involving the OH radical or ozone, of which the OH radical is the species of greatest reactivity (LeMaire, et al., 1982). Based on a first-order rate of reaction for vapor phase reactions with OH radical, the half life of a specific chemical species can be estimated if its OH radical reaction rate constant is known using:

$$t_{1/2} = 0.693/K_{OH}{}^{\circ}[OH^{\circ}]) \qquad (3\text{-}16)$$

where

$t_{1/2}$ = time to decrease component concentration by 50% (s)

$k_{OH^{\circ}}$ = OH radical reaction rate constant (cm^3/molecule-s)

$[OH^{\circ}]$ = atmospheric OH radical concentration (4×10^5 molecules/cm^3 = 6645×10^{-19} moles/cm^3)

A number of OH radical reaction rate constants as presented by Klopffer, (1980) and Cupitt, (1980), are given in Table 3-23. Table 3-24 presents additional constants, as given by Cupitt (1980), along with an estimation of the likelihood of a photolysis reaction occurring within the ambient atmosphere. Cupitt (1980) indicated that of all atmospheric removal mechanisms including physical, chemical, and photochemical, the photochemical reactions are of most significance for most classes of hazardous compounds and should be investigated further as a viable treatment option.

The use of photochemical reactions for the enhancement of compound biodegradation is an important area of interest for hazard mitigation from hazardous waste sites. Photolysis reactions are oxidative in nature and would be expected to aid in microbial degradation through the oxidation of resistant complex structures (Crosby, 1971; Sims and Overcash, 1983). Photoreactions are limited to soil surfaces due to light extinction within the soil system, but coupled to soil mixing, they may prove to be highly effective as an in-place treatment technique for relatively immobile chemical species.

TABLE 3-23. RATE CONSTANTS FOR THE HYDROXIDE RADICAL REACTION IN AIR WITH VARIOUS ORGANIC SUBSTANCES, $K_{OH}°$ IN UNITS OF $(MOLE\text{-}SEC)^{-1}$

Substance	$\log_{air} k_{OH}°$
Acetaldehyde	9.98
Acrolein	10.42
Acrylonitrile	9.08
Allyl chloride	10.23
Benzene	8.95
Benzyl chloride	9.26
Bis(chloromethyl)ether	9.38
Carbon tetrachloride	<5.78
Chlorobenzene	8.38
Chloroform	7.78
Chloromethyl methyl ether	9.26
Chloroprene	10.44
o-,m-,p-cresol*	10.52
p-cresol	10.49
Dichlorobromobenzene*	8.26
Diethyl ether	9.73
Dimethyl nitrosamine	10.37
Dioxane	9.26
Epichlorophydrin	9.08
1,2-epoxybutane	9.16
Epoxypropane	8.89
Ethanol	9.28
Ethyl acetate	9.06
Ethyl propionate	9.03
Ethylene dibromide	8.18
Ethylene dichloride	8.12
Ethylene oxide	9.08
Formaldehyde	9.78
Hexachlorocyclopentadiene	10.55
Maelic anhydride	10.56
Methanol	8.78
Methyl acetate	8.04
Methyl chloroform	6.86
Methyl ethyl ketone	9.32
Methylene chloride	7.93
Methyl propionate	8.23
Nitrobenzene	7.56
Nitromethane	8.81
2-nitropropane	10.52
n-nitrosodiethylamine	10.19
Nitrosoethylurea	9.89
n-propylacetate	9.41
Perchloroethylene	8.01
Phenol	10.01
Phosgene	nonreactive
Polychlorinated biphenyls	<8.78
Propanol	9.51
Propylene oxide	8.89
Tetrahydrofuran	9.95
Toluene	9.52, 9.56
Trichloroethylene	9.12
Vinylidene chloride	9.38
o-,m-,p-xylene*	9.98

Source: Adapted from Lemaire, et al. (1980) and Cupitt, (1980).

TABLE 3-24. ATMOSPHERIC REACTION RATES AND RESIDENCE TIMES OF SELECTED ORGANIC CHEMICALS

Compound	$k_{OH} \times 10^{12}$ (cm^3 $molecule^{-1}$ sec^{-1})	Direct Photolysis Probability	Physical Removal Probability	Residence Time (Days)	Anticipated Photoproducts
Acetaldehyde[c]	16	Probable	Unlikely	0.03-0.7[c]	H_2CO, CO_2
Acrolein	44[a]	Probable	Unlikely	0.2	OCH-CHO, H_2CO, HCOOH, CO_2
Acrylonitrile	2	–	Unlikely	5.6	H_2CO, HC(O)CN, HCOOH, CN^0
Allyl chloride	28[a]	Possible	Unlikely	0.3	HCOOH, H_2CO, $ClCH_2CHO$, chlorinated hydroxy carbonyls, $ClCH_2COOH$
Benzyl chloride	3[a]	Possible	Unlikely	3.9	OCHO, Cl; ring cleavage products chloromethyl-phenols
Bio(Chloromethyl) Ether	4[a]	Possible	Probable	0.02-2.9[d]	$HCl+H_2CO$, ClHCO, chloromethylformate
Carbon Tetrachloride	<0.001	–	Unlikely	>11,000	Cl_2CO, Cl^0
Chlorobenzene	0.4[a]	Possible	Unlikely	28	Chlorophenols, ring cleavage products
Chloroform	0.1	–	Unlikely	120	Cl_2CO, Cl^-
Chloromethyl methyl ether	3[a]	Possible	Probable	0.004-3.9[d]	chloromethyl and methyl formate, ClHCO
Chloroprene	46[a]	Probable	Unlikely	0.2	H_2CO, $H_2C{=}CClCHO$, OHCCHO, ClCOCHO, $H_2CCHCClO$, chloro-hydroxy acids, aldehydes
o,m,p-cresol[e]	55	–	Unlikely	0.2	hydroxynitrotoluenes, ring cleavage products
Dichlorobenzene[e]	0.3[a]	Possible	Unlikely	39	chlorinated phenols, ring cleavage products
Dimethyl Nitrosamine	39[a]	Probable	–	≤0.3	aldehydes, NO
Dioxane	3[a]	–	Unlikely	3.9	$OHCOCH_2CH_2OCHO$, OHCOCHO oxygenated formates
Dioxin	–	Probable	–	–	–
Epichlorohydrin	2[a]	Possible	Unlikely	5.8	H_2CO, OHCOCHO, $ClCH_2O(O)OHCO$
Ethylene Dibromide	0.25	Possible	Unlikely	45	Bi, $BrCH_2CH_2CHO$, H_2CO, Br HCO
Ethylene Dichloride	0.22	Possible	Unlikely	53	ClHCHO, $H_2CClCOCl$, H_2CO, $H_2CClCHO$
Ethylene Oxide	2[a]	–	Unlikely	5.8	OHCOCHO
Formaldehyde[c]	10	Probable	Unlikely	0.1-1.2[c]	CO, CO_2
Hexachlorocyclo-pentadiene	59[a]	Probable	–	0.2	Cl_2CO, diacylchlorides, ketones, Cl•
Maleic Anhydride	60[a]	Possible	Possible	0.1	CO_2, CO; acids, aldehydes and esters which should photolyze

TABLE 3-24. (Continued)

Compound	$k_{OH} \times 10^{12}$ (cm^3 $molecule^{-1}$ sec^{-1})	Direct Photolysis Probability	Physical Removal Probability	Residence Time (Days)	Anticipated Photoproducts
Methyl Chloroform	0.0[illegible]2	Possible	Unlikely	970	H_2CO, Cl_2CO, Cl•
Methylene Chloride	0.14	Possible	Unlikely	83	Cl_2CO, CO, ClHCO, Cl•
Methyl Iodide	0.004[a]	Possible	Unlikely	2900	H_2CO, I^0, ICHO, CO
Nitrobenzene	0.06[a]	Possible	Unlikely	190	Nitrophenols, ring cleavage products
2-Nitropropane	55[a]	Possible	Unlikely	0.2	H_2CO, CH_3CHO
N-Nitrosodi-ethylamine	26[a]	Probable	–	$\leqslant$0.4	aldehydes, nitroamines
Nitrosoethylurea	13[a]	Possible	–	$\leqslant$0.9	aldehydes, nitroamines
Nitrosomethylurea	20[a]	Possible	–	$<$0.6	aldehydes, nitroamines
Nitrosomorpholine	28[a]	Possible	–	$\leqslant$0.4	aldehydes, ethers
Perchlorethylene	0.17	Possible	Unlikely	67	Cl_2CO, $Cl_2C(OH)COCl$, Cl^0
Phenol	17[a]	–	Possible	0.6	dihydroxybenzenes, nitrophenols, ring cleavage products
Phosgene[f]	~0	–	Possible	–	CO_2, Cl^0, HCl
Polychlorinated Biphenyls	$<$1[a]	Possible	Unlikely	$>$11	hydroxy PCB's, ring cleavage products
POM (Benzo(a)-pyrene)	–	Possible	Probable	8	B(a)P-1,6-quinone
Propylene Oxide	1.3	–	Unlikely	8.9	$CH_3C(O)OCHO$, $CH_3C(O)CHO$, H_2CO, HC(O)OCHO
Toluene	6	–	Unlikely	1.9	Benzaldehyde, cresols, ring cleavage products, nitro compounds
Trichloroethylene	2.2	Possible	Unlikely	5.2	Cl_2CO, ClHCO, CO, Cl•
Vinyllidene Chloride	4[a]	Possible	Unlikely	2.9	H_2CO, Cl_2CO, HCOOH
o-,m-,p-xylene	16	–	Unlikely	~0.7	substituted benzaldehydes, hydroxy xylenes, ring cleavage products nitro compounds

a. Rate constant by method of Hendry and Kenley (1979).

b. Material is not expected to exist in vapor phase at normal temperatures. Residence time calculation assumes the chemical is substantially absorbed on aerosol particles and that the aerosol particles have a residence time of approximately 7 days.

c. The shorter residence time includes a photolysis rate as given in Graedel (1978).

d. Decomposition in moist air is expected. The shorter residence time includes the cited decomposition rate.

e. Values given are averages for the various isomers.

f. Reaction with O('D) is possible; $k = 3.6 \times 10^{-10}$ cm^3 $molecule^{-1}$ sec^{-1}, and $[O('D)] = 0.2$ molecules cm^{-3} implies a tropospheric lifetime of 440 years. In addition, slow hydrolysis is expected.

Source: Cupitt, L.T., 1980.

Photolysis of soil contaminants may be enhanced in 2 ways:

(1) by addition of proton donors, and
(2) by enhancing volatilization leading to photodegradation

3.4.3.1 Addition of Proton Donors

Description

Enhanced photodegradation of soil contaminants may be accomplished through the addition of various proton donor materials to the contaminated soils.

Photolysis of tetrachlorodibenz-p-dioxin (TCDD) on soil surfaces was reported by Crosby, et al. (1971) in the presence of suitable hydrogen sources in the form of polar solvents, and Plimmer, et al. (1973) indicated that methanol used as a solvent for TCDD photo-oxidation also acted as a hydrogen donor in the photolysis reaction.

Investigations of feasible in-place treatment methods for contaminated areas surrounding the TCDD release which occurred near Seveso, Italy, in 1976 have been reported by a number of authors. Wipf, H.K., et al. (1978) investigated the use of alternative hydrogen donors for the photo-oxidation of TCDD. Solutions of 80 percent olive oil and 20 percent cyclohexane at 350 L/ha and 40 percent aqueous emulsion with 4 percent biodegradable emulsifying agent at 400 L/ha were found to produce a thin film on vegetation and other smooth surfaces to provide a maximum reaction surface for TCDD photolysis. TCDD reductions in excess of 60 percent were observed within 48 hours after treatment. Under laboratory conditions, the oil and emulsion solutions reduced the half life of TCDD by a factor of 25 upon irradiation with simulated sunlight.

Liberti, et al. (1978) reported a 1:1 solution of ethyl oleate and xylene used as hydrogen donors also resulted in complete degradation of TCDD on building surfaces in approximately 1 hour at 2 mW/cm^2 and 72 hours at 20 $\mu W/cm^2$ light intensity.

Dehalogenation of kepone (Dawson, et al., 1978, in Dawson, et al., 1980) and enhanced PBB (Christensen and Weimer 1979) photolysis have been reported when hydrogen donors in the form of amino groups have been added to contaminated soils prior to irradiation with sunlight. However, no observable degradation of PCBs in soil was found with amine-enhanced soil (Meuser and Weimer, 1982).

Soil photodecomposition of PCBs was reported by Occhiucci and Patacchuiola, (1982) and was shown to be enhanced by the addition of a proton donor, triethylamine, to the waste/montmorillonite system. Addition of triethylamine resulted in a 2.5 to 5-fold increase in PCB degradation over a 100 hour irradiation period, providing 4 to 18 percent decomposition of the various chlorinated species tested.

The dechlorination reactions described above result from hydrogen abstraction by organic radicals formed upon irradiation as presented by Bunce (1982). Optimization of this process for soil systems has yet to be accomplished but appears to represent an area of potential for use in the in-place treatment of stable, nonmobile compounds.

Activated carbon adsorption of organics at hazardous waste sites followed by chemical addition and photolysis has been reported by React Environmental Crisis Engineers, St. Louis, MO (personal communication, 1983). The site is impregnated with activated carbon and soils are sampled. The most highly contaminated materials are physically removed, packaged, and disposed of in an approved hazardous waste disposal facility. The remaining material is mixed with sodium bicarbonate to increase soil pH, and the material is allowed to photochemically react resulting in the photolysis of the parent material. The level of treatment is expected to be high to medium. An increase in soil pH is the major secondary impact of the treatment method.

Wastes Amenable to Treatment

Photodegradable organic wastes are amenable to this treatment. Generally, this includes compounds with moderate to strong absorption in the >290-nm wavelength range. Such compounds generally have an extended conjugated hydrocarbon system or a functional group with an unsaturated hetero atom (e.g., carbonyl, azo, nitro). Groups that typically do not undergo direct photolysis include saturated aliphatics, alcohols, ethers, and amines. Tetrachlorodibenz-p-dioxin (TCDD), kepone, and PCBs have been treated with this method.

Status of Technology

In the field, several hazardous waste sites have been treated by this method. Laboratory studies have demonstrated the potential for photo-oxidation. However, the potential for the production of hazardous compounds from photodegradation needs to be further researched.

Ease of Application

Materials which are hydrogen donors are applied and depending on the depth of contamination, the soil is tilled to expose waste to the light. This may be easy or difficult depending on the trafficability of the site and the depth of contamination. Runon and runoff controls may be necessary to manage the drainage and erosion.

Potential Achievable Level of Treatment

The level of treatment achievable is potentially high, based on limited experimental data. Effectiveness also depends on the amount of tillage possible at the site and the depth of contamination.

Reliability of Method

Considering that hazardous compounds may be produced as a consequence of photodegradation, unless the constituents in the wastes and their photodegradation products are known, this is not a reliable method.

Secondary Impacts

Production of hazardous compounds from the photodegradation of pesticides has been documented, e.g., dieldrin formation from aldrin, paraoxon formation from parathion, phosgene formation from chloropicrin (Crosby 1971), and the formation of PCBs from the photoreaction of DDT (Woodrow et al., 1983). The potential for such occurrences with additional parent compounds is expected to be high and further research is needed to identify potential toxic product formation to ensure the safe application of this treatment methodology.

Equipment and Exogenous Reagents

Power implements, tillers, and applicators are required to apply the proton donors.

Information Requirements

- characterization and concentration of wastes, particularly organics at site;
- absorption spectra of waste constituents (at wavelength >290 nm, molecular absorptivities, absorption maxima, quantum yield);
- photolysis rate constant(s);
- products of photolysis (particularly hazardous products);
- volatility of organics (vapor pressure, Henry's Law Constant);

- depth, profile, and areal distribution of contamination;
- light intensity at site;
- trafficability of soil and site.

Sources of Information

Cupitt, 1980; Lemaire, et al., 1981; Spencer, 1980; Nilles and Zabik, 1975; Hautala, 1978; Burkhard and Guth, 1979; Crosby, 1971; Plimmer, 1971; Klopffer, 1980; Sims and Overcash, 1983; Plummer, 1973; Wiff, 1978; Liberte, et al., 1978; Dawson, et al., 1980; Christensen and Weimer, 1979; Meuser and Wiemer, 1982; Occhiucci and Patacchiola, 1982; Bunce, 1982; Woodrow, et al., 1983; Hendry and Kenley, 1979; Graedel, 1978.

3.4.3.2 Enhancement of Volatilization

Description

Enhancing volatilization of compounds from the soil which are susceptible to photodegradation may be a potential treatment technique. This method involves increasing the bulk density or drying of the soil system to increase soil vapor pore spaces and subsequently increase the vaporization rate of desired compounds, followed by photodegradation in air.

Wastes Amenable to Treatment

The technique is applicable to compounds of low water solubility, with low K_D values, low K_W values, and those that are highly photoreactive and that, once within the lower atmosphere, would have a relatively short half life (on the order of hours or preferably minutes). Generally, this includes compounds with moderate to strong absorption in the >290-nm wavelength range. Such compounds generally have an extended conjugated hydrocarbon system or a functional group with an unsaturated hetero-atom (e.g., carbonyl, azo, nitro). Groups that typically do not undergo direct photolysis include saturated aliphatics, alcohols, ethers, and amines.

Status of Technology

This technology is conceptual, based on observed laboratory reaction rates for waste constituents in simulated atmospheres. No information concerning the enhancement of volatilization to increase photolysis has been reported in the literature either on laboratory or pilot scale. Theoretically the process can be expected to be effective for compounds of high volatility and high photoreactivity. Further investigations concerning this process are needed.

Ease of Application

The soil may be tilled to enhance vaporization. Drying of the soil to increase volatilization may be accomplished by tilling, or by installation of a drainage system. This technology may be easy or difficult to apply, depending on the trafficability of the site and the depth of contamination. Controls for runon and runoff management may be necessary.

Potential Achievable Level of Treatment

The level of treatment achievable is potentially high for volatile, photoreactive compounds with short atmospheric half-lives.

Reliability of Method

As with the previous treatment technology, hazardous products may result from photodegradation. Unless there is sufficient certainty that the constituents in a waste will not produce hazardous photodegradative products, this technology should not be used.

Secondary Impacts

There is evidence of the production of degradation products that are hazardous from several pesticides. Potentially, this could also be true of other compounds. Tillage used to dry the soil could increase the susceptibility of the site to eorsion.

Equipment and Exogenous Reagents

Power implements, tillers, and a drainage system are required for this treatment technology.

Information Requirements

A determination of the volatility potential of the compounds of concern through an analysis of their partition coefficients, along with a determination of their half lives, will allow a determination of the applicability of this proposed treatment technique to a specific situation. The information requirements are:

- characterization and concentration of wastes, particularly organics at site;
- absorption spectra of waste constituents (at wavelengths >270 nm, molar absorptivities, absorption maxima, quantum yield);
- photolysis rate constant(s);
- products of photolysis (particularly hazardous products);
- volatility of organics (vapor pressure, Henry's Law Constant);
- depth, profile, and areal distribution of contamination;
- soil moisture;
- light intensity at site;
- trafficability of soil and site.

Sources of Information

Burkhard and Guth, 1979; Crosby, 1971; Crosby, et al., 1971; Cupitt, 1980; Hautala, 1978; Klopffer, 1980; Lemaire, et al., 1982; Nilles and Zabik, 1975; Plimmer, 1971; Sims and Overcash, 1983.

3.5 ATTENUATION

The basic principle of attenuation is the mixing of contaminated soil (or wastes) with clean soil to reduce the concentrations of hazardous components to acceptable levels. This process is potentially applicable to both inorganics and organics. However, acceptable concentration limits have been established only for heavy metals.

3.5.1 Attenuation of Metals

Description

Attenuation of metals which accumulate at the soil surface may be accomplished by either mixing subsurface soil with the top soil, or by applying uncontaminated soil from the adjacent area. Soil may also be purchased from local contractors, or pure soil materials (e.g., bentonite) may be obtained from commercial suppliers.

The treatment requires no chemical additives and is good for large-scale contaminated areas. It can easily be used with other treatments, such as pH adjustment, to maximize metal sorption to soils, to enhance its effectiveness. However, attenuation is generally not applicable for contamination below the plow layer (about 2 feet).

Wastes Amenable to Treatment

This treatment is applicable to all kinds of wastes containing metals. It is particularly applicable to wastes for which concentrations in soil have been defined. For those wastes which are extremely toxic, destructive treatment might be more acceptable.

Status of Technology

This technology has been applied in land treatment systems in which the addition of wastes to soil is controlled and monitored.

Ease of Application

Attenuation for metals may be easy or difficult to apply depending on site/soil trafficability considerations and depth of contamination. Thorough mixing is required. The site should be evaluated for erosion potential, and erosion controls and provisions for containment and treatment of runoff instituted as necessary. Controls for the prevention of runon of precipitation (both surface and subsurface) may also be necessary.

Potential Achievable Level of Treatment

The level of attenuation achievable depends primarily on the concentration and depth of contamination. It could be high with appropriate site/soil trafficability and if the contamination is not deep into the ground.

Reliability of Method

This method is reliable. However, the mixing of soil or addition of other soil into the site may alter the properties of the natural soil. Because of the increased mobility of metals with decreased pH, liming and reliming of the site may be necessary to maintain the metals in the soil if the pH is altered due to the treatment.

Secondary Impacts

The mixing of the soil profile or the addition of different soil may alter the physical, chemical, and biological properties of the original soil at the site. This may have adverse effects. Tilled soils are also usually more susceptible to water and wind erosion.

Equipment and Exogenous Reagents

Power implements and tillers are necessary. If new material is added, applicators will be necessary. Liming material and soil may also be needed.

Information Requirements

- characterization and concentration of metals in waste;
- depth, profile, and areal distribution of contamination;
- soil assimilative capacity of metals, both contaminated and uncontaminated soils;
- soil pH;
- trafficability of soil and site.

3.5.2 Attenuation of Organics

Description

Attenuation of organics in soil is performed in the same manner as attenuation for inorganics discussed above. In addition, the potential for a higher percentage of sorbed organics with an increased mass of soil may increase the level of attenuation for organics. The Freundlich Isotherm used to describe sorption is discussed under Immobilzation (Section 3.3).

As the amount sorbed per unit dry mass of soil (S in Equation 3.1) decreases, the percent sorbed increases. Therefore, the incorporation by mixing of endogenous soil layers, or the addition of exogenous soil to contaminated soil, represents another technology to increase the extent and effectiveness of immobilization of chemical contaminants at hazardous waste sites, and may also aid in decreasing toxicity of the contaminated soil to soil micro-organisms that is due to high concentrations of constituents.

Wastes Amenable to Treatment

This treatment is applicable to all organic wastes. However, organics that are very soluble in water may be more effectively treated by other methods since large amounts of soil may be required to reduce the mobility of the compound. If very toxic components are present in the waste, destructive treatment would be the preferable treatment alternative.

Status of Technology

There is limited field experience in this technology. Addition of dry soils for accomplishing treatment has been utilized for PCB wastes. Waste oil pits, that held transformer fluid containing PCB's ruptured causing a release of 740,000 liters of oil and acidic water. After dewatering, the oils remaining in the pits were treated by the addition of dry soil (Eckenfelder, 1970).

Ease of Application

The endogenous soil profile is tilled to mix uncontaminated soil with the contaminated layers. If the desired attenuation is not reached, imported soil or clay may be applied. This method may be easy or difficult depending on site/soil trafficability considerations for tillage and incorporation of added soil or clay. Erosion may be a problem due to tillage. Controls for erosion prevention and containment and runoff treatment may be necessary. Controls for ri prevention of runon of precipitation (both surface and/or subsurface) may also be necessary.

Potential Achievable Level of Treatment

The level of attenuation achievable is potentially high with suitable size, soil, and waste characteristics.

Reliability of Method

The mixing of soil or addition of other soil or clay to the soil system may alter the properties of the natural soil. As a result, the effectiveness of this method may vary for different compounds, and may not be as expected. However, this method should be reliable under most conditions.

Secondary Impacts

Toxic concentrations to soil organisms may be reduced and biodegradation may be enhanced by this treatment. If clay is added, clay catalyzed degradation and ion-exchange in the clay surface may also be enhanced. However, the mixing of the soil profile or the addition of different soil may alter the physical, chemical, and biological properties of the original soil at the site adversely. Tilled soils are usually more susceptible to water and wind erosion.

Equipment and Exogenous Reagents

Power implements and tillers are necessary. Applicators will be necessary for the addition of new soil or clay.

Information Requirements

- characterization and concentration of organic waste constituents;
- depth, profile, and areal distribution of contamination;
- soil organic carbon content;
- soil moisture;
- trafficability of soil and site.

Sources of Information

Eckenfelder, Jr., 1970.

3.6 REDUCTION OF VOLATILIZATION

Volatilization from a hazardous waste site may need to be controlled to reduce air emissions or to retain compounds within the soil system longer to allow for in-place treatment. There are three ways to reduce volatilization from a site: reduction of soil vapor pore volume, use of physical/chemical barriers, and soil cooling.

3.6.1 Reduction of Soil Vapor Pore Volume

Description

Reduction of soil vapor pore volume for volatilization control is accomplished by modifying the soil system to reduce the partitioning of a compound into the vapor phase and subsequently reduce its rate of volatilization. Such modifications include compaction and water addition to reduce the air-filled pore spaces within the soil. These techniques are especially applicable to those compounds found to represent a high vapor phase mobility potential and a low water phase partition potential (i.e., for compounds with a $K_D \times K_W$ product greater than about 50, and a K_D value greater than 10, where K_D is the soil:water partition coefficient and K_W is the air:water partition coefficient).

Under these circumstances, addition of moisture does not materially decrease the percent sorbed to the soil surface, but reduces greatly the soil vapor pore spaces. Air filled porosity is the soil parameter of major significance affecting volatilization because of increased partitioning onto the soil from the soil vapor phase as the volume of the vapor phase is reduced due to soil compaction and/or water addition (Farmer, et al., 1980).

Evaluation of the effectiveness of soil modifications on contaminant loss via vapor movement can be made rapidly using the technique described by Farmer et al. (1980), assuming steady-state vapor flux through a soil cover, i.e.:

$$J = -D_s (C - C_s)/L \qquad (3\text{-}17)$$

where

J = vapor flux (ng/cm^2/d),

D_s = apparent steady-state diffusion coefficient (cm^2/d),

C = concentration in the air at the soil surface (ng/cm^3),

C_s = concentration in the air at the bottom of the soil layer, i.e., saturation vapor pressure assuming steady-state (ng/cm^3),

L = depth of the soil layer, cm.

Millington and Quirk (1961) have suggested the use of the following equation for apparent diffusion in soil taking into account the porosity of the soil:

$$D_s = D_o (P_a^{10/3}/P_T^2) \tag{3-18}$$

where:

D_o = diffusion coefficient in air (cm^2/d),

P_a = soil air filled porosity, (cm^3/cm^3),

P_T = total soil porosity (cm^3/cm^3), 1 — soil bulk density/particle density,

θ = volumetric water content (cm^3/cm^3),

while Thibodeaux and Hwang (1982) model apparent soil diffusion using total soil porosity and tortuosity:

$$D_S = D_o P_T/t \tag{3-19}$$

where

t = soil tortuosity.

Equation (3-17) then becomes:

$$J = - D_o (P_a^{10/3}/P_T^2) (C - C_s)/L \tag{3-20}$$

or:

$$J = - D_o (P_T/t) (C - C_s)/L \tag{3-21}$$

Using oxygen as a reference, diffusion coefficients of other compounds can be estimated using:

$$D_A = 0.178 \text{ cm}^2/\text{s } (32/M_A)^{0.5} \tag{3-22}$$

where:

D_A = air diffusion coefficient for compound A (cm^2/s) in air at 0°C

M_A = molecular weight of compound A (μg/mole)

Temperature corrections are made to the above value according to:

$$D_{A2} = D_{A1}(T_2/T_1)^{0.5} \tag{3-23}$$

where:

D_{A1}, D_{A2} = diffusion coefficients at temperatures T_1 and T_2, respectively.

A compound's saturation concentration, C_s, may be calculated using the following relationship:

$$Cs = pM/RT \tag{3-24}$$

where:

p = compound vapor pressure (mm Hg),

M = compound molecular weight (μg/mole),

R = universal gas constant (L mm Hg/°K mole),

T = absolute temperature (°K).

Laboratory evaluation of hexachlorobenzene (HCB) vaporization from simulated landfill areas has been reported by Farmer et al. (1980). They found that increases in soil moisture content produced logarithmic decreases in HCB vaporation (an increase in θ from 17.24 to 19.58 percent resulted in HCB flux reductions of 20 percent). Water phase mobility of HCB was not of concern due to HCB's water solubility of only 6.2 μg/l.

Bulk density effects were also shown by Farmer, et al. (1980) to have a large impact on HCB vapor movement. An increase in bulk density from 0.96 to 1.15 g/cm^3 resulted in a 65 percent reduction in HCB flux from the laboratory landfill units.

Because of the insoluble nature of HCB, its volatilization flux through a water layer was shown to be reduced by a factor of 870 as compared to uncovered samples (Farmer, et al., 1980). Farmer, et al. indicated an expected similar reduction in HCB vapor flux to take place upon saturation of contaminated soil systems. Similar results would be expected for compounds with properties similar to those of HCB. Leaching would be minimized due to the low water solubility and the slight compound partitioning into the soil water phase.

Wastes Amenable to Treatment

This treatment is useful for most volatile organic (e.g., benzene, gasoline, phenol) and inorganic (e.g., H_2S, NH_3, Ra, methyl mercury) compounds. The nature of the waste is also very important. As discussed earlier, these control technologies are especially applicable to those compounds with a high vapor phase mobility potential and a low water phase partition potential.

Status of Technology

Reduction of soil vapor pore volume is at the laboratory stage for uncontrolled hazardous waste sites. Decreases in volatilization of compounds due to water addition and increase in bulk density have been demonstrated from simulated landfill areas.

Ease of Application

Irrigation water is applied to reduce vapor-filled pore space, but not in such amounts that would cause leaching of hazardous constituents. Frequent smaller applications are more desirable. This is comparatively easy. Soil may be compacted to increase bulk density so vapor-filled pore space is reduced.

Potential Achievable Level of Treatment

The achievable level of treatment is potentially low to medium, depending on the waste characteristics. This technology is most effective for constituents with a high vapor phase potential and a low water phase partition potential. The achievable level of treatment also depends on site and soil conditions that are necessary for the addition of soil moisture and for compaction to increase bulk density of the soil.

Reliability of Method

Since soil moisture evaporates, retreatment with irrigation water from time to time is essential for effective reduction in volatilization.

Secondary Impacts

There is a potential for increased liquid phase mobility when water is used if the constituent has a high K_w and low K_{oc}, i.e., it readily partitions into the water phase and is highly water-soluble. Compaction reduces aeration in the soil and could adversely affect degradative reactions, e.g., chemical oxidation or aerobic biodegradation. Similarly, reducing pore volume with increasing soil moisture content would reduce oxygen content.

Equipment and Exogenous Reagents

Irrigation and compaction equipment are required. Irrigation water is the reagent used in this technology.

Information Requirements

- characterization and concentration of waste, primarily organics at site;
- volatility of organic constituents (vapor pressure, Henry's Law Constant, air/water partition coefficient, solubility);
- sorption of organics in soil (K_{oc});
- depth, profile, and areal distribution of contamination;
- bulk density, particle density, tortuosity of soil pores, temperature, and organic matter content of soil;
- soil moisture content;
- precipitation at site;
- trafficability of soil and site.

Contaminant K_W and K_D or K_{OC} estimates will provide information necessary to indicate the relative importance of vapor and solution phase transport and will dictate appropriate control methods for the compounds.

Sources of Information

Farmer, et al., 1980; Millington and Quirk, 1961; Thibodeaux, and Hwang, (1982).

3.6.2 Soil Cooling

Description

Soil cooling may be used to decrease the temperature of soil systems to reduce the vapor pressure of volatile constituents and correspondingly reduce their volatilization rate. The technique may enhance treatment by other means through the retention of compounds for longer periods of time within the soil system.

One means of lowering soil temperature is by the use of cooling agents applied to the soil surface. Greer, and Gross, (1980) found solid carbon dioxide (dry ice) to be more effective than liquid carbon dioxide, liquid nitrogen, or ice in reducing ethyl ether vaporization from a liquid pool. They observed ethyl ether vapor concentration reductions from 8300 to 96 ppm with dry ice addition. Dry ice resulted in a liquid pool temperature of $-85°C$ for 80 minutes at an application rate of 250 kg/2.7 m^3. Liquid nitrogen produced $-120°C$ temperatures but required 1025 kg to reduce concentrations from 93,000 to 116 ppm, and was found more difficult to work with than the dry ice. Because of the effectiveness of solid carbon dioxide and its minimal risks for response personnel during its application, it was the cooling agent of choice (Greer and Gross, 1980).

Modifications to the soil surface may also result in cooler soil temperatures (Baver, et al., 1972). Vegetated soils are usually cooler in the summer but warmer in the winter. Soils with more moisture warm up more slowly in the spring. Irrigation can be used for cooling in the summer. Tillage of the soil creates a surface barrier reducing heat flow from the surface to the subsurface. Mulched soils are cooler in spring, winter, and fall, while summer temperatures are similar between mulched and bare plots.

Soil temperature control is discussed in greater detail in Section 4.

Wastes Amenable to Treatment

Volatile organics are the most suitable wastes for this technology.

Status of Technology

Experimental and limited field applications of cooling agents have been conducted only on liquid spills, and no reports are available on soil surface cooling for vapor mitigation. Soil cooling by soil surface modification is used in agricultural operations.

Ease of Application

Cooling agents are continually applied to soil surface with spray equipment for liquefied agents or with grinding and spreading equipment for solid agents. Mulch is applied by standard agricultural methods. Water is applied at frequent intervals through irrigation systems in small amounts. The use of cooling agents would be difficult due to the low temperature and the reactions that occur upon application of low-temperature materials. Soil modifications using mulches, tilling, and other techniques vary in difficulty depending on trafficability considerations.

Potential Achievable Level of Treatment

The effectiveness of soil cooling is related to the degree of temperature reduction possible. Cooling agents are more effective than soil modifications, but are not likely to be practical because of cost.

Reliability of Method

Soil modification techniques are more reliable than cooling agents. The latter would not be reliable for large areas or for long periods of time, and the extremely low temperatures obtainable would be unnecessary. Retreatment of soil surface modification techniques is necessary.

Secondary Impacts

Low temperatures may decrease or inhibit microbial activity. Extremely low temperatures may severely decrease microbial numbers and/or activity. The application of excessive irrigation water may result in leaching of hazardous constituents. Tilling will increase the potential for erosion at the site and actually increase the surface area available for volatilization.

Equipment and Exogenous Reagents

Mulch, applicators, grinders, tillers, and cooling agents are needed for the technology.

Information Requirements

- characterization and concentration of wastes, particularly organics at site;
- volatility of organic constituents (vapor pressure, Henry's Law Constant, air/water partition coefficient, solubility, and particularly their dependence on temperature);
- sorption of organics in soil (K_{oc});
- depth, profile, and areal distribution of contamination;
- soil moisture;
- effectiveness of cooling agents;
- trafficability of soil and site.

Sources of Information

Green and Soors, 1980; Baver, et al., 1972.

4. Modification of Soil Properties

4.1 INTRODUCTION

Implementation of in-place treatment techniques for contaminated soils often involves the modification of soil properties. Biological degradation, for example, can be enhanced by the addition of nutrients, and immobilization of heavy metals may require adjustment of soil pH. Soil properties discussed in this section include:

- oxygen content,
- moisture content,
- nutrient content,
- pH,
- temperature.

This Section emphasizes the mechanics of soil property modification, independent of the treatment technology. Table 4-1 lists the technologies of Section 3, indicating which soil properties may require modification as a part of treatment.

4.2 CONTROL OF OXYGEN CONTENT

Oxygen content in surface soils can be increased primarily through the use of tillage equipment which breaks, mixes and aerates the soil. Alternatively, oxygen content can be decreased by compaction or increased moisture content. Aeration of subsurface soils not accessible to tillage equipment can be accomplished using construction equipment, such as a backhoe, or using a well point injection system.

A variety of equipment is available to aerate surface soils. This equipment, grouped under the category of Tillers, is described in the appendix. Tilling equipment can also be used to mix wastes or reagents into the soil. Choice of equipment depends on the amount of soil disturbance or mixing desired, and on site characteristics such as the rockiness of the soil.

For some processes, such as anaerobic biological degradation, surface soil compaction may be desirable. By reducing pore sizes and restricting reaeration, anaerobic microsite frequency in the soil will increase. Compaction helps draw moisture to the soil surface. Thus, the problems of leaching that may occur if anaerobiosis were achieved by water addition would be lessened. If the compaction itself were not adequate to achieve the required degree of anaerobiosis, water could be added. Less water, however, should be required in a compacted soil than in an uncompacted soil; thereby minimizing the leaching potential. Volatilization may also be suppressed by surface soil compaction.

Aeration of soils deeper than about 2 feet can be accomplished by air injection through well points. In one case, air was injected into a series of 10 wells using diffusers attached to paint sprayer-type compressors. They delivered about 2.5 cfm to enhance microbial degradation. Various nutrients were added simultaneously. The diffusers were positioned 5 feet from the bottom of the well and below the water table (Raymond, et al., 1976). Aeration through well points has been primarily used for saturated soils and has been shown to be effective. Applicability of the technique for unsaturated soils is not certain.

TABLE 4-1. SOIL MODIFICATION REQUIREMENTS FOR TREATMENT TECHNOLOGIES

Technology	Oxygen Content	Moisture Content	Nutrient Content	pH	Temperature
EXTRACTION	–	–	–	X	X
IMMOBILIZATION					
Sorption (heavy metals)					
Agri. products	–	–	–	X	–
Activated carbon	–	–	–	X	–
Tetren	–	–	–	X	–
Sorption (organics)					
Soil moisture	–	X	–	–	–
Agri. products	–	–	–	–	–
Activated carbon	–	–	–	–	–
Ion exchange					
Clay	–	–	–	X	–
Synthetic resins	–	–	–	X	–
Zeolites	–	–	–	X	–
Precipitation					
Sulfides	X	X	–	X	–
Carbonates, phosphates and hydroxides	X	X	–	X	–
DEGRADATION					
Oxidation					
Soil-catalyzed reactions	X	–	–	X	–
Oxidizing agents	X	–	–	X	–
Reduction					
Reducing agents	X	X	–	X	–
Chromium	X	–	–	X	–
Selenium	X	–	–	X	–
PCBs and Dioxins	–	X	–	–	X
Polymerization	–	–	–	–	–
Modification of soil properties (for biodegradation)					
Soil moisture	–	X	–	–	–
Soil oxygen – aerobic	X	–	–	–	–
Soil oxygen – anaerobic	X	X	–	–	–
Soil pH	–	–	–	X	–
Nutrients	–	–	X	–	–
Nonspecific org. amendments	–	–	–	–	X
Analog enrichment for cometabolism	–	–	–	–	X
Exogenous acclimated or mutant micro-organism	–	–	X	–	X

(continued)

TABLE 4-1 (Continued)

Technology	Oxygen Content	Moisture Content	Nutrient Content	pH	Temperature
Cell-free enzymes	–	–	–	–	X
Photolysis					
Proton donors	–	–	–	–	–
Enhance volatilization	–	X	–	–	–
ATTENUATION					
Metals	–	–	–	–	–
Organics	–	–	–	–	–
REDUCTION OF VOLATILES					
Soil Vapor Volume	–	X	–	–	–
Soil Cooling	–	–	–	–	X

Source: Arthur D. Little, Inc.

4.3 MOISTURE CONTROL

Control of moisture content of soils at an in-place treatment site may be essential for control and optimization of some degradative and sorptive processes as well as for suppression of volatilization of some hazardous constituents. Moisture control may take the form of supplemental water to the site (irrigation), removal of excess water (drainage, well points), a combination of techniques for greater moisture control, or other methods, such as soil additives.

4.3.1 Irrigation

Irrigation may be accomplished by subirrigation, surface irrigation, or overhead (sprinkler) irrigation (Fry and Grey, 1971).

With subirrigation, water is applied below the ground surface and moves upward by capillary action. If the water has high salinity, salts may accumulate in the surface soil, resulting in an adverse effect on soil microbiological activity. The site must be nearly level and smooth, with either a natural or perched water table, which can be maintained at a desired elevation. The groundwater is regulated by check dams and gates in open ditches, or jointed perforated pipe to maintain the water level in soil. The use of such systems is limited by the restrictive site criteria. There may be situations in which a subirrigation system may be combined with a drainage system to optimize soil moisture content. However, at a hazardous waste site, raising the water table might result in undesirable groundwater contamination.

Trickle irrigation is a system of supplying filtered water directly on or below the soil surface through an extensive pipe network with low flow-rate outlets only to areas which require irrigation. It does not give uniform coverage to an area, but with proper management, does reduce percolation and evaporation losses. For most in-place treatment sites, this method would probably not be appropriate, but it may find application in an area where only "hot spots" of wastes are being treated.

Surface irrigation includes flood, furrow, or corrugation irrigation. Since the prevention of off-site migration of hazardous constituents to ground or surface waters is a primary restraint on in-place treatment technology, the use of surface irrigation should be viewed with caution. Contaminated water may also present a hazard to on-site personnel.

In flood irrigation, water covers the surface of a soil in a continuous sheet. Theoretically, water should stay at every point just long enough to apply the desired amount, but this is difficult or impossible to achieve under field conditions.

In furrow irrigation, water is applied in narrow channels or furrows. As the water runs down the furrow, part of it infiltrates the soil. Considerable lateral movement of the water is required to irrigate the soil between furrows. Salts also tend to accumulate in the area between furrows. Furrow irrigation frequently requires extensive land preparation, which usually would not be possible or desirable at a hazardous waste site due to contamination and safety considerations.

In corrugation irrigation, as with furrow irrigation, water is applied in small furrows from a head ditch. However, in this case, the furrows are used only to guide the water, and overflooding of the furrows can occur.

In general, control and uniform application of water is difficult with surface irrigation. Also, soils high in clay content tend to seal when water floods the surface, limiting water infiltration.

The basic sprinkler irrigation system consists of a pump to move water from the source to the site, a pipe or pipes leading from the pump to the sprinkler heads, and the spray nozzles. Sprinkler irrigation has many advantages. Erosion and runoff of irrigation water can be controlled or eliminated, application rates can be adjusted for soils of different textures, even within the same area, and water can be distributed more uniformly. Irrigation is also possible on steep, sloping land and irregular terrain. Usually less water is required than with surface flooding methods, and the amount of water applied can be controlled to meet the needs of the in-place treatment technique.

There are several types of sprinkler irrigation systems:

1) Permanent installations with buried main and lateral lines;

2) Semi-permanent systems with fixed main lines and portable laterals;

3) Fully portable systems with portable main lines and laterals, as well as a portable pumping plant.

The first two types (especially the first) would likely not be appropriate nor cost-effective for a hazardous waste site because of the required land disturbance for installation and the limited time period for execution of the treatment.

The fully portable systems may have hand-moved or mechanically moved laterals. To eliminate movement by hand, the system may have enough laterals to cover the whole area (a solid set system). Portable systems can be installed in such areas as forests in patterns such as to avoid interference with trees. Mechanically moved laterals may be divided into three categories: side-roll wheel move; center pivot systems; and traveling sprinklers. The amount of labor is considerably reduced compared to portable systems, but the cost of the equipment is higher. However, the health and safety of workers must be considered as well as cost in the choice of an appropriate system.

The side-roll wheel move is a lateral suspended on a series of wheels. The unit is stationary during operation and is moved while shut off by an engine mounted at the center of the line or an outside power source at one end of the line. A variation of this system is a continuous travel wheel with a flexible hose, which remains in operation as the wheel moves across the field.

The center pivot system is a pipeline suspended above ground with various sized sprinklers spaced along its length. The system is self-propelled and continuously rotates around a pivot point.

The traveling sprinkler consists of a single gun sprinkler mounted on a portable, wheeled unit which is self-propelled up and down the length of the field.

The choice of an appropriate irrigation system depends on site conditions, costs, and health and safety considerations for both on-site personnel and off-site populations. The system should be designed by a qualified specialist such as an agricultural engineer. Preliminary guidelines for designing an irrigation system can be found in the *Sprinkler Irrigation Handbook* (Fry and Grey, 1971) and *Planning for an Irrigation System* (Turner and Anderson, 1980). The latter publication, in addition to technical aspects of irrigation, also discusses sources of water, including legal rights, and methods of determining irrigation costs.

4.3.2 Drainage

A properly designed drainage system removes excess water and/or lowers the groundwater level to prevent waterlogging. Surface drainage is accomplished by open ditches and lateral drains, while subsurface drainage is accomplished by a system of open ditches and buried tube drains into which water seeps by gravity. The collected water is conveyed to a suitable disposal point. Subsurface drainage may also be accomplished by pumping from wells to lower the water table. Caution is required at a hazardous waste site to ensure that drainage water disposed off-site is not contaminated with hazardous substances. Provisions must be made to collect, store, treat, and/or recycle water that is not acceptable for off-site release. The drainage system should be managed to prevent or minimize contamination problems.

The design of a drainage system is affected by the topography, soil properties, and water source factors of a site. The two types of drainage systems are (Donnan and Schwab, 1974):

- surface drains — used where subsurface drainage is impractical (e.g., impermeable soils, excavation difficult), to remove surface water or lower water table;

- subsurface drains — used to lower the water table. Construction materials include clay or concrete tile, corrugated metal pipe, and plastic tubing. Selection depends on strength requirements, chemical compatibility, and cost considerations.

For the design and construction of a drainage system, a drainage engineer should be consulted. The American Society of Agronomy monograph, *Drainage for Agriculture* (Schilfgaarde, 1974) contains a complete discussion of drainage.

4.3.3 Well Points

Well points, like subsurface drains, can be used to lower the water table in shallow aquifers. They typically consist of a series of riser pipes screened at the bottom and connected to a common header pipe and centrifugal pump. Well point systems are practical up to 10 meters (33 feet) and are most effective at 4.5 meters (15 feet). Their effectiveness, however, depends on site-specific conditions, such as the horizontal and vertical hydraulic conductivity of the aquifer (Ehrenfeld and Bass, 1983).

4.3.4 Additives

Various additives are available to enhance moisture control. For example, the water-retaining capacity of the soil can be enhanced by adding water-storing substances. Three such synthetic substances were recently evaluated by Nimah, et al. (1983) for use in arid area soils. They found that available soil water content was increased by two of the products. Water-repelling agents are available which diminish water absorption by soils. On the other hand, water-repelling soils can be treated with surface-active wetting agents to improve water infiltration and percolation. Other soil characteristics which have been modified by surface active agents include acceleration of soil drainage, modification of soil structure, dispersion of clays, and soil made more compactable. Evaporation retardants are also available to retain moisture in a soil. Secondary effects of some of these amendments on soil biological activities, other soil physical properties, soil chemical properties and environmental effects, e.g., leachability and degradability, are discussed by Brandt, G.H., (1972).

4.4 NUTRIENT ADDITIONS TO THE SOIL

Degradation of organic compounds at a hazardous waste site requires an active population of micro-organisms. Among other environmetnal factors (e g., temperature, moisture, pH, etc.), adequate nutrition is vital to maintain the microbial population at an optimum level. The hazardous wastes being degraded may contribute some necessary nutrients, but may not supply all that are required or that may be beneficial (e.g., silicon and sodium). If the soil does not contain an adequate supply of nutrients, the soil must be supplied with the appropriate elements in the form of fertilizers. A fertilizer is any substance added to the soil to supply those elements required in plant nutrition (Tisdale and Nelson, 1975).

The number of substances suitable as fertilizers is very large, and their compositions and origins differ considerably. Classification systems incorporating many aspects of fertilizer origin, use, and characteristics are presented in Finck (1982). Because of the variety of possible classifications, the choice of an appropriate fertilizer can be complicated, and an agronomist should be consulted to develop a fertilization plan at a hazardous waste site. A plan may include types and amounts of nutrients, timing and frequency of application, and method of application. The nutrient status of the soil and the nutrient content of the wastes must be determined to formulate an appropriate fertilization plan. Basic textbooks on fertilization include *Soil Fertility and Fertilizers* (Tisdale and Nelson, 1975), *Fertilizers and Fertilization* (Finck, 1982), and *Fertilizers and Soil Amendments* (Follett, et al., 1981).

The development of a fertilization program not only includes the proper selection of fertilizer form and determination of correct fertilizer quantities, but also the selection of an application method. Fertilizers must be transported, stored, and applied so that no chemical or physical changes occur to decrease dispersibility and effectiveness. Improper handing during transportation and storage may result in the creation of safety hazards due to moisture absorption, such as increased flammability, explosiveness, and corrosiveness, or the formation of noxious gases. Improper mixing of fertilizer types before or after application may result in nitrogen losses, immobilization of water-soluble phosphate, or deterioration of distribution properties due to moisture absorption (Finck, 1982).

In agricultural application, fertilizers are either applied evenly over an area or concentrated at given points, such as banded along roots. However, at a hazardous waste site, fertilizer will likely be applied evenly over the whole contaminated area and incorporated by tilling, if necessary. Nutrients can also be injected through well points below the plow layer.

With broadcast fertilization, the fertilizer can be left on the surface or incorporated with a harrow (2 to 3 cm deep), a cultivator (4 to 6 cm deep), or with a plow (a layer at bottom of furrow, e.g., 15 cm deep). The depth of incorporation depends on the solubility of the fertilizer and the desired point of contact in the soil. In general, nitrate fertilizers move freely, while ammonia nitrogen is adsorbed by soil colloids and moves little until converted to nitrate. Potassium is also adsorbed and moves little except in sandy soils. Phosphorus does not move in most soils. Therefore, potassium and phosphorus need to be applied or incorporated to the desired point of use.

4.5 CONTROL OF SOIL pH

Control of soil pH at an in-place hazardous waste treatment site is a critical factor in several treatment techniques (e.g., metal immobilization, optimum microbial activity). The goal of soil pH adjustment in agricultural application usually is to increase the pH to near neutral values, since natural soils tend to be acidic.

The areas of the country in which the need for increasing soil pH is greatest are the humid regions of the East, South, Middle West, and Far West States. In areas where rainfall is low and leaching is minimal, such as parts of the Great Plain States and the arid, irrigated saline-alkali soils of the Southwest, Intermountain, and Far West States, pH adjustment is usually not necessary. Some soils, especially those high in carbonates, do require the pH to be lowered. However, a hazardous waste-contaminated soil may have substances high in pH, thus necessitating soil acidification.

The most common method of controlling pH is liming. Liming is the addition to the soil of any calcium or calcium-and magnesium-containing compound that is capable of reducing acidity (i.e., raising pH). Lime correctly refers only to calcium oxide, but is commonly used to refer to calcium hydroxide, calcium carbonate, calcium-magnesium carbonate, and calcium silicate slags (Tisdale and Nelson, 1975).

The benefits of liming to biological activity are several. At higher pH values, aluminum and manganese are less soluble. Both of these compounds are toxic to most plants. In addition, phosphates and most micro-elements necessary for plant growth (except molybdenum) are more available at higher pH. Microbial activity is greater at or near neutral pH, which enhances mineralization and degradation processes and nitrogen transformations (e.g., nitrogen fixation and nitrification).

A summary of commonly used liming materials is presented in Table 4-2. The choice of a liming material depends upon several factors. Calcitic and dolomitic limestones are the most commonly used materials. To be effective quickly, however, these materials must be ground, because the velocity of reaction is dependent on the surface in contact with the soil. The finer they are ground, the more rapidly they react with the soil. However, a more finely ground limestone product usually contains a mixture of fine and coarse particles in order both to effect a pH change rapidly and still be relatively long-lasting as well as reasonably priced. Many states require that 75 to 100 percent of the limestone pass an 8- to 10-mesh sieve and that 20 to 80 percent pass anywhere from an 8- to 100-mesh sieve (Tisdale and Nelson, 1975). Calcium oxide and calcium hydroxide are manufactured as powders and react quickly.

Other factors to consider in the selection of a limestone are neutralizing value, magnesium content, and cost per ton applied to the land.

Lime requirement for soil pH adjustment is dependent on several soil factors, including soil texture, type of clay, organic matter content, and exchangeable aluminum (Follett, et al., 1981). The buffering capacity of soil reflects the ability of soil components to hold a large number of ions in adsorbed or reserve form. Thus, adsorption or inactivation of H+ ions, or the release of adsorbed ions to neutralize OH- ions provides protection against abrupt changes in pH when acidic or basic constituents are added to soil. Differences among soils in their buffering capcity reflect differences in the soil cation exchange capacities and will directly affect the amount of lime required to adjust soil pH. The amount of lime required is also a function of the depth of incorporation at the site, i.e., volume of soil to be treated. The amount of lime required to effect a pH change in a particular site/soil/waste system is determined by a state experimental or commercial soil testing laboratory in short-term treatability studies or soil-buffer tests (McLean, 1982).

Lime requirements may also be affected by acid precipitation and acid-forming fertilizers. A field study in Connecticut showed that each year the acidity generated by acid precipitation would require 36 kg/ha (32 lb/A) of pure calcium carbonate for neutralization (Frink and Voight, 1976).

Lime is usually applied from a V-shaped truck bed with a spinner-type propeller in the back (Follett, et al., 1981). Uniform spreading is difficult with this equipment, and wind losses can be significant. A more accurate but slower and more costly method is a lime spreader (a covered hopper or conveyor) pulled by a tractor. Limestone does not migrate easily in the soil since it is only slightly soluble, and must be placed where needed. Plowing and/or discing surface-applied lime into the soil may therefore be required.

The application of fluid lime is becoming more popular, especially when mixed with fluid nitrogen fertilizer. The combination results in less trips across the soil, and the lime is available to counteract acidity produced by the nitrogen. Also, limestone has been applied successfully to a pharmaceutical wastewater land treatment facility through a spray irrigation system.

TABLE 4-2. LIMING MATERIALS

Liming Material	Description	Calcium Carbonate Equivalent[a]	Comments
Limestone, calcitic	$CaCO_3$, 100% purity	100	Neutralization value usually between 90-98% because of impurities; pulverized to desired fineness
Limestone, dolomitic	65% $CaCO_3$ + 20% $MGCO_3$, 87% purity[b]	89	Pure dolomite (50% $MgCO_3$ ano 50% $CaCO_3$) has neutralizing value of 109%; pulverized to desired fineness
Limestone, unslaked lime, burned lime, quick lime	CaO, 85% purity	151	Manufactured by roasting calcitic limestone; purity depends on purity of raw materials; white powder, difficult to handle – caustic; quick acting; must be mixed with soil or will harden and cake
Hydrated lime, slaked lime, builder's lime	$Ca(OH)_2$, 85% purity	85	Prepared by hydrating CaO; white powder, caustic, difficult to handle; quickly acting
Marl	$CaCO_3$, 50% purity	50	Soft, unconsolidated deposits of $CaCO_3$, mixed with earth, and usually quite moist
Blast furnace slag	$CaSi_2O_3$	75-90	By-product in manufacture of pig iron; usually contains magnesium
Waste lime products	Extremely variable in composition	?	––

a. Calcium carbonate equivalent (CCE): neutralizing value compared to pure calcium carbonate, which has a neutralizing value defined as 100.

State laws specify a calcium carbonate equivalent averaging 85%.

Source: Follett, R.H., et al., 1981, Tisdale S.L., and Nelson, W.L., 1975.

4.6 MODIFICATION OF SOIL TEMPERATURE

Soil temperature is one of the more important factors that controls microbiological activity and the rate of organic matter decomposition. Soil temperature is also important in influencing the rate of volatilization of compounds from soil. Soil temperature can be modified by regulating the incoming and outgoing radiation, or by changing the thermal properties of the soil (Baver, et al., 1972).

Vegetation plays a significant role in soil temperature because of the insulating properties of plant cover. Bare soil unprotected from the direct rays from the sun becomes very warm during the hottest part of the day, but also loses its heat rapidly during colder seasons.

However, a well-vegetated soil during the summer does not become as warm as a bare soil, and in the winter, the vegetation acts as an insulator to reduce heat lost from the soil. Frost penetration is more rapid and deeper under bare soils than under a vegetative cover.

Mulches can affect soil temperature in several ways. In general mulches reduce diurnal and seasonal fluctuations in soil temperature. In the middle of the summer, there is little difference between mulched and bare plots, but mulched soil is cooler in spring, winter and fall, and warms up more slowly in the spring. Mulches with low thermal conductivities decrease heat flow both into and out of the soil; thus, soil will be cooler during the day and warmer during the night. White paper, plastic, or other types of white mulch increases the reflection of incoming radiation, thus reducing excessive heating during the day. A transparent plastic mulch transmits solar energy to the soil and produces a greenhouse effect. A black paper or plastic mulch adsorbs radiant energy during the day and reduces heat loss at night. Humic substances increase soil temperature by their dark color, which increases the soil's heat adsorption.

The type of mulch required determines the application method. Mulches, in addition to modification of soil temperature, are also used to protect soil surfaces from erosion and to reduce water and sediment runoff, prevent surface compaction or crusting, conserve moisture, and help establish plant cover (Soil Conservation Service, 1979). A summary of mulch materials is presented in Table 4-3. Commercial machines for spraying mulches are available. Hydromulching is a process in which seed, fertilizer, and mulch are applied as a slurry. To apply plastic mulches, equipment which is towed behind a tractor mechanically applies plastic strips which are sealed at the edges with soil. For treatment of large areas, special machines that glue polyethylene strips together are available (Mulder, 1979).

Irrigation increases the heat capacity of the soil, raises the humidity of the air, lowers air temperature over the soil, and increases thermal conductivity, resulting in a reduction of daily soil temperature variations (Baver, et al., 1972). Sprinkle irrigation, for example, has been used for temperature control, specifically frost protection in winter and cooling in summer and for reduction of soil erosion by wind (Schwab, et al., 1981). Drainage decreases the heat capacity, thus raising the soil temperature. Elimination of excess water in spring causes a more rapid temperature increase. The addition of humic substances improves soil structure, thus improving soil drainability, resulting indirectly in increased soil temperature.

Several physical characteristics of the soil surface can be modified to alter soil temperature (Baver, et al., 1972). Compaction of the soil surface increases the density and thus the thermal conductivity. Tillage, on the other hand, creates a surface mulch which reduces heat flow from the surface to the subsurface. The diurnal temperature variation in a cultivated soil is much greater than in an untilled soil. A loosened soil is colder at night and more susceptible to frost.

TABLE 4-3. MULCH MATERIALS

	Organic Materials	Quality	Notes
a.	Small grain straw or tame hay	Undamaged, air dry threshed straw, free of undesirable weed seed	Spread uniformly – at least 1/4 of ground should be visible to avoid smothering seedling. Anchor either during application or immediately after placement to avoid loss by wind or water. Straw anchored in place is excellent on permanent seedings.
b.	Corn stalks chopped or shredded	Air dried, shredded into 8" to 12" lengths	Relative slow to decompose. Resistant to wind blowing.
c.	Wood Excelsior	Burred wood fibers approximately 4" long	A commercial product packaged in 80-90 lbs. bales. Apply with power equipment. Tie down usually.
d.	Wood Cellulose Fiber	Air dry, non-toxic with no growth inhibiting factors	Must be applied with hydraulic seeder.
e.	Compost or manure	Shredded, free of clumps or excessive coarse material	Excellent around shrubs. May create problems with weeds.
f.	Wood chips and bark	Air dried, free from objectionable coarse material	Most effective as mulch around ornamentals, etc. Resistant to wind blowing. May require anchoring with netting to prevent washing or floating off.
g.	Sawdust	Free from objectionable coarse material	More commonly used as a mulch around ornamentals, etc. Requires anchoring on slopes. Tend to crust and shed water.
h.	Pine straw	Air dry. Free of coarse objectionable material	Excellent around plantings. Resistant to wind blowing.

(continued)

TABLE 4-3 (Continued)

	Other Mulch Materials	Quality	Notes
a.	Asphalt Emulsion	Slow setting SS-1	Use as a film on soil surface for temporary protection without seeding. Requires special equipment to apply.
b.	Gravel or Crushed Stone		Apply as a mulch around woody plants. May be used on seeded areas subject to foot traffic. (Approximate weight – 1 ton per cu. yd.)
c.	Wood Excelsior Mats	Blanket of excelsior fibers with a net backing on one side	Roll 36" x 30 yards covers 161/2 sq. yds. Use without additional mulch. Tie down as specified by manufacturer.
d.	Jute, Mesh or Net	Woven jute yarn with 3/4" openings	Roll 48" x 75 yds. weighs 90 lbs. and covers 100 sq. yds.

Source: Soil Conservation Service, 1979.

Appendix: Cost Information

TABLE A-1. EQUIPMENT APPLICABLE TO TREATMENT OF HAZARDOUS-WASTE-CONTAMINATED SOILS

Function	Equipment	Principal Use	Examples of Capacity or Size Range	Approximate Cost[a]	Cost Units
Power Implements	Tractors, Crawler	May be needed when maximum traction and stability is needed. Especially useful on steep slopes.	Small, 28 maximum drawbar horsepower	$ 21,800	each
			Large, 300 maximum drawbar horsepower	210,000	each
	Tractors, wheel type, two-wheel drive	Adequate where traction or power requirements are less demanding.	Small, 12 maximum drawbar horsepower	6,000	each
			Large, 164 maximum drawbar horsepower	66,500	each
	Tractors, wheel type four-wheel drive	Better traction and higher horsepower available than with two wheel drive tractors.	Small, 12 maximum drawbar horsepower	6,800	each
			Very large, 552 maximum drawbar horsepower	275,000	each
Tillers: Loosening, aerating, and/or mixing the soil	Plows, Chisel	Loosens and aerates soil to 14 inch depth with minimum vertical mixing.	10 foot width	1,500	each
			41 foot width	18,700	each
	Plows, moldboard	Turn and aerate soil 8 to 12 inches deep. Poor mixing, but useful in rocky soils.	3-bottom, 2-way	4,800	each
	Rotary tillers[b]	Effective verticle mixing and aeration of surface 4 to 10 inches of soil. Combines effects of plowing and cultivation.	40 inch width, 15 inch rotor diameter	1,330	each
			300 inch width, 21 inch rotor diameter	16,600	each
	Subsoilers, chisel	Break up deep soil with little verticle mixing to 30 inches or more.	13 shank, 270 inch width	7,640	each
	Subsoiler, double tilling	Turns surface soil with moldboard plow then loosens and mixes subsoil to 20 inches with a rotary tiller. Fertilizer or other agent can be mixed into the subsoil.			

TABLE A-1. (Continued)

Function	Equipment	Principal Use	Examples of Capacity or Size Range	Approximate Cost[a]	Cost Units
Tillers (Continued)	Harrows, Disc	Loosen and aerate surface soil.	Small hitch mounted, 450 lb	450	each
		Provide more verticle mixing than most other harrows.	Large, pull type, 16,700 lb	32,100	each
	Harrows, power	Several varieties of power harrows available which use rotating (verticle or horizontal) oscillating, or reciprocating motion to loosen and aerate surface soil.	Small, 72 inch width, 10 inch working depth	5,000	each
			Large, 240 inch width, 8 inch working depth	19,500	each
	Harrows, spike	Break up clods formed when plowing sticky soil. Loosen and aerate shallow surface soil.	7 foot wide section	160-230	each
	Harrows, spring tooth	Loosen and aerate shallow surface soil, have vibrating action.	8 foot width	940	each
			20 foot width	2,900	each
Compactors: Compacting soil	Rollers	Compact soil surface, improve soil moisture retention, restrict gas diffusion.	4 foot width	586	each
			16 foot width	1,620	each
Applicators: Application of exogenous agent(s)	Sprayers, hydraulic	Treatment with relatively small amounts of fluid agents, e.g., 20 to 200 gallons per acre.	Small, hitch mounted, 7.5-foot treatment width	511	each
			Large, hitch mounted, 60-foot treatment width	9,330	each
			Small, self-propelled, 27-foot treatment width	6,730	each
			Large, self-propelled, 70-foot treatment width	144,000	each
	Spreaders, chemical fertilizer	Apply granular chemical fertilizers or other agents in similar form. Some fertilizer spreaders can be modified to apply agricultural limestone.	Small, hitch mounted 1 to 2 cu ft capacity	214	each
	Spreaders, manure or dried sewage sludge	Apply barnyard manure or dried sewage sludge.	Tractor drawn, 122 cu ft capacity	2,600	each
			Tractor drawn, 391 cu ft capacity	9,490	each
			Truck mounted	Bid	
	Spreaders, agricultural limestone	Apply ground lime or dolomite to soil for pH control.	Small, hitch mounted, 12 cu ft capacity	624	each

TABLE A-1. (Continued)

Function	Equipment	Principal Use	Examples of Capacity or Size Range	Approximate Cost[a]	Cost Units
			Large, truck mounted, 250 cu ft capacity	15,700	each
	Injectors, liquid	Inject liquid materials below the soil surface. Conventional equipment can inject 1100 gal per minute of liquid to 16 inches below the surface.	Self-propelled	Bid	
			Tractor-drawn	Bid	
Grinders: Grind plant Materials	Grinders, tub	Grind hay or similar plant material to be used as organic matter amendment or mulch.	8 ton per hour	12,500	each
			15 to 25 ton per hour	41,000	each
Covers, Mulches: Soil coverings and applicators	Plastic sheeting	Cover soil to manage soil temperature or to suppress volatilization.	2 mil thick 10.5 ft wide, 1400 ft. long	90-95	roll (140 lb)
	Plastic laying machine	Applies plastic sheeting over the soil by unrolling it, burying, and/or gluing the edges. Mounted on tractor hitch.	10.5 ft wide (sizes 9 to 20 ft wide available)	3,300	each
	Hydromulching	Ground plant materials (frequently wood fiber) are sprayed onto the soil in aqueous slurry. Chemical binders may be added to stabilize the material against the wind.		1,500 to 1,700 (Includes mulch, equipment and labor)	pr acre
Irrigation	Sprinkler, hand move	Apply water to manage soil moisture and/or soil temperature. Fertilizers or other treatment agent may be applied with the irrigated water.	Portable, 1 to 40 acres	200-500	pr acre
			Solid set, 1 or more acres	800-2000	pr acre
	Sprinkler, self-move		Motor driven, side roll, 20 to 80 acres	300-500	pr acre
	Sprinker, self-propelled		Center pivot driven by water, 40-240 acres	325-450	pr acre
			Water driven, side roll, 80-160 acres	275-450	pr acre
Drainage	Perforated pipe drains	Lower shallow water table to improve soil aeration.	Costs depend on depths of drains, soil hydraulic conductivity, and site parameters	350-500	pr acre

a. Manufacturers suggested retail prices are used for most agricultural tractors and implements.

b. Larger rotary tillers can be modified to carry fertilizer or other chemical application equipment so that treatment can be done in a single pass over the soil.

Source: Utah Water Research Laboratory.

TABLE A-2. EXOGENOUS AGENTS, EXCAVATION' AND HAULING COSTS

Identifying Term	Agent	Approximate Cost	Cost[a] Unit
Acidifying Agents	Aluminum sulfate	$235	Ton
	Ferrous sulfate	130	Ton
	Ferric sulfate	108	Ton
	Liquid ammonium sulfide	235	Ton
	Sulfur (crude)	109-126	Ton
	Sulfuric acid	20-96	Ton
Activated Carbon	Activated carbon, powder	0.55	lb
	Activated carbon, granular	1.05	lb
Carbonates/Phosphates	Calcium corbonate (limestone)	6.50-35	Ton
	Triple superphosphate	127-165	Ton
Cooling Agents	Liquid and gas carbon dioxide	0.12	lb
	Liquid nitrogen	0.28	lb
	Solid carbon dioxide	0.16	lb
	Water ice	0.03	lb
Excavation and Hauling	Excavation	0.77-4.58[b] (Average ~ 2.10)	yd^3
	Hauling less than 5 miles	0.50	yd^3-mile
	Hauling more than 5 miles	0.25-0.30	yd^3-mile
Fertilizer	Ammonia, anhydrous	135-180	Ton
	Ammonia, aqueous	210	Ton
	Ammonium nitrate	91-115	Ton
	Ammonium sulfate	74-79	Ton
	Diammonium phosphate	165	Ton
	Phosphoric acid, 52-54%	165	Ton
	Phosphate, rock	23	Ton
	Potassium chloride, 60-62.4%	57	Ton
	Potassium nitrate	277-284	Ton
	Sodium nitrate	130	Ton
	Superphosphate, triple	160-165	Ton
	Urea	200-215	Ton
	Blended fertilizers (N-P-K)		
	16-16-8%	220	Ton
	16-16-16	230	Ton
	18-46-0	260	Ton
	29-14-0	230	Ton
Flushing Agents	Caustic soda, liquid, 50%	150-230	Ton
	Citric acid	0.81	lb
	Hydrochloric acid, 20 Be[c]	55-115	Ton
	Nitric acid, 36°, 38°, 40° Be	195	Ton
	Sodium lauryl sulfate, 30%	0.29-0.32	lb
	Sulfuric acid	20-96	Ton

(Continued)

TABLE A-2. (Continued)

Identifying Term	Agent	Approximate Cost	Cost[a] Unit
Liming Material	Agricultural limestone or dolomite	6.50-34	Ton
	Lime (85% CaO)	31.25-32.50	Ton
	Hydrated lime (85% CaOH)	32.50-34.50	Ton
Organic Materials	Animal Tankage	55	Ton
	Alfalfa hay	80-120	Ton
	Bone meal, steamed	300	Ton
	Castor pomace	149	Ton
	Cottonseed meal, 41%	215	Ton
	Horse feed (grains and molasses)	180	Ton
	Manure, dairy cattle	0-1	Cubic Foot
	Molasses	125	Ton
	Peanut meal, 50%	235	Ton
	Sewage sludge, activated	80	Ton
Oxidizing Agents	Hydrogen peroxide, 35-70%	0.22-0.43	lb
	Ozone generator, 22 lb/day	40,000-45,000	Each
	Potassium permanganate	2	Kilogram
Precipitating Agent	Ferrous sulfate	130	Ton
Proton Donors	Methanol	0.48	Gallon
	Mineral oil	2.69-2.72	Gallon
	Vegetable oils	0.22-0.47	lb
	Xylene	1.20-1.60	Gallon
Reducing Agents	Acetic acid	0.23	lb
	Iron powder	1.00	lb
	Soda caustic (NaOH-Tech):		
	Liquid, 50%	150-230	Ton
	Flake, 76%	500-570	Ton
	Granular, 75%	520	Ton
	Sodium borohydride powder	18-19	lb
	Sodium borohydride stabilized solution, 12%	16	lb
Resins	Anion Exchange	191-197	Cubic Foot
	Cation exchange	211-217	Cubic Foot
Soil/Clay	Bentonite, industrial grade	94	Ton
	Kaolin, uncalcined	58	Ton
	Top soil	4-10	Cubic Yard

(Continued)

TABLE A-2. (Continued)

Identifying Term	Agent	Approximate Cost	Cost[a] Unit
Sulfides	Ammonium sulfide liquid	235	Ton
	Hydrogen sulfide liquid	0.11	lb
	Sodium sulfide	410-470	Ton
Tetren	Tetraethylenepentamine	1.70-1.78	lb
Zeolites	Clinoptilolite	45-50	Ton

a. Most costs are wholesale, bulk in train car, tank car, or truckload quantities FOB factories or ports.

b. Depends on equipment type and size. Add 60% for hard or rocky soil, deduct 15% in light soil or sand (Godfrey, R.S., ed. 1982. Building construction cost data 1983. Robert Snow Means Co., Inc., Kingston, MA. 421 p.).

c. Be = Baumé.

Source: Utah Water Research Laboratory.

Bibliography

Ahrens, J.F., and J.B. Kring. 1968. Activated carbon adsorbs pesticides. Front. Plant Sci. 20:13.

Alexander, M. 1977. Introduction to soil microbiology. Second Edition. John Wiley and Sons, New York, NY. 467 p.

Alexander, M. 1981. Biodegradation of chemicals of environmental concern. Science 211:132-138.

American Public Health Association. 1975. Standard Methods for Examination of Water and Wastewater. Fourteenth edition. Wash., D.C.

Anderson, A.H. 1968. The inactivation of simazine and liuron in soil by charcoal. Weed Res. 8:58-60.

Anderson, J.R. 1978. Pesticide effects on non-target soil micro-organisms. pp. 313-533. *In* pesticide microbiology. Hill, I.R. and Wright, S.J.L. (eds.) Academic Press, New York, NY.

Anonymous. 1981. It's hard to find jobs for industrious bugs. Chem. Week 128(2):40-41.

Anonymous. 1982. Bugs tame hazardous spills and dumpsites. Chem. Week 130(18):49.

Arthur D. Little, Inc. 1976. State-of-the-art survey of land reclamation technology. EC-CR-76-76. Department of the Army, Edgewood Arsenal, Aberdeen Proving Ground, MD. 105 p.

Artiole, J., and W.H. Fuller. 1979. Effect of crushed limestone barriers on chromium attenuation in soils. J. Environ. Qual. 8:503-510.

Atlas, R.M., and R. Bartha. 1981. Microbial ecology, fundamentals and applications. Addison-Wesley Publishing Company, Reading, MA. 560 p.

Baas Becking, L.G.M., I.R. Kaplan, and D. Moore. 1960. Limits of the natural environment in terms of pH and oxidation-reduction potentials. J. Geology 68:243-284.

Bailey, G.W., J.L. White, and T. Rothberg. 1968. Adsorption of organic herbicides by montmorillonite: Role of pH and chemical character of adsorbate. Soil Sci. Soc. Am. Proc. 32:222-234.

Baker, M.D., and C.I. Mayfield. 1980. Microbial and non-biological decomposition of chlorophenols and phenol in soil. Water Air Soil Poll. 13:411-424.

Bartha, R. 1980. Pesticide residues in humus. ASM News 46:356-360.

Baver, L.D., W.H. Gardner, and W.R. Gardner. 1972. Soil physics. Fourth edition. John Wiley and Sons, Inc., New York, NY: 498 p.

Bhattacharyya, D., A.B. Jumawan, G. Sun, C. Sund-Hagelberg, and K. Schwitzgebel. 1981. Precipitation of heavy metals with sodium sulfide: Bench scale and full-scale experimental results. *In* Water-1980. AIChE 209(77).

Biddappa, C.C., M. Chino, and K. Kumazawa. 1982. Migration of heavy metals in two Japanese soils. Plant and Soil 66:299-316.

Black, C.A. (ed.). 1965. Methods of soil analysis. Part 2. Chemical and microbiological properties, Agronomy No. 9, Amer. Soc. Agron. Madison, WI.

Bollag, J.-M. 1983. Cross-coupling of humus constituents and xenobiotic substances. pp. 127-141. *In* Aquatic and terrestrial humic materials. Christman, R.F., and Gjessing, E.T. (eds.). Ann Arbor Science Publishers, Ann Arbor, MI.

Bollag, J.-M., P. Blattman, and T. Laanio. 1978. Adsorption and transformation of four substituted anilines in soil. J. Agric. Food Chem. 26:1302-1306.

Bollag, J.-M., S.-Y. Liu, and R.D. Minard. 1980. Cross-coupling of phenolic humus constituents and 2,4-dichlorophenol. Soil Sci. Soc. Am. J. 44:52-56.

Bollag, J.-M., and S.-Y. Liu. 1983. Incorporation of pentachlorophenol into humic acid polymers by a fungal laccase. p. 273. *In* Abstracts of the Annual Meeting of the ASM. Amer. Soc. Mocribiol., Wash., DC.

Bollag, J.-M., R.D. Minard, and S.-Y. Liu. 1983. Cross-linkage between anilines and phenolic humus constituents. Environ. Sci. Tech. 17:72-80.

Bonazountas, M., and J. Wagner. 1981. SESOIL — A seasonal soil compartment model. Arthur D. Little., Inc., Cambridge, Mass. Prepared for U.S. Environmental Protection Agency Office of Toxic Substances, Wash., D.C.

Botre, C., A. Memoli, and F. Alhaique. 1978. TCDD solubilization and photodecomposition in aqueous solutions. Environmental Science and Technology 12(3):335-336.

Bowers, W. 1970. Modern concepts of farm machinery management. Stipes Publishing Co., Champaign, IL.

Brady, N.C. 1974. The nature and properties of soil. Eighth edition. Macmillan Publishing Co., Inc., New York, NY.

Brandt, C.H. 1972. Soil physical property modifiers. *In* organic chemicals in the soil environment. Vol. 2. Goring, C.A.I., and Hamaker, J.W. (eds.). Marcel Dekker, Inc., New York, NY. pp 691-729

Breck, D.W. 1974. Zeolite molecular sieves, structure, chemistry, and use. Wiley Interscience. New York, NY.

Brown, D.P. 1978. Proposed volatile spill suppression concepts. pp. 303-308. *In* National Conference on Control of Hazardous Material Spills.

Brown, L.S. 1972. A physical barrier system for control of hazardous material spills in waterways. Proceedings of the 1972 National Conference on Control of Hazardous Material Spills, Houston, TX.

Brunner, W., and D.D. Focht. 1983. Persistence of polychlorinated phenyls (PCB) in soils under aerobic and anaerobic conditions. p. 266. *In* Abstracts of the annual meeting of the ASM. American Society for Microbiology, Wash., DC.

Bunce, N.J. 1982. Photodechlorination of PCB's: Current status. Chemosphere 11(8):701-714.

Burkhard, N., and J.A. Guth. 1979. Photolysis of organophosphorus insecticides on soil surfaces. Pest. Sci. 10:313.

Camoni, I., A. DiMuccio, D. Pontecovo, F. Taggi, and L. Vergoni. 1982. Laboratory investigation for microbiological degradation of 2,3,7,8-TCDD in soil by addition of organic compost. *In* Chlorinated dioxins and related compounds impact on the environment. Hutzinger, O. (ed.) Pergamon Press, New York, NY.

Cerniglia, C.E., and D.T. Gibson. 1979. Oxidation of benzo(a)pyrene by the filamentous fungus *Cunninghamella elegans*. J. Biol. Chem. 254(23):12174.

Cerniglia, C.E., R.L. Herbert, R.H. Dodge, P.J. Szaniszlo, and D.T. Gibson. 1979. Some approaches to studies on the degradation of aromatic hydrocarbons by fungi. *In* Microbial degradation of pollutants in marine environments. Bourquin, A.L., and Pritshard, H. (eds.). EPA-600/9-79-012, 360.

Chakrabarty, A.M. 1980. Plasmids and dissimilation of synthetic environmental pollutants. pp. 21-30. *In* Plasmids and transposons, environmental effects and maintenance mechanisms. Stuttard, C., and Rozee, K.R. (eds.). Academic Press, New York, NY.

Chakrabarty, A.M. 1982. Genetic mechanisms in the dissimilation of chlorinated compounds. pp. 127-139. *In* Biodegradation and detoxification of environmental pollutants, Chakrabarty, A.M. (ed.). CRC Press, Inc., Boca Raton, FL.

Chowdhury, A., D. Vockel, P.N. Moza, W. Klein, and F. Korte. 1981. Balance of conversion and degradation of 2,6-dichlorobenzonitrile - 14C in humus soil. Chemosphere 10:1101-1108.

Christensen, D.C., and W.C. Weimer. 1979. Enhanced photodegradation of persistent halogenated organic materials. Presented at the 34th Annual Purdue Industrial Waste Conference, Purdue University, Lafayette, IN., pp 160-166.

Coffey, D.L., and G.F. Warren. 1969. Inactivation of herbicides by activated carbon and other adsorbents. Weed Sci. 17:16-19.

Council for Agricultural Science and Technology (CAST). 1975. Utilization of animal manures and sewage sludges in food fiber production. Report No. 41.

Crosby, D.G. 1971. Environmental photooxidation of pesticides. pp 260-290. *In* Degradation of synthetic organic molecules in the biosphere. National Academy of Sciences, Wash., DC.

Crosby, D.G., A.S. Wong, J.R. Plimmer, and E.A. Woolson. 1971. Photodecomposition of chlorinated dibenzo-p-dioxins. Science 73:748.

Cupitt, L.T. 1980. Fate of toxic and hazardous materials in the air environment. EPA-600/53-80-084. U.S. Environmental Protection Agency, Athens, GA.

Damanakis, M., D.S. Drennan, J.D. Fryer, and K. Holly. 1970. The adsorption and mobility of praquat and different soil and soil constituents. Weed Res. 10:264-277.

David, D.J. 1978. Analysis of solids, plants, fertilizers, and other agricultural materials. Progress in Analytical Atomic Spectroscopy 1(3).

Dawson, G.W., B.W. Mercer, and C.H. Thompson. May 1980. Strategy for the treatment of spills on land. National Conference on Control of Hazardous Materials Spills.

Deere and Company. 1957. The operation, care and repair of farm machinery. 28th edition. Deere and Company, Moline, IL.

Department of the Army, Corps of Engineers. 1982. Preliminary guidelines for selection and design of remedial systems for uncontrolled hazardous waste sites. EM1110-2-(Draft). Department of the Army, Corps of Engineers, Wash., D.C.

Dobbs, R.A., and J.M. Cohen. 1980. Carbon adsorption isotherms for toxic organics. EPA-600/8-80-023. Municipal Environmental Research Laboratory, U.S. Environmental Protection Agency, Cincinnati, OH.

Donnan, W.W., and G.O. Schwab. 1974. Current drainage methods in the U.S.A. pp. 93-114. *In* Proceedings of the First International Symposium on Acid Precipitation and the Forest Ecosystem. U.S. Forest Service General Technical Report NE-23, USDA — Forest Service, Upper Darby, PA.

Doyle, R.C., and J.D. Isbister. 1982. Treatment of TNT and RDX contaminated soils by composting. pp. 209-213. *In* Managing uncontrolled hazardous waste sites. Hazardous Materials Control Research Institute, Silver Spring, MD.

Dragun, J., and D.E. Baker. 1979. Electrochemistry. pp. 130-135. *In* The encyclopedia of soil science, Part I: Physics, chemistry, biology, fertility, and technology. Dowden Hutchinson, and Ross, Inc., Stroudsburg, PA.

Dragun, J., and C.S. Helling. 1982. Soil- and clay-catalyzed reactions: I. Physicochemical and structural relationships of organic chemicals undergoing free-radical oxidation. *In* Land disposal of hazardous waste. Proc. Eighth Annual Research Symposium. EPA-600/9-82-002.

Eckenfelder, Jr., W.W. 1970. Water quality engineering for practicing engineers. Barnes and Noble, Inc., New York, NY.

Edgehill, R.U., and R.K. Finn. 1983. Microbial treatment of soil to remove pentachlorophenol. Appl. Environ. Microbiol. 45:1122-1125.

Ehrenfeld, J.R., and J.M. Bass. 1983. Handbook for evaluating remedial action technology plans. U.S. Environmental Protection Agency, EPA-600/2-83-076, MERL, Cincinnati, OH.

Farmer, W.J., M.S. Yang, J. Letey, and W.F. Spencer. 1980. Land disposal of hexachlorobenzene wastes-controlling vapor movement in soils. EPA-600/2-80-119. Municipal Environmental Research Laboratory, ORD, U.S. EPA, Cincinnati, OH.

Finck, A. 1982. Fertilizers and fertilization. Verlag Chemie, Deerfield Beach, FL.

FMC Corporation. 1979. Industrial waste treatment with hydrogen peroxide. Industrial Chemicals Group, Phil., PA.

Focht, D.D., and M. Alexander. 1970. DOT metabolites and analogs: Ring fission by *Hydrogenomonas*. Science 170:91-92.

Follett, R.H., L.S. Murphy, and R.L. Donahue. 1981. Fertilizers and soil amendments. Prentice-Hall, Inc., Englewood Cliffs, NJ.

Franklin Institute Research Laboratory, Inc., 1981. The Franklin Institute chemical method for detoxifying polychlorinated biphenyls (PCBs) and other toxic waste. Phila., PA.

Frink, C.R., and G.K. Voight. 1976. Potential effects of acid precipitation on soils in the humid temperature zone. pp. 685-709. *In* Proceedings of the First International Symposium on Acid Precipitation and the Forest Ecosystem. U.S. Forest Service General Technical Report NE-23, USDA — Forest Service, Upper Darby, PA.

Fry, A.W., and A.S. Grey. 1971. Sprinkler irrigation handbook. Rain Bird Sprinkler Mfg. Corporation, Glendora, CA.

Furukawa, K. 1982. Microbial degradation of polychlorinated biphenyls (PCBs). pp. 33-57. *In* Biodegradation and detoxification of environmental pollutants. Chakrabarty, A.M., (ed.). CRC Press, Boca Raton, FL.

Furukawa, T., and G.W. Brindley. 1973. Adsoption and oxidation of benzidine and aniline by montmorillonite and hectorite. Clay Miner. 21:279-288.

Graedel, T.E. 1978. Chemical compounds in the atmosphere. Academic Press, New York, NY.

Gray, T.R.G. 1978. Microbial aspects of the soil. plant, aquatic, air, and animal environments: Soil and plants. pp. 19-38. *In* Pesticide microbiology. Hill, I.R., and Wright, S.J.L. (eds.) Academic Press, New York, NY.

Greer, J.S., and S.S. Gross. 1980. The practicality of controlling vapor release from spills of volatile chemicals through cooling. pp. 130-135. *In* National Conference on Control of Hazardous Material Spills.

Griffin, R.A., and N.F. Shimp. 1978. Attenuation of pollutants in municipal landfill leachate by clay minerals. U.S. Environmental Protection Agency-600/2-78-157.

Grove, J.H., and B.G. Ellis. 1980. Extractable chromium as related to soil pH and applied chromium. Soil Sci. Soc. Am. J. 44:238-242.

Guidelines for data acquisition and data quality evaluation in environmental chemistry. 1980. Anal. Chem. 52:2242-2249.

Gupta, O.P. 1976. Adsorbents and antidotes. World Crops. 28:134-138.

Hautala, R. 1978. Surfactant effects on pesticide photochemistry in water and soil. EPA-600/3-78-060. U.S. Environmental Protection Agency, Athens, GA.

Hazardous Materials Control Research Institute. 1983. Hazardous Materials Control Directory, 1983-1984. Silver Spring, MD.

Henderson, R.W., G.R. Lightsey, and N.A. Poonawala. 1977b. Competitive adsorption of metal ions from solutions by low-cost organic materials. Nature 162:746-747.

Henderson, R.W., D.S. Andrews G.R. Lightsey, and N.A. Poonawala. 1977a. Reduction of mercury, copper, nickel, cadmium, and zinc levels in solution by competitive adsorption onto peanut hulls, and raw and aged bark. Bull. Environ. Contam. Toxicol. 17:355-359.

Hendry, D.G., and R.A. Kenley. 1979. Atmospheric reaction products of organic compounds. Environmental Protection Agency Report EPA-560/12-79-001. 81p.

Hiltz, R. 1982. Vapor hazard control. pp. 10-21 to 10-33. *In* Hazardous Spills Handbook. Bennett, G.F., Featers, F.S. and Wilder, I. (eds.) McGraw-Hill Book Co., New York, NY.

Hirschler, A.E. 1966. Proton acids as electron acceptors on aluminosilicates. J. Catalysis 5:196-197.

Horowitz, A., J.M. Sulflita, and J.M. Tiedje. 1983. Reductive dehalogenation of halobenzoates by anaerobic lake sediment microorganisms. Appl. Environ. Microbiol. 45:1459-1465.

Huang, C.P., and F.B. Ostoric. 1978. The removal of cadmium(II) from dilute aqueous solution by activated carbon adsorption. J. Environ. Eng. Div., ASCE 104.

Huang, C.P., and E.H. Smith. 1981. Removal of cadmium(II) from plating waste water by an activated carbon process. *In* W.J. Cooper (ed.) Chemistry in water reuse, 2:355-400. Ann Arbor Sci. Publ., Ann Arbor, MI.

Implement and Tractor Red Book. 1983. Intertee Publishing Corp., Overland Park, KS.

Jacobs, L.W. 1977. Utilizing municipal sewage wastewaters and sludges on land for agricultural production. North Central Regional Extension Publication No. 52.

Jensen, R.A. 1982. United States industrial plans. *In* Hazardous materials spills handbook. Bennett, G.F., Feates, F.G., and Wilder, I. (eds.). McGraw-Hill Book Company, New York, NY. pp. 7-43 to 7-55.

Johnston, J.B., and S.B. Robinson. 1982. Opportunities for development of new detoxification processes through genetic engineering. pp. 301-314. *In* Detoxication of hazardous waste. Exner, J.H (ed.). Ann Arbor Science, Ann Arbor, MI.

JRB. 1982. Techniques for evaluating environmental processes associated with the land disposal of specific hazardous materials. Vol. I: Fundamentals. Report to U.S. Environmental Protection Agency.

JRB, Inc. 1982. Techniques for evaluating environmental processes associated with the land disposal of specific hazardous materials. II — Incompatible Wastes.

Jurinak, J.J., and N. Bauer. 1956. Thermodynamics of zinc adsorption on calcite, colomite, and magnesite-type minerals. Soil Sci. Soc. Am. Proc. 20:466-471.

Kaplan, D.L., and A.M. Kaplan. 1982a. Thermophilic biotransformations of 2,4,6-trinitrotoluene under simulated composting conditions. Appl. Environ. Microbiol. 44:757-760.

Kaplan, D.L., and A.M. Kaplan 1982b. Composting industrial wastes — biochemical consideration. Biocycle May/June:42-44.

Khan, S.U. 1982. Studies on bound 14C-prometryn residues in soil and plants. Chemosphere 11:771-795.

Khan, S.U., and K.C. Iverson. 1982. Release of soil bound (non-extractable) residues by various physiological groups of microorganisms. J. Environ. Sci. Health B17:737-749.

Kilbane, J.J., D.K. Chatterjee, and A.M. Chakrabarty. 1983. Detoxification of 2,4,5-trichlorophenoxyacetic acid from contaminated soil by *Pseudomonas cepacia*. Appl. Environ. Microbiol. 45:1697-1700.

Kim, B.E. 1981. Treatment of metal containing wastewater with calcium sulfide. *In* Water-1980, AIChE 209 (77).

Kim, B.E. 1981. Treatment of metal containing wastewater with calcium sulfide. *In* Water-1980. AIChE (209) 77.

Kirk Othmer Encylopedia of Chemical Technology. 1982. Vol. 18. John Wiley & Sons, New York, NY.

Klopffer, W. Dec. 2-4, 1980. Rapid test for simulation of photo-oxidative degradation in the gas phase. *In* Int. Workshop on Test Methods and Assessing Procedures for the Determination of the Photochemical Degradation Behavior of Chemical Substances, Berlin.

Ko, W.H., and J.L. Lockwood. 1968. Accumulation and concentration of chlorinated hydrocarbon pesticides by micro-organisms in soil. Can. J. Microbiol 14:1075-1078.

Kobayashi, H., and B.E. Rittman. 1982. Microbial removal of hazardous organic compounds. Environ. Sci. Technol. 16:170A-183A.

Konetzka, W.A. 1977. Microbiology of metal transformations. pp. 317-342. *In* Microorganisms and minerals. Weinberg, E.D. (ed.). Marcel Dekker, Inc., New York, NY.

Korte, N.E., J. Skopp, W.H. Fuller, E.E. Niebla, and B.A. Alesii. 1976. Trace element movement in soils influence of soil physical and chemical properties. Soil Sci. 121:350-359.

Kowalenko, C.G. 1978. Organic nitrogen, phosphorus and sulfur in soils. pp. 95-135. *In* Soil organic matter. Schnitzer, M. and Khan, S.U. (eds.). Elsevier Scientific Publishing Co., New York, NY.

Larsen, V.J., and H.H. Schierup. 1981. The use of straw for removal of heavy metals from wastewater. J. Environ. Qual. 10:188-193.

Lemaire, J., I. Campbell, H. Hulpke, J.A. Guth, W. Merz, L. Philip, and C. Von Waldow. 1982. An assessment of test methods for photodegradation of chemicals in the environment. Chemosphere 11(2):119-164.

Liang, L.N., J.L. Sinclair, L.M. Mallory, and M. Alexander. 1982. Fate in model ecosystems of microbial species of potential use in genetic engineering. Appl. Environ. Microbiol. 44:708-714.

Liberti, A., D. Broco, A. Allegrini, and G. Bertoni. 1978. Field photodegradation of TCDD by ultra-violet radiation. pp. 195-200. *In* Dioxin: Toxilogical and chemical aspects. Cattabeni, F., Cavallero, A., and Galli, G. (eds.). Spectrum Publications, Inc., New York, NY.

Lichtenstein, E.P., T.W. Fuhremann, and K.R. Schulz. 1968. Use of carbon to reduce the uptake of insecticidal soil residues by crop plants. J. Agr. Food Chem. 16:348-355.

Liu, S.-Y., R.D., Minard, and J.-M. Bollag. 198 . Coupling reactions of 2,4-dichlorophenol with various anilines. J. Agric. Food Chem. 29:253-257.

Lyman, W.J., W.F. Reehl, and D.H. Rosenblatt. 1982. Handbook of chemical property estimation methods. McGraw-Hill Book Company, New York, NY.

Mahmood, R., R. C. Sims. 1984. Management of toxic organic compounds in soil systems. Dissertation prepared for Utah State University, Dept. of Civil and Environmental Engineering. To be submitted August, 1984.

Marinucci, A.C., and R. Bartha. 1979. Biodegradation of 1,2,3- and 1,2,4-trichlorobenzene in soil and in liquid culture. Appl. Environ. Microbiol. 38:811-817.

Mattigod, S.W., G. Sposito, and A.L. Page. 1981. Factors affecting the solubility of trace metals in soils. *In* Chemistry in the Soil Environment, ASA Special Publication 40.

McBride, M.B. 1980. Chemisorption of Cd^{2+} on calcite surfaces. Soil Sci. Soc. Am. J. 44:26-28.

McBride, M.B. 1979. Chemisorption and precipitation of Mn^{2+} at $CaCO_3$ surfaces. Soil Sci. Soc. Am. J. 43:693-698.

McLean, E.O. 1982. Soil pH and lime requirement. pp. 199-224. *In* Methods of soil analysis. Part 2 — chemical and microbiological properties. Second Edition. Page, A.L. (ed.). American Society of Agronomy, Inc., Madison, WI.

McLean, E.O., S.W. Dumford, and F. Coronel. 1966. A comparison of several methods of determining lime requirements of soil. Soil Sci. Soc. Am. Proc. 30:26-30.

Mercer, B.W., A.J. Shuckrow, and L.L. Ames. 1970. Fixation of radioactive wastes in soil and salt cores with organic polymers. BNWL-1220.

Metry, A.A. 1980. The handbook of hazardous waste management. Technomic Publ., Westport, PA.

Meuser, J.M. and W.C. Wiemer. 1982. Amine-enhanced photodegradation of polychlorinated biphenyls. EPRI Research Project 1263-6. Battelle Pacific Northwest Laboratories.

Millington, R.J., and J.P. Quirk. 1961. Permeability of porous solids. Trans. Faraday Soc. 57:1200-1207.

Mulder, D. (ed.). 1979. Soil disinfection. Elsevier Scientific Publishing Co., Amsterdam.

Mulle, G.J. 1981. Chemical decomposition of PCBs in transformer fluids. The Acurex Process. Acurex Waste Technologies, Inc., Mountain View, CA.

Munnecke, D.M. 1980. Enzymatic detoxification of waste organophosphate pesticides. J. Agric. Food Chem. 28:105-111.

Munnecke, D.M., L.M. Johnson, H.W. Talbot, and S. Barik. 1982. Microbial metabolism and enzymology of selected pesticides. pp. 1-32. *In* Biodegradation and detoxification of environmental pollutants. Chakrabarty, A.M. (ed.). CRC Press, Inc., Boca Raton, FL.

Nagel, G., et al. 1982. Sanitation of groundwater by infiltration of ozone treated water. GWF-Wasser/Abwasser 123(8):399-407.

Nakayama, S., et al. Sept. 15, 1979. Ozone. Science and Engineering 1(2):119-132.

Nilles, G.P., and M.J. Zabik. 1975. Photochemistry of bioactive combasagran. J. Agr. Food Chem. 23:410.

Nimah, M.H., J. Ryan and M.A. Chaudhry. 1983. Effect of synthetic conditioners on soil water retention, hydraulic conductivity, pososity, and aggregation. Soil. Sci. Soc. Am. J. 47:742-745.

Occhiucci, G., and A. Patacchiola. 1982. Sensitized photodegradation of adsorbed polychlorobiphenyls (PCBs). Chemisphere 11(3):255-262.

O'Sullivan, D. 1981. Strong growth ahead for industrial enzymes. Chem. Eng. News 59(3):37-38.

Ou, L.-T., P.S.C. Rao, and J.M. Davidson. 1983 Methyl parathion degradation in soil: influence of soil-water tension. Soil Biol. Biochem. 15:211-215.

Overcash, M.R., and D. Pal. 1979. Design of land treatment systems for industrial wastes —theory and practice. Ann Arbor Science, Ann Arbor, MI.

Page, J.B. 1941. Unreadability of the benzidine color reactions as a test for montmorillonite. Soil Sci. 51:133-140.

Pal, D., J.B. Weber, and M.R. Overcash. 1980. Fate of polychlorinated biphenyls (PCBs) in soil-plant systems. Residue Reviews 74:45-98.

Peech, M. 1961. Lime requirements vs. soil pH curves for soils of New York State. Cornell University, Ithaca, NY.

Pfaender, F.K., and M. Alexander. 1972. Extensive microbial degradation of DDT in vitro and DDT metabolism by natural communities. J. Agric. Food Chem. 20:842-846.

Pillie, R.J., R.E. Baier, R.C. Ziegler R.P. Lenard, J.C. Micholovic, S.L. Pek, and P.H. Bock. 1975. Methods to treat, control and monitor spilled hazardous materials. EPA-670/2-75-042.

Plimmer, J.R. 1971. Principles of photodecomposition of pesticides. pp. 279-290. *In* Degradation of synthetic organic molecules in the biosphere. National Academy of Sciences, Wash., D.C.

Plimmer, J.R., and U.I. Klingebiel. 1973. Photochemistry of dibenzo-p-dioxins. pp. 44-54. *In* Chlorodioxins — origin and fate. Blair, E.H. (ed.). American Chemical Society, Wash., DC.

Prengle, Jr., H.W. 1977. Evolution of the ozone/UV process for wastewater treatment. Presented at Seminar on Wastewater Treatment and Disinfection with Ozone. Cincinnati, OH. International Ozone Association, Vienna, VA.

Prior, L.A. 1975. Land availability, crop production, and fertilizer requirements in the United States. EPA/530/Sw-166. U.S. Environmental Protection Agency, Wash., D.C.

Purdue University. 1977. Soil characterization in Indiana: I. Field and laboratory procedures. Research Bulletin No. 943, Department of Agronomy.

Quastel, J.H., and P.G. Scholefield. 1953. Arsenite oxidation in soil. Soil Sci. 75:279-285.

Raymond, J.L., V.W. Jamison, and J.O. Hudson. 1976. Beneficial stimulation at bacterial activity in groundwaters containing petroleum products. AIChE, (73), 166:390-404.

Rice, R.G. 1981. Ozone for the treatment of hazardous materials. Water — 1980. Symposium series, American Institute of Chemical Engineers, 209 (77):79-107.

Santillan-Medrano, J., and J.J. Jurinak. 1975. The chemistry of lead and cadmium in soil solid phase formation. Soil Sci. Soc. Am. J. 39:851-856.

Saxena, A., and R. Bartha. 1983a. Modeling of the covalent attachment of chloroaniline residues to quinoidal sites of soil humus. Bull. Environ. Contam. Toxicol. 30:485-491.

Saxena, A., and R. Bartha. 1983b Microbial mineralization of humic acid — 3,4-dichloroaniline complexes. Soil Biol. Biochem. 15:59-62.

Saxena, J., and P.H. Howard. 1977. Environmental transformation of alkylated and inorganic forms of certain metals. pp. 185-226. *In* Advances in applied microbiology., Perlman, D. (ed.). Academic Press, New York, NY.

Schilfgaarde, J.V. 1974. Drainage for agriculture. Agronomy Monograph No. 17, Amer. Soc. of Agron., Inc., Madison, WI.

Schwab, G.O., R.K. Frevert, T.W. Edminster, and K.K. Barhes. 1981. Soil and water conservation engineering. Third edition. John Wiley and Sons, New York, NY.

Sharma, S., and R. Singh. 1983. Selenium in soil, plant, and animal systems. CRC Critical Reviews in Environmental Control 13(1):23-50.

Sherman, J.D. 1978. Ion exchange separation with molecular sieve zeolites. *In* Adsorption and Ion Exchange Separation. Sherman, J.D. (ed.). AIChE Symposium Ser. No. 179:98-116.

Shin, Y., J.J. Chodan, and A.R. Wolcott. 1970. Adsorption of DDT by soils, soil fractions, and biological materials. J. Agr. Food Chem. 18:1129-1133.

Shippen, J.M., C.R. Ellin, and C.H. Clover. 1980. Basic farm machinery. Third edition, Pergamon Press, New York, NY.

Sikora, L.J., D.D. Kaufman, M.A. Ramirez, and G.B. Willson. March 8-10, 1982. Degradation of pentachlorophenol and pentachloronitrobenzene in a laboratory composting system. *In* Proc. Eighth Ann. Symp. on Disposal of Hazardous Wastes. U.S. Environmental Protection Agency, Cincinnati, OH

Sims, R.C., and M.R. Overcash. 1981. Land treatment of coal conversion wastewaters. *In* Environmental aspects of coal conversion technology VI, a symposium on coal-based synfuels. EPA-600/9-82-017, U.S. Environmental Protection Agency, Wash., DC.

Sims, R.C. and M. Overcash. 1983. Fate of polynuclear aromatic compounds (PNAs) in soil-plant systems. Residue Reviews 88:1-68.

Sjobland, R.D., and J.-M. Bollag. 1981. Oxidative coupling of aromatic compounds by enzymes from soil microorganisms. pp. 113-152. *In* Soil biochemistry. Vol. 5. Paul, E.A., and Ladd, J.N. (eds.). Marcel Dekker, Inc., New York, NY.

Skujins, J., S.O. McDonald, and W.G. Knight. 1983. Metal ion availability during biodegradation of waste oil in semi-arid soils. Environmental Biogeochemistry 35:341-350.

Smeulders, F., A. Maes, J. Sinnaeve, and A. Cremers. 1983. In situ immobilization of heavy metals with tetraethylenepentamine (tetren) in natural soils and its effect on toxicity and plant growth. Plant and Soils. 70:37-47.

Soil Conservation Service. 1972. Soil survey laboratory methods and procedures for collecting soil samples. Soil Survey Investigation Report No. 1, U.S. Department of Agriculture, Wash., D.C.

Soil Conservation Service. 1979. Guide for sediment control on construction sites in North Carolina. U.S. Dept. of Agriculture, Soil Conservation Service, Raleigh, NC.

Solomon, D.H. 1968. Clay minerals as electron acceptors and/or electron donors in organic reactions. Clay Miner. 16:31-39.

Sommers, L.E., C.M. Gilmour R.E. Wildung, and S.M. Beck. 1981. The effect of water potential on decomposition processes in soils. pp. 97-117. *In* Water potential relations in soil microbiology. SSA Special Publication Number 9. Soil Sci. Soc. of Amer., Madison, WI.

Spencer, W.F., J.D. Adam, T.D. Shoup, and R.C. Spear. 1980. Conversion of parathion to paraoxon on soil dusts and clay minerals as affected by ozone and UV light. J. Agr. Food Chem. 28:369.

Staiff, D.C., L.C. Butler, and J.E. Davis. 1981. A field study of the chemical degradation of paraquat dichloride following simulated spillage on soil. Bull. Environ. Contam. Toxicol. 26:16-21.

Stevenson, F.J. 1972. Role and function of humus in soil with emphasis on adsorption of herbicides and chelation of micronutrients. Bioscience 22(1):643-650.

Stotsky, G., and V.N. Krasovsky. 1981. Ecological factors that affect the survival, establishment, growth, and genetic recombination of microbes in natural habitats. pp. 31-42. *In* molecular biology, pathogenicity and ecology of bacterial plasmids. Levy, S.F., Clowes, R.C., and Koenig E.L. (eds.). Plenum Press, New York, NY.

Strek, H.J., J.B. Weber, P.J. Shea, E. Mrozek, Jr., and M.R. Overcash. 1981. Reduction of polychlorinated biphenyl toxicity and uptake of carbon-14 activity by plants through the use of activated carbon. J. Agr. Food Chem. 29:288-293.

Sulflita, J.M., and J.-M. Bollag. 1980. Oxidative coupling activity in soil extracts. Soil Viol. Biochem. 12:177-183.

Sulflita, J.M., A. Horowitz, D.R. Shelton, and J.M. Tiedje. 1982. Dehalogenation: a novel pathway for the anaerobic biodegradation of haloaromatic compounds. Science 218:1115-1117.

Sulflita, J.M., J.A. Robinson, and J.M. Tiedje. 1983 Kinetics of microbial dehalogenation of haloaromatic substrates in methanogenic environments. Appl. Environ. Microbiol. 45:1466-1473.

Sulflita, J.M., and J.M. Tiedje. 1983. The anaerobic degradation of 2,4,5-trichlorophenoxyacetic acid by a dehalogenating bacterial consortium. p. 266. *In* Abstracts of the Annual Meeting of the ASM. Amer. Soc. for Microbiol., Wash., D.C.

Sweeney, K. 1981. Reductive treatment of industrial wastewaters. I. Process and description and II. Process application. pp. 67-78. *In* Water 1980. G.F. Bennett (ed.). American Institute of Chemical Engineers Symposium, Series 209 (77).

Tames, R.S., and R.J. Hance. 1969. The adsorption of herbicides by roots. Plant and Soil. 30:221-226.

Taylor, J.M., J.F. Parr, L.J. Sidora, and G.B. Wilson. 1980. Consideration in the land treatment of hazardous wastes: Principles and practices. *In* 2nd Oil and Hazardous Spills Conference and Exhibition.

Taylor, S.A., and G.L. Ashcroft. 1972. Physical Edaphology. W.H. Freeman and Co., San Francisco, CA. 533 p.

Theng, B.K.G. 1974. The chemistry of clay-organic reactions. Adam Hilger, London.

Thibault, G.T., and N.W. Elliott. 1980. Biological detoxification of hazardous organic chemical spills. pp. 398-402. *In* Control of hazardous material spills. Vanderbilt University, Nashville, TN.

Thibault, G.T., and N.W. Elliott. 1979. Accelerating the biological clean-up of hazardous materials spills. pp. 115-120. *In* Oil and hazardous material spills: Prevention-control-cleanup-recovery-disposal. Information transfer, Inc., Silver Spring, MD.

Thibodeaux, L.J., and S.T. Hwang. 1982. Landfarming of petroleum wastes modeling the air emission problem. Env. Progress 1(1):42-46.

Turner, J.H., and C.L. Anderson, 1980. Planning for an irrigation system. American Association for Vocational Instruction Materials, Athens, GA.

U.S. Environmental Protection Agency. 1976. Application of sewage sludge to cropland: Appraisal of potential hazards of the heavy metals to plants and animals. MCD — 33 EPA 430/9-76-013.

U.S. Environmental Protection Agency. 1977. Environmental assessment of subsurface disposal of municipal wastewater treatment sludge. Interim report. EPA/530/SW-547.

U.S. Environmental Protection Agency. 1977. Municipal sludge management: Environmental factors. MCD-28, EPA 430/9-9-77-004.

U.S. Environmental Protection Agency. 1978. Municipal sludge agricultural utilization practices. An environmental assessment. 1: SW-709.

U.S. Environmental Protection Agency. 1978. Applications of sludges and wastewaters on agricultural land: A planning and educational guide. MCD-35.

U.S. Environmental Protection Agency. 1980. A guide to regulations and guidance for utilization and disposal of municipal sludge. MCD-72. EPA 430/9-80-015.

U.S. Environmental Protection Agency. 1981. Effects of sewage sludge on the cadmium and zinc content of crops. EPA-600/8-81-003.

U.S. Environmental Protection Agency. 1982. Handbook for remedial action at waste disposal sites. EPA-625/6-82-006. Municipal Environmental Research Laboratory, Office of Research and Development. Wash., D.C.

U.S. Environmental Protection Agency. 1982. Test methods for evaluating solid waste. SW-846, 2nd Ed.

U.S. Environmental Protection Agency. 1983. Hazardous waste land treatment. SW-874 (Revised edition). Wash., D.C.

Voerman, S., and P.M. Tammes. 1969. Adsorption and desorption of lindane and dieldrin by yeast. Bull. Environ. Contam. Toxicol. 4:271-277.

Walker, A., and D.V. Crawford. 1968. The role of organic matter in adsorption of the triazine herbicides by soil. *In* Isotopes and radiation in soil organic matter studies. Proc. 2nd Symp. Int. Atomic Energy Agency. Vienna, Austria.

Wallnöfer, P.R., W. Ziegler, and G. Engelhardt. 1981. Bacterial oxidation of 4-methylthio-substituted veratrylglycerol-B-aryl-ethers, model compounds for soil bound pesticide residues. Chemosphere 10:341-346.

Walsh, L.M., and J.D. Beaton (ed.). 1973. Soil testing and plant analysis. Soil, Sci. Soc. Amer., Inc., Madison, WI.

Walton, G.C., and D. Dobbs. 1980. Biodegradation of hazardous materials in spill situations. pp. 23-29. *In* Control of hazardous material spills. Vanderbilt University, Nashville, TN.

Weast, R.C., 1983. Handbook of chemistry and physics. CRC Press, Cleveland, OH.

Weber, J.B., and E. Mrozek, Jr. 1979. Polychlorinated biphenyls: Phytotoxicity, adsorption and translocation by plants, and inactivation by activated carbon. Bull. Environ. Contam. Toxicol. 23:412-417.

Weber, J.B., P.W. Perry, and R.P. Upchurch. 1965. The influence of temperature and time of the adsorption of paraquat, diquat, 2,4-D, and prometone by clays, charcoal, and an anion exchange resin. Soil Sci. Soc. Am. Proc. 29:678-688.

Williams, E.B. May 26-28, 1982. Contamination containment by in-situ polymerization. Second National Symposium on Aquifer Restoration and Groundwater Monitoring. National Water Well Association. Worthington, OH.

Wilson, L.G. 1980. Monitoring in the vadose zone: A review of technical elements and methods. EPA-600/7-80/134. U.S. Environmental Protection Agency, Las Vegas, NV.

Wipf, H.K., E. Homberger, N. Neuner, and F. Schenker. 1978. Field trials on photodegradation of TCDD on vegetation after spraying with vegetable oil. pp. 201-217. *In* Dioxin: Toxilogical and chemical aspects. Cattabeni, F., Cavallero, A., and Galli, G. (eds.). Spectrum Publications, Inc., New York, NY.

Woodrow, J.E., D.G. Crosby, and J.N. Seiber. 1983. Vapor-phase photochemistry of pesticides. Residue Reviews 85:111-125.

Woolson, E.A. 1977. Fate of arsenicals in different environmental substrates. Environ. Health Perspect. 19:73-81.

Yagi, O., and R. Sudo. 1980. Degradation of polychlorinated biphenyls by microorganisms. JWPCF. 52:1035-1043.

Index

Copyright Notice

Figure 3-3	From Huang, C.P., and E.H. Smith. 1981. Removal of cadmium(II) from plating waste water by an activated carbon process. *In* W.J. Cooper (ed.) Chemistry in water reuse, 2:355-400. Ann Arbor Sci. Publ., Ann Arbor, MI. Used by permission.
Figure 3-4	From Huang C.P., and E.H. Smith. 1981. Removal of cadmium(II) from plating waste water by an activated carbon process. *In* W.J. Cooper (ed.) Chemistry in water reuse, 2:355-400. Ann Arbor Sci. Publ., Ann Arbor, MI. Used by permission.
Table 3-2	From Huang, C.P., and E.H. Smith. 1981. Removal of cadmium(II) from plating waste water by an activated carbon process. *In* W.J. Cooper (ed.) Chemistry in water reuse, 2:355-400. Ann Arbor Sci. Publ., Ann Arbor, MI. Used by permission.
Table 3-6	From Coffey, D.L., and G.F. Warren. 1969. Inactivation of herbicides by activated carbon and other adsorbents. Weed Sci. 17:16-19. Used by permission.
Table 3-8	From Coffey, D.L., and G.F. Warren. 1969. Inactivation of herbicides by activated carbon and other adsorbents. Weed Sci. 17:16-19. Used by permission.
Table 3-9	From Overcash, M.R., and D. Pal. 1979. Design of land treatment systems for industrial wastes — theory and practice. Ann Arbor Science, Ann Arbor, MI. Used by permission.
Table 3-10	From Overcash, M.R., and D. Pal. 1979. Design of land treatment systems for industrial wastes — theory and practice. Ann Arbor Science, Ann Arbor, MI. Used by permission.
Table 3-11	From Artiole, J., and Fuller, W.H., 1979. Effect of Crushed Limestone Barriers on Chromium Attenuation in Soils. J. Environ. Qual. 8:503-510.
Table 3-12	From Mattigod, S.W., G. Sposito, and A.L. Page. 1981. Factors affecting the solubility of trace metals in soils. *In* Chemistry in the Soil Environment, ASA Special Publication 40. Used by permission.
Table 3-15	From Rice, R.G 1981. Ozone for the treatment of hazardous materials. Water — 1980. Symposium series. American Institute of Chemical Engineers, 209 (77):79-107. Used by permission.
Table 3-18	From Staiff, D.C., L.C. Butler, and J.E. Davis. 1981. A field study of the chemical degradation of paraquat dichloride following simulated spillage on soil. Bull. Environ. Contam. Toxicol. 26:16-21. Used by permission.

☆ U.S. GOVERNMENT PRINTING OFFICE: 1984 - 759-102/10712

II. Background Information

1. Introduction

To support the methodology presented in Volume 1 monitoring information and data on the behavior and fate of hazardous constituents in soil systems are presented here in Volume 2 of the Manual. This additional information is intended to provide the manual user with more information and background material for making more complex decisions and analyses concerning site, soil, and waste interactions than can be presented in Volume 1.

Section 2, "Monitoring," includes a discussion of the factors required for a comprehensive and effective monitoring program including soil care, soil pore liquid, unsaturated zone, groundwater and atmosphere sampling. A monitoring program is necessary to assure that the objectives of the in-place treatment program are being met.

Section 3, "Characterization and Evaluation of Fundamental Processes in Soil/Waste Systems," presents information on the behavior and fate of hazardous constituents in soil systems. The information presented served as a basis for selecting treatment techniques and developing the methodology for in-place treatment. Elements of soil/waste systems considered included site and soil factors important in influencing in-place treatment, soil mobilization processes for leaching and volatilization control, biodegradation processes and transformation processes. A brief discussion addressing the modeling of waste constituents in soil systems is presented.

The Appendix provides compound properties and adsorption, degradation and volatilization parameters for various chemical compounds. A glossary of terms is also included.

The information in Volume 2 represents an effort to present the manual user with the most current information and research activities that directly or indirectly influence the in-place processes and treatment techniques. The topics addressed present areas of critical concern and provide fundamental principles for characterizing the basic processes in soil/waste systems. Since most in-place treatment techniques have not been thoroughly tested, the information in Volume 2 will provide the background or framework for planning and interpreting in-place treatment techniques specifically addressed in Volume 1.

2. Monitoring

INTRODUCTION

During the execution of an in situ remedial action, the primary objectives are to prevent off-site migration of hazardous constituents via migration to ground or surface waters or transmission through the atmosphere, as well as to render the wastes nonhazardous through degradation, detoxification, or immobilization. To assure that these objectives are being met, a monitoring program must be established.

Specific objectives of a monitoring program are to: (K. W. Brown and Associates 1980, U. S. Environmental Protection Agency 1983):

1. Assure that the hazardous or toxic constituents of the waste are being degraded, detoxicated, or inactivated as planned.

2. Monitor degradation rates of degradable constituents.

3. Assure that waste constituents are not entering runoff or leachate water and leaving the area in unacceptable concentrations.

4. Determine whether adjustments in treatment management are needed to maintain the treatment process (e.g., Is soil pH within desirable range? Are adequate nutrients available for biological degradation of organic constituents? Does soil moisture require adjustment? Are additional chemical amendments needed to complete chemical treatment of the wastes?).

A complete monitoring program would include the media listed in Table 2-1. The waste constituents to be monitored in the various media are those determined to be hazardous in the initial site/waste characterization study as well as expected important degradation or transformation products. Nonhazardous constituents and/or their transformation/degradation products and soil properties which might affect treatment processes should also be monitored. The monitoring program may include substances which are required for treatment, whether native to the soil or added as a treatment agent.

STATISTICAL CONSIDERATIONS

A limitation in field sampling is the large number of samples required to ensure that the measured mean value of a particular parameter is within a given range of a true value (Wilson 1980). This limitation is due to the natural spatial variability of soil properties. It is therefore necessary to

TABLE 2-1. REQUIREMENTS OF A COMPLETE MONITORING PROGRAM (ADAPTED FROM U.S. ENVIRONMENTAL PROTECTION AGENCY 1983)

Media to be Monitored	Purpose
Soil in the treatment zone	Determine extent of degradation, transformation, and immobilization pH and nutrient status, and any other factors or substances affecting treatment execution and effectiveness.
Soil cores (unsaturated zone)	Determine slow movement of hazardous constituents.
Soil-pore liquid in the unsaturated zone	Determine highly mobile constituents.
Groundwater	Determine mobile constituents.
Runoff water	Determine migration off-site of soluble, suspended, or adsorbed constituents.
Air	Determine personnel and population health hazards.

take a significant number of replicate samples which are representative of the area being sampled. Realistic, unbiased data are required so that valid comparisons can be made between the values of monitored parameters and background values.

The hazardous waste site should be divided into uniform areas for sampling purposes. These areas may be according to soil series or to phases based on surface texture within a series. If known "hot spots" of wastes or significant differences in types of wastes occur at the site, these areas should also be monitored separately. For each medium sampled, appropriate background sampling sites should be monitored.

Guidelines for statistical sampling procedures within the selected areas are given in Hazardous Waste Land Treatment (U. S. Environmental Protection Agency 1983) and Test Methods for Evaluating Solid Waste (U.S. Environmental Protection Agency 1982c). Professionals who may be consulted for assistance may include certified professional soil scientists, statisticians, and environmental analytical chemists.

SOILS SAMPLING IN THE TREATMENT ZONE AND IN THE UNDERLYING UNSATURATED ZONE

Soil cores and borings are used to measure the vertical movement of waste constituents as well as to determine the progression of treatment in the soil treatment zone. Soil core sampling in the treatment zone is used as a management tool. The extent of treatment (i.e., degradation, transformation, and/or immobilization) of the hazardous constituents can be monitored as well as factors which may affect the execution of the treatment, such as nutrients and pH. Soil core sampling below the treatment zone is used to determine whether significant concentrations of hazardous constituents are moving below the treatment zone. If unacceptable leaching of contaminants is occurring, contingency plans should provide means to prevent groundwater contamination, such as the use of a grout bottom seal.

Soil sampling methods used by soil scientists and irrigation and drainage engineers to evaluate physical properties of soils are also suitable for determining chemical constituents in the upper layers of soils (Wilson 1980). Several types of equipment which have been used are listed in Table 2-2. The split spoon sampler is a barrel-type auger, with one side which pivots on a hinge. A tube-type sampler consists of a tube bevelled and sharpened on one end to aid insertion in the soil. A drive hammer is used to force the tube to the desired depth, and an intact soil core is removed from the tube.

Wilson (1980) identified several problems associated with the use of soil samplers. With augers and tube-type samplers, dry soil may fall out of the unit when it is withdrawn from the soil, which requires the use of a core catcher. Also, if a soil is wet, chemical interactions occurring between the soil water and metal parts of the sampler may introduce metal contamination into the sample. Contamination is an especially important problem in the case of organic chemicals and microorganisms. Techniques for microbial sampling have been described by Bordner, Winter, and Scarpino (1978) and for organic and microbial sampling by Dunlap et al. (1977). One of the most promising methods appears to be the use of a combination auger/dry tube corer technique. A hole is augered to the top of the desired sampling depth; then the soil is sampled with a core sampler forced into the sampling region. Wilson (1980) suggests that a lucite plastic insert in the dry tube coring barrel would minimize contamination with metals.

TABLE 2-2. EQUIPMENT FOR FIELD COLLECTION OF SOIL SAMPLES (WILSON 1980)

Hand-driven Equipment	Power-driven Equipment
Screw-type auger	Continuous flight power auger (hollow-stemmed)
Post-hole auger	Core sampler
Barrel auger	Bucket auger
Dutch auger	Cable-tool drill rig
Split-spoon sampler	Rotary drill rig
Tube-type sampler	
Auger/dry-tube corer	

Soil samples from lower soil depths are more easily obtained by the use of powered coring or drilling equipment. The U.S. Environmental Protection Agency (1983) suggests that even for samples at shallower depths, powered equipment may be less costly due to clean, minimally disturbed samples for analysis.

Several types of power-driven sampling units are listed in Table 2-2. Coring type machines are generally more desirable than drilling rigs since the addition of water during the drilling operation, to remove cuttings and in some cases mud to hold open the hole, may change the soil water chemistry and may result in sample contamination. Air rotary drilling rigs, which use air to bring cuttings to the surface, do not have the problem of contamination. Spiral type augers, unless hollow-stemmed (which allows for the insert of a core sampler), are not recommended because of difficulty in determining the depth the sample was taken (Wilson 1980).

After soil core sampling, holes should be backfilled with native soil (compacted to field bulk density), clay slurry, or other suitable materials in order to prevent the channelling of hazardous constituents down the holes U.S. Environmental Protection Agency 1983).

Procedures for handling, preserving, and shipping soil solids samples may be found in Test Methods for Evaluating Solid Waste (U.S. Environmental Protection Agency 1982c).

SOIL PORE LIQUID SAMPLING IN THE UNSATURATED ZONE

Water added to a site by precipitation or irrigation while percolating through the treatment zone may rapidly transport some mobile hazardous constituents or transformation products through the unsaturated zone to the groundwater (U.S. Environmental Protection Agency 1983). The purpose of soil-pore liquid monitoring is to detect these rapid pulses of contaminants. Heavy precipitation, snow melt, and irrigation events are often responsible for such pulses, and sampling periods should be scheduled to correspond to such events. Soil texture and structure and other soil properties, as they affect infiltration and percolation rates, also determine to what extent precipitation events cause rapid water movement in a particular soil and can be used as predictive tools to schedule sampling events.

Soil pore liquid sampling may also indicate the amount of materials leaching to the groundwater. Samples of groundwater do not provide this information due to the diluting effect of the groundwater. In addition, analysis of the soil pore liquid can provide an early warning signal that remedial action is required for the management of the treatment process.

Since water in the unsaturated zone is held under negative pressure (suction), wells and open cavities cannot be used to collect the flowing water. Therefore samplers in the unsaturated zone are called suction sampling devices. A summary of various types of suction samplers are presented in Table 2-3. The reader is referred to Wilson (1980) for a comprehensive discussion of these samplers.

TABLE 2-3. SOIL-PORE LIQUID SAMPLING DEVICES (WILSON 1980)

Sampling devices
A. Ceramic-type samplers
1. Suction cup
a. Vacuum-operated soil-water samplers
b. Vacuum-pressure samplers
c. Vacuum-pressure samplers with check valves
2. Filter candle
B. Cellulose-acetate hollow fiber samplers
C. Membrane filter samplers

WATER SAMPLES FROM THE SATURATED ZONE
(i.e., groundwater)

Groundwater sampling will indicate whether hazardous waste constituents have indeed migrated to the groundwater and may present a public health hazard, depending on the hydrogeological characteristics and uses of the groundwater system. If contamination has occurred and is deemed unacceptable, provisions must be made to recover and treat or otherwise handle the contaminated groundwater.

Sophisticated groundwater sampling equipment and procedures are not desirable for monitoring programs (U. S. Army Corps of Engineers 1982). Rather, to ensure long-term, efficient operation of the monitoring system, the devices should be simple, rugged, foolproof, and operable by trained, but not necessarily educationally skilled personnel (Vanhoff, Weyer, and Whitaker 1979). In Table 2-4, a list of saturated zone sampling methods and sample extraction techniques are given. Again, the reader is referred to Wilson (1980) for a discussion of these methods. In addition, the U. S. Environmental Protection Agency (1983) has suggested the following documents as sources of information on groundwater monitoring:

1. Manual of Ground-Water Sample Procedures (Scalf et al. 1981);

2. Ground-Water Manual (U.S. Department of the Interior, Bureau of Reclamation 1977);

3. Procedures Manual for Ground-Water Monitoring at Solid Waste Disposal Facilities (U.S. Environmental Protection Agency 1977);

TABLE 2-4. GROUNDWATER SAMPLING METHODS (WILSON 1980)

<u>Sampling devices</u>
Tile lines
Collection pans and manifolds
Wells
Piezometers
Multilevel samplers
Groundwater profile samplers
<u>Sample extraction methods</u>
Hand bailers
Air-lift and gas-lift pumps
Suction lift pumps
Piston pumps
Centrifugal pumps
Submersible pumps

4. Ground-water Monitoring Systems, Technical Resource Document (U.S. Environmental Protection Agency, in preparation); and

5. Ground-water Monitoring Guidance for Owners and Operators of Interim Status Facilities (U.S. Environmental Protection Agency 1982b).

Cost estimates prepared by the U.S. Environmental Protection Agency (1982) for several types of monitoring techniques are presented in Table 2-5.

Caution should be exercised in the choice of well casing material. The selection of casing depends on the constituents being monitored. Steel casing may contribute such contaminants as zinc and iron; therefore the use of PVC, fiberglass, or teflon is recommended. Wilson (1980) discusses the controversy of whether PVC is appropriate for sampling of organic pollutants. Dunlap et al. (1977) feel that PVC adsorbs organic constituents as well as contributes such contaminants as phthalic acid esters to samples. However, Geraghty and Miller, Inc. (1977) indicate that PVC does not leak organic compounds and is more inert than steel casing, which develops an iron oxide coating which has an unpredictable and changeable adsorptive capacity. With PVC, once the adsorption sites are saturated, the concentration of organics in the water remain in equilibrium with the adsorbed organics. For microbial sampling, Wilson (1980) suggests the use of PVC pipe, because metal constituents in steel wells could inhibit microbial growth.

RUNOFF WATER MONITORING

To ensure the health and safety of off-site populations, runoff water, if any occurs, should be monitored for hazardous constituents or byproducts. If

TABLE 2-5. COST ESTIMATES FOR VARIOUS MONITORING TECHNIQUES AND CONSTRUCTION METHODS IN THE ZONE OF SATURATION (U.S. ENVIRONMENTAL PROTECTION AGENCY 1982)

Monitoring technique & construction method	Price per installation well diameter		
	51 mm (2-inch)	102mm (4-inch)	152mm (6-inch)
Screened over a single interval (plastic screen and casing)			
1. Entire aquifer	$1,600-$3,700	$2,300-$4,500	$6,400-$7,500
2. Top 3 meters (10 feet) of aquifer	600- 1,050	700- 1,150	-
3. Top 1.5 meters (5 feet) of aquifer with drive point	100- 200	-	-
Piezometers (plastic screen & casing)			
1. Entire aquifer screened			
a. Cement grout	2,100- 4,700	2,800- 5,500	6,900- 8,500
b. Bentonite seal	1,850- 4,150	2,350- 4,950	6,650- 7,950
2. Top 3 meters (10 feet) of aquifer screened			
a. Cement grout	1,150- 2,050	1,200- 2,150	-
b. Bentonite seal	900- 1,500	950- 1,600	-
Well clusters			
1. Jet-percussion			
a. Five-well cluster, each well with a 6-meter (20-foot) long plastic screen	2,500- 3,800	-	-
b. Five-well cluster, each well with only a 1.5 meter (5-foot) long plastic screen	1,700- 2,300	-	-
2. Augers			
a. Five-well cluster, each well with a 6-meter (20-foot) long stainless steel wire-wrapped screen	4,600- 5,300	-	-
b. Five-well cluster, each well with only a 1.5 meter (5-foot) long gauze wrapped drive points	1,800- 2,600	-	-
3. Cable tool			
a. Five-well cluster, each well with a 6-meter (20-foot) long stainless steel, wire-wrapped screen	-	-	9,850-14,150

TABLE 2-5. CONTINUED

Monitoring technique & construction method	Price per installation well diameter		
	51 mm (2-inch)	102mm (4-inch)	152mm (6-inch)
4. Hydraulic rotary			
a. Five-well cluster, each well with a 6 meter (20-foot) long plastic screen casing grounded in place	-	9,050-14,900	13,800-19,400
b. Five-well cluster, completed in a single large-diameter borehole 4.5 meter (15-foot) long plastic screens, 1.5-meter (5-foot) seal between screens	4,240- 5,800	8,250-11,000	-
5. Single well/multiple sampling point	-	-	3,000- 4,700
a. 33.5-meters (110-foot) deep well with 1-foot long screens separated by 1.2 meters (4-feet) of casing starting at 3 meters (10-feet) below ground surface			
Sampling during drilling	-	3,000- 4,700	3,300- 5,200

Note: Cost estimates are for an aquifer composed of unconsolidated sand with a depth to water of 3 meters (10 feet) and a total saturated thickness of 30 meters (100 feet). Cost estimates are based on rates prevailing in the Northeast in Autumn, 1975. Actual costs will be lower and higher depending upon conditions in other areas. Therefore, while the costs presented will become outdated with time, the relative cost relationships among the monitoring techniques should remain fairly constant.

significant concentrations are found, provisions should be available for collecting, storing, treating, and/or recycling the water through the site.

AIR MONITORING

Air monitoring at hazardous waste sites is essential for the protection of health and safety of the remedial action team due to the volatile nature of many hazardous compounds anticipated at hazardous waste sites. An adequate air monitoring program allows the evaluation of the relative importance of vapor transport from the site, provides a means for evaluating the effectiveness of vapor suppression techniques, permits the identification of volatile daughter products from the various treatment techniques utilized, and is a requirement if vapor phase photolysis is used as a treatment alternative.

In addition to personal monitoring equipment, a perimeter sampling network should be established to detect off-site migration of gaseous and/or particulate emissions. Upwind and downwind sampling sites should also be used to determine background air quality as well as the extent of off-site contamination, if any. A large number of sources are available concerning the development of air sampling networks from both the USEPA and professional organizations such as the Air Pollution Control Association, and these entities should be contacted for specifics.

Both high-efficiency particle filter samplers and gas/vapor samplers should be used for contaminant collection (Cheremisinoff et al. 1982). Solid sorbent traps have become the standard sampling medium for volatile organic air pollutants (Seiber and Woodrow 1983) and are suggested for materials with boiling points greater than 100°C (Cheremisinoff et al. 1982).

Continuous air monitoring during normal working hours, suggested for industrial and hazardous waste impoundments (Cheremisinoff et al. 1982), would be a minimum requirement at hazardous waste sites with potentially significant vapor phase hazards. Sites with known significant vapor phase hazard would require continuous air monitoring until hazard mitigation is complete.

Cheremisinoff et al. (1982) indicate that a minimum air monitoring program should include the following analyses:

1. Organic compounds with boiling points greater than 300°C, from particulate filter samplers.

2. Total Chromatographable Organic (TCO) analysis for compounds with boiling points from 100 to 300°C from solid sorbent samplers.

3. Mass Spec analysis for inorganic/element determinations of materials collected on particulate filter samplers.

4. Analysis for specific compounds of concern based on specific site information via acceptable EPA or NIOSH standard procedures.

Specific constitutent analysis would be a requirement for treatment effective evaluation and would indicate parent compound destruction and reaction product formation during implementation of on-site hazard mitigation procedures.

REFERENCES

Bordner, R., J. Winter, and F. Scarpino. 1978. Microbiological methods for monitoring the environment. EPA-600/8-78-017. U.S. Environmental Protection Agency, Washington, D.C.

Brown, K. W. and Associates, Inc. 1980. Hazardous waste land treatment. SW-874 (Draft) U.S. Environmental Protection Agency, Cincinnati, OH.

Cheremisinoff, N. P., P. N. Cheremisinoff, F. Ellerbush, and A. J Perna. 1982. Industrial and hazardous waste impoundments. Ann Arbor Science Publishing, Inc., Ann Arbor, MI.

Dunlap, W. J., J. F. McNabb, M. R. Scalf, and R. L. Cosby. 1977. Sampling for organic chemicals and microorganisms in the subsurface. EPA-600/2-77-176. U.S. Environmental Protection Agency, Washington, D.C.

Geraghty and Miller, Inc. 1977. The prevalence of subsurface migration of hazardous chemical substances at selected industrial waste land disposal sites. EPA/530/SW-634, U.S. Environmental Protection Agency, Washington, D.C.

Scalf, M. R., J. F. McNabb, W. J. Dunlap, R. J. Cosby and J. Fryberger. 1981. Manual of ground-water sampling procedures. National Water Well Association., Worthington, OH: 93 p.

Seiber, J. N., and J. E. Woodrow. 1983. Methods for studying pesticide atmospheric dispersal and fate at treated areas. Pesticide Reviews 85:217-229.

U.S. Army Corps of Engineers. 1982. Preliminary guidelines for selection and design of remedial systems for uncontrolled hazardous waste sites. EM 1110-2-(Draft). Department of the Army, Corps of Engineers, Washington, D.C.

U.S. Department of the Interior, Bureau of Reclamation. 1977. Groundwater manual. U.S. Government Printing, Washington, D.C.

U.S. Environmental Protection Agency. 1977. Procedures manual for groundwater monitoring at solid waste disposal facilities. SW-616. U.S. Environmental Protection Agency, Washington, D.C.

U.S. Environmental Protection Agency. 1982a. Handbook for remedial action at waste disposal sites. EPA-625/6-82-006. U.S. Environmental Protection Agency, Cincinnati, OH.

U.S. Environmental Protection Agency. 1982b. Ground-water monitoring guidance for owners and operators of interium status facilities. SW-963. U.S. Environmental Protection Agency, Washington, D.C.

U.S. Environmental Protection Agency. 1982c. Test methods for evaluating solid waste. SW-846. U.S. Environmental Protection agency, Washington, D.C.

U.S. Environmental Proection Agency. 1983. Hazardous waste land treatment. SW-874, U.S. Environmental Protection Agency, Cincinnati, OH.

Vanhoff, J. A., K. U. Weyer, and S. H. Whitaker. 1979. Discussion of "A multilevel device for ground-water sampling and piezometric monitoring by J. F. Pickens, J. A. Cherry, G. E. Grisak, W. F. Merritt, and B. A. Rizto." Groundwater 17(4):391-393.

Wilson, L. G. 1980. Monitoring in the vadose zone: A review of technical elements and methods. EPA-600/7-80-134, U.S. Environmental Protection Agency, Las Vegas, NV.

3. Characterization and Evaluation of Fundamental Processes in Soil/Waste Systems

SITE AND SOIL FACTORS RELATED TO IN SITU TREATMENT

Introduction

Identification of site characteristics is necessary before the initiation of in situ remedial actions for treatment of hazardous waste contaminated soils for three reasons. First, residual hazardous constituents are of public health concern because of their ultimate ability to contaminate the atmosphere, through volatilization or resuspension, and the hydrosphere, through leaching and runoff (Dawson and Brown 1981). Thus to protect the public health, both route characteristics by which the contaminant migrates off-site, and off-site receiver characteristics must be considered. Route characteristics determine the potential for contamination, while receiver characteristics and the corresponding degree of public health hazard indicate the time frame in which the remedial action must be performed. Site characterization also may serve to elucidate aspects of site modification or management to improve protection of human health.

In determining the public health hazard potential, both site and waste characteristics must be integrated. For example, if site characteristics indicate an immediate public health hazard due to high potential for off-site migration of an extremely hazardous chemical, an expensive but rapidly acting chemical in situ treatment technique may be more appropriate than a slower acting, less expensive technique utilizing natural soil processes.

Figure 3-1 depicts transport, decomposition, and immobilization processes influencing the migration of hazardous constituents in the environment. Hazardous compounds can be dispersed through the atmosphere via airborne particles or as gases. Human exposure occurs directly through dermal contact and inhalation of particles or gas or indirectly through deposition on crops or bioaccumulation in grazing game and agricultural animals, either or both of which may be ingested by humans. Hazardous compounds may reach surface waters in runoff, either dissolved or suspended in water or adsorbed to eroding soil particles. Movement through the soil may occur as liquid or gas or dissolved in soil water both in lateral and vertical directions to ground and surface waters. Human contact occurs through ingestion of the contaminated water. For purposes of this manual, detoxification by plants or removal in vegetation are not considered as means of in situ treatment.

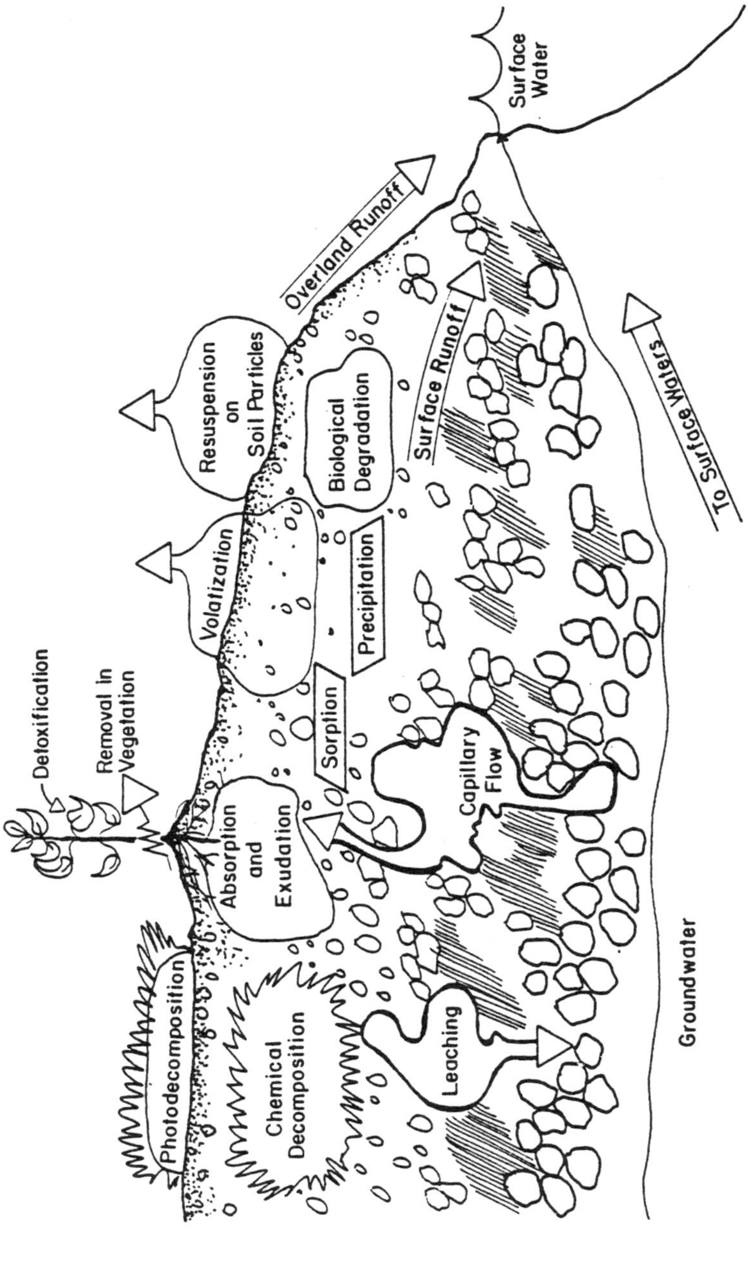

Figure 3-1. Processes influencing the migration of hazardous constituents in the terrestrial environment (adapted from Weber et al. 1973).

Decomposition and transformation of compounds may occur by biological processes, by chemical means (e.g. oxidation, reduction, hydrolysis) or by photodecomposition at the soil surface. Degradation refers to changes from hazardous to less hazardous or innocuous substances, while transformation refers to changes from hazardous to less toxic or innocuous substances. Compounds may be also immobilized by adsorption to soil particles, which may either be organic materials, such as humus or inorganic materials, such as the clays montmorillonite, vermiculite, or the hydrous oxides.

Secondly, the site must be evaluated in terms of its potential for decreasing the degree of hazardous of contaminants by degradation, transformation and immobilization. This aspect of site characterization focuses on soil physical, chemical, and biological properties and on climatological variables. The choice of a particular in situ treatment technology or train of treatments and its (their) effectiveness is often highly dependent on site/soil characteristics as well as on waste characteristics.

The third aspect of site characterization concerns the actual physical execution of the treatment techniques. Trafficability of the site (steep slopes, excessively wet conditions, soils high in sticky, plastic clays, etc.) may affect the choice of an appropriate treatment or the manner in which the in situ treatment is conducted, or may even preclude the use of any in situ treatment method. Appropriate site modifications and management options to enhance treatment are dependent upon the existing site conditions.

In many instances, data needs for off-site migration potential, treatment choice and effectiveness, and treatment execution may overlap, but the use of particular types of data should be kept in mind while investigating and characterizing a site. This will ensure that data are collected in an integrated manner and that necessary data are collected. At sites where previous remedial actions have removed acute hazards but contaminated soils with low levels of hazardous constituents remain, a preliminary assessment of available information may indicate that existing data are sufficient and few or no new characterization efforts are required.

<u>Site Characterization Related to Off-site Migration</u>

Site characteristics that relate to potential for off-site migration of hazardous compounds include a) site location and topography; b) soil properties; c) geological factors; d) hydrogeological factors; and 3) climatological factors.

Site Location and Topography--

Potential for contaminant migration due to soil permeability, depth to groundwater, erodibility, and flooding potential can sometimes be predicted by knowledge of the type of landform on which the hazardous waste site is located (Table 3-1). The type of landform may also indicate possible site modifications necessary to minimize migration (Phung et al. 1978). For example, on upland crests and valley side landforms, surface water is limited to incident precipitation and controllable off-site runoff. These landforms may require the diversion of surface waters to reduce the amount of water entering and infiltrating the site. However, upland crests or valley sides

TABLE 3-1. LANDFORMS AND TOPOGRAPHY OF HAZARDOUS WASTE SITES AS RELATED TO POTENTIAL FOR MIGRATION OF HAZARDOUS CONSTITUENTS (ADAPTED FROM RYAN AND LOEHR 1981)

Landscape Position: Topography	Migration Potential	
	Groundwater	Erosion
Eskers: long, low, narrow steep-sided ridges	high permeability	steep slopes
Kames: low, long, steep-sided ridges	high permeability	steep slopes
Outwash plains: near level broad tracts gently sloping from origin	potential high watertable, especially on fringe	
Terraces: flat areas with stair-stepped development, commonly between river and upland:		
glacial	high permeability	
marine	high to moderate permeability	
lake	low permeability	
Lake beds: broad, exceptionally flat	low permeability high watertable	
Till plains: young-broad, gently rolling	high watertable	
old-broad, level areas	moderately deep watertable	
Alluvial fans: smooth moderate slopes, transitional area between highlands and lowlands	high permeability potential high watertable at bottom of fan	moderate slopes
Playas: broad, exceptionally flat surfaces	low permeability	
Loess: undulating topography with smoothly rounded convex hills	moderate permeability	high erosion potential
Moraine	highly heterogeneous - more investigation required	

TABLE 3-1. (continued)

Landscape Position: Topography	Migration Potential Groundwater	Erosion
Floodplain		flooding
Delta		flooding
Beach ridges	high groundwater	
Coastal plains	highly heterogeneous - more investigation required	
Tidal flat		flooding
Sand dunes		wind erosion potential

may pose a high hazard to groundwater since they are often in groundwater recharge areas.

Upland flat areas with fine-grained soils of low permeability pose less risk of high groundwater and erosion and have greater attenuation capacities than terrace landforms. Terrace landforms are often underlain by highly permeable coarse-grained soils, sometimes at shallow depths. Contamination from these sites may occur at nearby surface expressions of underlying groundwater. The probability of groundwater intersecting a terrace site increases as the site position approaches either the valley wall or the level of the modern floodplain.

Warner (1976) describes four site conditions where pollution potential is especially high. They are as follows:

1. Sloping sites with relatively impermeable bedrock (e.g. shale, dense limestone, crystalline igneous rock) 0.6 m or less from the surface --high potential for erosion, seepage and overland flow of contaminated effluent.

2. Sites located in karst topography, with clayey residual soils overlying limestone or dolomite with fracture and solution porosity and permeability - high potential for contamination of groundwater, for though infiltration into soil itself is slow, liquids can rapidly enter bedrock where soil is absent, creating sinkholes and paths for direct flow into groundwater systems.

3. Sites with little topographic relief where groundwater table is at or very near the surface (e.g. old lake beds, floodplains) - high groundwater pollution potential.

4. Sites with fractured bedrock and a shallow soil depth (e.g. in granitic areas) - high groundwater pollution potential.

Soil Characteristics--

Important soil hydraulic, physical, and chemical properties that affect the migration through the vadose zone of hazardous constituents to groundwater or off-site in runoff must also be characterized. The vadose zone is the region extending from the ground surface to the upper surface of the principal water-bearing formation (Everett et al. 1982). In this zone, water in pore spaces primarily coexists with air, though saturated regions may occur. Perched water tables may develop at interfaces of layers with greatly different textures. Prolonged infiltration may also result in saturated conditions. The vadose zone usually consists of topsoils, which are weathered geological materials usually 3 to 6 feet deep, arranged in more or less well-developed profiles. Water movement in the topsoil is usually unsaturated, with soil water at less-than atmospheric pressure. Weathered topsoil materials gradually merge with underlying earth materials, which may include residual or transported clays or sands. The topsoil differs from the material lying below it in that it is more weathered, contains organic matter and the biological life associated with organic matter, and is the zone of plant-root growth. The entire vadose zone may be hundreds of feet thick and the travel time of pollutants hundreds or thousands of years, while other regions may be underlain by shallow potable aquifers which are especially susceptible to contamination due to short travel times and reduced potential for pollutant attenuation.

Those soil characteristics that affect water movement, i.e. infiltration and permeability, and those factors that affect contaminant mobility are the most important. It must be noted, however, that certain waste characteristics may affect natural infiltration and permeability of the soil and such interactions must be considered (see section 4). The site/soil properties that should be characterized are given in Table 3-2. If such a complete soil description can be obtained, predictions for potential migration can be fairly accurately defined.

Soil Classification--

There are three systems under which soils are most likely to have been classified in the United States (Fuller 1978): the Unified Soil Classification System (USCS), the 1938 U. S. Department of Agriculture System (USDA) and the present (1960, 1968) U. S. Department of Agriculture System.

The USCS was developed to describe engineering properties of soils. (Fuller 1978). Classification of soil types are based on particle (grain) sizes and response to physical manipulation at various water contents. An abbreviated description of the system (not including information on manipulation (liquid limit and plasticity index)) is given in Table 3-3 (Fuller 1978).

TABLE 3-2. SITE AND SOIL CHARACTERISTICS IDENTIFIED AS IMPORTANT IN IN SITU TREATMENT

Site location/topography and slope
Soil type, and extent
Soil profile properties
boundary characteristics
depth
texture*
amount and type of coarse fragments
structure*
color
degree of mottling
bulk density*
clay content
type of clay
cation exchange capacity*
organic matter content*
pH*
Eh*
aeration status*
Hydraulic properties and conditions
soil water characteristic curve
field capacity/permanent wilting point
water holding capacity*
permeability* (under saturated and a range of unsaturated conditions)
infiltration rates*
depth to impermeable layer or bedrock
depth to groundwater,* including seasonal variations
flooding frequency
runoff potential*
Geological and hydrogeological factors
subsurface geological features
groundwater flow patterns and characteristics
Meteorological and climatological data
wind velocity and direction
temperature
precipitation
water budget

*Factors that may be managed to enhance soil treatment.

TABLE 3-3. MAJOR DIVISIONS, SOIL TYPE SYMBOLS, AND TYPE DESCRIPTIONS FOR THE UNIFIED SOIL CLASSIFICATION SYSTEM (USCS)(FULLER 1978)

Major Divisions			Symbol	Description
Coarse-grained soils More than half of material is larger than No. 200 sieve size	Gravels More than half of coarse fraction is larger than No. 4 sieve size	Clean Gravel (little or no fines)	GW	Well graded gravels, gravel-sand mixtures, little or no fines.
			GP	Poorly graded gravels or gravel-sand mixtures, little or no fines.
		Gravels with Fines (appre-ciable fines)	GM	Silty gravels, gravel-sand-silt mixture.
			GC	Clayey gravels, gravel-sand-clay mixtures.
	Sands More than half of coarse fraction is smaller than No. 4 sieve size	Clean Sands (little or no fines)	SW	Well graded sands, gravelly sands, little or no fines.
			SP	Poorly graded sands or gravelly sands, little or no fines.
		Sands with Fines (appre-ciable fines)	SM	Silty sands, sand-silt mixtures.
			SC	Clayey sands, sand-clay mixtures.
Fine-grained soils More than half of material is smaller than No. 200 sieve size	Silts and Clays Liquid limit is less than 50		ML	Inorganic silts & very fine sands, silty or clayey fine sands or clayey silts with slight plasticity.
			CL	Inorganic clays of low to medium plasticity, gravelly clays, sandy clays, silty clays, lean clays.
			OL	Organic silts and organic silty clays of low plasticity.
	Silts and Clays Liquid limit is greater than 50		MH	Inorganic silts, micaceous or diatomaceous fine sandy or silty soils, elastic silts.
			CH	Inorganic clays of high plasticity fat clays.
			OH	Organic clays of medium to high plasticity, organic silts.
Highly Organic Soils			P	Peat & other highly organic soils.

Notes: ML includes rock flour. The No. 4 sieve opening is 4.76 mm (0.187 in.); the No. 200 sieve opening is 0.074 mm (0.0029 in.).

The USDA system was developed for agricultural and other land management uses and is based on both chemical and physical properties of the soil. The first system (1938-1960) was based on soil genesis, i.e., how soils formed or were thought to have formed, while the present system is based on quantitatively measurable properties of soils as they exist in the field. The present system is being constantly refined but is in general use by U. S. soil scientists. The highest level of the present USDA system is described in Table 3-4, and the approximate equivalents in the 1938 USDA system are shown in Table 3-5.

Fuller (1978) developed a comparison of the USDA classification and the USCS. The part of the USDA system that can be compared most directly in the USCS system is soil texture and associated modifiers (e.g. gravelly, mucky). The size ranges for the USDA and the USCS particle designations are shown in Table 3-6. The two systems are not directly comparable. The soil texture designation in the USDA system is based only on the amounts of sand-silt-and clay-sized particles in the soil (Figure 3-2). In the USCS, the soil type is determined both on the amounts of certain sizes of soil particles and on the response of the soil to physical manipulation at varying water contents. Correlations of USDA soil textures and USCS soil types are presented in Tables 3-7 and 3-8. Correlations between USCS soil types and other parts of the USDA system would not be possible. Texture is a major criterion in the USCS but only a minor criterion in the USDA system. A soil of a given texture can be classified into only a limited number of the 15 USCS soil types. However, in the USDA system, soils of the same texture may be found in many of the 10 orders and 43 suborders of the system because of differences in their chemical properties or the climatic areas in which they are located.

The complete USDA soil classification system follows the pattern of (Anderson 1977):

1. orders
2. suborders
3. great groups
4. subgroups
5. family
6. series
7. types

Each level is a subdivision of the preceding classification level. The orders are based on soil morphology and development similarity. Suborders emphasize genetic homogeneity. Great groups are based on similarity in diagnostic surface and subsurface horizons. A subgroup is based on general similarities of profiles of soils located within a large area. Families are based on properties important to the growth of plants, such as texture, mineral composition, and soil temperature. A soil series is composed of soils with similar but not identical profiles but with different surface layer textures. A soil type consists of soils with similar surface texture, and may be divided into phases based on some prominent deviation (e.g. slope, stoniness, erosion, or soluble salt content).

TABLE 3-4. DESCRIPTIONS OF SOILS IN THE HIGHEST (MOST GENERAL) CATEGORIES OF THE PRESENT USDA CLASSIFICATION SYSTEM (FULLER 1978)

ORDER	DESCRIPTION
Alfisols	Alfisols have a clayey subsoil horizon and moderate to high base (cation) saturation. Water is held above the wilting point during at least three of the warm months of the year. Alfisols are higher in hydroxy-oxides (sesquioxides) than most soils, as the name implies, and therefore may have a fragipan, duripan, sodium horizon, petrocalcic (lime), and plinthite (iron oxides or sesquioxides) or other similar features which separate them from other soils. Where the temperature is moderate to cool, Alfisols form a belt between the Mollisols of the grasslands, and Spodosols and Inceptisols of very humid climates. In warmer climates, Alfisols form a belt between Aridisols of arid regions and Inceptisols, Ultisols, and Oxisols of warmer climates. Leaching of bases from the soil may occur almost every year or may be infrequent.
Aridisols	Aridisols occur in arid climates. They do not have water available to mesophytic plants for long periods as do the Alfisols. Water is held at less than 15 bars or it is salty. A few Aridisols occur in semiarid climates because they take up water slowly and most of the rainfall runs off. Aridisols have one or more pedogenic horizons that may have formed in the present dry environment or that may be relics of a former pluvial period. The pedogenic horizon may be the result of translocation and accumulation of salts, lime, or silicate clays or of cementation by carbonates or silica. The pH usually is neutral alkaline, sometimes highly alkaline.
Entisols	Entisols do not have horizon or profile development, or at least no evidence of such. In many of the soils time has been too short for distinct horizons to differentiate. Other Entisols are on soil slopes too steep for water to pentrate well or where erosion rate exceeds development rate, still others are on flood or glacial outwash plains which continuously accumulate new alluvium. Some are wind-moved sand. Not all Entisols are young. Some are actually very old, consisting mostly of quartz sand which weathers very slowly. Such materials as organic matter, lime, gypsum, iron oxides, and clays do not accumulate or at least only to a very small extent.
Histosols	Most Histosols are saturated or nearly saturated with water most of the year. They are high in organic matter and represent what is often described as mucks, peat bogs, high moors, or raised peats.

TABLE 3-4. (CONTINUED)

ORDER	DESCRIPTION
Inceptisols	By USDA definition (1968), in part at least, "Inceptisols are soils of humid regions that have altered horizons which have lost bases or iron and aluminum but retain some weatherable minerals. They do not have an illuvial horizon enriched either with silicate clay that contains aluminum or with an amorphous mixture of aluminum and organic matter." This is a difficult soil order to visualize from the description. It represents more what other soils are not than what they are. Inceptisols develop mainly in the more clayey parent materials, in contrast to Spodosols, which develop in materials which have little clay.
Mollisols	Very dark colored soils, rich in bases and naturally covered by grass (steppe land) are called Mollisols. Many soils of this order accumulate lime and/or sodium, and clay. Mollisols occupy extensive subhumid to semiarid areas of the grass plains in the USA. They are located generally between the Aridisols of arid climates and the Spodosols or Alfisols of humid climates. These soils are highly productive, constituting rich crop land in the breadbasket areas of central and west central U.S. Areas of Mollisols appear in nearly every state. Luxuriant, perennial grass seems to be essential to their formation. Where waste waters and leachates are applied to the surface, the high soil organic matter plays an important part in mobility of trace and heavy elements. Where leachate and aqueous wastes pass through subsoils only, lime and bases influence mobility in addition to the effect of clay minerals.
Oxisols	Those soils which once were called red and yellow laterites are now named Oxisols. Reddish, yellowish, or grayish soils of tropical and subtropical climates that form on mostly gentle slopes on surfaces of great age are Oxisols. They are mixtures of quartz, kaolin, free oxides and organic matter. The boundaries of horizons blend into each other so gradually they are generally arbitrary. Weathering has proceeded to great depths. Water moves through these soils rapidly. Because of the high oxide (primarily iron) content that coats particles and forms granular particulates, these soils attenuate the trace and heavy metals very well. Oxisols occur in Hawaii.
Spodosols	Two well defined and obvious-to-the-eye horizons distinguish Spodosols. Just below the surface layer of forest litter and partly decomposed dark organic matter is a bleached layer of uncoated quartz sand. The second layer, just below, usually is coffee color. Organic matter and iron complexes accumulate to give the dark brown color. However, this spodic is one in which amorphous mixtures of organic matter and aluminum may also occur with or without iron.

TABLE 3-4. (CONTINUED).

ORDER	DESCRIPTION
	Spodosols generally are coarse textured, containing only small quantities of clay, if any, and usually permit rapid water movement. These soils occur under high rainfall conditions and coniferous forest, though sometimes hardwoods are present. Attenuation is poor despite the spodic horizon where aluminum, and often iron, complexes with organic matter. Usually the iron and aluminum content is low, though, even when a fragipan (soft when wet, brittle when dry) is present. The textures are mostly sandy, sandy-skeletal, coarse loam, loamy-skeletal, and coarse-silty. New England, New York, Northern lake states, and Alaska are most noted for spodic (Podzolic) soils.
Ultisols	The concept of Ultisols is that of soils of mid-to-low latitudes that have a horizon that contains an appreciable amount of translocated silicate clay but few bases. Highly humid conditions sometime during the year cause water to move through them. Ultisols are most commonly found in warm-humid climates that have a seasonal deficit of rain and on older surfaces. They develop from a wide variety of parent materials. Kaolin, gibbsite, and aluminum-interlaid clays are common in the soil clay fraction. They usually form under coniferous and hardwood-forest vegetation in the United States.
Vertisols	Vertisols are clayey soils which crack severely when dry and have high bulk densities between the cracks. The clay minerals are dominated by montmorillonite. Most are found under warm climatic conditions, i.e., thermic or warmer. In arid regions they form in closed depressions or playas. Vertisols often are referred to as churning soils because during swelling, pressure is exerted, causing them to heave and recycle the soil.

TABLE 3-5. ORDERS IN THE PRESENT USDA SOIL CLASSIFICATION SYSTEM AND APPROXIMATE EQUIVALENTS IN THE 1938 USDA system (Fuller 1978)

Present Order[a]	Approximate Equivalents[b]
1. Entisols	Azonal soils, and some Low-Humic Gley soils
2. Vertisols	Grumusols
3. Inceptisols	Ando, Sol Brun Acide, some Brown Forest, Low-Humic Gley, and Humic Gley soils
4. Aridisols	Desert, Reddish Desert, Sierozem, Solonchak, some Brown and Reddish Brown soils, and associated Solonetz
5. Mollisols	Chestnut, Chernozem, Brunizem (Prairie), Rendzinas, some Brown, Brown Forest, and associated Solonetz and Humic Gley soils
6. Spodosols	Podzols, Brown Podzolic soils, and Groundwater Podzols
7. Alfisols	Gray-Brown Podzolic, Gray Wooded soils, Non-calcic Brown soils, Degraded Chernozem, and associated Planosols and some Half-bog soils
8. Ultisols	Red-yellow Podzolic soils, Reddish-brown Lateritic soils of the U.S., and associated Planosols and Half-bog soils
9. Oxisols	Laterite soils, Latosols
10. Histosols	Bog soils

[a]Present (1960, 1968) USDA comprehensive soil classification.

[b]Old (1938) USDA soil classification.

TABLE 3-6. U.S. DEPARTMENT OF AGRICULTURE (USDA) AND UNIFIED SOIL CLASSIFICATION SYSTEM (USCS) PARTICLE SIZES (FULLER 1978)

USDA		USCS	
Particle	Size Range (mm)	Particle	Size Range (mm)
Cobbles	76.2-254	Cobbles	>76.2
Gravel	2.0-76.2	Gravel	4.76-76.2
Coarse gravel	12.7-76.2	Coarse gravel	19.1-76.2
Fine gravel	2.0-12.7	Fine gravel	4.76-19.1
Sand	0.05-2.0	Sand	0.074-4.76
Very coarse sand	1.0-2.0	Very coarse sand	
Coarse sand	0.5-1.0	Coarse sand	2.0-4.76
Medium sand	0.25-0.5	Medium sand	0.42-2.0
Fine sand	0.1-0.25	Fine sand	0.074-0.42
Very fine sand	0.05-0.1	Very fine sand	
Silt	0.002-0.05	Fines[a]	<0.074
Clay	<0.002	(Silt and clay)	

[a]USCS silt and clay designations are determined by response of the soil to manipulation at various water contents rather than by measurement of size.

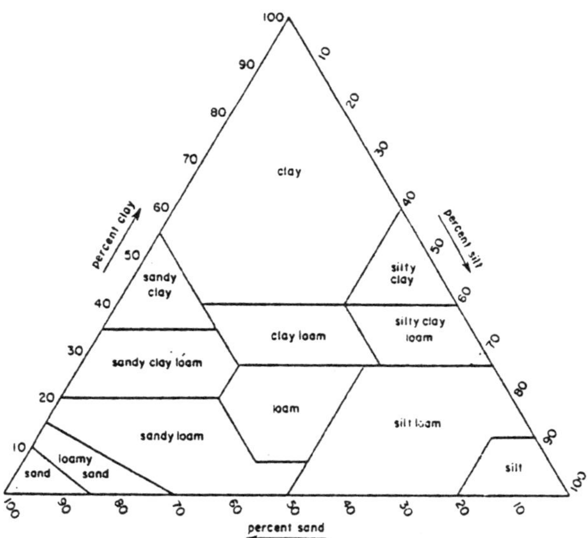

Figure 3-2. USDA soil textural classification (Fuller 1978).

TABLE 3-7. CORRESPONDING USCS AND USDA SOIL CLASSIFICATIONS (FULLER 1978)

Unified Soil Classification System (USCS) Soil Types	Corresponding United States Department of Agriculture (USDA) Soil Textures
1. GW	Same as GP--gradation of gravel sizes not a criteria.
2. GP	Gravel, very gravelly[a] sand less than 5% by weight silt and clay.
3. GM	Very gravelly[a] sandy loam, very gravelly[a] loamy sand very gravelly[a] silt loam, and very gravelly[a] loam[b].
4. GC	Very gravelly clay loam, very gravelly sandy clay loam, very gravelly silty clay loam, very gravelly silty clay, very gravelly clay[b].
5. SW	Same--gradation of sand size not a criteria.
6. SP	Coarse to fine sand; gravelly sand[c] (less than 20% very fine sand).
7. SM	Loamy sands and sandy loams (with coarse to fine sand), very fine sand; gravelly loamy sand[c] and gravelly sandy loam[c].
8. SC	Sandy clay loams and sandy clays (with coarse to fine sands); gravelly sandy clay loams and gravelly sandy clays[c].
9. ML	Silt, silt loam, loam very fine sandy loam[d].
10. CL	Silty clay loam, clay loam, sandy clays with <50% sand[d].
11. OL	Mucky silt loam, mucky loam, mucky silty clay loam, mucky clay loam.
12. MH	Highly micaceous or diatomaceous silts, silt loams -- highly elastic.
13. CH	Silty clay and clay[d].
14. OH	Mucky silty clay.
15. PT	Muck and peats.

[a]Also includes cobbly, channery, and shaly.
[b]Also includes all of textures with gravelly modifiers where >1/2 of total held on No. 200 sieve is of gravel size.
[c]Gravelly textures included if > 1/2 of total held on No. 200 sieve is of gravel size.
[d]Also includes all of these textures with gravelly modifiers where >1/2 of the total soil passes the No. 200 sieve.

TABLE 3-8. CORRESPONDING USDA AND USCS SOIL CLASSIFICATIONS (FULLER 1978)

United States Department of Agriculture (USDA) Soil Textures	Corresponding Unified Soil Classification System (USCS) Soil Types
1. Gravel, very gravelly loamy sand	GP, GW, GM
2. Sand, coarse sand, fine sand	SP, SW
3. Loamy gravel, very gravelly sandy loam, very gravelly loam	GM
4. Loamy sand, gravelly loamy sand, very fine sand	SM
5. Gravelly loam, gravelly sandy clay loam	GM, GC
6. Sandy loam, fine sandy loam, loamy very fine sand, gravelly sandy loam	SM
7. Silt loam, very fine sandy clay loam	ML
8. Loam, sandy clay loam	ML, SC
9. Silty clay loam, clay loam	CL
10. Sandy clay, gravelly clay loam, gravelly clay	SC, GC
11. Very gravelly clay loam, very gravelly sandy clay loam, very gravelly silty clay loam, very gravelly silty clay and clay	GC
12. Silty clay, clay	CH
13. Muck and peat	PT

Identification of kinds of soils at a site in terms of the higher categories of the classification system can provide information that is relevant to identifying pollutant attenuation and migration. The names of soils from the order to subgroup level are composed of a series of formative elements, which can be used to predict many soil properties relevant to in situ teatment of wastes. Table 3-9 illustrates the kinds of information that can be inferred from soil names using the order Mollisol as an example. The reader is referred to Soil Taxonomy (Soil Survey Staff 1975) for a comprehensive description of the USDA soil classification system.

Slope--
The type and degree of slope indicates surface drainage problems. Concave slopes cause surface runoff to converge while convex slopes disperse runoff (U.S. Environmental Protection Agency 1980).

Soil Profile Properties--
In climates where there is sufficient rainfall, soils become highly organized into layers (horizons). A soil profile is a vertical cross-section of the soil, made up of several horizons, each having its own distinctive characteristics. Descriptions of soil profile characteristics made in the field are especially valuable, because predictions about waste management can be made more accurately.

Soil Horizons--
Soil horizons are designated A, B, and C to represent the surface soil, subsoil, and substratum, respectively. The A and B horizons are formed by weathering and other soil-forming processes. The C horizon is usually the parent material or undifferentiated geological deposits from which the A and B horizons developed and is unaltered by soil forming processes (e.g. sediment from ancient sea and lake beds, loess blown from dry floodplains, rocks and rock powder released from melting glaciers, alluvium from flooding streams). Not all soils have all three horizons, while many soils show variations within each master horizon. A hypothetical soil profile with all the common soil horizons is illustrated in Figure 3-3. Each horizon can have significantly different characteristics such as depth, texture, structure, bulk density and chemical properties which will result in differing drainage and pollutant attenuative characteristics.

Boundary Characteristics--
Boundary characteristics, especially abrupt changes in texture and structure (e.g. clay or sand layers, hard pans) can adversely affect vertical percolation of water. Horizontal flow may result from textural discontinuities, thus resulting in contamination of adjacent areas.

Depth--
The depth of the soil profile is important for attenuation of hazardous constituents moving through the soil zone, with greater depths being more desirable. K.W. Brown and Associates, Inc. (1980) suggest a depth of three times the depth of incorporation of a hazardous waste or 1.4 meters, whichever is greater, as appropriate for treatment of hazardous wastes. Federal regulations for the land treatment of hazardous wastes also require 1.4 m (Federal Register 1982). In general, the greater the depth of the soil profile, the

TABLE 3-9. INFORMATION THAT CAN BE INFERRED FROM THE USDA SOIL CLASSIFICATION SYSTEM, USING THE ORDER MOLLISOL AS AN EXAMPLE. (FORMA- ELEMENTS FOR THE CLASSIFICATION LEVEL ARE UNDERLINED) (ADAPTED FROM AGRICULTURAL RESEARCH SERVICE 1974)

Soil Classification Level	Example	Information Relevant to In Situ Treatment
Order	Mollisol	Fertile, high in organic matter
Suborder	Boroll	Long cold, winters; soils frozen for extended periods; only short time periods that soil is warm enough for biological activity
	Aquoll	Naturally wet; often develops in low places where water collects and stands
	Xeroll	Dry for extended periods in summer; in most, moisture moves through soil in underlying layers in winter
Great Group	Calcixeroll	Strong concentration of calcium carbonate or gypsum at some depth above 60 inches; calcareous in all parts above that depth unless texture is sandy
	Argixeroll	Horizon of clay accumulation in subsoil; most with horizon of carbonate accumulation below that
Subgroup	Aquic Argixeroll	Moderately shallow groundwater at some times of the year unless artificially drained
	Lithic Argixeroll	Hard rock at shallow depths
	Vertic Argixeroll	Fine-textured soils with deep, wide cracks at some period in most years
	Typic Argixeroll	Lacking features or combination of features specified for other subgroups

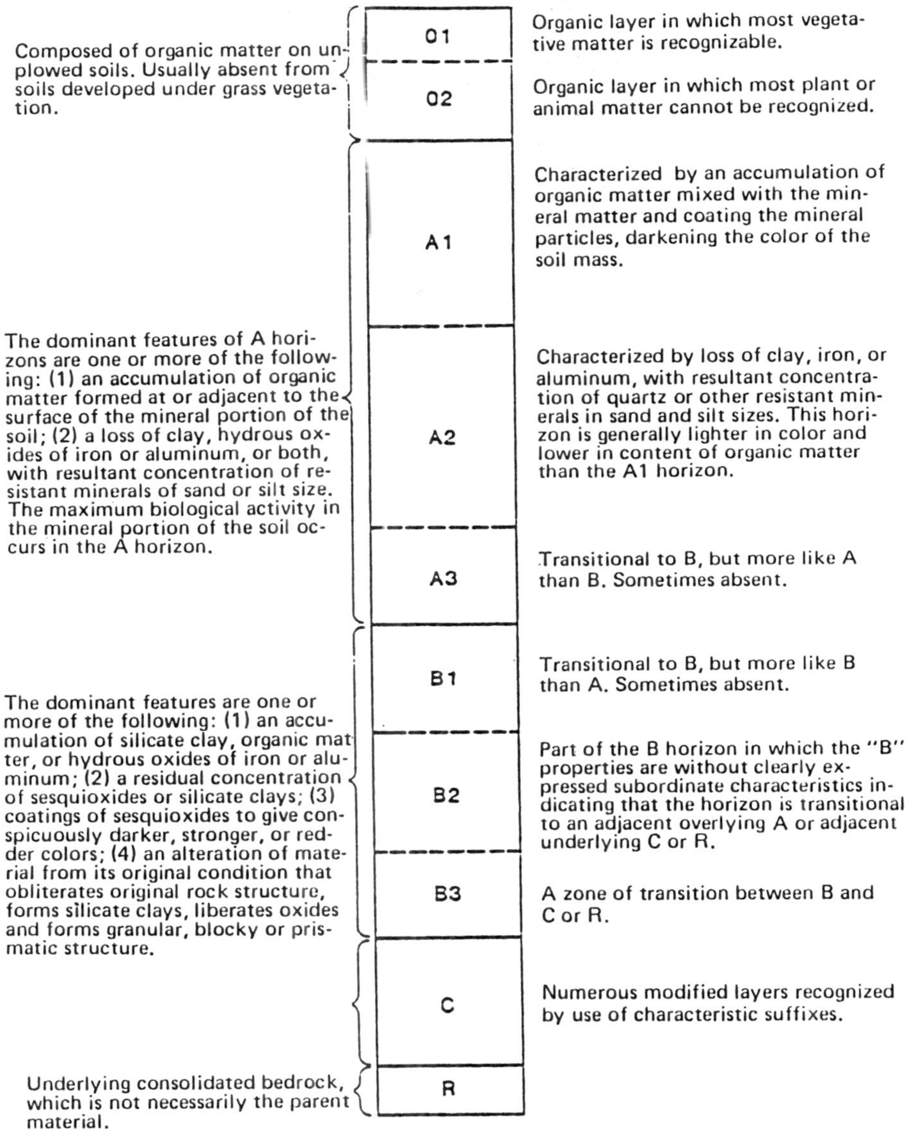

Figure 3-3. A hypothetical soil profile illustrating the common soil horizons (Otis 1983). Used by permission, see Copyright Notice.

greater the protection against migration of hazardous constituents to groundwater and greater the soils attenuation properties.

Soil Texture--

Soil texture is a description of the particle size distribution of various soil separates (particles less than 2 mm). The USDA soil texture system was shown in Figure 3-2, while a comparison of the USDA system and USCS system was shown in Table 3-8. K. W. Brown and Associates Inc. (1980) described soil characteristics important to hazardous waste treatment based on soil texture in the following manner:

> The sand and gravel particles are normally thought of as the coarse separates, and soils dominated by these coarse separates are usually of an open character, have low water-holding capacity, possess good drainage, high permeability and aeration, and are generally in a loose friable condition. In contrast to the coarse separates, the silt and clay particles are considered the fine separates. Silt particles are micro-sand particles predominantly composed of quartz and have some plasticity, cohesion, and adsorption. A high percentage of silt is undesirable and leads to physical problems such as soil crusting. Clay particles are very small, less than 0.002 mm in diameter, and therefore have a very high surface area. Clays are commonly flat or plate-like, highly plastic, cohesive, and have a very high adsorptive capacity for water, ions and gases. In general soils high in silt and clay are called fine textured soils and are very plastic, cohesive, high in water holding capacity, and have very slow water and air permeabilities. Each type of soil, fine or coarse, has various advantages and disadvantages for use in a waste disposal system. Coarse textured soils will remain oxidized while fine soils may often become reduced due to wetness. The oxidation state will play an important role in how quickly, by what pathway, and in what form various waste constituents will be degraded. Ammonia, for example, will be rapidly converted to nitrate in an oxidized coarse soil and, due to the potential for rapid water infiltration, may leach to the groundwater supply. Oxidation also affects the degradation of organics by regulating which and how many microbes can exist. Therefore, a given hydrocarbon will degrade via one pathway at a rapid rate when well oxidized, a slower rate when less well oxidized, and via a second pathway and rate when under reduced conditions. This is exemplified by the data of Gibson and Yeh (1973) who show that under aerobic conditions the primary pathway for hydrocarbon degradation is oxidation to form epoxides while under anaerobic conditions organic materials are mainly fermented into organic acids such as methanol. The high adsorptive capacity of the fine textured soils may, however, be very useful in holding various ions, such as heavy metals, in an immobile form and preventing their movement to water supplies or other undesirable areas.... In general, it can be suggested that hazardous waste land treatment facilities not be established on extremely deep sandy soils due to the potential for leaching to groundwater. Similarly, silty soils having a severe problem with crusting should not be selected due to the extreme

potential for runoff. In general, the loam, silt loam, clay loam, sandy clay loam, silty clay loam, silty clay, and sandy clay soils are best suited for land treatment of hazardous wastes.

A summary of the advantages of each type of soil texture in the USDA system is given in Table 3-10.

Soil Structure--

Soil structure refers to the aggregation of primary soil particles (sand, silt, and clay) into compound particles or clusters of primary particles, which are called peds. The aggregates, separated by surfaces of weakness or open planar voids, are often seen as cracks in the soil. The structures of the different horizons of a soil profile are essential characteristics of the soil profile just as are texture or chemical composition. Seven basic structure types are defined by the shape and arrangement of peds. These are illustrated in Figure 3-4. Soils may also have no structure: coarse textured soils with no texture are called single grain while fine textured soils are called massive (soil material clings together in large uniform masses). Aggregate formation is thought to result from the cementing action of soil colloidal matter (clay minerals, colloidal oxides of iron and manganese, and colloidal organic matter). Aggregation may vary in stability, changing in response to moisture content, chemical composition of the soil solution, biological activity, and management practices. Soils high in shrink/swell clays show particularly dramatic structural changes in response to changes in water content.

Structure may modify the influence of texture in regard to moisture and air relationships. Interpedal voids are often large and continuous compared to voids between the primary particles within the peds. For example, a soil with a high content of plastic clays would exhibit limited permeability if it did not have a well-developed structure which facilitates water and air movement. Aggregates in coarse textured soils stabilize the surface and increase water retention in the soil.

The type of structure determines the dominant direction of the pores and thus the direction of water movement. Platy structures restrict vertical percolation, prismatic and columnar enhance vertical percolation, and blocky and granular enhance percolation both vertically and horizontally (Otis 1983). Structural units that can withstand moderate handling without disintegrating will provide better hydraulic properties.

Color--

Color and color patterns in soil are good indicators of drainage characteristics of soil. Soil colors can be described in general terms (e.g. brown, gray, yellow, etc.) or by the Munsell color system, which characterizes a color in terms of hue, value, and chroma.

Uniform red, yellow, or brown colors indicate that a soil is well-drained and seldomly or never saturated with water. Gray or blue colors indicate that the soil is saturated continuously or for extended periods. Soils with spots or streaks of red, yellow, or black in a gray matrix ("mottled") are usually periodically saturated with water.

TABLE 3-10. SUITABILITY OF VARIOUS TEXTURED SOILS FOR LAND TREATMENT OF HAZARDOUS INDUSTRIAL WASTES (K. W. BROWN AND ASSOCIATES 1980)

Texture	Advantages	Disadvantages
Sand	Very rapid infiltration Usually oxidized & dry Low runoff potential	Very low CEC Very high hydraulic conductivity rate Low available water Little soil structure
Loamy sand	High infiltration Low to medium runoff	Low CEC Moderate to high hydraulic conductivity rate Low to medium available water
Loam	Moderate infiltration Fair oxidation Moderate runoff potential Generally accessible Good CEC	Fair structure
Silt loam	Moderate infiltration Fair oxidation Moderate runoff potential Generally accessible Good CEC	Some crusting Fair to poor structure
Silt	Low infiltration Fair to poor oxidation Good CEC Good available water	High crusting potential Poor structure High runoff
Silty clay loam	Medium-low percolation Fair structure High CEC	Med-low infiltration Some crusting potential
Silty clay	Good to high available water	Moderate runoff Often wet Fair oxidation
Clay loam	Medium-low percolation Good structure Med-poor aeration High CEC High available water	Med-low infiltration Mod-high runoff Often wet
Clay	Low percolation High CEC High available water	Low infiltration Often massive structure High runoff Sometimes low aeration
Sandy clay	Med-low percolation Med-high CEC	Fair structure Moderate-high runoff
Sandy clay loam	Med-high available water Good aeration	Medium infiltration

Structure Type	Aggregate Description	Diagrammatic Aggregate	Common Horizon Location
Granular	Relatively nonporous, small and spheroidal peds; not fitted to adjoining aggregates		A horizon
Crumb	Relatively porous, small and spheroidal peds; not fitted to adjoining aggregates		A horizon
Platy	Aggregates are platelike. Plates often overlap and impair permeability		A2 horizon in forest and claypan soils
Blocky	Blocklike peds bounded by other aggregates whose sharp angular faces form the cast for the ped. The aggregates often break into smaller blocky peds		Bt horizon
Subangular blocky	Blocklike peds bounded by other aggregates whose rounded subangular faces form the cast for the ped		Bt horizon
Prismatic	Columnlike peds without rounded caps. Other prismatic aggregates form the cast for the ped. Some prismatic aggregates break into smaller blocky peds		Bt horizon
Columnar	Columnlike peds with rounded caps bounded laterally by other columnar aggregates that form the cast for the peds		Bt horizon

Figure 3-4. Diagrammatic definition and location of various types of soil structure (Foth 1978). Used by permission, see Copyright Notice.

Mottles--

Mottles result from chemical and biochemical reactions when saturated conditions, organic matter, and temperatures above 4°C occur together in the soil. Bacteria utilizing organic matter deplete oxygen present in the soil. Other bacteria continue the organic decomposition using oxidized iron and manganese compounds instead of oxygen in their metabolism. The insoluble iron and manganese are reduced to soluble forms, causing the soil to lose its red, yellow, and brown color and turning it gray. When the soil drains, the soluble iron and manganese are carried with the water to the larger pores in the soil. When they, contact air-filled pores, they are reoxidized, forming insoluble compounds that accumulate as red, yellow or black spots near pore surfaces. The soil from which the compounds were removed remain gray (Otis 1983).

Bulk Density--

Bulk density is the mass of dry soil per unit bulk volume. As the density of a soil increases, the volume of pore space decreases. Density and pore space relationships determine the ease and amount of air and water stored in and moving through pore spaces. Of soils with the same texture, those with higher bulk densities are more dense with less pore volume and thus less permeable. Good structure and lower bulk densities promote good aeration and drainage. Soil bulk densities usually increase with depth due to less organic manner, less aggregation, and compression from the weight of the overlying soil (K.W. Brown and Associates, Inc. 1980). Sandy soils, with particles lying close together, exhibit high bulk densities, ranging from 1.2 to 1.8 g/cm^3. Finer textured soils with usually higher organic matter contents and more structure have more pore space and lower bulk densities, generally ranging from 1.0 to 1.6 g/cm^3 (Brady 1974).

Type and Amount of Soil Colloids--

Clays and organic matter--The colloidal fraction of a soil is of primary importance in the sorption and immobilization of hazardous organic and inorganic compounds. The colloidal fraction, composed of organic and inorganic particles with a maximum size of 0.001 mm, is the most chemically active portion of a soil. These particles are characterized by a large exposed surface or interface, a capacity to adsorb and hold solids, gases, ions, and polar compounds, and a tendency to hasten or retard chemical reactions by catalytic action (Anderson 1977). Of the mineral, inorganic portion of soils, only the clay particles are colloidal in size and even some clay particles are too large to be classified as colloidal. However, recognizing the exceptions, the term clay is used for the inorganic colloidal portion of soils. The organic colloidal fraction is composed of amorphous humus, derived from organic materials during their breakdown by microoganisms.

Inorganic soil colloids--The formation of soil inorganic colloids results from the weathering of minerals, which in mixtures form the rocks of the earth. Since oxygen and silicon make up about 75 percent of the earth's crust on a weight basis, most minerals are primary or secondary silicates. The average mineralogical composition of igneous and sedimentary rocks is shown in Table 3-11.

Weathering is basically a combination of destruction and synthesis (Brady 1974). Both mechanical factors (e.g. strains from temperature changes,

TABLE 3-11. AVERAGE MINERALOGICAL COMPOSITION OF IGNEOUS AND SEDIMENTARY ROCKS (FOTH 1978)
Used by permission, see Copyright Notice

Mineral Constituent	Origin	Igneous Rock, %	Shale, %	Sand-stone, %
Feldspars	Primary	59.5	30.0	11.5
Amphiboles and pyroxenes	Primary	16.8	-	a
Quartz	Primary	12.0	22.3	66.8
Micas	Primary	3.8	-	a
Titanium minerals	Primary	1.5	-	a
Apatite	Primary or secondary	0.6	-	a
Clay	Secondary	-	25.0	6.6
Limonite	Secondary	-	5.6	1.8
Carbonates	Secondary	-	5.7	11.1
Other minerals	-	5.8	11.4	2.2

[a]Present in small amounts.

pressures of freezing water, erosive action of water, wind, and ice and exfoliation) and chemical processes (e.g. hydration hydrolysis oxidation, solution, and carbonation) act in the process of weathering. Rocks are broken into smaller rocks and eventually into their individual minerals, many of which are primary silicates (i.e. silicate minerals not altered chemically because of deposition and crystallization from molten lava). At the same time, the rock fragments and minerals are changed to new secondary minerals either by minor alterations or complete chemical changes. These changes are accompanied by a continued decrease in particle size, increase in specific surface area (area per unit weight) and by the release of soluble constituents, most of which may be lost in drainage. An illustration of the weathering process is shown in Figure 3-5. The size, number, and surface area of soil particles from very coarse sand to clay are shown in Table 3-12.

The minerals which are synthesized are primarily layer silicate clays and very resistant end products, usually iron and aluminum oxides. Soils are predominately composed of these synthesized secondary minerals and very resistant primary minerals such as quartz. The sand and coarse silt fractions of soils are composed mainly of quartz and generally quite inactive chemically (Figure 3-6). Primary silicates such as the feldspars, amphiboles, and micas are present in sands but decrease in amount in silts. Secondary silicates dominate the clay fraction, while oxides of iron and aluminum are important in the fine silt and coarse clay fractions. The silicate clays are characteristic of temperate regions while the hydrous oxides are characteristic of the tropic

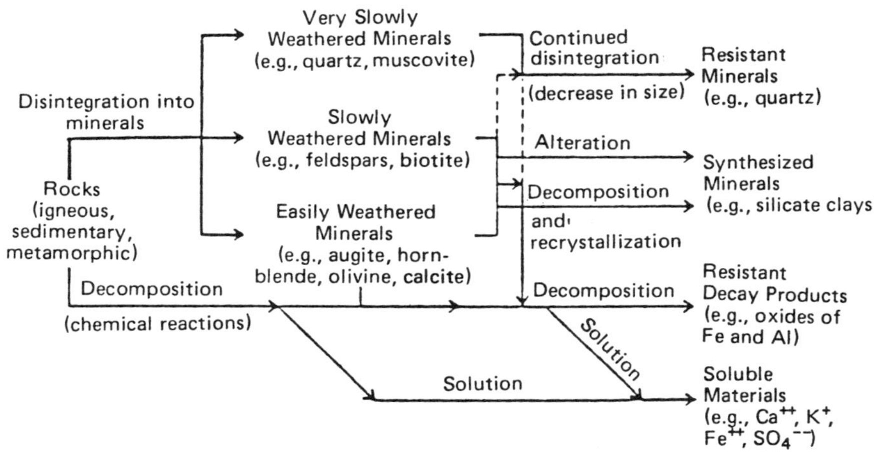

Figure 3-5. Weathering pathways which take place under moderately acid conditions common in humid temperature regions. (Major paths of weathering are indicated by the heavier arrows, minor pathways by broken lines.) (Brady 1974). Used by permission, see Copyright Notice.

TABLE 3-12. THE SIZE, NUMBER, AND SURFACE AREA OF SOIL PARTICLES (FOTH 1978)
Used by permission, see Copyright Notice

Particle type	Diameter (mm)	Number of particles/g[a]	Surface area (sq cm/g)
Very coarse sand	2.00-1.00	90	11
Coarse sand	1.00-0.50	720	23
Medium sand	0.50-0.25	5,700	45
Fine sand	0.25-0.10	46,000	91
Very fine sand	0.10-0.05	722,000	227
Silt	0.05-0.002	5,776,000	454
Clay	below 0.002	90,260,853,000	8,000,000

[a]Assumed to have spherical shapes, based on maximum diameter of the particle type.

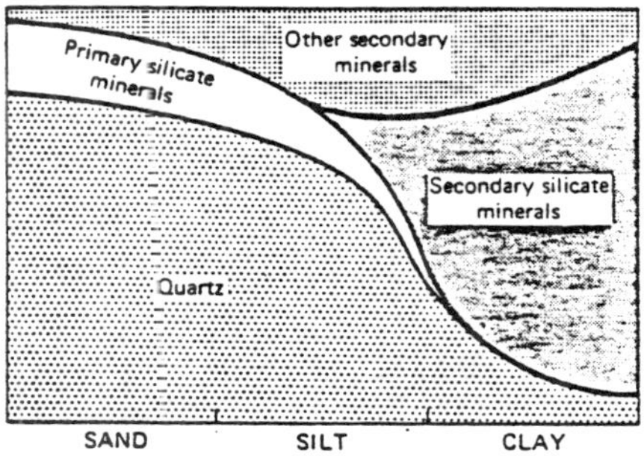

Figure 3-6. General relationship between particle size and kinds of minerals present (Brady 1974). Used by permission, see Copyright Notice.

and semitropics, though they do occur in temperate regions intermixed with silicate clays.

The layer silicates have a planar geometry, very large specific surface areas, and very high residual negative charge densities which are neutralized by a large external swarm of cations, thus resulting in a capacity for strong adsorption of and catalytic action towards hazardous compounds (Ahlrichs 1972).

Two basic sheet-like molecules make up the structure of silicate clays. A tetrahedral sheet is composed of series of tetrahedrons with four oxygen atoms surrounding a central cation, which is usually silicon (Si^{+4}), but may be aluminum (Al^{+3}), in a close-packed arrangement (Figure 3-7). An octahedral sheet is composed of a series of octahedrons with six oxygen atoms forming the corners around a large cation, which is usually Al^{+3}, but may be magnesium (Mg^{+2}) or iron (Fe^{+2} or Fe^{+3}). The sheets are formed by the sharing of corner oxygens (Figure 3-8).

The sheets may be joined in one of two ways, 1:1 or 2:1 arrangements. In 1:1 arrangements, one tetrahedral layer is connected to one octahedral layer by sharing of a common oxygen. Repeats of this 1:1 unit produces the clay kaolinite. The 2:1 arrangement has single tetrahedral layers joined to each side of the octahedral layer by sharing of oxygen atoms. This 2:1 unit produces the basic layer of the clays montmorillonite, chlorite, vermiculite, and micas. The 1:1 and 2:1 units are shown in Figure 3-9.

Some 2:1 clays exhibit a very high negative charge due to substitution of low valency for high valency cations within normally neutral crystals. During formation, ions of similar radi, such as Al^{+3} for some Si^{+4} in the tetrahedral layer and Mg^{+2} or Fe^{+2} for Al^{+3} or Fe^{+3} in the octahedral layer, may be substituted. The lower valency of the substituting cation results in a residual negative charge, which must be balanced by a cation external to the

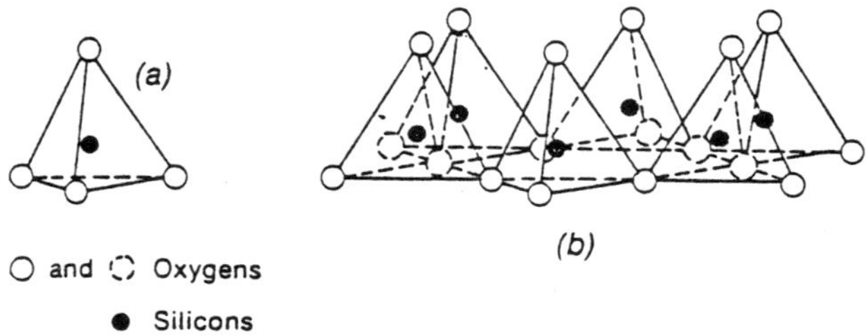

Figure 3-7. (a) Single silica tetrahedron. (b) Sheet structure of silica tetrahedrons arranged in a hexagonal network (Foth 1974). Used by permission, see Copyright Notice.

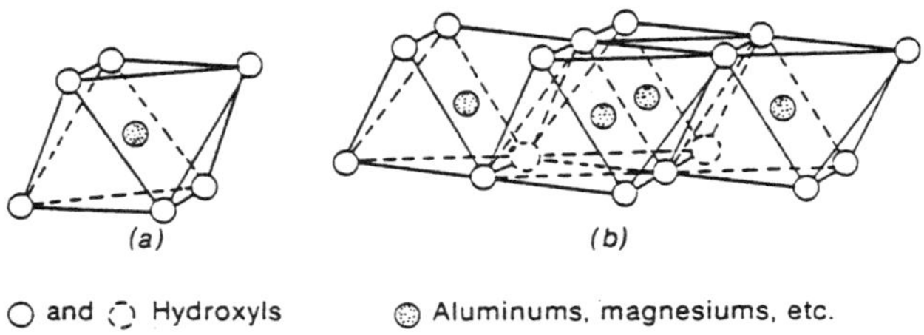

Figure 3-8. (a) Single octahedral unit. (b) Sheet structure of octahedral units (Foth 1974). Used by permission, see Copyright Notice.

2:1 unit either on its edge or in the interlayer surface. Cation exchange capacity is an expression of the number of cation adsorption sites per unit weight of soil. It is defined as the sum total of exchangeable cations adsorbed, expressed in milliequivalents per 100 grams of oven dry soil.

There are a number of silicate clays but most are variations of a few major types. Kaolinite is the most common 1:1 type clay. The 1:1 units are bonded together by hydrogen bonding between hydroxyl groups. There is no interlayer surface area and kaolinite does not swell with the addition of water. There is also little substitution within the tetrahedral or octahedral layers so the charge deficit is low. Thus kaolinite has quite low reactivity and sorptive capacity.

Montmorillonite is a common 2:1 type clay which has a negative charge of 30 to 120 milliequivalents per 100 grams of soil and swells and shrinks as water moves freely between the weakly bonded 2:1 units. Water is attracted to the oxygen surface of the clay and to the neutralizing cations in the

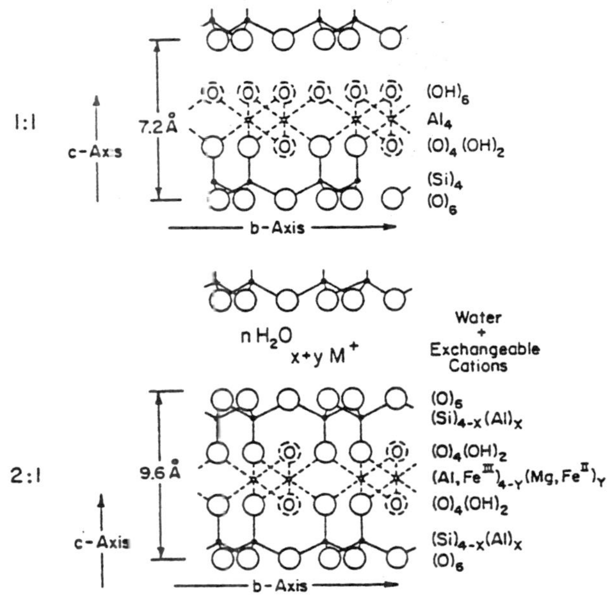

Figure 3-9. Sketch showing an edge view of the crystal structure of a 1:1 and a 2:1 type clay mineral (Ahlrichs 1972). Used by permission, see Copyright Notice.

interlayers. The swelling caused by the water exposes the large surface area and the large charge deficit of montmorillonite.

Illite or hydrous mica possesses the same 2:1 structure as montmorillonite but differs in that adjacent 2:1 units are tightly bonded by a potassium bridge. Illite therefore does not swell in water, and most of its charge is neutralized by potassium. Illite is expected to exhibit minimal interaction with hazardous organic compounds added to the soil (Ahlrichs 1972).

Vermiculite is similar to illite, except that potassium is not present, and the charge deficit is somewhat lower in the crystal. Vermiculite usually has hydrated Mg^{+2} or calcium (Ca^{+2}) in the interlayers neutralizing the charge. A summary of the overall characteristics of these common clay minerals is given in Table 3-13.

The hydrous oxide clays are oxides containing associated water molecules. They are formed by intense weathering, where silicon has been removed from the silicate clays, leaving iron and aluminum hydroxides in a highly colloidal

TABLE 3-13. SUMMARY OF CHARACTERISTICS OF SOIL COLLOIDS (COOK AND PACE 1978)
Used by permission, see Copyright Notice

Type	Example	Structure	Interlayer Bonding	Size (μ)	Charge (meq/100 g)	Source	Shrink-Swell	Weathering / Geography	Parent Material
A. Silicates 1:1	Kaolinite	Crystalline, platelike	-H. . .O-	0.1–5.0	3–16	H-Dissociation	Low	Intermediate / Southeast U.S.	Alumino-silicates
2:1 Expanding	Montmorillonite	Crystalline, platelike	None	0.01–1.0	80–100	Mg–Al (octahedral)	High	Low / Not Limiting	High Mg (basic), poorly drained
Partially-Expanding	Vermiculite	Crystalline, platelike	$Mg(H_2O)_6$	—	150	Al–Si (tetrahedral)	Low	Low / Northern U.S., Canada	Mica
Non-Ex.	Illite	Crystalline, platelike	K	0.1–2.0	15–40	Al–Si (tetrahedral)	Low	Low / Northern U.S., Canada	Mica
B. Hydrous Oxides	$Al_2O_3 \times H_2O$ $Fe_2O_3 \times H_2O$	Amorphous / Crystalline	N.A.	Colloidal	pH-dependent	H Dissociation	N.A.	High / Southeast U.S., Tropics	Fe, Al containing minerals
C. Organic	Humic Acid	Amorphous, Organic structure (C,H,O,N,P,S)	N.A.	Colloidal	pH-dependent (200+)	–COOH –OH 	Low (can lose volume by burning)	Low lying, poorly drained areas	plant (animal)

state. Their important sites for adsorption are -OH and $-OH_2$ on the outer surfaces, which may become charged in the presence of excess H+ or OH- ions. The pH values at which the charge on the ions changes ranges from 7.5 to 9 for the iron and aluminum oxides. Below these values, the compounds are positively charged, thus providing adsorption capacity for anions.

The surface geometry of the silicate clays is important in the attenuation of hazardous compounds. The interlayer surfaces provide most of their adsorptive surface area (Ahlrichs 1972). The source of the charge deficit in a clay has been suggested as important in determining the bonding action between layers and consequently the amount of swelling. Marshall (1964) suggests that tetrahedral sources of charge deficit give greater bonding energies than octahedral sources since the distance from the cause of charge deficiency is closer to the planar surface in the tetrahedra.

The size of the crystal c-axis spacing (see Figure 3-9) may determine which compounds enter the interlayer of a clay. For example, montmorillonite will allow almost any size molecule or compound to enter if there is an attraction for it. When neutralized with sodium (Na^+) or other similar monovalent ions as the exchangeable cations, these clays are especially free swelling. Less swelling is exhibited with multivalent ions. The interlayer surfaces of vermiculite, with a more restrained swelling potential then montmorillonite, are small enough to prevent the entrance of many compounds.

The importance of the spatial geometry of negative charge sites in clays was demonstrated by Weed and Weber (1968), using two divalent cationic organic compounds. The average distance between charge sites in a montmorillonite is 11-12 Å, while in micas it is 6-8 Å. Diquat, with charge sites 3-4 Å apart on ring nitrogen groups, preferentially sorbed to the higher charge density (closer spacing) of the micas. In paraquat, the charge sites were 7-8 Å apart on the ring nitrogen groups, and paraquat was sorbed preferentially by montmorillonite, with its lower charge density.

Anion exchange on colloidal clay minerals also occurs but apparently to a much lesser extent in most soils than that of cation exchange. Anions may replace the hydroxyl groups in the clays, and as these groups are more numerous in kaolinte than in other silicate clays, kaolinite is considered to be primarily responsible for anion exchange in temperature or arid region soils. Aluminum and iron oxides become protonated in the acid environments which are normal for highly weathered tropical soils and thus are also important in anion exchange reactions, as discussed previously.

Organic colloids--Soil contains many organic compounds in various stages of decomposition. Soil organic matter is derived from: 1) plant material; 2) animal matter; 3) microorganisms, both living and dead; 4) synthesized and secreted products of living plants and microorganisms; and 5) decomposition products of organic debris (Anderson 1977).

Schnitzer (1978) estimated that 65 to 75 percent of the organic matter in mineral soils consists of humic materials, i.e., humic acid (HA), fulvic acid (FA), and humin. They are amorphous, dark-colored, hydrophilic, acidic, partly aromatic, chemically complex organic substances ranging in molecular weight from hundreds to several thousand, with very large surface areas (500 - 800 m^2/g) and high cation exchange capacities. The remainder of the organic matter is composed primarily of polysaccharides and protein-like substances (Flaig et al. 1975). These include substances with still recognizable physical and chemical characteristics, such as carbohydrates, proteins, peptides, amino acids, fats, waxes, alkanes and low molecular weight organic acids (Schnitzer 1982). They are readily decomposed by microorganisms and have a short lifetime in the soil. Schnitzer (1982) identified the following important characteristics of all humic materials:

1. Ability to form water-soluble and water-insoluble complexes with metal ions and hydrous oxides.

2. Ability to interact with minerals and a wide variety of organic compounds, including alkanes, fatty acids, dialkyl phthalates, pesticides, herbicides, carbohydrates, amino acids, peptides, and proteins.

The formation of water soluble complex of FA with metals and toxic organics can increase the concentrations of these constituents in soil solutions and natural waters to levels much greater than their calculated solubilities. The oxygen-containing functional groups are important for the reactions of humic materials with metals, minerals, and organic compounds. Humic matter is somewhat organophilic, which is very important for adsorption of some hazardous nonionic organic molecules. A summary of types of functional groups and distribution in humic materials is presented in Table 3-13a. The functional groups contribute to a high cation exchange capacity (200-400 milliequivalents/100 g), thus acting similarly to clays in preventing cations from leaching. A summary of characteristics of soil humus is previously included in Table 3-6.

Humus increases the water holding capacity of a soil, since it swells when wet and can adsorb two to six times its own weight in water. However, it does rewet slowly if thoroughly dried.

Humus and other organic matter, because of their chemical composition, can add to the nutrient status of the soil, thus increasing microbial activity which may be responsible for attenuation of hazardous soil contaminants. Nitrogen, phosphorus, and some of the minor nutrients, such as sulfur, zinc, and boron all can be contributed by organic matter.

The amount of nitrogen in decomposing organic matter relative to the amount of carbon is especially important, for insufficient nitrogen may limit the rate of degradation. Organic matter with a low nitrogen content (or wide C:N ratio) is often associated with a slow rate of decomposition. Materials which contain more than 1.5 to 1.7 percent nitrogen probably do not need additional fertilizer or soil nitrogen to meet the requirements of microorganisms during decomposition. This corresponds to a threshold carbon:nitrogen ratio of 25 to 30 (Taylor et al. 1980). However, carbon:nitrogen ratios should be used with caution, for the ratio does not indicate the availability of the carbon or nitrogen to microorganisms.

In summary, humified soil organic matter, because of its surface area, surface properties, and functional groups, can serve as a buffer, an ion exchanger, a surfactant, a chelating agent, and a general sorbent, all of which are important in the attenuation of hazardous compounds in soils (Ahlrichs 1972).

TABLE 3-13a. OXYGEN-CONTAINING FUNCTIONAL GROUPS IN HUMIC SUBSTANCES (SCHNITZER 1975).

Type of Material	meq/g					
	CO_2H	Phenolic OH	Alcoholic OH	Ketonic C=O	Quinonoid C=O	Methoxyl
HA	4.4	3.3	1.9	1.2	1.0	0.3
Humin	3.1	2.2	N.D.[a]	3.1	2.0	0.4
FA	8.1	3.9	4.0	1.4	0.6	0.4

[a]N.D. = not determined.

Soil pH--Soil pH determines in part the degree of surface charge on colloidal-sized soil particles. At high pH values, negatively charged surfaces develop, while at low pH values, positively charged surfaces occur. The tendency for adsorption of anions or cations is thus dependent on the pH of the soil water in the vadose zone.

Soil pH also has major effects on biological activity in the soil. Some organisms have small tolerances to variations in pH while others can tolerate a wide pH range. The optimum pH range for rapid decomposition of most wastes and residues is 6.5 to 8.5. Bacteria and actinomycetes have pH optima near neutrality and do not compete effectively with fungi under acidic conditions (Taylor et al. 1980). Soil pH also affects the availability of nutrients, as shown in Figure 3-10.

There are several sources of hydrogen (H^+) and hydroxyl (OH^-) ions in soil solutions. The hydrolysis of exchangeable bases (Ca^{+2}, Mg^{+2}, Na^+, and K^+) which dissociate from cation exchange surfaces contribute OH^- ions. Exchangeable hydrogen (H^-) which has dissociated contributes H^+. Exchangeable H^+ is the principal source of H^+ until the pH of the soil goes below 6, when aluminum in the octahedral sheet of clays becomes unstable and is adsorbed as exchangeable Al. Upon hydrolysis, each Al ion becomes the source of three H^+ ions.

The pH of a calcareous (containing $CaCO_3$) soil or a calcareous soil horizon ranges from 7 to a maximum of 8.3. A calcareous soil horizon may greatly affect migration of hazardous compounds by an abrupt change in soil pH which may affect solubility or ionization states of the compounds.

A soil high in sodium (sodic soil) has an even higher pH than a calcareous soil, due to the hydrolysis of sodium carbonate and the formation of a strong base, sodium hydroxide. When the cation exchange capacity is 15 percent or more saturated with sodium, or a significant amount of sodium carbonate exists in the soil, the pH values range from 8.5 to 10.

Saline soils are high in soluble salts. Plant growth may be impaired due to the osmotic pressure of the soil solution restricting water uptake. Saline soils tend to have a pH at or near 7 due to hydrolysis of soluble salts (Foth 1978). The pH of soils can be adjusted by several means. Sulfur compounds can be used to lower pH or liming compounds can be used to raise pH.

Oxidation-reduction potential--In conjunction with measurement of pH in the soil solution, a measurement of the oxidation-reduction potential or Eh of the soil solution will add valuable information. Eh is an expression of the electron density of a system. As a system becomes reduced, there is a corresponding increase in electron density, resulting in a progressively increased negative potential (Taylor et al. 1980). With Eh and pH known, Eh-pH diagrams can be constructed showing stability fields for major dissolved species and solid phases. These diagrams can be useful in understanding the occurrence and mobility of hazardous compounds in soils (Everett et al. 1982).

The maximum rate of decomposition of degradable hazardous compounds is correlated with a continuous supply of oxygen. Excessive levels of

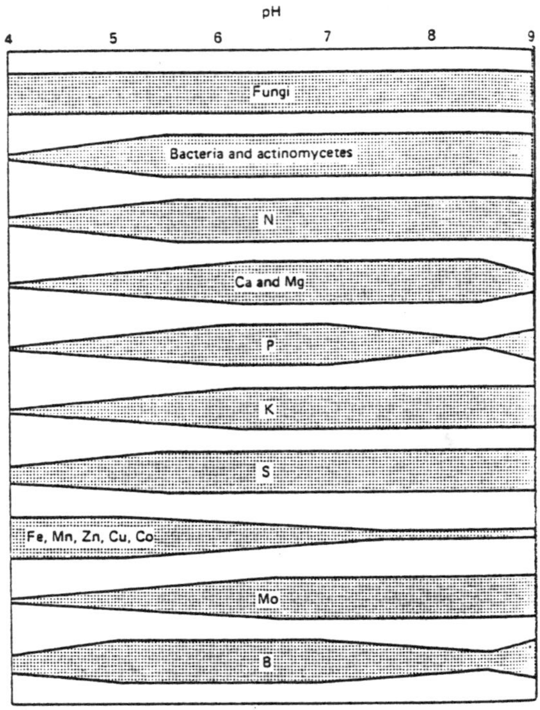

Figure 3-10. Relationships in mineral soils between pH and the activity of microorganisms and the availability of plant nutrients (Brady 1974). Used by permission, see Copyright Notice.

degradable materials may lead to depletion of O_2 in soil and anaerobiosis, which slows the rate and extent of decomposition and may produce some reduced compounds which are odorous and toxic to micrognanisms and plants. Table 3-14 shows the succession of microbial events relative to soil redox potential.

The degradative pathways for some hazardous compound may involve some critically reductive steps. An important initial step in the degradation of DDT is a reductive one, which requires anaerobiosis (Guenzi and Beard 1967). Parr and Smith (1973, 1976) have shown that toxaphene and trifluralin degrade more rapidly under anaerobic conditions. Thus an engineering management tool to maximize detoxification and degradation of some compounds may be the alternation of aerobic/anaerobic conditions (Guenther 1975) by adjusting Eh

TABLE 3-14. SUCCESSION OF EVENTS RELATED TO THE REDOX POTENTIAL WHICH CAN OCCUR IN WATERLOGGED SOILS, OR POORLY DRAINED SOILS RECEIVING EXCESSIVE LOADINGS OF ORGANIC CHEMICAL WASTES OR CROP RESIDUES (TAKAI AND KAMURA 1966)

Period of Incubation	State of Reduction	System	Redox Potential (Millivolts)	Nature of Microbial Metabolism	Formation of Organic Acids
Early	First Stage	Disappearance of O_2	+600 to +400	Aerobes	None
		Disappearance of NO_3^-	+500 to +300	Facultative anaerobes	Some accumulation after addition of organic matter
		Formation of Mn^{2+}	+400 to +200		
		Formation of Fe^{2+}	+300 to +100		
Later	Second Stage	Formation of S^{2-}	0 to -150	Obligate anaerobes	Rapid accumulation
		Formation of H_2	-150 to -220		Rapid decrease
		Formation of CH_4	-150 to -220		

through flooding or cultivation. Anaerobic conditions may be maintained by keeping the soil saturated with water and limiting aeration. Regular cultivation of soil can be used to maintain aerobic conditions.

Nutrient status--The biological degradation of hazardous compounds requires the presence of nutrients for optimum biological growth (Table 3-15). Three of the major nutrients, nitrogen, phosphorus, and potassium, can be supplied in common inorganic fertilizers. Calcium deficiencies usually occur only in acid soils and can be corrected by liming. If the soil is deficient in magnesium, the use of dolomitic lime is advised. A high level of exchangeable bases (calcium, magnesium, sodium, and potassium) on the surface exchange sites of the soil is also desirable for good microbial activity and to prevent excessively acid conditions.

Though sulfur levels in soils are usually sufficient, sulfur is also added as a constituent of most inorganic fertilizers. Micronutrients also occur in adequate amounts in most soils. At hazardous waste sites, the primary danger may be in overloading of the soil with one of these elements

TABLE 3-15. ESSENTIAL ELEMENTS FOR BIOLOGICAL GROWTH (BASED ON REQUIREMENTS FOR PLANT GROWTH) (TISDALE AND NELSON 1975)
Used by permission, see Copyright Notice

Elements		Source
Major nutrients:	Carbon, Hydrogen, Oxygen	Air and Water
	Nitrogen, Phosphorus, Potassium, Sulfur	Soil, inorganic fertilizers, or in waste
	Calcium, Magnesium	Soil liming materials, or in waste
Minor nutrients:	Iron, Manganese, Boron, Molybdenum, Copper, Zinc, Chlorine, Sodium, Cobalt, Vanadium, Silicon	Soil, soil amendments, or in waste

which may have been in the waste, thus causing toxicity and leaching problems. The pH of the soil is also important, for it determines for some of the elements their solubilities and availabilities and thus toxicity and leaching potential (Figure 3-10) (K. W. Brown and Associates 1980).

Hydraulic Properties of the Soil Profile--

Soil/water relationships and associated soil hydraulic properties affect both the movement of hazardous compounds through the soil and the soil processes acting within the soil profile to effect attenuation of waste compounds. Biodegradation of waste chemicals requires water for microbial growth and for diffusion of nutrients and by-products during the breakdown process. Soil hydraulic properties are those properties whose measurement involves the flow or retention of water within the soil profile (U.S. EPA 1977).

The total volume of a soil consists of about 50 percent pore space and 50 percent solid matter. Water entering the soil fills the pore spaces until they are all full. The water then continues to move down into the subsoil, displacing air as it travels; this flow, when the soil is at its maximum retentive capacity, is said to be saturated. After water input to the soil ceases, the water drains from the pores, and the soil becomes unsaturated. Water in the soil below the saturation level is held in the soil against the force of gravity. The forces that hold the water in the soil result from the surface tension of water, the cohesion of water molecules, the adhesion of water molecules to soil surfaces, and other electrical forces at the molecular level. The terms soil water pressure potential or matric potential are used to describe the energy required to remove water from an unsaturated soil. This energy may be expressed as a potential, (e.g., erg/g), as pressure (e.g., dyne/cm^2 or bar), or as pressure head (e.g., cm). All of these terms are negative quantities, since water is held in the soil pores at less than atmospheric pressure. Soil tension and suction are also used to describe the energy of soil water retention, but are reported as positive quantities and are not precise in regard to units.

The force by which water is held in the soil pores is approximately inversely proportional to the pore diameter. As water evaporates or drains, the larger, or macropores, drain first, while the smaller, or micropores, are still filled with water. Therefore, as soil water content decreases, the absolute value of the matric potential increases. A graphical representation of such a relationship is known as a soil moisture characteristic curve and is illustrated in Figure 3-11 for several different soil textures. The shape of the soil water characteristic curve is strongly dependent on soil texture and structure. Soils with primarily large pores, such as sands, lose nearly all their water at a very small (absolute value) matric potential. However, soils with a mixture of pore sizes, such as loamy soils, hold more water at saturation due to a greater porosity and lose water more slowly as the absolute value of the matric potential increases.

The terms field capacity and permanent wilting point are qualitative descriptions of soil water content. Field capacity refers to the percentage of water remaining in a soil after having been saturated and after free gravitational drainage has ceased. Gravitational water movement is significant in migration due to leaching of hazardous compounds and nutrients for use

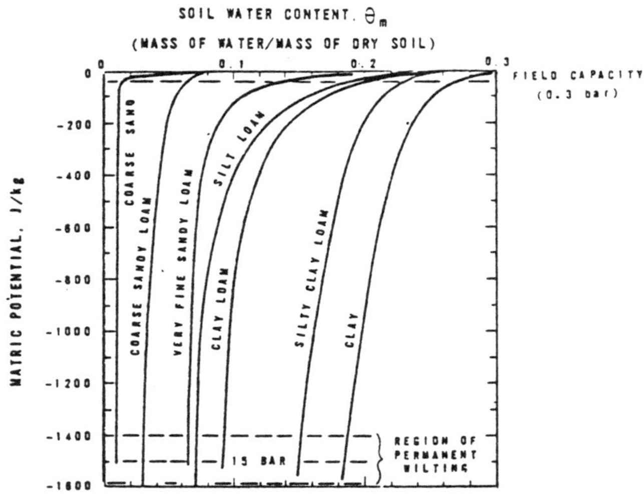

Figure 3-11. Soil-water characteristic curves for several soils (Taylor and Ashcroft 1972). Used by permission, see Copyright Notice.

in microbial degradation of the compounds. Also, if water does not drain quickly from the soil, it may have a harmful effect on microbial activity due to poor aeration. It may also affect solubility of compounds due to an alteration in oxidation-reduction potential. Field capacity of a particular soil is not a unique value but represents a range of water contents. In sands, moisture content at field capacity corresponds to matric potentials in the range of -0.10 to -0.15 bar, while in medium to fine-textured, potentials range from -0.3 to -0.5 bar, with -0.3 bar most commonly used.

Drainage does not cease at field capacity but continues at a reduced rate due to movement of water through micropores by capillarity. Adhesive attraction between the water and the walls of the micropores causes water to move through the pores, pulling along other water due to the cohesive attraction between water molecules. Capillary water can move in any direction in the soil, following micropore channels.

When moisture in a soil is no longer in adequate supply to meet the demands of plants growing in the soil, and plants wilt and remain wilted, the moisture content is said to be at the permanent wilting point. This moisture content occurs in most soils when the matric potential is in the range of -15 bars. The amount of water held in a soil between field capacity and the permanent wilting point is known as available water. This is the water available for plants and for soil microbial and chemical reactions. Information on optimal and marginal water potentials for growth, reproduction, and survival of individual species of microorganisms in soils is limited (Taylor et al. 1980). Bacterial activity is highest in wet conditions, but noticeably decreases by about -3 bars (Clark 1967). Some fungi can grow and survive in soils under dry conditions. Fungal growth is often decreased in wet soils, suggesting that bacteria may be antagonistic to fungi under these conditions. At low potentials, bacteria are less active, thus allowing fungal growth (Cook and Papendick 1970).

As seen in Figure 3-12, though fine-textured soils have the maximum total water-holding capacity, medium textured soils have the maximum available water due to favorable pore size distribution. Even at the permanent wilting point, soil contains a considerable amount of water, though it is unavailable for use. This water, which is bound tightly to individual soil particles, is known as hygroscopic water.

Permeability describes the ease with which liquids pass through the soil. Knowledge of permeability is necessary to predict the rate of movement of hazardous compounds through a soil profile.

Water moves through soils according to Darcy's Law:

$$q = K \; dH/dL \tag{3-1}$$

where

q = flux of water per unit cross sectional area (cm/h)

K = permeability of hydraulic conductivity (cm/h)

dH/dL = total head (or hydraulic) gradient (m/m)

The total head is the sum of the soil water pressure head (h) and the head due to gravity (z), or $H = h + z$. The path length of water is L.

The hydraulic conductivity constant K is not a true constant but changes rapidly as a function of water content. Even under conditions of constant water content, K may change due to swelling of clay particles or changes in the chemical nature of the soil water. Due to their negatively charged

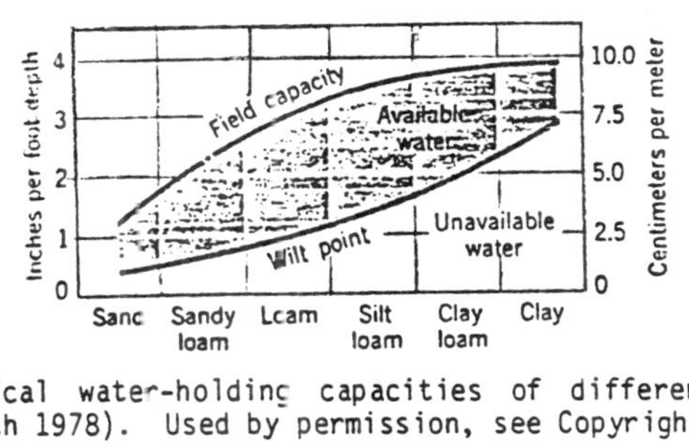

Figure 3-12. Typical water-holding capacities of different textured soils (Foth 1978). Used by permission, see Copyright Notice.

nature, soil clay particles tend to repel each other and remain dispersed, resulting in low permeability. Positively charged cations in the soil water neutralize the negative charges and allow the soil particles to come close enough together so that flocculation can occur, which increases soil pore size and permeability. Thus a water low in salts may result in permeability problems.

However, sodium in its hydrated state is much larger than the other common cations and tends to keep clay particles dispersed. If sufficient salt is present, the layer of sodium ions will be suppressed so the clay particles can flocculate. However, the required salt concentration may be so high as to restrict microbial activity and degradative processes in the soil. The sodium adsorption ratio (SAR) is used as a measurement of the degree to which sodium will be adsorbed by soils from a solution in equilibrium with the soil. Specifically:

$$SAR = \frac{Na^{+}}{\sqrt{Ca^{+2} + Mg^{+2}/2}} \qquad (3.2)$$

where the ionic concentrations are expressed in milliequivalents per liter.

As a general rule, an SAR of 15 or greater is considered unacceptable. However, there is a difference depending on the dominant clay type. With kaolinate, the SAR may be as high as 20 before serious swelling problems occur. However, with soils in which montmorillonite is dominant, SAR values of 8-10 may cause serious swelling problems (Ferguson 1976).

Other chemicals, especially organic compounds, may also alter soil permeability. Potential for such alteration should be investigated in order to predict and minimize leaching of the compound, if it is hazardous itself, or to determine if it would increase the leaching of associated hazardous compounds.

Soil permeability is primarily determined by soil texture, with coarser materials usually having higher conductivities. However, soil structure may also be important by increasing macropore space in a finer-textured soil. A listing of permeability values for soils classified by the USCS is shown in Table 3-16.

Hydraulic conductivity decreases greatly as water content decreases below saturation. In sandy soils, though permeability is much higher at saturation than in loamy soils, permeability decreases more rapidly as the matric potential becomes more negative, eventually becoming lower than in medium textured soils.

Drainability is a term used to describe the relative rapidity and extent of removal of water from a soil profile. Drainability is dependent upon the permeability (i.e., K, the hydraulic conductivity) and groundwater relationships that are controlled by soil properties and the position of the site on the landscape (i.e., the hydraulic gradient, dH/dL). A well-drained soil (e.g. a loamy soil) is one in which water is removed readily but not rapidly; a

TABLE 3-16. PERMEABILITY VALUES FOR SOILS CLASSIFIED IN THE UNIFIED SOIL CLASSIFICATION SYSTEM (YANG AND BYE 1979)

Earth Material Category	Unified Soil Classification System Designation	Permeability Range (cm/sec)
Gravel	GW, GP	
Medium to Coarse Sand	SW, SP	Permeable
Fine to Very Fine Sand	SW, SP	>10^{-4} cm/sec
Sand with ≤15% Clay, Silt	GM, SM, SC	Semi-permeable
Sand with ≥15% but ≤50% Clay	GM, SM, ML	10^{-2} to 10^{-6} cm/sec
Clay with <50% Sand	OL, MH	Relatively impermeable
Clay	CL, CH, OH	<10^{-6} cm/sec

poorly drained soil (e.g., a poorly structured fine soil) remains waterlogged for extended periods of time, resulting in reducing conditions and insufficient oxygen for biological activity; an excessively drained soil (e.g. a sandy soil) is one in which water is removed so completely that droughty conditions occur.

For in situ treatment of hazardous waste contaminated soils, the most desirable soil would be one in which permeability was only large enough to maximize soil attenuation processes (e.g. adequate aeration for aerobic microbial degradation) while still minimizing leaching (assume lower permeabilities protect against leaching).

Infiltration rate refers to the rate at which water enters the soil from the surface. When the soil profile is saturated, the infiltration rate is equal to the saturated hydraulic conductivity. However, when the soil initially is relatively dry, the infiltration rate is higher as water enters large pores and cracks. The infiltration rate is reduced rapidly to a near steady-state value as the large pores fill and clay particles swell. Infiltration rates are affected by soil texture and structure, ionic composition of the applied liquid, condition of the soil surface, and type of vegetation.

Another soil characteristic important in terms of potential migration of hazardous compounds is the depth to an impermeable layer, bedrock, or groundwater (including seasonal variations). The depth affects the drainability of the soil and the effective depth for pollutant attenuation. It may also indicate whether leaching to groundwater poses an acute hazard.

Ryan and Loehr (1981) reported that with depths of less than 1.5 m, horizontal flow predominates and the saturated hydraulic conductivity can be assumed to be equal to the permeability of the saturated horizon with the highest permeability. The hydraulic gradient is assumed to be equal to the slope of the limiting layer and can be approximated by the slope of the soil surface. When the depth to an impermeable layer, bedrock, or groundwater is greater than 1.5 m, vertical flow is predominant. The saturated hydraulic conductivity of the soil can be assumed to be equal to the permeability of the most limiting horizon, and the hydraulic gradient is assumed to be one.

An assessment of the flooding frequency of the site should be made to determine potential for off-site migration in floodwaters. Only slight hazards exist if the soil is usually not flooded any part of the year, moderate hazards if occasional flooding occurs (10-50 percent chance of flooding once every two years) and severe hazards if frequent flooding occurs (greater than 50% chance of flooding every two years) (Ryan and Loehr 1981).

Potential for off-site migration in runoff can be predicted by site/soil characteristics and waste characteristics. Various runoff models have been developed and can be used as predictive tools. (See Huber and Heaney (1981) and Foster (1981) for reviews of available models.) If the situation warrants, precautionary methods to reduce erosion should be included as an important and necessary part of an in situ remedial treatment plan.

Runoff is the portion of precipitation that appears in surface waters. Technically runoff not only includes surface and subsurface runoff but also movement of water vertically to groundwater and then lateral movement of the groundwater to surface receiving waters (baseflow). Surface runoff is water that travels over the ground surface to reach a lake or stream (overland flow). Subsurface runoff is water that has infiltrated the surface soil and moved laterally through the vadoze zone to the receiving water as shallow flow above the groundwater (interflow). Because the pathway that water takes to the receiving water determines what type and how much of a chemical is transported to that water, the three types of runoff need to be estimated separately to assess potential problems. Surface runoff may carry chemicals in solution, in suspension, or adsorbed to suspended soil particles. Subsurface runoff and groundwater carry primarily soluble chemicals not strongly adsorbed to soil particles.

Steenhuis and Walter (1979) describe a method of categorizing pollutants with regard to potential losses in soil and water according to their relative concentrations in water and on soil particles as indicated by adsorption-desorption isotherms. An adsorption partition coefficient, k_s, for a given solution concentration is a calculated as the ratio of amount adsorbed to that in solution:

$$k_s = \frac{\text{concentration of substance adsorbed to soil particles (ppm; mg/kg)}}{\text{concentration of substance in solution (ppm; mg/l)}}$$

Group I pollutants are those with k_s values approximately equal to 1000. These include the strongly adsorbed and solid phase pollutants. The loss of these pollutants in baseflow and interflow is small. Their losses in overland flow are high and are related to the amount of sediment in the soil and the amount of substance in the soil. Organic matter, which may be an important sorbent of hazardous compounds, is more easily eroded than most mineral particles and tends to remain in suspension because of its low density. Silt and clay are also more erodible than sand and are usually higher in organic matter content, which may be adsorbed on the clay particles. Clay is also an important sorbent of hazardous compounds. Therefore, an "enriching process" occurs in overland runoff in which the concentration of a hazardous compound in an eroded sediment may be much higher than in the original soil. Loss of Group I pollutants can be decreased by erosion control practices which minimize sediment detachment and transport.

Group II pollutants, with k_s values of about 5, include the moderately adsorbed pollutants (e.g. most pesticides). Their loss in overland flow has been shown to be related to the amount of runoff water and not to the amount of soil loss. Erosion control practices which prevent sediment detachment and transport are not as effective as practices which reduce the total amount of runoff volume. Transport of adsorbed substances by water passing through the soil matrix is much slower than transport by surface flow. An equilibrium exists between substances dissolved in the soil water and those adsorbed to the soil. The greatest pollution of sub-surface flow water comes from those substances which are weakly adsorbed or those slow to degrade.

The Group III pollutants have k_s values of about 0 to 0.5 and are non-adsorbed or soluble pollutants. Their primary pathway of loss is through interflow and baseflow. Losses in surface runoff are small and therefore are not greatly affected by practices to reduce runoff. In fact, reduced runoff might increase subsurface flow and thus increase their losses to interflow and groundwater.

Moderately and weakly adsorbed substances usually migrate fairly rapidly after application to the soil with initial precipitation events. Strongly adsorbed substances, depending on their recalcitrance, may pose hazards for years due to movement in overland flow.

The control of runoff can be accomplished by several means (Steenhuis and Walter 1979). Decreasing runoff velocity will reduce both surface runoff volume and sediment loss. More water remains on the soil for a longer period of time, thus permitting increased infiltration. Runoff velocity may be reduced in several ways: 1) by forcing water to move laterally rather than straight down slopes; 2) by reducing the slope of the land through land-forming; or 3) by increasing the roughness of the soil surface to dissipate the kinetic energy of the water.

An increase in surface storage will remove trapped water from total surface runoff volume, thereby resulting in decreased runoff velocity and a reduced sediment carrying capacity. Surface storage can be increased by various engineering and agricultural practices (e.g., creation of ridges of soil or vegetation) which allow water to pool. Moisture storage capacity in the soil itself can be increased by addition of organic matter or by draining or evaporating moisture already in the soil profile.

Infiltration governs the amount of water that will enter a soil. Thus engineering and agronomic practices that affect the physical, chemical, and biological soil characteristics may alter infiltration rates. Changes in bulk density, porosity, and percent of water-stable aggregates all affect a soil's capacity for infiltration and its erodibility. Infiltration rates can be increased by lower bulk densities and higher porosity or by an increase in the number of macropores connecting to the soil surface.

Frozen soil often has a lower infiltration rate than unfrozen soil, especially if the soil was frozen when moist. Since frost usually penetrates deeper if soil is bare than if it is snow covered, practices that prevent snow from blowing away may lessen frost penetration. However, the additional snow may increase surface runoff. Snowmelt occurring in the spring, often on frozen ground, can carry a higher contaminant load than rainfall runoff that has infiltrated the ground. Also the time frames in which rainfall runoff and snow melt runoff occur are different. Rainfall runoff occurs when the infiltration capacity of the soil is exceeded by the rate of precipitation. The infiltration capacity may be exceeded by the intensity of the rainfall event and/or its duration. Spring snowmelt occurs over a period of time interrupted periodically by subfreezing weather and continuing until all the snow melts.

Runoff can also be controlled by reduction of the splash-energy of falling rain. Raindrop impact on bare soil may break soil aggregates to component particles. These smaller particles may be carried by water into larger pores, thus forming a thin surface layer with low hydraulic conductivity. Dissipating raindrop energy by use of a plant canopy or mulch or promoting aggregate stability with organic matter addition may greatly reduce this surface sealing effect.

Off-site Migration in Air--

Soil properties that affect off-site migration of hazardous compounds via air transmission also need to be characterized. Again, the degree of migration in air is an interaction between soil/site characteristics and waste characteristics.

Important soil properties that determine the extent and rate of volatilization of hazardous compounds are those related to soil permeability and soil moisture. The total porosity of the soil, the distribution of macro and micropores, and the tortuosity of the soil pores should be characterized. The range of air-filled porosities exhibited by the soils under moisture regimes encountered at the contaminated site also should be investigated, for wetter soils are less permeable to gases than dry soils. At lower moisture contents, there is also an increase in sorption of compounds. Volatilization of compounds from soils is more completely discussed in section 6.

Wind erosion, unlike water erosion, is not divided into types but varies only by degree. The major factors affecting erosion by wind are climate, soil, and vegetation. Specific factors are shown in Table 3-17. A wind erosion model has been developed by Chepil and Woodruff (1963) to predict soil loss from wind erosion.

Geological and Hydrogeological Factors--

In addition to soil characteristics, geological information in the form of subsurface geological characteristics and hydrogeological factors are

TABLE 3-17. FACTORS AFFECTING EROSION OF SOIL BY WIND (SCHWAB ET AL. 1966)
Used by permission, see Copyright Notice

Climatic Variables	Soil Factors	Vegetative Factors
Precipitation	Texture	Height and density of cover
Temperature	Structure	Type of vegetation
Wind	Particle density	Seasonal distribution
Humidity	Bulk density	
Viscosity of air	Organic matter	
Density of air	Moisture content	
	Surface roughness	

important in determining potential for offsite migration via transmission by water to ground or surface receiving waters. The nature of groundwater flow systems within the subsurface structure is an important factor which determines the impact hazardous wastes may have on the environment.

The geological framework of the site consists of the rocks or sediments in the formations beneath the site. Information is required on the extent, composition, stratification, and thickness of the layers. For example, sedimentary layers (e.g., limestones, sandstones, and shales) tend to channel flows along bedding planes. Thus, flow directions may be determined by dips in the strata. In humid climates, solution channels may form in limestones, which may allow very rapid transport of pollutants over long distances with little attenuation. Fracture zones that occur in igneous and metamorphic bedrock (e.g. granites, diorite, marble, quartzite, slate, gneiss, and schist), may also permit rapid transport of polluted groundwater (Blackman et al. 1980). The most favorable areas would be those covered by thick deposits of unconsolidated low permeability materials overlying shales or undisturbed fine-grained sedimentary bedrock formations which have no major structural variations or fractures affecting formation stability (Corbin 1980). Large thicknesses of unconsolidated materials allow opportunities for natural attenuation while providing a protective barrier to any usable aquifer system.

Hydrogeological factors relating to groundwater are also required to assess potential for pollution from hazardous waste contaminated soils. For groundwater in unconsolidated formations, less hazard exists if there is no connection with surficial or buried drift aquifers, especially if the hazardous waste site is overlain with lower permeable materials to bedrock. For groundwater in bedrock formations, more favorable conditions to minimize pollution potential from hazardous waste sites exist if the site is away from any recharge areas to major freshwater aquifers or there is no direct connection to a usable bedrock aquifer (Corbin 1980). Confined groundwater, which is isolated from the surface by a relatively impermeable bed consisting of clay, shale, or dense limestone, is not easily contaminated, nor is it affected much by local sources of recharge (Warner 1976).

Knowledge of the nature of groundwater flow patterns is also critical. Localized, rather than regional, groundwater flow patterns, preferably with known discharge points and a large distance to the water table are the most favorable conditions. Information required to assess hazard might include:

1. Elevations of water table and potentiometic surface (groundwater gradients).

2. Fluctuations of groundwater levels due to natural inputs and outputs of water.

3. Drawdowns of groundwater levels from wells (cones of depression caused by pumping can alter groundwater levels from those that would naturally exist).

4. Effects on groundwater flow patterns from changes in surface water flows or levels.

5. Hydraulic characteristics of the aquifer including transmissivity, specific yield, and specific retention.

Specific yield and specific retention are measures of the amount of groundwater an aquifer will yield upon pumping. Specific yield is the amount of water that will drain by gravity from a saturated aquifer divided by the bulk volume of the aquifer. Specific retention is equal to the porosity minus the specific yield under saturated conditions. Transmissivity is the rate at which water is transmitted through a unit width of the aquifer under a unit hydraulic gradient. It is equal to the permeability multiplied by the aquifer thickness (U. S. Environmental Protection Agency 1977).

Meteorological and Climatological Data--

Meteorological and climatological data are required to assess the public health hazard from migration of hazardous compounds to receiving ground and surface waters and via air transmission. These factors also impact upon the attenuation of hazardous compounds in the soil environment.

The dispersion characteristics of the area are an important component of the air migration potential. Greater dispersion associated with open lands is more favorable than areas with channel type dispersion, such as in valley and depressional areas. Determination of prevailing wind directions and wind velocities will give an indication of the direction and extent of migration.

Temperature of both the air and soil affect the rate of biological and chemical attenuation processes in the soil, the volatilization of compounds, and the soil moisture budget. In general, temperature is difficult to control in a field situation, but may be affected by the use of mulches of natural or artificial materials and soil moisture control.

Most soil microorganisms are mesophiles, i.e. they exhibit maximum growth and activity in the 20 to 35°C temperature range. Soils also contain some microorganisms which grow best at temperatures below 20°C (psychropiles) and some which exhibit maximum growth rates between 50 to 60°C (thermophiles). In general, organic matter decomposition increases with increasing temperature. The influence of temperature on microorganism activity should be considered to estimate the time for site recovery.

Soil temperature data are not as extensive as air temperature data. K. W. Brown and Associates, Inc. (1980) discuss a method for predicting annual soil temperature cycles developed by Fluker (1958). They also present isotherm maps of soil temperature data at the 4-inch depth for the spring of 1979 in the United States.

The preparation of a water budget for the contaminated hazardous waste site will aid in indicating leaching potential and predicting the type and extent of attenuation processes that will act upon the waste (biodegradation, sorption, etc.). Inputs to the site include precipitation and any added water that may be necessary for execution of the appropriate in situ treatment technique. Factors which may limit required water addition must also be considered, such as the total amount of precipitation relative to evapotranspiration, the distribution of precipitation during the year, and changes in precipitation from year to year. Outputs include evaporation, transpiration, percolation to the groundwater, and subsurface and surface runoff.

Precipitation data may be based on real measured rainfall events or on frequency analyses, (e.g. amount of precipitation expected in a 10-year, 24-hour storm, 25-year, 1-hour storm, etc.). Of importance is the total rainfall a site receives as well as the intensity, duration, and frequency of single precipitation events.

Evaporation is the transfer of liquid water into the atmosphere. Factors affecting the rate of evaporation are the nature of the evaporating surface and the vapor pressure differences as affected by temperature, wind, atmospheric pressure, quality of water, and available energy (Schwab et al. 1966). In saturated soils, evaporation is expected to be the same as from open freewater surfaces. However, in unsaturated soils, below field capacity, evaporation is very low, as soil moisture movement is slow when the soil is relatively dry.

Transpiration is the process by which water vapor passes into the atmosphere through the tissues of living plants. Loss of soil moisture by transpiration is often a substantial portion of the total water available during a growing season. Transpiration is dependent on the moisture available, the kind and density of plant growth, the amount of sunshine, and soil fertility and structure.

The measurement of evaporation and transpiration is usually combined and referred to as evapotranspiration. Evapotranspiration can either be measured directly or predicted using various models (e.g. see Schwab et al. 1966 for a discussion of methods).

The Receiver System--

The receiver system around the hazardous waste contaminated site may be an important site charactertistic, which, when evaluated in conjunction with waste characteristics, may determine precautions necessary to prevent migration of hazardous constituents and also may dictate the remedial action appropriate for the specific site/soil/waste situation. For example, a contaminated site located in a sparsely populated arid region over a deep unusable aquifer may not require the use of a quick-acting remedial in situ treatment technique. However, a site in an urban area, with high rainfall and soils of high erosion potential, overlying a heavily used aquifer, may require the choice of an in situ treatment technique with little or no secondary health impacts and which can treat the wastes in a shorter span of time.

The characteristics listed in Table 3-18 can be used to give an assessment of hazard potential. Information from organized hazard ranking systems, such as the Mitre Hazard Ranking System Model, the LeGrand Model, the Surface Impoundment Assessment Model, the EPA Solid and Hazardous Waste Research Division Model, and the Rating Metholodogy Model (Caldwell et al. 1981) may be available or be used to obtain an indication of the urgency of clean-up and required precautionary methods.

TABLE 3-18. IMPORTANT RECEIVER CHARACTERISTICS

Migration to Groundwater
Groundwater use - present and potential
Groundwater quality-usability of aquifer
Distance to nearest downgradient well-treatment of water from the well
Population served
Discharges to surface waters - uses of surface waters
Recharge zone for freshwater aquifer
Migration in Runoff to Surface Waters
Surface water use - drinking, recreation, fishing, irrigation, livestock watering
Population served
Distance to a sensitive environment - floodplain, wetlands, etc.
Migration via Air Transmission
Distance to nearest human population
Population within 1 mile radius
Population downwind from site
Land use - crops, forestry products, livestock, urban, schools, parks, playgrounds, industrial, residential, etc.
Distance to a sensitive environment

Site Characteristics with Regard to In Situ Treatment Techniques

Soil and site characteristics necessary for the choice and design of in situ treatment techniques overlap significantly with the data required to assess the potential for off-site migration of hazardous compounds from contaminated sites. Many of the natural attenuation processes which determine migration potential may be used as in situ techniques but under engineering management control and monitoring. Therefore, the characteristics listed in Table 3-2 are also relevant for choosing appropriate treatments for degrading, transforming, or immobilizing hazardous compounds. Those characteristics which are subject to some degree of engineering control are indicated in Table 3-2. The remainder of this chapter is devoted to the discussion of soil attenuation processes related to hazardous waste treatment. In Section 2 of this manual, various treatment processes based on soil attenuation properties are discussed.

Site Characterization Related to Physical Execution of In Situ Treatment Technology

Site requirements for the physical execution of an in situ treatment technique may be a determining factor in whether the technique may actually be utilized at a particular site or what precautions or operational controls need to be incorporated into the execution plan.

The trafficability of the soil under different climatological and soil moisture conditions needs to be assessed. There may be restrictions on the type of equipment that can be used and times when the equipment can be used (e.g., presence of boulders, steep slopes, excessively wet conditions in clayey soils).

Trafficability refers to the capability of a soil to permit the movement of a vehicle over the land surface (Reeve and Fausey 1974). In military operations, trafficability is defined as whether or not vehicles of various kinds can pass over a given terrain without regard for the final condition of the soil. However, in agriculture, the primary concern is for successfully performing given operations on the land without damaging the soil. Such damage might include decreased permeability to air and water, altered thermal relations, and resistance to root penetration. Generally, trafficability means being able to perform required operations in such a way as to create a desired soil condition or to get an operation completed expediently. Operations that require manipulation (tillage) require a different interpretation of trafficability than do operations in which soil is used as a surface on which to operate (non-tillage).

The U. S. Army Engineer Waterways Experiment Station (1956) identified four soil characteristics that relate to trafficability of soils: 1) bearing capacity, 2) traction capacity, 3) slipperiness, and 4) stickiness. Any one or a combination of these may cause vehicle immobilization.

The trafficability of a soil is considered adequate for a vehicle if it has sufficient bearing capacity to support the vehicle and sufficient traction to develop the forward thrust necessary to overcome the rolling resistance. Bearing and traction capacities are related to soil strength or shear resistance. Soil strength can be determined by laboratory tests (direct shear, triaxial shear, and unconfined compression) or by a field test using a cone penetrometer.

Slipperiness is the condition of deficient traction capacity in a thin surface layer of soil which is otherwise trafficable. When soils adhere and build up on the running gears of a vehicle, increasing rolling resistance and making steering difficult, the condition is called sticky. Soil stickiness and slipperiness usually occur on soils high in clay. When the soil surface is cooler than the underlying soil, moisture migrates from the lower layers to the surface. If the evaporative demand is not great, the moisture accumulates at the surface and causes decreased traction. This condition is not a problem which can be alleviated by drainage, but by water management. It is especially a problem in seasons when the radiant energy input is low.

Damage to the soil by vehicular traffic usually results from compressing and puddling the soil. To avoid such damage, the soil must be manipulated or traversed when the soil is below some critical moisture level, which is dependent on the type of soil. Wet soils are easily compacted by both tillage and nontillage operations. Clay soils are especially a problem, since they hold a large amount of water that must be removed by internal drainage or evaporation before tillage is possible.

Soil compaction can be reduced in several ways. Reducing the load intensity on a soil or reducing the number of trips over a soil can be accomplished by changing machinery configurations or tractor tire designs. Subsurface and surface drainage systems can also be used to reduce soil moisture content.

Other site conditions which may affect trafficability include slopes and the presence of large coarse fragments, such as boulders. Reeve and Fausey (1974) present a review of methods regarding the determination of soil trafficability using predictive equations and empirical rating systems.

If the treatment itself requires a certain site condition, modifications must be made, if possible, to achieve that condition. An assessment must be made to determine if required modifications are feasible at that particular site. For example, the initial steps in biodegradation of a chlorinated organic compound may require anaerobic conditions followed by aerobic condition. Thus the site/soil infiltration, permeability, and drainability characteristics all determine whether anaerobic/aerobic conditions can be achieved.

Sources of Information

Many hazardous waste sites for which in situ treatment may prove to be an appropriate remedial response may have been subjected to earlier acute emergency response cleanup actions or preliminary investigations to determine the necessity of remedial action, such as a Superfund site. Therefore, there may be significant existing data on site and soil characteristics.

The necessary data required to assess the choice and execution of an appropriate treatment or series of treatments has been discussed generally in this section and will be discussed in greater detail in the following sections relating to soil attenuation processes. Existing data should be assessed to determine adequacy for meeting information requirements and to design any subsequent data collection activities. If data are sufficient, no new characterization efforts are needed. If further data collection is needed, trained soil scientists, geotechnical engineers, geologists, and other persons trained in appropriate disciplines should be hired to generate the required information.

Some sources of existing information include the following:

1. Government investigative reports (e.g., Field Investigation Team (FIT) reports, Hazard Assessment Reports (Mitre Model)

2. Engineering data from other public or private agencies or firms (e.g., university geology departments, state water resource agencies, state geological surveys, city water departments)

3. County soils maps (general background information--on-site investigation also required)

4. Aerial photographs

5. Water well borings logs

6. Geotechnical reports from nearby facilities

It is expected that data related to potential migration from the site (route characteristics) and potential public health hazards (waste and receiver characteristics) will have been generated for many sites where in situ treatment is being considered. Most new data collection efforts will be expected to emphasize information required for the choice and execution of appropriate successful in situ treatment technology, i.e., the information will center around soil characterization. Therefore, persons knowledgeable in soil science should be utilized for sampling and analysis of soil samples and for the interpretation of sample results.

WASTE CHARACTERIZATION RELATED TO IN SITU SOIL TREATMENT

Introduction

The hazardous-waste contaminated site must be thoroughly characterized in terms of waste characteristics in order to assess the degrees of hazard to both on-site workers and to off-site populations and to determine applicable and appropriate types of in situ treatment technology. Together with knowledge of soil and site characteristics, knowledge of waste characteristics will enable the selection of a particular treatment technique specifically designed for the site/soil/waste situation or will indicate whether in situ treatment is not appropriate for the specific situation.

Since the wastes at a contaminated site do not exist separately from the soil, measurement of waste types, amounts, and distributions must be assessed as part of the soil, soil pore liquid, groundwater, or surface or subsurface runoff. Thus analysis techniques will usually include both extraction procedures to separate waste constituents from the environmental matrix as well as chemical analyses of the constituents. It is not practical to use identification based on observable characteristics, such as odor, color, reaction, etc. In general, analysis of substances classified as hazardous requires more sophisticated methods and instrumentation than traditional waste parameters in pollution control, such as chemical oxygen demand, nitrogen forms, phosphorus, etc.

Definition of Hazard and Degree of Hazard

A waste stream is characterized as hazardous if it is defined as hazardous, or if it contains substances which are defined as hazardous (40 CFR 261), or if it exhibits one or more of the characteristics of hazardous waste: ignitability, corrosivity, reactivity, or toxicity.

Ignitability refers to wastes that either present fire hazards during routine storage, transportation, or disposal or are capable of severely exacerbating a fire once started. Corrosivity identifies wastes which have the ability to mobilize toxic metals when discharged to a land environment, corrode handling, storage, and management equipment, or destroy human or animal tissue in the event of inadvertent contact. Reactivity refers to wastes which have any of the following properties: 1) readily undergo violent chemical change; 2) react violently or form potentially explosive mixtures with water; 3) generate toxic fumes when mixed with water or, in the case of cyanide or sulfide-bearing wastes, when exposed to mild acidic or basic conditions; 4) explode when subjected to a strong initiating force; 5) explode at normal temperatures and pressures; or 6) fit within the Department of Transportation's list of forbidden explosives (49 CFR 173.53), or Class B explosives (49 CFR 173.88).

Toxicity, as used in defining a hazardous waste, refers to a reading of toxicity in the Extraction Procedure (EP) Toxicity Test. The EP is designed to simulate leaching of waste in a sanitary landfill. It is a laboratory test in which a representative sample of a waste is extracted with distilled water with pH maintained at 5 by acetic acid. The EP extract is analyzed to determine if threshold levels for eight metals, four pesticides, and two herbicides (Table 3.19) have been exceeded. If the EP extract contains one or more of the substances in amounts equal to or exceeding the specified levels, the waste is classified as hazardous according to the Extraction Procedure Toxicity Test.

Definitions of methods to determine these characteristics are presented in Test Methods for Evaluating Solid Waste: Physical/Chemical Methods (U.S. Environmental Protection Agency 1982b). In a contaminated environmental sample, it may be difficult to determine if the waste substances exhibit these hazardous characteristics. Chemical identification may be the only way to determine which waste constituents are undesirable and thus require in situ treatment.

TABLE 3-19. MAXIMUM CONCENTRATION OF CONTAMINANTS FOR CHARACTERISTIC OF EP TOXICITY (U.S. ENVIRONMENTAL PROTECTION AGENCY 1982b)

EPA Hazardous Waste Number	Contaminant	Maxiumum Concentration (mg/1)
D004	Arsenic	5.0
D005	Barium	100.0
D006	Cadmium	1.0
D007	Chromium	5.0
D008	Lead	5.0
D009	Mercury	0.2
D010	Selenium	1.0
D011	Silver	5.0
D012	Endrin (1,2,3,4,10,10-Hexachloro-1 7-epoxy-1,4,4a,5,6,7,8,8a-octahydro-1 4-endo, endo-5,8-dimethanonaphthalene)	0.02
D013	Lindane (1,2,3,4,5,6-Hexachlorocyclohexane, gamma isomer)	0.4
D014	Methoxychlor (1,1,1-Trichloro-2,2-bis (p-methoxyphenyl)ethane)	10.0
D015	Toxaphene ($C_{10}H_{10}Cl_8$, Technical chlorinated camphene, 67-69% chlorine)	0.5
D016	2,4-D (2,4-Dichlorophenoxyacetic acid)	10.0
D017	2,4,5-TP (Silvex) (2,4,5-Trichlorophenoxypropionic acid)	1.0

Other waste characteristics are important in determining degree of hazard, though not legally defined as such. These include various types of human toxic effects, animal toxic effects, persistence/biodegradability, infectiousness, volatility (affecting migration via atmospheric route), and solubility (affecting migration via water routes) (U. S. Environmental Protection Agency 1982b). These must also be assessed in order to select an appropriate treatment technology which is effective and ensures safety for both on-site workers and off-site populations.

Toxicity information is available from many different sources. The following is a partial listing of reference publications and systems which may be consulted for human and/or ecological toxicity data and chemical hazards:

1. Chemical Hazards Response Information System (CHRIS), U.S. Coast Guard, consists of four handbooks (A Condensed Guide to Chemical Hazards, Hazardous Chemical Data Manual, Hazard Assessment Handbook, and Response Methods Handbook), data bases for regional contingency planning, and the Hazard Assessment Computer System.

2. Oil and Hazardous Materials Technical Assistance Data System (OHM-TADS) is an automated information retrieval file designed to facilitate the rapid retrieval of information on approximately 1000 chemicals. Includes information on physical, chemical, biological, toxicological, air, land, and water effects and commercial data.

3. Chemical Transportation Emergency Center (CHEMTREC) serves as clearinghouse by providing an emergency 24-hour telephone number for chemical emergencies. Provides warnings and limited guidance on hazards of spills, fire, or exposure, and contacts shipper of the chemical.

4. TOXLINE (Toxicity Information On-Line), National Library of Medicine, is a computerized data resource for health and toxicological effects information.

5. NIH/EPA Chemical Information System (CIS), Interactive Sciences Corporation, Washington, D.C, contains information on toxicological, analytical, and physical properties of chemical substances.

6. Organic Chemical Producers Data Base, EPA/ORD Cincinnati Research Laboratory, contains data on physical/chemical properties, manufacturing processes, uses, and by-products.

7. Karnofsky, B., J. King, P. Thielmann, K. Gleason, and M. Baer. 1981. Chemical information resources handbook: Toxics integration information series. EPA-560/TIIS-81-001. Office of Pesticides and Toxic Substances, Washington, D.C. This publication describes chemical information resources dealing with chemical toxicology, environmental effects, spill responses, disposal methods, ambient air and water concentrations, control technologies, and existing regulations.

8. Sax, Irving N. 1975. Dangerous properties of industrial materials. Van Nostrand Reinhold Co., New York, NY.

9. National Fire Protection Association. Hazardous chemicals data. Boston, MA.

10. Milson, M. et al. (Eds.). 1976. The Merck Index, 9th Ed. Merck and Co., Inc. Rahway, NJ.

11. Chemical Manufacturers Association. Chemical safety data sheets. (SD-1-SD-96). CMA, Washington, D.C.

12. Fawcell, H. H., and W. S. Woods (Eds.). 1981. Safety and health in chemical operations. 2nd ed., Wiley-Interscience, New York, NY.

13. Bretherick, L. 1979. Handbook of reactive chemical hazards. Butterworths, Boston, MA.

14. National Institute for Occupational Safety and Health (NIOSH). Registry of Toxic Effects of Chemical Subtances. Annual publication. U.S. Government Printing Office, Washington, D.C.

15. Survey of Compounds Which have been Tested for Carcinogenic Activity. U.S. Government Printing Office, Washington, D.C. Seven-volume series.

16. American Industrial Hygiene Association. Hygienic guide series. Detroit, MI.

17. Patty, F. A. Industrial hygiene and toxicology. Wiley Interscience, New York, NY.

18. National Institute for Occupational Safety and Health. NIOSH/OSHA pocket guide to chemical hazards. Publ. No. 78-210, NIOSH, Washington D.C.

A comprehensive list of biological test systems which may be used to detect genetic toxicity of hazardous wastes is presented in Hazardous Waste Land Treatment (U.S. Environmental Protection Agency 1983).

Toxicity of waste constituents to soil microorganisms is especially of concern because of the effects on biological in situ treatment processes. Acute microbial toxicity can be evaluated using a pour plate method which enumerates total viable heterotrophic and hydrocarbon-utilizing microorganisms. A overview of this method, which involves a series of soil/waste dilutions in phosphate buffer, is presented in Table 3-20.

Another method is currently being tested by the U.S. Environmental Protection Agency (1983) for evaluation of acute microbial toxicity in hazardous waste land treatment systems. The Beckman Microtox™ measures the light output of a suspension of marine luminescent bacteria before and after a sample is added. A reduction in light output signifies the presence of toxicants in the waste sample (Beckman Instruments, Inc. 1982).

Soil respirometry (monitoring CO_2 evolution from soil/waste samples) may also be used to determine effects of the complex waste in the soil system on

TABLE 3-20. METHOD FOR DETERMINATION OF MICROBIAL TOXICITY OF A WASTE/SOIL MIXTURE (U.S. ENVIRONMENTAL PROTECTION 1983)

Microorganism to be Enumerated	Type of Media for Enumeration
Total viable heterotrophs	Soil extract agar, with amphoteracin B.
Soil fungi	Potato dextrose agar or soil extract agar with rose bengal and streptomycin.
Hydrocarbon - utilizing bacteria and fungi	Soil extract agar, with carbon source replaced with silica gel oil. (Prepared by mixing fumed silica gel with waste.)

microbial activity. In addition to monitoring toxicity, the technique can also be used to study the effects of various soil treatments and amendments on degradation of organic compounds (e.g., levels and timing of nutrient additions, temperatures, etc.). A description of the technique is presented in Hazardous Waste Land Treatment (U.S. Environmental Protection Agency 1983).

Preliminary Waste Identification

At a contaminated site, it may be possible to obtain a preliminary assessment of expected waste constituents if the industrial source or sources of the wastes stored at the site are known. Unique hazardous waste streams are produced from a number of industries, including textiles, paper, leather products, primary metals, etc. (U.S. Environmental Protection Agency 1983). The draft edition of Hazardous Waste Land Treatment (K. W. Brown and Associates, Inc. 1980) contains a section dealing with hazardous wastes by specific industry. An update of this section is currently being prepared by K. W. Brown and Associates, Inc.

Several hazardous waste generating activities are not specific to a particular industry but are used in many industries to perform certain processes. Examples of these activities include painting, electroplating, and cleaning, which produce paint residues, metals and cyanide, and solvents, respectively (U.S. Environmental Protection Agency 1983). Therefore, the presence of specific hazardous substances may be predicted if the industry or industrial process responsible for the wastes is known. Knowledge of the feedstocks used and expected industrial products and by-products will indicate what substances may be present. Chemical analyses are still required, however, to confirm the presence and the degree of contamination.

Another source of information on waste constituents at a site may be preliminary site assessments, such as those performed by Field Investigation Teams (FIT) for the U. S. Environmental Agency. From such reports, information on type and amount of constituents in soil, groundwater, and surface water samples may be available.

Chemical Analysis of the Wastes

In order to effectively design an in situ treatment technology, the types and amounts, and distribution of wastes present must be accurately known. Knowledge of waste source will only give an estimate of specific substances that may be present. Reactions between different wastes and in the soil matrix may have created new hazardous substances which must also be treated. In addition to the hazardous components of the wastes which require treatment (i.e., by degradation, transformation, or immobilization), other waste substances which may impact on the effectiveness of the treatment process or have adverse effects on the soil system must also be characterized. Treatment effectiveness and completeness also can only be determined by knowledge of the waste constituents originally present at the site.

Identification of hazardous constituents in wastes is usually accomplished by instrumental methods rather than the standard types of analyses commonly used for pollution analysis. Instrumental methods require more costly equipment, a well-equipped analytical laboratory, and trained, skilled personnel.

The instrumental methods used for analysis are:

1. Gas chromatography (GC)

2. High-performance liquid chromatography (HPLC)

3. Gas chromatography/mass spectrometry (GC/MS)

4. Atomic absorption spectrometer (AAS)

5. Inductively-coupled, argon-plasma spectrophotometry (ICAP)

GC and HPLC are used to quantitatively determine contamination levels of known specific organic materials. Concentration levels in the parts per billion range (ppb) can be determined. They are not suitable for qualitative analysis of unknown substances. Different GC detectors are available for analysis of different classes of organic compounds (Table 3-21).

GC/MS is primarily used for qualitative analysis but recent improvements permit quantitative measurements as well. A large number of substances can be simultaneously analyzed at ppb or sub-ppb levels.

TABLE 3-21. TYPES OF GC DETECTORS (U.S. ENVIRONMENTAL PROTECTION AGENCY 1983)

Type of Detector	Type of Compounds Detected
Flame ionization detector	Hydrocarbons
Flame photometric detector	Sulfur and/or phosphorus containing compounds
Electron capture detector	Halogenated hydrocarbons and phthalates
Nitrogen-phosphorus detector	Nitrogen and/or phosphorus containing compounds

AAS and ICAP are used for quantitative determination of metals in the parts per million (ppm) or ppb range. AAS can quantify amounts of individual metals, while ICAP can quantify amounts of several dozen metals simultaneously.

Tables summarizing U.S. Environmental Protection Agency-approved analytical procedures for 16 classes of organic chemicals and approved analytical procedures for 80 hazardous materials are given in EPA Field Guide for Scientific Support Activities Associated with Superfund Emergency Response (U.S. Environmental Protection Agency 1982a).

The U.S. Environmental Protection Agency (1983) has defined waste characteristics important for hazardous waste land treatment in terms of determination of hazardous constituents of the wastes and constituents which may affect the soil as an effective treatment medium. These characteristics are presented in Table 3-22. The reader is referred to the referenced document for a thorough discussion of the significance of the various constituents.

Waste Characteristics Related to Soil Treatment

Wastes identified at uncontrolled soil contaminated sites must also be characterized with respect to properties that affect the behavior and fate of chemicals in soil systems. Knowledge of properties that affect the behavior and fate of chemicals also directly affect pathways of treatment, or assimilation, of waste constituents in soil systems. Pathways of treatment include 1) degradation, 2) transformation, and/or 3) immobilization of chemical constituents in a soil treatment zone, with subsequent protection of groundwater, surface water, and atmosphere.

Degradation and transformation reactions include chemical and biological reactions which breakdown waste constituents resulting in nontoxic

TABLE 3-22. METHODS FOR ANALYZING WASTE CONSTITUENTS IMPORTANT TO IN SITU TREATMENT (ADAPTED FROM U.S. ENVIRONMENTAL PROTECTION AGENCY 1983)

Element/Compound/ Property	Possible Analysis Method(s)	Reference(s)
Inorganic chemicals/ properties		
Metals	Sample digestion followed by atomic absorption spectrophotometry or inductively coupled arc spectrometry	U.S. EPA 1982b U.S. EPA 1979 Page et al. 1982
Halides (bromine, chlorine, and fluorine)	Various methods	Adriano and Doner 1982 U.S. EPA 1982b U.S. EPA 1979
Nitrogen forms	Various methods	Bremner and Mulvaney 1982 Stevenson 1982 Keeney and Nelson 1982 U.S. EPA 1982b U.S. EPA 1979
Electrical conductivity	Saturation extracts and other aqueous extracts	Rhoades 1982
pH	Colorimetric or potentiometric	McLean 1982 U.S. EPA 1979 U.S. EPA 1982b
Titratable acids and bases	Aqueous waste suspensions	McLean 1982 U.S. EPA 1979
Organic chemicals		
Total organic carbon	Dry or wet combustion with CO_2 determinations; dichromate oxidation techniques	Nelson and Sommers 1982
Volatile organic compounds	Purge and trap or head space determinations; analysis with GC or GC-MS	U.S. EPA 1982b

TABLE 3-22. CONTINUED.

Element/Compound/ Property	Possible Analysis Method(s)	Reference(s)
Extractable organics organic acids (e.g., carboxylic acids, quaiacols, and phenols)	Liquid/liquid acid/ base extraction method: Analysis with GC with capillary or packed columns	U.S. EPA 1983
organic bases (e.g., alkyl, aromatic, and aza-heterocyclic amines)	Analysis with GC with capillary or packed columns	U.S. EPA 1983
neutrals (e.g., aliphatic and aromatic hydrocarbons, and oxygenated and chlorinated hydrocarbons)	Analysis with GC or HPLC	U.S. EPA 1983
Water solubles	Variable; further study needed	U.S. EPA 1983
Residual solids (e.g., inorganics and relatively nondegradable forms of carbon such as coke, charcoal, and graphite)	Evaporation of water from aqueous fraction of acid-base extraction procedure	U.S. EPA 1983

products. Immobilization reactions include adsorption and chemical reactions which result in accumulation and termination of constituent mobility in the soil treatment zone. Since many organic compounds and organo-metallic complexes exhibit simultaneous decomposition and migration, the relative rates of degradation and mobilization must be considered.

Characterization of chemical contaminants with respect to soil treatment pathways (degradation, transformation, and immobilization) provides the basis for establishing qualitative and calculational procedures for determining soil assimilative capacities for hazardous waste constituents or classes of constituents.

Utilizing this approach for hazardous waste contaminated soils management, all constituents and constituent classes identified as associated with a specific uncontrolled site can be classified functionally into three pathways by which soil treatment occurs. This approach also provides endpoints for field verification of treatment effectiveness and protection of groundwater, surface water, and atmosphere.

Specific parameters important for determining the behavior and fate, and therefore treatment pathways, for waste constituents in soil are listed in Table 3-23. For each chemical, or chemical class, the information required can be summarized as: 1) characteristics related to potential leaching, e.g., water solubility, octanol/water partition coefficient, solid sorption coefficient; 2) characteristics related to potential volatilization, e.g., vapor pressure, relative volatilization index; 3) characteristics related to potential decomposition, e.g., half-life, degradation rate biodegradability index; and 4) characteristics related to chemical reactivity e.g., oxidation, reduction, hydrolysis potential. Interfacing these "soil-based behavorial characteristics" with specific site and soil properties allows a determination of the potential for: 1) soil treatment and 2) offsite contamination. This information, then, provides a rational basis for selection of treatment techniques that augment natural soil processes to accomplish complete treatment of the waste.

Statistical Considerations

A sampling and analysis plan for characterization of wastes at a hazardous waste contaminated site must be based on fundamental statistical concepts such that the uncertainty of general conclusions based on partial knowledge can be evaluated. Table 3-24 presents basic statistical terminology associated with a sampling/analysis strategy. The primary objectives of such a strategy are to collect samples that will allow sufficiently accurate (closeness of a sample to its true value) and precise (closeness of repeated sample values) measurements of the chemical properties of the wastes. A complete discussion of a statistical strategy to determine if chemical contaminants are present at hazardous levels is presented in Test Methods for Evaluating Solid Waste: Physical/Chemical Methods (U.S. Environmental Protection Agency 1982b).

IMMOBILIZATION OF CHEMICAL CONSTITUENTS AS RELATED TO IN SITU TREATMENT

Inorganics

Rather than discuss all possible inorganic pollutants, this section will be limited to those inorganics which have been the subject of concern resulting from their toxicity and those that, because of their use in industrial processes, are likely to be involved in an accidental spill. The listing of inorganic priority pollutants as defined by the U.S. EPA (1976), is given in Table 3-25. Figure 3-13 illustrates the frequency of occurrence of inorganics in soils found at Superfund and non-Superfund sites (F.I.T. Reports). Table 3-26 lists the percent occurrence for each element. The concentration range of inorganics found at Superfund plus non-Superfund sites,

TABLE 3-23. SOIL-BASED WASTE CHARACTERIZATION

Chemical Class	Soil Sorption Parameters	Soil Degradation Parameters	Chemical Properties
Acid	Freundlich sorption constants (K, N)	Half-life ($t_{1/2}$)	Molecular weight
Base	Sorption based on organic carbon content (K_{oc})	Rate constant (first-order)	Melting point
Polar Neutral	Octanol water partition co-efficent (K_{ow})	Relative biodegradability	Specific gravity
Nonpolar Neutral			Structure
Inorganic			Water solubility

Volatilization Parameters	Chemical Reactivity	Soil Contamination Parameters
Air:water partition coefficient (K_w)	Oxidation	Concentration in soil
Vapor pressure	Reduction	Depth of Contamination
Henry's law constant ($1/K_w$)	Hydrolysis	
Sorption based on organic carbon content (Koc)	Precipitation	
Water solubility	Polymerization	

TABLE 3-24. BASIC STATISTICAL TERMINOLOGY APPLICABLE TO SAMPLING PLANS FOR SOLID WASTES (U.S. EPA 1982b)

Terminology	Symbol	Mathematical equation	(Equation)
• Variable (e.g., barium or endrin)	X	---	
• Individual measurement of variable	X_i	---	
• Mean of all possible measurements of variable (population mean)	μ	$\mu = \frac{\sum_{i=1}^{N} X_i}{N}$, with N = number of possible measurements	(1)
• Mean of measurements generated by sample (sample mean)	$\bar{x}$	Simple random sampling and systematic random sampling	
		$\bar{x} = \frac{\sum_{i=1}^{n} X_i}{n}$, with n = number of sample measurements	(2a)
		Stratified random sampling	
		$\bar{x} = \sum_{k=1}^{r} W_k \bar{x}_k$, with $\bar{x}_k$ = stratum mean and W_k = fraction of population represented by Stratum k (number of strata [k] ranges from 1 to r)	(2b)
• Variance of sample	s^2	Simple random sampling and systematic random sampling	
		$s^2 = \frac{\sum_{i=1}^{n} X_i^2 - (\sum_{i=1}^{n} X_i)^2/n}{n - 1}$	(3a)
		Stratified random sampling	
		$s^2 = \sum_{k=1}^{r} W_k s_k^2$, with s_k^2 = stratum variance and W_k = fraction of population represented by Stratum k (number of strata [k] ranges from 1 to r)	(3b)

TABLE 3-24. CONTINUED

Terminology	Symbol	Mathematical equation	(Equation)
• Standard deviation of sample	s	$s = \sqrt{s^2}$	(4)
• Standard error (also standard error of mean and standard deviation of mean) of sample	$s_{\bar{x}}$	$s_{\bar{x}} = \frac{s}{\sqrt{2}}$	(5)
• Confidence interval for μ[a]	CI	$CI = \bar{x} \pm t_{.20}\, s_{\bar{x}}$, with $t_{.20}$ obtained from Table 2 in this section for appropriate degrees of freedom	(6)
• Regulatory threshold[a]	RT	Defined by EPA (e.g., 100 ppm for barium in elutriate of EP toxicity test)	(7)
• Appropriate number of samples to collect from a solid waste (financial constraints not considered)	n	$n = \frac{t_{.20}^2 s^2}{\Delta^2}$, with $\Delta = RT - \bar{x}$	(8)
• Degrees of freedom	df	$df = n - 1$	(9)
• Square root transformation	---	$\sqrt{X_i + 1/2}$	(10)
• Arcsin transformation	---	Arcsin $\sqrt{p}$; if necessary, refer to any text on basic statistics; measurements must be converted to percentages (p)	(11)

[a]The upper limit of the CI for μ is compared to the applicable regulatory threshold (RT) to determine if a solid waste contains the variable (chemical contaminant) of concern at a hazardous level. The contaminant of concern is not considered to be present in the waste at a hazardous level if the upper limit of the CI is less than the applicable RT. Otherwise, the opposite conclusion is reached.

TABLE 3-25. LIST OF INORGANIC PRIORITY POLLUTANTS (U.S. EPA 1976)

Antimony	Lead
Arsenic	Mercury
Asbestos	Nickel
Beryllium	Selenium
Cadmium	Silver
Chromium	Thallium
Copper	Zinc
Cyanide	

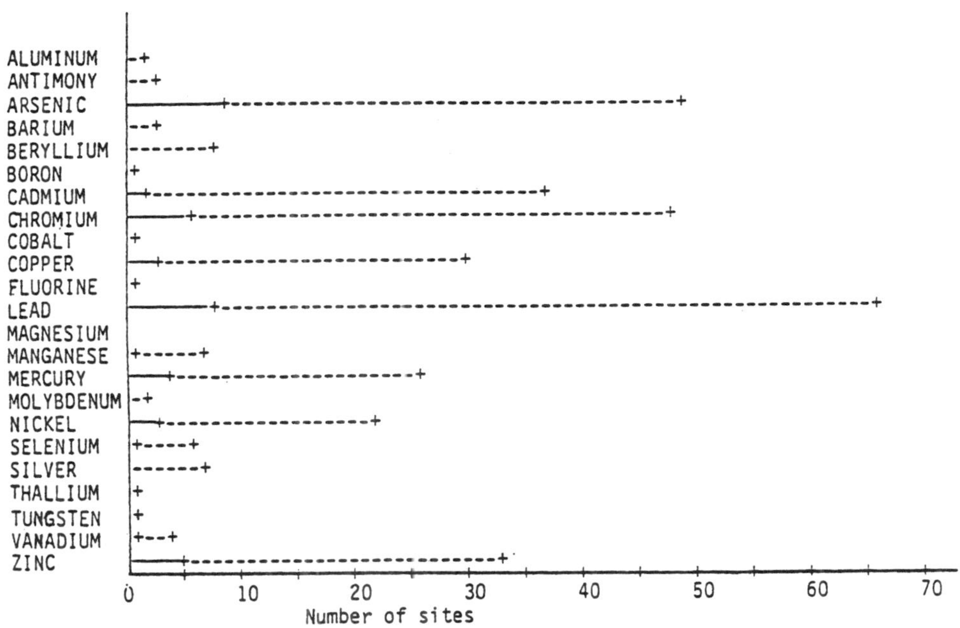

Figure 3-13. Frequency of occurrence of inorganic constituents in soil at FIT sites.

TABLE 3-26. PERCENT OCCURRENCE OF INORGANICS IN SOILS--SUPERFUND AND NON-SUPERFUND SITES (TOTAL NUMBER OF SITES=436)

Pb	15	Hg	6	Al	1	Mo	1
As	11	Ni	5	Sb	1	Tl	1
Cr	11	Ag	2	Ba	1	W	1
Zn	9	Be	2	B	1	V	1
Cd	8	Se	1	Co	1		
Cu	7			F	1		

along with the content of these elements in uncontaminated soils, is listed in Table 3-27. Barium, boron, cobalt, and vanadium concentrations at these polluted sites were within background levels.

Considering inorganics relative toxicity (Table 3-25) and their concentrations and frequency of occurrence in present disposal sites, the following discussion on the behavior of inorganics in soils will be limited to: As, Be, Cd, Cr, Cu, Pb, Hg, Ni, Se, Ag, and Zn. Common industrial use of these elements is listed in Table 3-28.

All the elements under consideration are metals. Arsenic, selenium, and chromium are the only metals listed that can exist as anions in nature. Because of their anionic nature, their behavior in soil will differ from the other heavy metals. A discussion of their unique behavior will follow a general overview of the fate of metals in soils.

Fate of Metals in Soils

The fate of metals added to soil will be controlled by a complex and dynamic system of physical, chemical and biological reactions. Metals, unlike many hazardous organic constituents, cannot be readily degraded or detoxified. Toxic metals represent a long term threat in the soil environment. This threat can be reduced considerably if the heavy metals can be permanently immobilized by either chemical or physical methods.

The heavy metal-soil interaction is such that an accumulation of metals normally occurs on the soil surface and downward transport does not occur to any great extent unless the buffer capacity of the soil is overcome. Soils can be regarded as having a finite loading, or buffer capacity for metals. This capacity is intimately related to the solution and surface chemistry of the soil matrix with reference to the heavy metal in question. Because of the wide range of soil characteristics and various forms by which heavy metals can be added to soil, heavy metal contamination presents a major problem which is highly site-specific.

A schematic diagram illustrating the multiphase equilibria which must be considered when defining the soil solution is shown in Figure 3-14. As Figure 3-14 indicates, at any given time, heavy metal concentrations in the soil

TABLE 3-27. CONTENT OF VARIOUS ELEMENTS IN SOILS (LINDSAY 1979) AND IN FIT SITES

Element	Common Range for Soils (ppm)	Average for Soils (ppm)	Range found in FIT sites (ppm)
Sb	2-10		15[a]
As	1-50	5	0.02-11,700
Ba	100-3,000	430	0.02-0.2
Be	0.1-40	6	2-38
B	2-100	10	0.3[a]
Cd	0.01-0.70	0.06	0.3-118,300
Cr	1-1,000	100	0.04-136,000
Cu	2-100	30	2.2-186,300
Co	1-40	8	2.6[a]
Pb	2-200	10	0.16-466,000
Hg	0.01-0.3	0.03	0.04-83,200
Mo	0.2-5	2	322[a]
Ni	5-500	40	0.4-8,800
Se	0.1-2	0.3	0.9-1.3
Ag	0.01-5	0.05	1.2-18
W	NG	NG	b
V	20-500	100	0.04-2
Zn	10-300	50	0.03-38,000
F	10-4,000	200	110,000[a]
Tl	NG	0.1	b

[a]One site reporting concentration.
[b]No sites reporting concentration.

TABLE 3-28. INDUSTRIAL USE OF SELECTED METALS

As	Pesticides, pigments, glass, textiles, wood preservatives, fireworks, printing, tanning, antifouling paints, enamels, ceramics, lubricating oil, alloys, oil cloth, linoleum, semiconductors, photoconductors.
Be	Rocket fuel, alloys, ceramics
Cd	Electroplating, pigments, alloys, enamels, batteries, rubber, plastics, fungicides, motor oil, textiles
Cr	Pigments, chrome tanning, electroplating, chrome-plating, corrosion inhibitor, varnishes, dye fixers, photography emulsion, defolient
Cu	Brass, dyes, wire, fungicides, alloys, plating, pipes, roofing, paints
Pb	Batteries, paints, glass, insecticides, gasoline additive, ammunition, solder, brass and bronze, pigments
Hg	Paints, catalysis, fungicides, pharmaceutical, plastics, paper products, batteries, electrical apparatus manufacturing
Ni	Steel and alloys, pigments, cosmetics, batteries, electroplating, electrical contacts, gasoline, spark plugs, paints, laquers, cellulose compounds
Se	Glass, photocopy, pigments, electrical industry, paints and inks, cosmetics, paint remover
Ag	Photography, electroplating, mirror manufacturing
Zn	Alloys, metal coating, inks, copying paper, cosmetics, paints, rubber and linoleum, glass

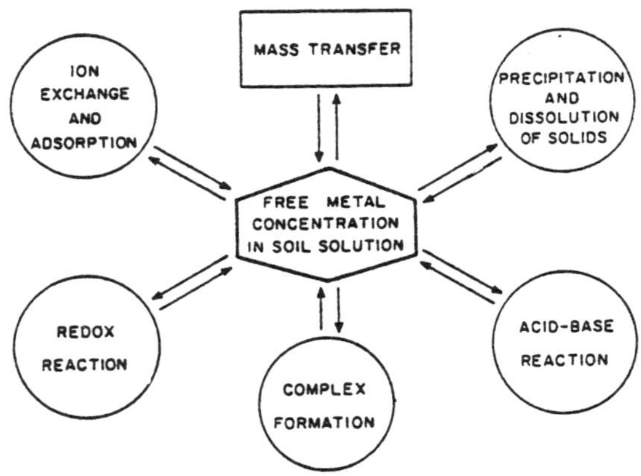

Figure 3-14. Principal controls on free trace metal concentrations in soil solutions (Mattigod et al. 1981). Used by permission, see Copyright Notice.

aqueous phase are governed by a number of interrelated processes, e.g., inorganic and organic complexation, acid-base reactions, redox, precipitation/ dissolution, and interfacial interaction. The ability to predict the concentration of a given heavy metal in the soil solution depends, to a large degree, on the accuracy with which the multiphase equilibria can be calculated.

In terms of industrial spills involving heavy metals, an obvious strategy is to apply a treatment(s) which will minimize the concentration of heavy metal(s) in the soil solution. Ideally the treatment will reduce the metal's aqueous concentration to essentially zero, thus resulting in a leakage (leaching) rate from the site to a biologically insignificant level.

The soil chemistry of heavy metals in soils can be divided for convenience into two interdependent but separate categories: 1) solution chemistry and 2) interfacial chemistry. Both are important in defining the status of heavy metals in soils. However, since one of the critical parameters in evaluating detoxification procedures will be the concentration of a given metal in solution, the topic of soil solution chemistry will be addressed first. The following brief discussion will be directed toward the general principles which will affect the dissolution/precipitation of the solid phase since solid phase formation to scavenge metals from solution is a primary objective of treatment.

Solution Chemistry

The kinetic aspects of dissolution and precipitation reactions involving heavy metals in the soil matrix suffer from the lack of published data. Thus the kinetic component, which in many cases can be critical to decontamination procedures, cannot be assessed easily. The currently acceptable alternative is to assume localized equilibrium occurs in the soil profile (Stumm and Morgan 1981) and then establish the boundary conditions toward which the system is moving. The application of equilibrium thermodynamics to the natural soil system will not only allow one to predict if a given treatment reaction is possible, but it will also provide a prediction of the solution concentration when equilibrium is attained. This approach relies heavily on the accuracy of thermodynamic data that can be found in the literature.

Solid Phase Formation

Consider the dissolution/precipitation of a generic solid phase compound containing a heavy metal cation Me and ligand L in an aqueous solution

$$Me_jL_k = j\ Me^{Z+} + k\ L^{Z-} \quad (3\text{-}3)$$

where j and k are stoichiometric numbers which numerically represent the valence Z of L and Me, respectively. Assuming constant pressure and temperature and ignoring the charge designation of the ions, the solubility equilibrium constant K_{so} is

$$K_{so} = (Me)^j(L)^k \quad (3\text{-}4)$$

where ( ) represent activities and the pure solid phase Me_jL_k is considered to be at unit activity. The value of K_{so} is related to the Gibbs free energy of dissolution by

$$\Delta G = RT \ln Q/K_{so} \quad (3\text{-}5)$$

Where Q, the reaction quotient, has the exact form as K_{so} (see Equation 3-4) but represents the actual ion activity product IAP of Me^{Z+} and L^{Z-} species in solution. When $Q = K_{so}$ the system is at equilibrium and $\Delta G=0$; if $Q/K_{so} > 0$, the solution is oversaturated and precipitation will occur, i.e., $\Delta G > 0$ and Equation 3-3 will proceed in the reverse direction; if $Q/K_{so} < 1$ the solution is undersaturated and dissolution occurs, i.e., $\Delta G < 0$ and Equation 3-3 proceeds as written.

The concept of ionic activity is important in the development of any model of the effect of treatment on heavy metal equilibria. The relationship between the molar concentration of the ith specie m_i in solution and its activity is

$$r_i = a_i/m_i \quad (3\text{-}6)$$

where r_i is the activity coefficient which has the property that $\lim r_i=1$ as $m_i \rightarrow 0$. Since natural systems are not infinitely dilute, r_i cannot be assigned a value of unity. The calculation of r_i for an ionic specie of valence Z_i is given by the Davies equation (Davies 1962)

$$-\log r_i = A\ Z_i^2\ [(\sqrt{I}/1+\sqrt{I}) - 0.3I] \quad (3\text{-}7)$$

where $I = 1/2 \sum_i m_i Z_i^2$ and is summed over all charged species in solution and A is the Debye-Huckel constant which has an approximate value of 0.5 in an aqueous solution of a mixed electrolyte (Stumm and Morgan 1981). Equation 3-7 can be used for solution concentrations of $I \leq 0.5\ ML^{-1}$.

Equation 3-4 can now be written

$$K_{so} = {}^cK_{so}\, r^j\ \ r^k \tag{3-8}$$

where ${}^cK_{so} = m^j\ M^k$, the ion product in terms of molar concentration. Using the Davies equation and rearranging, Equation 3-8 can be written

$$pK_{so} = p^cK_{so} + (j\ Z_{Me}^2 + k\ Z_L^2)\ \left(\frac{\sqrt{I}}{1+\sqrt{I}} - 0.3I\right) \tag{3-9}$$

where pK = -log K. Equation 3-9 shows that the solubility of any solid phase increases with increasing I at moderate salinities. This assumes, of course, that I is the result of the addition of an inert electrolyte. In more concentrated systems ($I > 1ML^{-1}$) the possibility exists that the value of r_i may become greater than 1 and the solubility of the solid phase may decrease in the presence of an inert electrolyte (Stumm and Morgan 1981).

The use of solubility diagrams is a convenient technique of showing how the solubility of heavy metals compounds varies with pH and at the same time it allows some prediction as to what solid phase regulates heavy metal activity in the soil solution. An example is the solubility diagram for lead in soils (Lindsay 1979) which shows the relative solubilities of lead silicate, phosphate, and carbonate as predicted in pure water in equilibrium with atmospheric CO_2. This is shown in Figure 3-15 which also incorporates the solubility isotherm for tricalcium phosphate (TCP), dicalcium phosphate dihydrate (DCPD), and hydroxyapatite (HA). The solubility data presented supports the idea that soil phosphorus may be a factor in regulating Pb^{2+} ion activity as originally suggested by Nriagu (1974).

Santillan-Medrano and Jurinak (1975) obtained experimental data from soil column studies using both Pb and Cd. The data in Figure 3-16 show that in the calcareous Nibley soil: 1) the solubility of Pb decreases with increasing soil pH, which is the usual trend for most heavy metals, 2) Pb phosphate compounds could be regulating the activity of Pb^{++} ion in solution, and 3) mix compound precipitation cannot be precluded between pH 7.5-8.0 because of the convergence of the solubility isotherms for $PbCO_3$, $Pb(OH)_2$, $Pb_3(PO_4)_2$, and $Pb_5(PO_4)_3Cl$. Similar comments can be made for the solubility of Cd in Nibley soil (Figure 3-17) with the additional comment that the solubility of Cd is considerably higher than Pb at any pH and at the high pH values the soil solution is undersaturated with respect to the compounds considered.

Heavy Metal Complexation in Soil Solution

Heavy metal ions form many soluble complexes with both organic and inorganic ligands. The effect of complexation is to increase the solubility

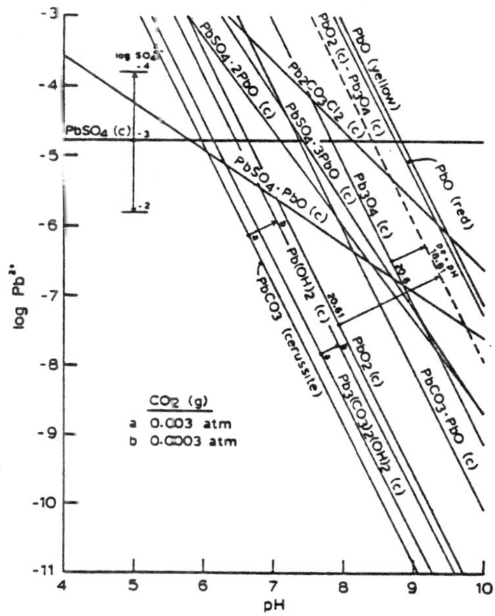

Figure 3-15. The solubility of vairous lead oxides, carbonates, and sulfates when SO_4^{2-} and Cl^- are 10^{-3} M and CO_2 is 0.003 atm or as specified (L'ndsay 1979). Used by permission, see Copyright Notice.

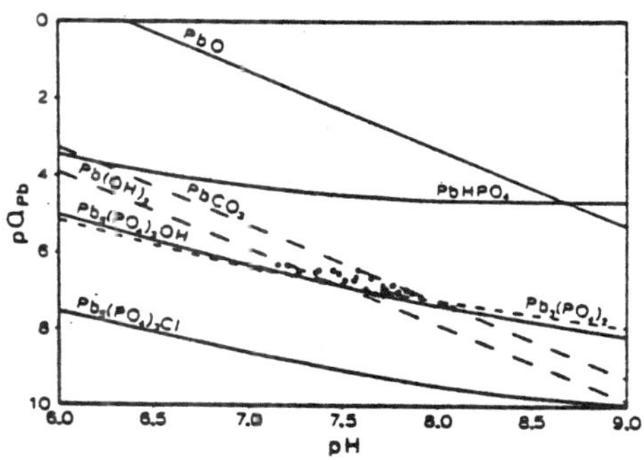

Figure 3-16. The solubility diagram for Pb in Nibley clay loam soil (Santillan-Medrano and Jurinak 1975). Used by permission, see Copyright Notice.

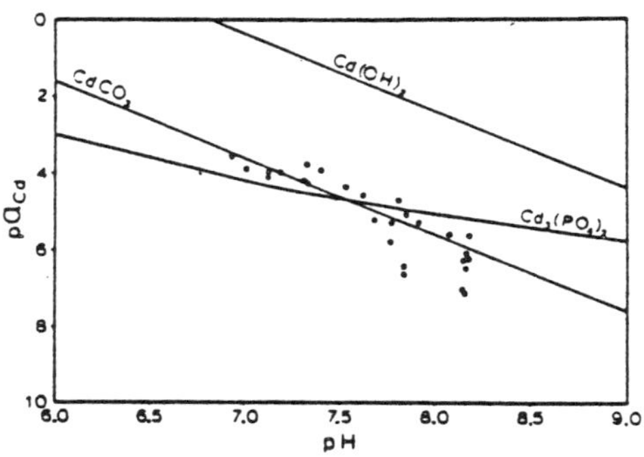

Figure 3-17. The solubility diagram for Cd in Nibley clay loam soil (Santillan-Medrano and Jurinak 1975). Used by permission, see Copyright Notice.

of the solid phase of which the complex ions are constituent. At any time, the total concentration of metal Me_T in the soil solution is the sum of the free ion concentration $[Me^{Z+}]$ and the concentrations of all organic and inorganic metal complexes. A general expression considering complex formation with any type of ligand L or its protinated form H_xL plus hydroxo complexes is given (Stumm and Morgan 1981)

$$Me_T = [Me^{Z+}]_{free} + \Sigma[Me_m\ H_k\ L_n\ (OH)_i] \qquad (3\text{-}10)$$

where m, n, i or K $\geq$ 0. In Equation 3-10 ion pairs are included in the complex designation.

The dominant complex specie in the soil solution can be significant when dealing with the transport of heavy metal through the matrix. The factors involved are: 1) the free metal concentration, 2) the dominant complex specie, and 3) the charge on the complexes. Inorganic complexes which can be expected to be formed with bivalent metals are given in Table 3-29.

Doner (1978), using soil columns, studied Cd, Cu, and Ni transport as affected by Cl^- and ClO_4^- ions. This study clearly showed the effect of the chloride complex on heavy metal movement in soils. The breakthrough curves for the three metals are shown in Figures 3-18, 3-19, and 3-20. The mobilities were found to be in order with the dissociation constants of the chloro-complexes of the metal ions, i.e., Cd > Cu > Ni.

An example of how complexation affects solubility is shown in Figure 3-21 where the solubility of CuO is plotted as a function of (H^+). The observed solubility, in the pH range of most natural systems, is attributed to the Cu^{2+} ion and the two carbonate complexes.

TABLE 3-29. SOME PROBABLE BIVALENT METAL COMPLEXES WITH INORGANIC LIGANDS IN SOIL SOLUTIONS (FROM MATTIGOD ET AL. 1981)
Used by permission, see Copyright Notice.

OH	CL	SO_4	CO_3	PO_4
MOH^+	MCL^+	$MHSO_4^+$	$MHCO_3^+$	$MH_2PO_4^+$
$M(OH)_2^0$	MCL_2^0	MSO_4^0	$M(HCO_3)_2^0$	$MHPO_4^0$
$M(OH)_3^-$	MCL_3^-	$M(SO_4)_2^{2-}$	MCO_3^0	MPO_4^-
$M(OH)_4^{2-}$	MCL_4^{2-}		$M(CO_3)_2^{2-}$	

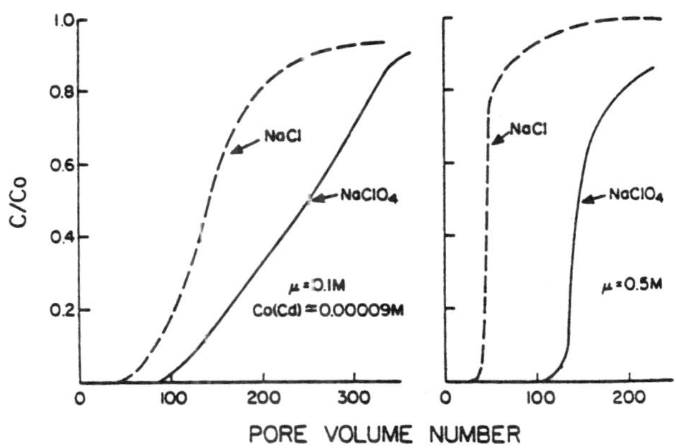

Figure 3-18. Breakthrough curves for Cd as affected by Cl^- and ClO_4^- ions (Doner 1978). Used by permission, see Copyright Notice.

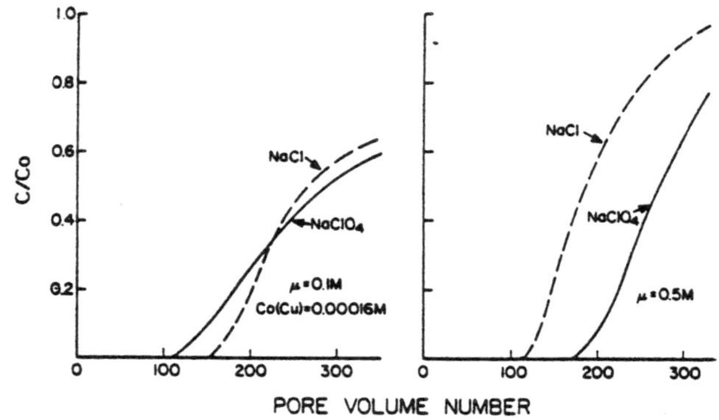

Figure 3-19. Breakthrough curves for Cu as affected by Cl^- and ClO_4^- ions (Doner 1978). Used by permission, see Copyright Notice.

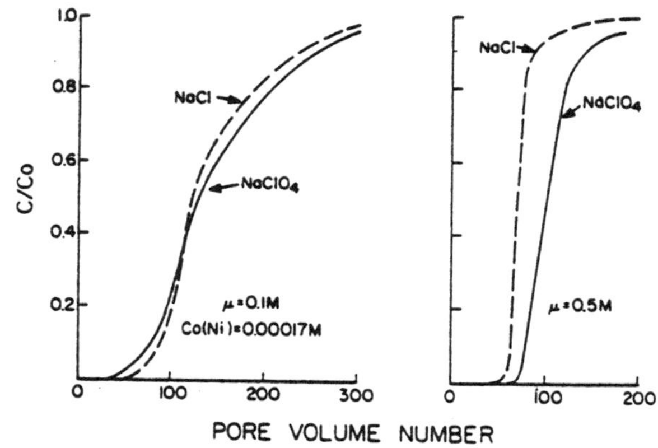

Figure 3-20. Breakthrough curves for Ni as affected by Cl^- and ClO_4^- ions (Doner 1978). Used by permission, see Copyright Notice.

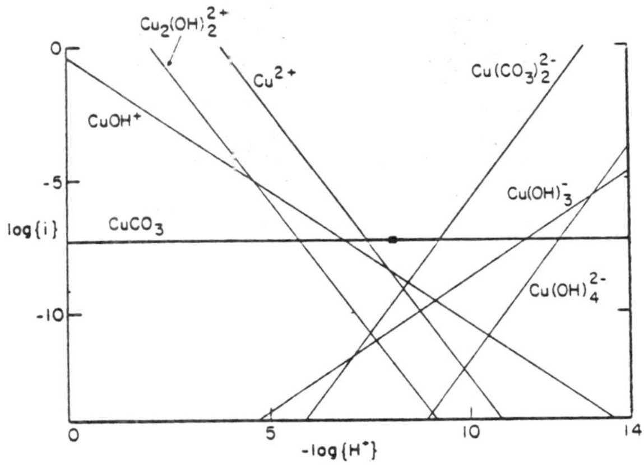

Figure 3-21. Solubility of CuO as a function of log $\{H^+\}$ (25°C., I = 0, log P_{CO_2} = -3.52) (Schindler 1967). Used by permission, see Copy-Right Notice.

Organic complexation of heavy metals in soil is not as well defined as inorganic complexation because of the difficulties of identifying the organic ligands. The recent disposal technique of land application of sewage sludge has fostered considerable interest in heavy metal complexation by fulvic acid extracted from sludge (Sposito et al. 1977, 1979, 1981; Behel et al. 1983, and Baham and Sposito 1983).

Sposito and co-workers have ceveloped an interesting conceptual fulvic acid model to interpret potentiometric titration data involving heavy metals. In brief, a fulvic acid extract is treated as an assembly of hypothetical macro molecules (similar to proteir molecules). This system has been determined to contain four classes of ac-dic functional groups which can react with heavy metals through proton exchange (Sposito and Holtzclaw 1977). Of these four groups, only two appear to be important in complexing the metals studied. Table 3-30 gives experimental anc estimated values of ^cK_i for i = 1, 2. These studies were conducted in an ionic medium of 0.1M $KClO_4$. K_1 and K_2 refer to the stability constarts of the two functional groups of the fulvic acid model. It is of interest to note that log ^cK_i are comparable to the log K values for metal complexes with alphatic acids such as citric and acetic (Martell and Smith 1977).

Computer Simulation of the Soil Solution

The multiphase equilibria and the myriad of potential interactions involving heavy metals make the soi- solution a prime candidate for computer simulation. However, because of the limitations in analytical techniques, the

TABLE 3-30. ESTIMATED VALUES OF LOG $^cK_1$ AND LOG $^cK_2$ (MODIFIED FROM SPOSITO ET AL. 1981)
Used by permission, see Copright Notice

Metal cation	$\log {}^cK_1$	$\log {}^cK_2$
Mg^{2+}	2.71	0.69
Ca^{2+}	(3.12)[a]	1.23
Mn^{2+}	3.93	2.23
Fe^{2+}	3.96	2.28
Ni^{2+}	3.81	2.08
Cu^{2+}	(3.88)	(2.11)
Zn^{2+}	3.54	1.74
Cd^{2+}	(3.04)	(2.27)
Pb^{2+}	(4.22)	(2.62)

[a]Parentheses denote measured values.

complete validation of the free ion and metal complex concentrations predicted is difficult.

An excellent example of an equilibrium model which has been used extensively in soil chemistry is GEOCHEM. This model (Mattigod and Sposito 1979) was developed from the program REDEQL2 (McDuff and Morel 1974). GEOCHEM expands REDEQL2 to account for ion exchange and expands the base for equilibrium constants particularly with regards to inorganic and organic metal complexes.

Solid-aqueous Interface Exchange
(Outer Sphere Complexes)

As cations, metals are capable of participating in cation exchange reactions on negatively charged surfaces of layer silicates (clays), oxides, and organic matter. Negatively charged mineral surfaces arise from either substitution of structural cations by ions of lower charge (permanent negative charge) or by dissociation of hydrogen ions of surface edges (pH-dependent charge). The negative charge of organic matter is totally pH dependent and arises from the dissociation of functional groups. Cations are held loosely in the vicinity of these negative charged sites by electrostatic forces.

These sites are "non-specific" (outer sphere complexes) and cations held at these sites can be displaced by other cations.

Specific Adsorption (Inner Sphere Complexes)

Specific adsorption forming inner sphere complexes at the interface is distinguished from the exchangeable state in terms of the increase in binding energy between the metal and the surface. Metals specifically adsorbed are apparently held by electrical forces as well as by additional forces including covalent bonding, Van der Waals forces, and steric fit at the site. The term specific implies that other metal cations do not effectively compete for the surface site occupied by the specifically adsorbed metal cation. This is the practical distinction between the non specific exchangeable and specifically adsorbed states.

Specific adsorption is the most important interfacial mechanism controlling soil-water concentration of a particular metal ion at low concentrations. As more of this metal ion is added, a specific adsorption capacity or limit is apparently reached and additional metal ions enter less energetically bound exchange positions. Specifically adsorbed metals are relatively immobile and unaffected by moderate concentrations of "macro" salt cations, i.e., Ca, Mg, Na and K. Metals held by exchange mechanisms will, however, be subject to exchange and subsequent leaching as "macro" cation concentrations increase.

The sorption capacity (exchange and adsorption) of a soil is determined by the number and kinds of sites available. Sorption of metals has been correlated with such soil properties as clay content, soil organic matter, Fe and Mn oxides and calcium carbonate, all of which contribute to the soils cation adsorption potential. Sorption processes are affected by these various soil factors, by the form of the metal added, and by the nature of the solvent introduced along with the metal. The result of these interactions may increase or decrease the movement of heavy metals with the soil water.

Factors Affecting Sorption

Effect of Anions--

Metal cations will form ion-pairs, complexes and chelates with inorganic and organic anions. Anionic ligands form metal associations that have a lower positive charge than the "free" metal cation. The resulting association may be uncharged or it may have a net negative charge. Benjamin and Leckie (1982) stated that the interaction between metal ions and complexing ligands may result in either a complex that is weakly sorbed to sorbent surfaces or in a complex that is sorb more strongly than the "free" metal ion. In general, the decrease in positive charge on the complexed metal reduces sorption to a negatively charged surface. One noted exception is the preferential sorption of hydrolyzed metals (M_eOH^+) versus the "free" bivalent metal (James and Healy 1972).

Within normal concentration of electrolytes in soil solution, Elrashidi and O'Connor (1982) found no measurable changes in Zn sorption by soils due to complex formation of Zn with Cl^-, NO_3^{2-}, or SO_4^{2-} ions. Under these conditions, anion complex formation did not compete with the highly selective sorption sites for Zn. The highest anion concentration studied was 0.1 ML^{-1}.

Doner (1978) using 0.1 to 0.5 M NaCl leaching solutions, concluded that the increased mobility, relative to $NaClO_4$ leachate, of Ni, Cu, and Cd through a soil column was due to complex formation of the metals with Cl^-. The mobility of Cd increased more than that of Ni and Cu, Ni being the least mobile. These observed mobilities are in the same order as that of the stability constants of the chloride complexes of these metals (Table 3-31). Complex ion formations may be represented by the equation:

$$Me^{2+} + nCl^- = MeCl_n{}^{2-n} \tag{3-11}$$

where n is the number of Cl^- ions complexed with the metal ion, Me^{2+}. The stability constant, K, is defined as:

$$K = (MeCl_n{}^{(2-n)})/(Me^{2+})\ (Cl^-)^n \tag{3-12}$$

where ( ) denotes activity. The larger the log K value, the more stable the complex. Doner et al. (1982), using geothermal brine solutions, again found increased mobility of Cu, Pb, Zn, and Cd due to complex formation with Cl^-. Mercuric-chloride complexes are not as strongly adsorbed as the Cl^- free species (Kinniburgh and Jackson 1978). The effect is pH-dependent, with the greatest complexing of Cl^- and Hg(II) at low pH (4.5). These authors caution against generalization regarding Cl^- inhibitory effect on sorption of other metal complexes.

Complex formation between metals and organic ligands may also affect metal sorption and hence mobility. This complexation may be of particular importance when organic waste is involved at the same spill site. Bowman et al. (1981) demonstrated drastic reduction in Ni sorption in the presence of EDTA, a synthetic chelating agent. Elrashidi and O'Connor (1982) likewise found that EDTA and DTPA can compete effectively with the soil sorbing sites for Zn. Both papers emphasize the great potential for leaching of metals in the presence of organic ligands.

Application of organic matter to soils, e.g. sludge, may, in some cases, increase the mobility of metals if this material contains significant levels of dissolved organic matter. The fulvic acid fraction of sludges has been shown to complex metals (Sposito et al. 1979; Sposito et al. 1981; Boyd et al. 1979; Sposito et al. 1982). Fulvic acid is soluble between pH 1 to 11 (Holtzclaw et al. 1976). Baham's et al. (1978) study illustrated the relative order of affinity of fulvic acid fraction of sludge for metals: Cu $\geq$ Fe > Zn $\geq$ Ni > Cd. Stability constants of metal organic complexes may govern their relative mobilities. Increased mobility of metals in a sandy soil and a sandy loam soil with the addition of sludge was also observed by Gerritse et al. (1982). The authors attributed this increase not only to complexation by dissolved organic compounds, but also to high ionic strength of the soil solution. Khan et al. (1982) showed that the mobility of metals through soil followed the order: Cu > Ni > Pb > Ag > Cd. The high mobility of Cu and Ni was attributed to their high complexing nature with soluble soil organic matter. Overcash and Pal (1979) reported that the order metal-chelate stabilities is Hg > Cu > Ni > Pb > Co > Zn > Cd.

TABLE 3-31. STABILITY CONSTANTS FOR Cl COMPLEXES TO Ni(II), Cu(II), AND Cd(II). FROM MATTIGOD ET AL. (1981)

Metal	Stability constant, log
Ni(II)	-0.43
Cu(II)	0.40
Cd(II)	1.98

The extent of complexation between a metal and soluble organic matter, and hence mobility, depends on the competition between metal-binding surface sites and the organic ligand for the metal. Metals with a high affinity for organic chelates, such as Cu and Ni, will readily be complexed by the organic matter and will be subject to leaching. Metals such as Cd and Zn that do not form highly stable complexes with organic matter will be less affected by its presence in the soil solution. Elliott and Denneny (1982) stated that soil metal binding sites should outcompete organic ligands for Cd if the log stability constant for the metal-organic complex is less than six (Table 3-32). Their study showed that only NTA and EDTA reduced Cd sorption by soil.

Effect of pH on Sorption--

Most fuctional groups of chelates, as well as most complex and ion-pair formers are weak acids, thus the stability of the metal complex is pH dependent with little association in acid media. The degree of association inreases with pH to a maximum often determined by some competing alternative reaction such as precipitation. Elliott and Denneny's (1982) study illustrated this effect of soil solution pH on organic ligands. As soil solution became increasing acidic, the influence of the organic ligands (Ox^0) on metal sorption diminished (Figure 3-22) due to a decreasing ability of the ligands to bind the metal relative to H^+. As the pH increases, the concentration of metal-ligand complex again decreases due to the hydrolysis of Cd ($CdOH^+$, $CdOH_2^0$) at pH>8.

pH also directly effects sorption (McBride and Blasiak 1979, Kuo and Baker 1980, Harter 1983). Many sorption sites in soil are pH dependent so that as the pH decreases, the number of possible negatively charged sites diminishes. In acid media, metals face competition for available permanent charged sites by H^+ and Al^{3+}. Figure 3-23 shows the impact of soil pH on Cd sorption by three soils. As is true for all cationic metals, Cd sorption increased with pH.

Cavallaro and McBride (1980) found that soils of higher pH retain more Cd and Cu. Copper sorption showed a stronger pH dependence than Cd. This finding is consistent with the hypothesis that hydrolysis of Cu at pH 6 increases its retention by soil. Cadmium does not hydrolyze until pH 8 (see

TABLE 3-32. OVERALL STABILITY CONSTANTS OF Cd COMPLEXES (FROM ELLIOTT AND DENNENY 1982)
Used by permission, see Copyright Notice

Ligand	Log of equilibrium constant[a] K_1	K_2	K_3	K_4
OH^-	3.81	9.52	8.70	8.85
Ac^-	1.66	2.75	2.73	
Ox^{2-}	3.46	5.24		
NA^{3-}	10.3			
$EDTA^{4-}$	17.2			

[a]At 25°C, adjusted to an ionic strength of 0.025 M with the Davies Equation: $K_n = (CdL_n)(L^n(Cd^{2+})$, where L is the ligand.

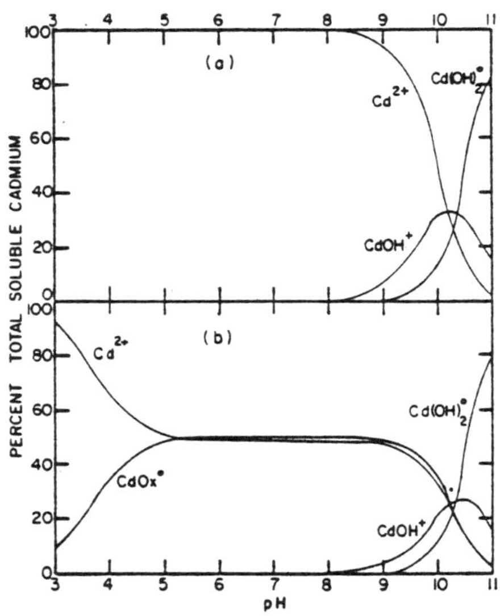

Figure 3-22. The pH-dependent speciation of Cd in: (a) the absence of organic ligands Ox°, and (b) the presence of 3.5×10^{-4} M Ox. Determined using constants in Table 3-32 (Elliott and Denneny 1982). Used by permission, see Copyright Notice.

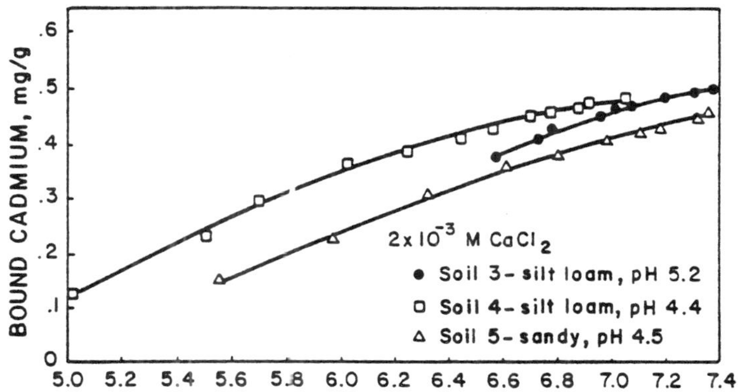

Figure 3-23. Cadmium binding as a function of pH (Tirsch et al. 1979). Used by permission, see Copyright Notice.

Note: The sandy textured soil, #5, sorbed the least Cd over this pH range. Differences in sorption of Cd between soils 3 and 4 may be due to the higher organic matter content in soil 4.

Figure 3-22). Table 3-33 shows the predicted percent distribution of metal species with pH (Harter 1983). The increasing importance of the hydrolyzed metal species with pH should effect metal sorption.

Jenne (1968) stated that hydrous oxides of Fe and Mn play a principal role in the retention of metals in soils. Solubility of Fe and Mn oxides is pH-related. Below pH 6, the oxides of Fe and Mn dissolve releasing sorbed metals ions to solution (Essen and El Bassam 1981).

Effect of Other Cations Present in Solution--

For specific adsorption sites, heavy metals are preferentially adsorbed over alkaline metals (Na, K, Ca, Mg) (Milberg et al. 1978; Kinniburgh and Jackson 1982). However, when specific adsorptions sites become saturated, exchange reactions dominate and competition for these sites with soil "macro" cations become important. Cavallaro and McBride (1978) found that sorption of Cu and Cd decreased in the preserce of 0.01M $CaCl_2$. They attributed this decrease, not to complexation with Cl, but to competition with Ca for sorption sites (Figure 3-24). Cadmium sorption (not shown) was more affected by the presence Ca than was Cu. The mobility of Cd may be greatly increased due to such competitions. Likewise, Harter (1979) indicated the Ca in solution had a greater affect on Pb sorption than on Cu.

TABLE 3-33. EFFECT OF pH ON PROBABLE SOLUTION PERCENT COMPOSITION OF DIFFERENT ION SPECIES[a] (FROM HARTER 1983)
Used by permission, see Copyright Notice

pH	4	5	6	7	8
Pb^{2-}	100	100	98	83	33
$PbOH^-$			2	17	66
$Pb(OH)2^0$					1
CU^{2-}	100	100	96	33	1
$CuOH^-$			2	7	1
$Cu(OH)2^0$			2	56	92
$CuCO_2^0$				4	6
Zn^{2-}	100	100	98	83	31
$ZnOH^-$			2	17	64
$Zn(OH)_2^0$					5
Ni^{2-}	100	100	100	100	99
$NiOH^-$					1

[a]The solution is assumed to be in equilibrium with atmospheric CO_2 and it is further assumed that anions other than OH^-, HCO_3^-, and $CO_3^=$, are not present in large enough quantities to make a contribution to solution complexes of these metals.

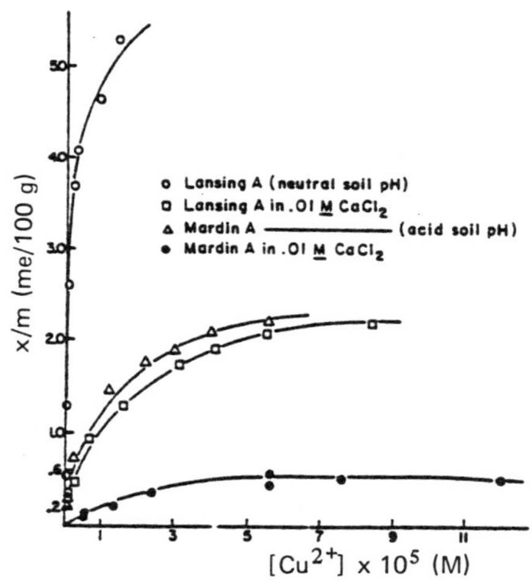

Figure 3-24. Cu^{2+} adsorption isotherms on Lansing and Mardin A horizon soils in the absence and presence of 0.01M $CaCl_2$ (Cavallaro and McBride 1978). (Note: under the same conditions the neutral Lansing A soil sorbed more Cu^{+2} than the acid Mardin A soil.) Used by permission, see Copyright Notice.

Oxidation-reduction

The metals under discussion at the present time, Cd, Zn, Pb, Ni, and Cu, normally have one oxidation state in the soil environment and are not directly affected by changes in the oxidation-reduction (redox) potential in soils. The redox potential does affect the solid phase of soils, in particular Fe and Mn oxides. Under reduced conditions Fe and Mn oxides are solubilized, releasing associated heavy metals. Sims and Patrick (1978), however, found that Zn and Cu did not remain in solution but became associated with exchange and organic sorption sites. The solubility and decomposition of organic matter is also affected by soil redox.

Mercury, As, Se and Cr all have multiple oxidation states in soils. Their reaction in soil is directly dependent on the redox potential of the soil. (This subject will be discussed in greater detail in a later section.)

Adsorption Isotherms--

Langmuir and Freundlich equations have become popular as a means of describing solid-solution reactions in soils. The Langmuir expression provides a theoretical adsorption maximum and a coefficient which is theoretically related to bonding energy. The linearized form of the Langmuir equation:

$$\frac{C}{x/m} = \frac{C}{(x/m)_{max}} + \frac{1}{(x/m)_{max}\ b} \qquad (3\text{-}13)$$

where C is the concentration or activity of the free metal in solution, x/m is the quantity (meq/100 g) of the metal ion adsorbed by the soil, $(x/m)_{max}$ is the maximum adsorption capacity of the soil, and b is the coefficient related to bonding energy. When $\frac{1}{x/m}$ is plotted as a function of C, the slope will be the reciprocal of the adsorption capacity, $(x/m)_{max}$, and the intercept will be $\frac{1}{(x/m)max\ b}$. A typical linear Langmuir adsorption isotherm is given in Figure 3-25.

The Freundlich expression is essentially empirical but has the virtue of being of appropriate form to fit the graphical shape (Figure 3-26) of many metal sorption data. The linearized form of the Freundlich isotherm equation is:

$$\log (x/m) = 1/n \log C + \log K \qquad (3\text{-}14)$$

where x/m and C have defined previously and 1/n and K are constants fitted from the experimental data. A typical Freundlich adsorption isotherm is given in Figure 3-27. The two sections of this plot, the linear form of the Freundlich equation followed by the curvilinear section, has been interpreted to indicate two different mechanisms of retention of Ni by this soil.

Cavallaro and McBride (1978) used the Langmuir expression to describe sorption of Cu and Cd by two soils. Table 3-34 gives values for the maximum

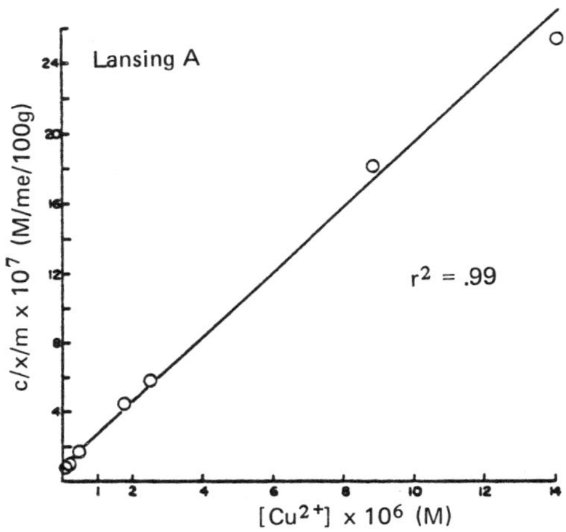

Figure 3-25. Langmuir adsorption isotherm for Cu^{2+} adsorption on the Lansing A soil (Cavallaro and McBride 1978). Used by permission, see Copyright Notice.

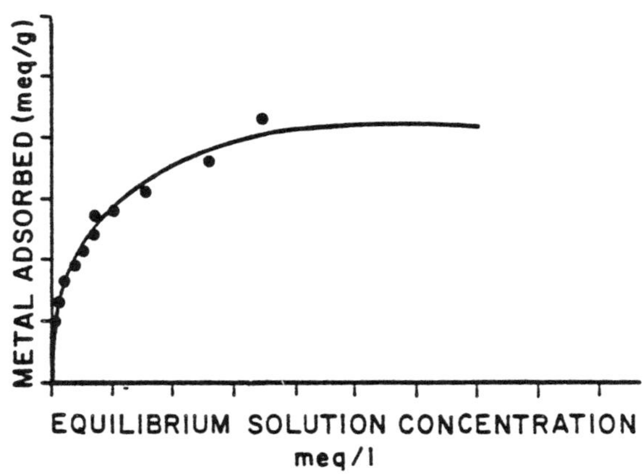

Figure 3-26. Typical adsorption isotherm for metals and soil.

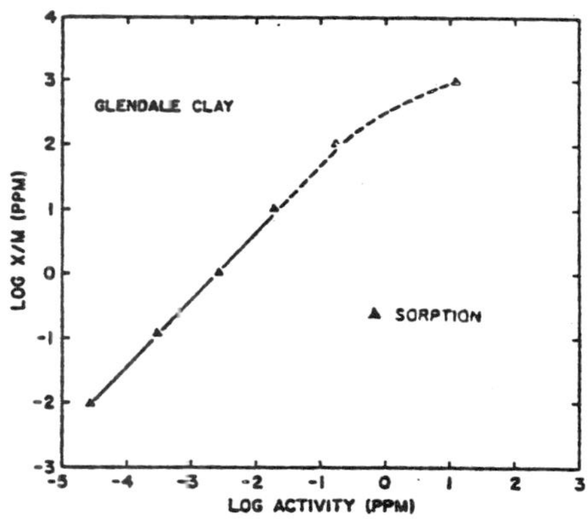

Figure 3-27. Nickel sorption by Glendale soil, 0.01N $CaCL_2$.

TABLE 3-34. CALCULATED LANGMUIR PARAMETERS FROM SOIL ADSORPTION OF Cu^{2+} AND Cd^{2+} (FROM CAVALLARO AND MCBRIDE 1978)
Used by permission, see Copyright Notice

	$(x/m)_{max}$(meq/100 g)		$b\times10^{-5}$ (M^{-1})		r^2	
	No Salt	0.01 M $CaCl_2$	No Salt	0.01 M $CaCl_2$	No Salt	0.01 M $CaCl_2$
Lansing A-Cu^{2+}	5.6 ± 0.4	2.2 ± 0.2	15.6	3.2	0.99	0.99
Lansing A-Cd^{2+}	4.0 ± 0.5	0.2 ± 0.05	1.7	∼0.2	0.97	a
Lansing C-Cd^{2+}	4.1 ± 0.8	1.2 ± 0.1	1.3	2.3	0.95	0.99
Mardin A-Cu^{2+}	2.3 ± 0.2	0.6 ± 0.1	1.5	0.5	0.99	0.94
Mardin A-Cd^{2+}	1.1 ± 0.1	0.0	1.9	a	0.99	a

[a]Adsorption was too low to permit calculation.

adsorption capacity ($(x/m)_{max}$) and for the "bonding constant" (b). The authors determined that the sorption maximum for Cu^{2+} was slightly less than the CEC of the Lansing A soil (neutral soil pH), while the maximum Cu^{2+} adsorbed by Mardin soil (acid soil pH) was less than half the soil's CEC. This indicates that Cu was not held as strongly on the Mardin soil. The adsorption maximum for Cd was less than Cu. The bonding constant (b) reflected the weaker retention of Cd relative to Cu in the Lansing A soil, although this effect is not consistent for Mardin A soil. The presence of 0.01M $CaCl_2$ lowered the values of $(x/m)_{max}$ and b by introducing Ca^{2+} ions competition for exchange sites.

Bowman et al. (1981), using K values derived from the Freundlich equation, compared the sorbing tendencies of 11 soils for Ni (Table 3-35). The Tuff sample (unweathered rock) showed much lower affinity for Ni than did the soils studied. Table 3-36 (Bowman et al.. 1981) illustrates that the effect of Ca concentration and the effect of anions, Cl^- and ClO_4^- on the sorption resulted in a lowering of sorption. Sorption in the non-complexing perchlorate (ClO_4^-) system was greater than in the 0.1N $CaCl_2$ system, but was much less than that in the 0.01N $CaCl_2$ system. The authors concluded that Ca competition for sorbing sites, rather than Cl^- complex formation, played the greater role in reducing Ni sorption.

Elrashidi and O'Connor (1982) compared the affinity of Zn on nine soils (Table 3-37) using Freundlich parameters. The authors also used these parameters in a correlation of sorption affinity with various soil properties. Clay percent, CEC, and specific surface were all found to be significantly correlated to Zn sorption.

From the above examples it is evident that the mathematical form of the Langmuir and Freundlich isotherm expressions, have proven valuable in interpreting metal behavior in soils. At present, however, the applicability of adsorption isotherm equations to the interpretation of soil chemical phenomena is a subject of controversy. For further discussion of this controversy see Elprince and Sposito (1981); Griffin and Au 1977; Veith and Sposito 1977; and Sposito (1979).

The preceding discussion dealt with the general behavior of divalent heavy metals in the soil environment. Although principles of sorption, precipitation, etc., are applicable to the fate of all ions in soil, some specific aspects of the chemistry of As, Se, Hg, Cr, Ag, and Be must be discussed individually.

Anions in the Soil Environment--

Of the metals under consideration in this text, arsenic, selenium and chromium, in one of its oxidation states, exist as anions in soils. Like cationic metals, these anions will participate in precipitation and sorption reactions.

Soil particles, though predominantly negatively charged, may also carry some positive charges. The oxide surfaces, notably iron and aluminum hydrous-oxides, can generate a significant number of positive charges as the pH decreases. For example, at pH 8-8.5 the positive and negative charges on

TABLE 3-35. FREUNDLICH PARAMETERS AND CORRELATION COEFFICIENTS FOR SORPTION OF Ni IN 0.01N $CaCl_2$ BY 12 SOILS. DATA FOR INITIAL NI CONCENTRATIONS OF 100 AND 1,000 ppm OMITTED (FROM BOWMAN ET AL. 1981)
Used by permission, see Copyright Notice

Soil	K	1/n	R^2
Carjo	262	0.95	0.999
Puye	74.0	0.87	0.993
Tuff	8.23	0.92	0.997
Chem Bottom	517	1.18	0.999
R-28	534	1.02	1.000
R-30	380	1.01	0.999
Glendale	650	1.03	0.998
Reagen	541	1.02	0.999
Palouse	396	0.96	0.999
Doak	199	0.99	0.999
Harvey	233	0.97	0.997
Lea	441	0.97	0.998

TABLE 3-36. FREUNDLICH PARAMETERS FOR SORPTION OF Ni IN DIFFERENT Ca SOLUTIONS BY FOUR SOILS. DATA FOR INITIAL Ni CONCENTRATIONS OF 100 AND 1,000 ppm OMITTED (FROM BOWMAN ET AL. 1981)
Used by permission, see Copyright Notice

Matrix	0.01N	$CaCl_2$	0.1	$CaCl_2$	0.1N	$Ca(ClO_4)_2$
Soil	K	1/n	K	1/n	K	1/n
Carjo	262	0.95	43.8	0.91	70.2	0.94
Glendale	650	1.03	534	0.95	691	0.98
Doak	199	0.99	78.2	0.81	98.6	0.85
Tuff	8.23	0.92	3.45	0.87	5.05	0.87

TABLE 3-37. FREUNDLICH PARAMETERS FOR SORPTION OF ZN IN 0.01N $CACL_2$ BY THE SOILS STUDIED[a] (FROM ELRASHIDI AND O'CONNOR (1982))
Used by permission, see Copyright Notice

Soil	Freundlich parameters			
	Part 1[b]		Part 2[b]	
	K	1/n	K	1/n
Carjo silt loam	134	0.84	92.8	0.47
Puye sandy loam	61.0	0.88	29.9	0.52
Tuff loamy sand	29.2	0.84	17.2	0.45
Chem B sand	1434	1.04	108	0.42
R28 sand	829	0.91	102	0.44
Glendale clay	2732	0.94	469	0.44
Reagan clay loam	808	0.97	302	0.53
Lea sandy loam	1318	1.03	293	0.46
Harvey sandy loam	557	0.97	189	0.41

[a]Correlation coefficients for all Freundlich adsorption equations are > 0.98.
[b]Parts 1 and 2 refer to linear isotherms at low and high Zn solution concentrations, respectively.

the surface of goethite (FeOOH) are balanced and the surface exhibits no net charge. This pH region is known as the zero point of charge (pzc). At pHs above the pzc the goethite surface will have a net negative charge and at pHs below the pzc the surface will have a net positive charge (Figure 3-28). Oxides can differ widely in their pzc.

It has also been suggested (Van Olphen 1963) that broken bonds at the edges of the particles of clay minerals, particularly of kaolin types, may carry localized positive charges. This might be expected, particularly at low pH values.

Clay minerals and oxides exert a strong preference for some anions in comparison to other anions, indicating the existence of chemical bonds between the solid surface and the specific anion. Phosphate has been the most extensively studied anion that exhibits this specific sorption phenomena. Phosphate ions are bonded strongly, or "fixed," by clay minerals and oxides. The sorption capacity for phosphate and other anions is, however, very small relative to cation sorption capacity of soils.

Chromium--Chromium exists in four possible forms in soils--trivalent forms, the Cr^{3+} cation and the CrO_2^- anion and hexavalent forms, $Cr_2O_7^{2-}$,

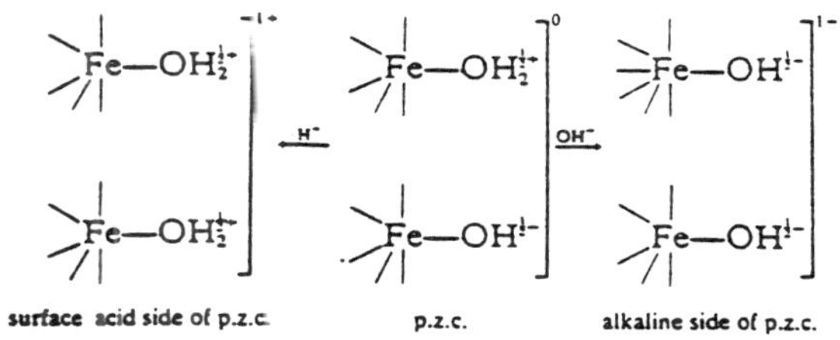

Figure 3-28. Zero point of charge (pzc) on an iron hydrous oxide (Greenland and Hayes 1981). Used by permission, see copyright notice.

and CrO_4^{2-}. Figure 3-29 illustrates the relationships between pH and redox potential and speciation of chromium in water.

Hexavalent chromium is the major chromium species used in industry. Forms of Cr(VI) in soils are as chromate ions CrO_4^{2-}, predominant at pH 6, and as dichromate ions $Cr_2O_7^{2-}$, predominant at pH 2-6 (Figure 3-29). The dichromate ions possess a greater health hazard than chromate ions. Both Cr(VI) ions are more toxic than Cr(III) ions.

Because of Cr(VI) anionic nature, its association with soil surface is limited to positively charged exchange sites, the number of which decreases with increasing soil pH (Figure 3-30, Bloomfield and Pruden 1980; James and Bartlett 1983). No precipitates of any hexavalent compounds of chromium were observed in the pH range 1.0 to 9.0 (Griffin and Shimp 1978). Thus Cr(VI) is highly mobile in soil.

Chromium (III) is considered the stable form of chromium in soil. The predominant cationic form of Cr(III) in aqueous solution is $[Cr(H_2O)_6]^{3+}$. Increasing pH results in the formation of hydroxy species such as $Cr(OH)^{2+}$, $Cr(OH)_2^{+}$, ... $Cr_6(OH)_{12}^{6+}$. All species of Cr(III) are readily sorbed by soils. In a study of the relative mobility of metals in soils of pH 5, Cr(III) was found to be the least mobile (Griffin and Shimp 1978). Hydroxy species of Cr(III) precipitate at pH 4.5. Complete precipitation of the hydroxy species occurs at pH 5.5 (Figure 3-31).

At most soils' pH and redox, Cr(VI) is reduced to Cr(III). Soil organic matter and reduced Fe can serve as electron donors in the reduction of Cr(VI) (Artiole and Fuller 1979). Bartlett and James (1979), however, demonstrated that under conditions prevalent in many soils, Cr(III) can be oxidized. The presence of oxidized Mn, which serves as an electron acceptor, was determined as an important factor in this reaction.

Industrial use of chromium also includes organic complexed Cr(III). Organically complexed Cr(III) may remain mobile in soil (James and Bartlett 1983). In addition to decreased Cr(III) sorption, added organic matter may

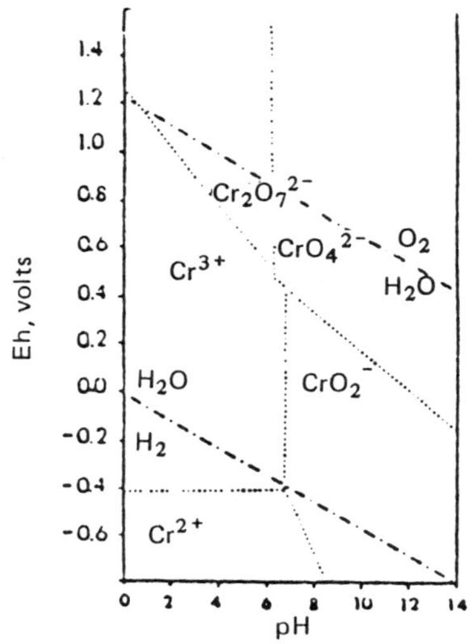

Figure 3-29. Eh-pH diagram of Cr species in water at 25°C calculated (Bartlett and Kimble 1976a). Used by permission, see Copyright Notice.

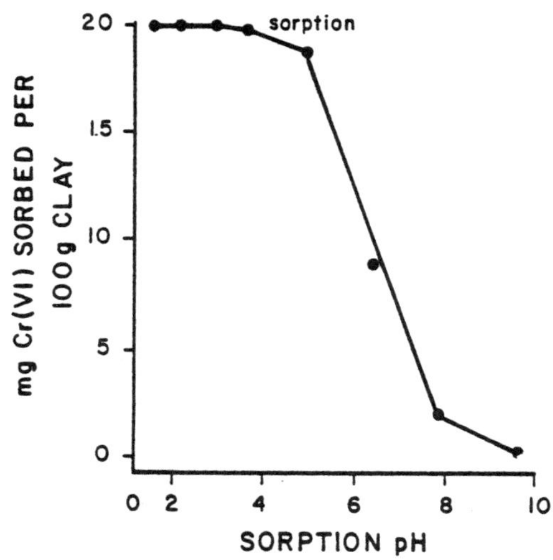

Figure 3-30. Sorption of Cr(VI) on NaOH-extracted soil. (Modified from Bloomfield and Pruden 1980.)

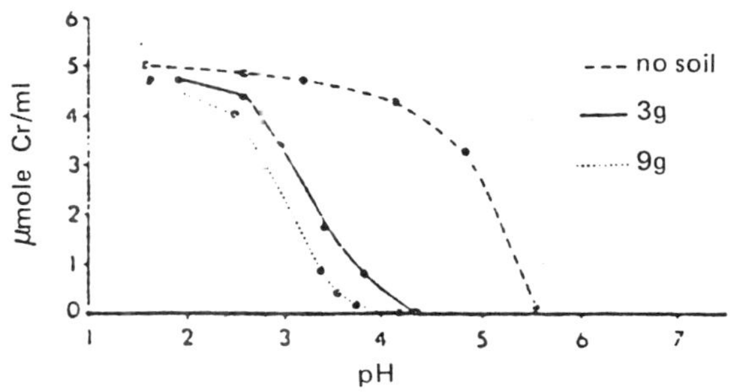

Figure 3-31. Concentration of Cr(III) in 24-hour equilibrium solutions as a function of pH (HCl-$CaCO_3$) and presence of Marlow Ap soil. Initial concentration was 5 μmole $CrCl_3$/ml (Bartlett and Kimble 1976a). Used by permission, see Copyright Notice.

also facilitate the oxidation of Cr(III) to Cr(VI). Under different conditions, this organic matter may facilitate the reduction of Cr(VI) (James and Bartlett 1983).

Mercury--The distribution of mercury species in soils, elemental mercury (Hg^0), mercurous ions (Hg_2^{2+}) and mercuric ions (Hg^{2+}), is dependent on soil pH and redox. Cationic mercury forms may be sorbed earily onto soil constituents. Divalent mercury is rapidly and strongly complexed by covalent bonding to organic matter and inorganic particles. Walters and Wolery (1974) stated that as much as 62 percent of the mercury in surface soils was bound to these particles. Mercuric ions are also bound to exchange sites of clays and hydrous oxides of iron and manganese. Sorption is pH dependent, increasing with increasing pH.

Mercurous and mercuric mercury are also immobilized by precipitation with phosphate, carbonate, and sulfide. Precipitates of $Hg(OH)_2$, $HgSO_4$, $HgNO_3$, and $Hg(NH_3)_4$ are insoluble at high pH. Insoluble HgS and $HgCl_3$ occur at all pH ranges. Divalent mercury will also form complexes with soluble organic matter, chlorides, and hydroxides which may contribute to its mobility.

The major loss of mercury from soils is through volatilization. Griffin and Shimp (1978) estimated that the removal of Hg from a leachate was not due to sorption by clay, but are due to volatilization and/or precipitation. This removal of mercury increased with pH. Rogers (1979) also found large amounts of mercury volatilized from soils. Amounts of mercury volatilized appeared to be affected by the solubility of the mercury compound added to soil (Table 3-38). Volatilization was also found to be inversely related to soil sorption capacity, thus losses followed sand > loam > clay (Table 3-38).

Under mildly reducing conditions, both organically bound mercury and inorganic mercury compounds may be degraded to the elemental form of mercury, Hg^0. Mercury metal can readily be converted to methyl or ethyl mercury by microbial transformation. These are the most toxic forms of mercury. Both

TABLE 3-38. PERCENT OF APPLIED Hg EVOLVED FROM SOILS WITHIN 144 HOURS (ROGERS 1979).
Used by permission, see Copyright Notice

Hg Compound	% Hg Evolved		
	Sand	Soil Loam	Clay
$HgCL_2$	38.3	32.9	14.2
$Hg(NO_3)_2$	36.5	24.1	13.4
$Hg(C_2H_3O_2)_2$	26.4	30.5	12.1
HgO	19.6	15.0	6.4
HgS	0.2	0.3	0.2

methyl and ethyl mercury are volatile and soluble in water. The formation of methyl mercury occurs primarily under acid conditions, while dimethyl mercury is produced at near neutral pH.

Arsenic--In the soil environment, arsenic exists as either arsenate, As(V) or as arsenite, As(III). Arsenite is the more toxic form of arsenic.

The behavior of arsenate in soil has been assumed to be analogous to that of phosphate, because of their chemical similarity. Like phosphate, arsenate is fixed to soils and is thus relatively immobile. Iron, Al, and Ca influences this fixation by forming insoluble precipitates with arsenate. Texture is often related to As(V) fixation. The reason for this relationship is that both reactive Fe and Al usually vary directly with clay content of the soil. Woolson et al. (1971) stated that arsenate may be leached from soil if the level of reactive Fe, Al, and Ca in soil are low. The solubility product for $FeAsO_4$ is reported as 5.7×10^{-21}, whereas for $Ca_3(AsO_4)_2$ it is 6.8×10^{-19}. The presence of Fe in soils is therefore most effective in controlling arsenates' mobility. Arsenite compounds are reported to be 4-10 times more soluble than arsenate compounds.

Griffin and Shimp (1978), in a sorption study of arsenate by kaolinite and montmorillonite, found maximum sorption of As(V) to occur at pH 5 (Figure 3-32). In comparing these sorption curves with the distribution diagram of As(V) (Figure 3-33), they concluded that $H_2AsO_4^-$ ion was the principal As(V) ion being sorbed by the clays. The non-sorption of $HAsO_4^{2-}$ was apparently due to repulsion of the ion from the clay surface resulting in decreased sorption of As(V) above pH 5.

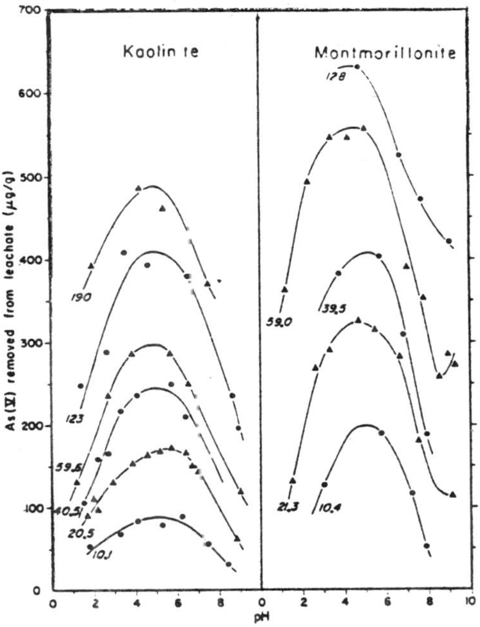

Figure 3-32. The amount of As(V) removed from DuPage leachate solutions by kaolinite and montmorillonite at 25°C plotted as a function of pH. Initial solution concentrations of As(V) are given in ppm. Each data point was obtained by using either 4 g of kaolinite or 1 g of montmorillonite in a total solution volume of 52.5 ml (Griffin and Shimp 1978).

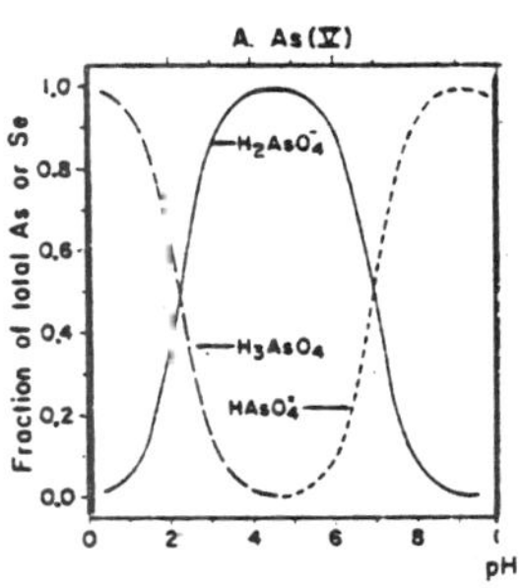

Figure 3-33. Diagram showing distribution of forms of As(V) (Griffin and Shimp 1978).

The sorption of As(III) is also strongly pH-dependent. Figure 3-34 shows increased sorption with pH in the pH range 3-9 (Griffin and Shimp 1978). These authors concluded that this increased sorption with pH was due to an increase in concentration of monovalent As(III) species with pH.

Under anaerobic conditions, arsenate may be reduced to arsenite. The reduced form of arsenic is more subject to leaching because of its increased solubility. Arsenite may also be volatilized from soils. High organic matter, warm temperatures, adequate moisture, and other conditions conducive to microbial activity drive the reaction sequence toward methylation and volatilization which reduces arsenic residues in soil (Woolson 1977). Woolson's (1977) study showed that only 1-2 percent of the arsenate applied at a rate of 10 ppm was volatilized as arsine (AsH_3) in 160 days.

Selenium--Forms and concentrations of selenium in soil solution are governed by various physical and chemical factors expressed in terms of pH, dissolution constants, solubility products, and redox potential. Figure 3-35 illustrates a phase diagram for selenium. Selenate, Se(VI) is the predominant form of selenium in calcareous soils and selenite, Se(IV), is the predominant form in acid soils.

Selenite can be bound to sesquioxides, especially to Fe oxides. Griffin and Shimp (1978) found maximum sorption of selenite on montmorillinite and kaolinite to occur at pH 2-3 (Figure 3-36).

Selenium is least soluble under acid conditions. Precipitates of selenite by ferric hydroxide in acid soils is presumed to be the factor in reducing selenium's mobility. In the solubility diagram (Figure 3-37, Geering et al. 1968), selenous acid potential (pH + $pHSeO_3$) is plotted against the ferric hydroxide potential (pH-1/3pFe). The solid line represents an estimated, average formation potential for the compounds $Fe_2(SeO_3)_3$ and $Fe_2(OH)_4SeO_3$. For the soils used in this study, all fall along the $Fe(OH)_3$ line and below the ferric selenite lines, indicating that crystalline ferric selenites are not governing the observed selenium solubility in these soils. Selenite probably forms sorption complexes with ferric oxides in soils rather than crystalline ferric selenite.

In contrast to selenite, selenate, Se(VI), is highly mobile in soils. Selenate dominates under alkaline conditions. Unlike most metals, selenium is more mobile at higher pHs.

Under reduced conditions, selenium is converted to elemental form (Figure 3-35). This conversion can provide an effective mechanism for attenuation since mobile selenate occurs only under well aerated, alkaline conditions.

Silver--Silver is very strongly sorbed by clay and organic matter. Precipitate of silver, AgCl, Ag_2SO_4, and Ag_2CO_3, are highly insoluble. Silver is highly immobile in the soil environment. Published data concerning the interaction of Ag with soil are rare.

Beryllium--Although beryllium is considered the most toxic metal in the environment, few studies have been done on the chemistry of beryllium in

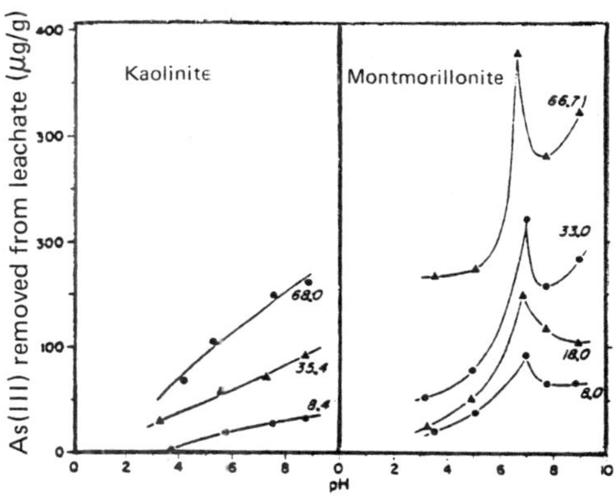

Figure 3-34. The amount of As(III) removed from DuPage leachate solutions by kaolinite and montmorillonite at 25°C plotted as a function of pH. Numbers give the initial solution concentration of As(III) in ppm. Each data point was obtained by using 4 g of clay in a total solution volume of 52.5 ml (Griffin and Shimp 1978).

soil. Berillium is known to be strongly adsorbed at exchange sites and is readily precipitated by liming. Beryllium is an alkaline earth metal (the same group as Ca and Mg) and presumably its activity in soils would be controlled by the same solution chemistry as other metals. Generalizations are, however, impossible.

SOIL SORPTION - ORGANICS

Soil sorption is perhaps the most important soil-waste process affecting the toxic and recalcitrant fractions of hazardous waste. The influence of soil sorption on the extent and rate of leaching, and also on biological decomposition of these fractions must be understood and described in order to effectively use the sorption reaction as a treatment process. Understanding the effect of different solid surfaces on hazardous waste constituents provides a mechanism for rationally selecting additional sorbents for use in augmenting the natural ability of a soil system for immobilizing hazardous chemicals. Also, understanding the relationship between soil water content and extent of sorption of hazardous chemicals provides the hazardous waste manager with a process for controlling the potential release and migration of constituents through leaching. Thus, this section of the manual describes the factors involved in soil sorption of chemical constituents and the basic parameters influencing the sorption process that may be used in treatment processes for immobilizing specific hazardous waste fractions.

Soil sorption is a physical/chemical process which involves the increase of solute at the soil water interface. The terms sorption and adsorption will

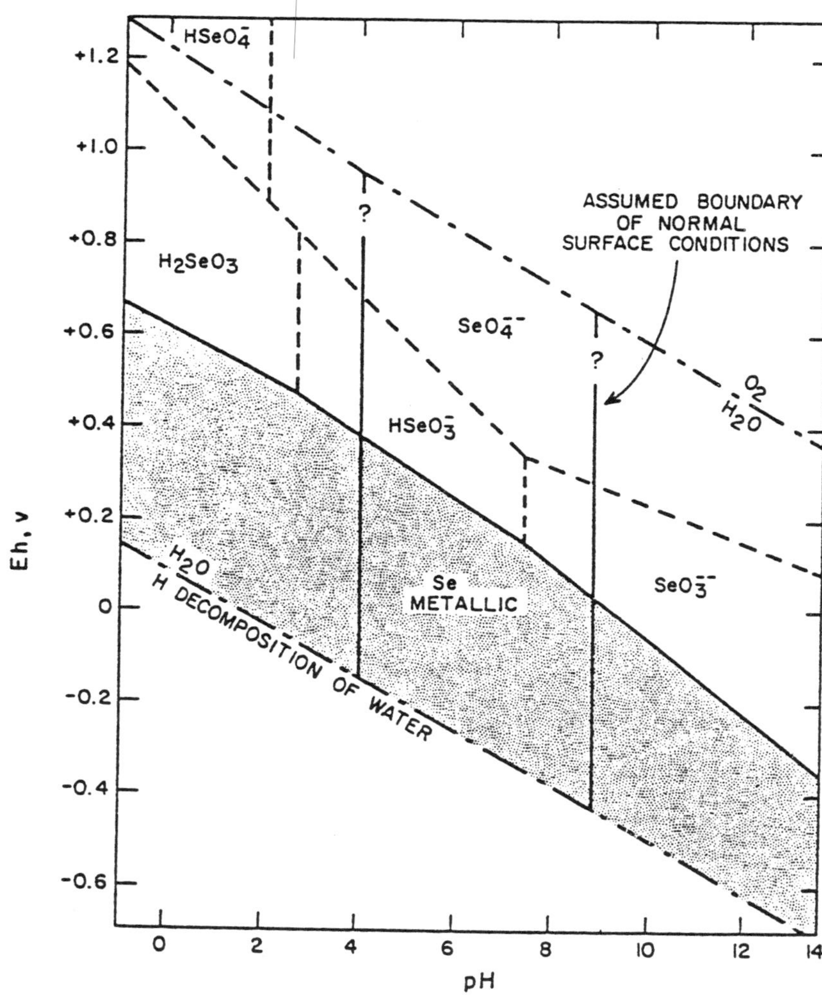

Figure 3-35. Stable fields of selenium (Brown and Associates, Inc. 1980).

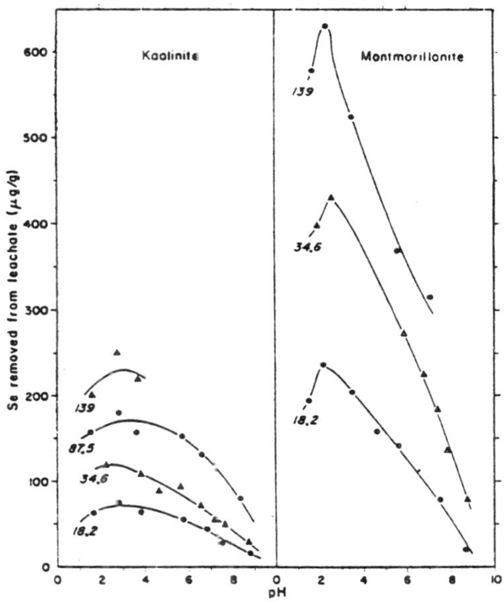

Figure 3-36. The amount of Se(IV) removed from DuPage leachate solutions by kaolinite and montmorillonite at 25°C, plotted as a function of pH. Numbers are the initial solution concentration of Se(IV) in ppm. Each data point was obtained by using either 5 g of kaolinite or 1 g of montmorillonite in a total solution volume of 52.5 ml (Griffin and Shimp 1978).

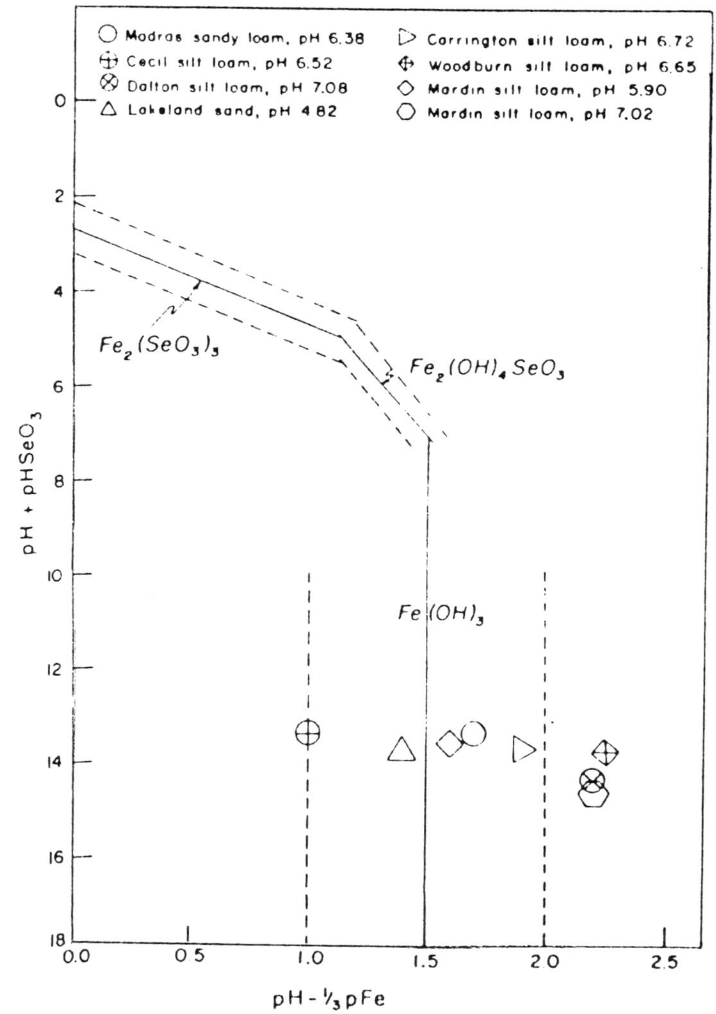

Figure 3-37. Solubility diagram for the ferric selenites, and the solubility data obtained from 1:10 soil-0.01M Ca $(NO_3)_2$ extracts of several soils (Geering et al. 1968). Used by permission, see Copyright Notice.

be used interchangeably to include both adsorption and absorption, unless otherwise stated. Sorption represents an important soil mechanism for removal of pollutants from the soil solution. The pollutants are "immobilized" by the soil, hence preventing polluted groundwater. In this section adsorption and factors affecting adsorption that may be used for controlling, managing, and treating hazardous constituents are discussed. Thorough descriptions of adsorption can be found in comprehensive reviews by Davies and Rideal (1963), Kipling (1965), Ekwall et al. (1963), Bailey and White (1970), and Hamaker and Thompson (1972).

Sorption can be "specific" or "nonspecific." Specific sorption occurs when specific sites on the surface exert forces on a particular unit of a molecule at a certain configuration, for example, adsorption of fluoride anion on geothite (Bohn et al. 1979). Nonspecific sorption is more general, and it precedes the specific sorption due to lower heat of adsorption. The major forces that make sorption possible are: 1) Van der Waals-London forces, 2) hydrophobic bonding, 3) hydrogen bonding, 4) charge transfer, 5) ligand exchange, and 6) chemisorption.

Van der Waals-London forces are weak electrostatic forces which are caused by uneven distribution of electrons of molecules due to the circulation in their orbits. This motion causes an instantaneous dipole which results in attraction. Van der Waals forces are weak forces; however, they are additive forces. Van der Waals forces become significant for large molecules when they interact with the soil surface at a few sites. Van der Waals bonding is important for organic molecules.

Hydrophobic bonding is actually a partitioning between a polar solvent (e.g. water) and a nonpolar adsorbent surface (e.g. soil humus). Hydrophobic bonding is a phenomenon related to entropy at low temperatures (Hamaker and Thompson 1972). The entropy is decreased as a result of hydrophobic compound dissolution in water. Hence, the molecules tend to leave water to increase the entropy and accumulate on hydrophobic regions of the adsorbent (organic matter of the soil). At higher temperatures, the enthalpy of transfer of molecules from adsorbent surface to the water increases which leads to the unfavorable condition where the molecule reenters the soil water. Hydrophobic bonding was used to explain the high correlation observed between the soil adsorption coefficients normalized for percent organic carbon (K_{OC}), and octanol/water partition coefficient (K_{OW}) for nonionic and polar organic compounds (Karickhoff et al. 1979, Karickhoff 1981, and Chiou et al. 1983). This correlation will be discussed in detail.

Hydrogen bonding is a special case of charge transfer. The hydrogen bond is a weak electrostatic bond and occurs between hydrogen and two atoms of high electronegativity (e.g., F, O, and N). Energy of adsorption in hydrogen bonding ranges from 0.5 - 15 kcal/mole. Hydrogen bonding becomes more significant with larger molecules, as in the case of Van der Waals bonding. Hydrogen bonding has been suggested for the adsorption of s-triazine herbicide molecules (Hamaker and Thompson 1972).

Charge transfer is a partial overlap of molecular orbitals of two molecules in an electron donor-acceptor system. Charge transfer complexes are formed between bonds or lone pairs of electrons. Charge transfer adsorption is an important mechanism in adsorption of organics (Hamaker and Thompson 1972).

Ligand exchange is a replacement of one or more ligands by an adsorbent molecule. This replacement occurs only if an adsorbent molecule is a stronger chelating agent than the ligand. For example, bipyridines and organophosphates can be adsorbed by ligand exchange (Hamaker and Thompson 1972).

Ion exchange is a process in which a cation is taken from a solution to replace another cation adsorbed on the soil surface. Cation exchange involves Coloumbic forces which reach up to 50 kcal/mole. Many organic molecules will be exchanged upon protonation in soil including alcohols, amines, and carbonyl groups. After protonation these molecules can be adsorbed on clay surfaces (Hamaker and Thompson 1972).

Chemisorption is an exothermic process with an energy range of 30-190 kcal/mole (Morrill et al. 1982). Usually, chemisorption involves a chemical bonding between adsorbent and adsorbate. In short-term sorption studies (<12 hours) the occurrence of chemisorption is low. However, chemisorption might be very important in the long-term immobilization of organic constituents and in influencing their fate in soil systems (Osgerby 1970).

The sorption behavior of organic compounds can be simplified by categorizing organics into functional classes (Weber 1972). The two major classes are: 1) ionic and 2) nonionic compounds.

Ionic Compounds

Organic ionic compounds consist of basic, cationic and acidic compounds. Organic basic compounds represent organic constituents that are capable of accepting a proton to become an ionized cation. Properties of basic pesticides, taken from Weber (1972), are shown in Table 3-39. Three mechanisms that function in adsorption of basic compounds by soil organic matter include: 1) ion exchange, depending on protonation, 2) hydrogen bonding, and 3) hydrophobic bonding, which is greatest at higher pH values when the molecules are not protonated. Acidic groups of soil organic matter have an average pK_a value of approximately 5.5 (Hayes 1970). Some acidic functional groups of fractions of soil organic matter have pK_a values less than three (fulvic acid) and 3.4 to 3.6 (humic acids) (Gamble 1970 and Gilmour and Coleman 1971). Adsorption of s-triazine molecules to soil organic matter is attributed to the complexing of the s-triazine molecule by ionizable H^+ ions on functional groups of organic colloids and to adsorption of protonated species by ion exchange (Weber et al. 1969). Factors shown to influence the adsorption of the s-triazine class of organic bases in soil systems include: 1) molecular structure, basicity, and solubility; 2) type of clay mineral; 3) acidity; 4) type and concentration of ions in solution and on clay surface; and 5) temperature.

Talibudeen (1955) reported that organic nitrogen bases react directly with H-montmorillonite through a cation-exchange mechanism which is pH dependent. Strong bases were adsorbed more strongly than weak bases.

TABLE 3-39. PROPERTIES OF BASIC PESTICIDES (WEBER 1972)
Used by permission, see Copyright Notice

Common Name or Designation	Trade Name	Chemical Name	pK_a	Water Solubility 20-25°C,ppm pH3	pH7	Vapor Pressure mm Hg 20°C (x10-6)
Atrazine	AAtrex	2-chloro-4-ethylamino-6-isopropyl-amino-s-triazine	1.68	31	35	0.3
Propazine	Milogard	2-chloro-4,6-bisisopropyl-amino s-triazine	1.05	4.8	4.8	0.029
Simazine	Princep	2-chloro-4,6-bisisopropylamino-s-triazine	1.65	5.8	5.0	0.0061
SD-15418	Bladex	2-chloro-4-(1-cyano-1-methylethylamino)-6-ethylamino-s-triazine	1.1		160	0.01 @30°
Ametryne	Evik	2-methylthio-4-ethylamino-6-isopropylamino-s-triazine	3.93	404	194	0.84
Prometryne	Caparol	2-methylthio-4,6-bisisopropyl-amino-s-triazine	4.05	206	40	1.0
Desmetryne	Semeron	2-methylthio-4n-methylamino-6-isopropylamino-s-triazine	4.0		501	1.0
Terbutryne	Igran	2-methylthio-4-ethylamino-6-tert-butylamino-s-triazine	4.07		58	0.96
Atratone	Gesatamin	2-methoxy-4-ethylamino-6-isopropylamino-s-triazine	4.20	1900	1600	2.9
Prometone	Pramitol	2-methoxy-4,6-bisisopropylamino-s-triazine	4.28	1000	677	2.3
Hydroxy Propazine	"Degradation product"	2-hydroxy-4,6-bisisoproplamino-s-triazine	5.20	326	41	
Menazon	Saphos	S-(4,6-diamino-s-triazin-2-ylmethyl)-O,O-dimethylphosphorodithioate	3.8		250	
Amitrole	Amino Triazole	3-amino-s-triazole	4.17	280,000		

He suggested that the basic property of organic bases was the most important property influencing their adsorption. Frissel (1961) also reported the pH-dependency of organic bases (s-triazine herbicides) on adsorption and concluded that triazines are adsorbed as neutral molecules in neutral and basic environments and as positively charged ions in acidic solutions. Weber (1966) demonstrated that maximum adsorption of 13 s-traizines on Na-montmorillonite clay occurred in the vicinity of the pK_a of each s-triazine (Figure 3-38). The more basic compounds were adsorbed in greater amounts than less basic compounds. H- and Al-montmorillonite may be acidic enough to hydrolyze atrazine to hydroxyatrazine (Weber 1970). Generally, adsorption of s-triazines increases as acidity, organic matter, and clay content of soils increase (Lavy 1968).

The mechanism of adsorption of basic compounds to soil systems can be illustrated using the s-triazine class of organic bases as a model. The following processes, taken from Weber (1970), will be considered:

$$R + H^+ = RH^+ \tag{3-15}$$

$$R + X\text{-mont} = RX\text{-mont} \tag{3-16}$$

$$RH^+ + X\text{-mont} = RH\text{-mont} + X^+ \tag{3-17}$$

$$H^+ + RH\text{-mont} = H\text{-mont} + RH^+ \tag{3-18}$$

$$H^+ + X\text{-mont} = H\text{-mont} + X^+ \tag{3-19}$$

$$R + H\text{-mont} = RH\text{-mont} \tag{3-20}$$

where R = triazine compound; RH^+ = triazine cation; X-mont = montmorillonite with X = exchangeable cations Na, Li, 1/2Ca, 1/2Mg, etc.; H-mont = hydrogen (aluminum) montmorillonite; and H^+ = hydrogen ion, as the hydrated H_2O species.

Equation 3-15 is related to the basicity of the compound, indicated by the equilibrium constant K_a, given by:

$$pK_a = \log [(RH^+)/(R)] + pH \tag{3-21}$$

As indicated previously, maximum adsorption occurs at a pH equal to the pK_a of the compound.

Equations 3-16 and 3-17 represent adsorption of netural and cationic s-triazines by montmorillonite. Equations 3-17, 3-18, and 3-19 indicate competition of s-triazine molecules with other cations for soil sites. Equation 3-20 indicates s-triazine adsorption through direct complexation with hydrogen ions on H-montmorillonite.

The effects of pH on organic base adsorption is illustrated as follows, and is taken from Weber (1970). As the pH is decreased, the s-triazine molecules become protonated (Equation 3-15), and these cations are adsorbed by clay minerals such as montmorillonite in exchange for X (Equation 3-17). The

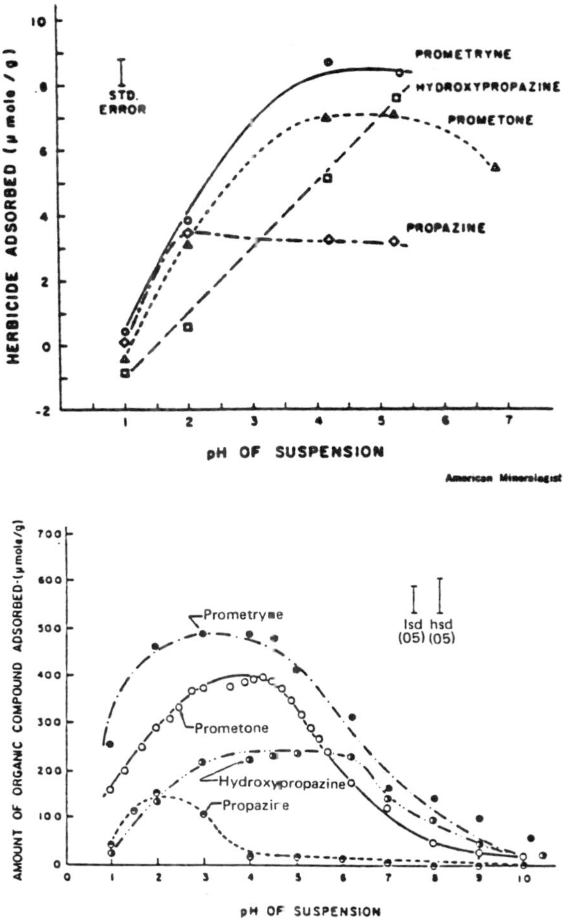

Figure 3-38. Effect of pH on adsorption of four related s-triazines on (top) soil organic matter and (bottom) Na-montmorillonite (Weber 1972). Used by permission, see Copyright Notice.

result is an increase in adsorption. Addition of acid also causes some of the X cations on the clay to be replaced by hydrogen (Equation 3-19) and the resulting H-montmorillonite may absorb neutral R molecules (Equation 3-20). When the pH equals approximately the pK_a, adsorption reaches a maximum. As pH is increased further, adsorption is decreased. The decrease in adsorption is attributed to competition from the H^+ ions (Equation 3-18). It is also known (Weber 1970) that salts such as barium chloride (0.1m) cause large decreases in adsorption of these basic molecules because Equation 3-17 is driven to the left, releasing the s-triazine cations to the solution.

Adsorption of s-triazine bases has also been reported to be relatively independent of soil pH. The s-triazine molecules atrazine, propazine, prometone, and prometryne were found to adsorb to 30 soils varying in pH from 4.9 to 7.6 according to the linear Freundlich isotherm:

$$Y = Kd\ C \tag{3-22}$$

For soils containing greater than 5 percent carbon, adsorption was independent of clay content and directly dependent on soil organic matter. Aqueous solubility did not influence sorption. Thus, adsorption of basic compounds may also be due to a partitioning out of aqueous solution onto nonpolar hydrophobic surfaces, in addition to specific charge related sorption sites (Walker and Crawford 1968).

Desorption of s-triazine molecules from soil minerals and soil organic matter occurs relatively easily in the presence of distilled water. Strong bonding of these organic bases to soil components is not indicated.

Cationic compounds generally are characterized by high-water solubility and by strong adsorption to clay fractions of soils. Examples of strongly adsorbed cations include the compounds paraquat and diquat. The mechanism of adsorption involves cation exchange. That is, a positively charged cation associated with a clay mineral is displaced by the positively charged organic cation. Cation properties are shown in Table 3-40 for several compounds.

Acidic compounds (anionic compounds) contain functional groups which may ionize to produce organic anions. The tendency to ionize is described by the dissociation constant, K_a:

$$RCOOH = RCOO^- + H^+ \tag{3-23}$$

$$pK_a = PH - \log [(RCOO^-)/(RCOOH)] \tag{3-24}$$

where RCOOH is the undissociated molecule and $RCOO^-$ is the anion species. Properties of acidic compounds are shown in Table 3-41. Weak acids, with relatively higher dissociation constants than strong acids, are in the free acid form at pH values below their pK_a values and are adsorbed to a much greater extent than in dissociated (anion) form. Due to the net negative charge associated with clay minerals, there is a net repulsion between the negatively charged organic anion and soil fractions including minerals and organic matter. However, Harter and Ahlrichs (1967) demonstrated that the pH of water at the clay surface is lower than in an aqueous soil bulk solution.

TABLE 3-40. PROPERTIES OF CATIONIC PESTICIDES (WEBER 1972)
Used by permission, see Copyright Notice

Common Name or Designation	Trade Name	Chemical Name	Molecular Weight	Water Solubility 20°C,%
Diquat	Ortho diquat	6,7-dihydrodipyrido[1,2-a:2',1'-c]pyrazidiinium dibromide	344	70
Paraquat	Ortho paraquat	1,1'-dimethyl-4,4'-bypyridinium dichloride	257	70
Chlormequat	Cycocel	(2-chloroethyl)-trimethyl-ammonium chloride		74
Morfamquat	Ceroxone	1,1-bis(3,5-dimethylmorpholino-carbonylmethyl)-4,4'-bipyridylium dichloride		
Methylene blue		3,7-bis(dimethylamino)-phenazathionium chloride	374	4
Phosphon	Phosfon	tributyl-2,4-dichlorobenzyl phos-phonium chloride	398	high
Hyamine	Hyamine 10-X	di-isobutyl cresoxyethoxy-ethyldimethyl-benzyl ammonium chloride, mono-hydrate	480	high
Phenacridane chloride	Acrizane	9-(p-n-hexyloxyphenyl)-10-methyl-acridinium chloride	406	high
Et. Pyr. Br.		1-ethylpyridinium bromide	188	high
Pyr. Pyr. Cl.		N-(4-pyridyl)-pyridinium chloride	193	high

TABLE 3-41. PROPERTIES OF ACIDIC PESTICIDES (WEBER 1972)
Used by permission, see Copyright Notice

Common Name or Designation	Trade Name	Chemical Name	pK_a	Water Solubility 20-25°C, ppm	Vapor Pressure mm Hg 35°C (x10-6)
2,4-D	Weedone 638	2,4-dichlorophenoxyacetic acid	2.80	650	
2,4,5-T	Brush Killer	2,4,5-trichlorophenoxyacetic acid	2.84	238	
MCPA	Methoxone	(4-chloro-o-tolyloxy)acetic acid	3.11	550	
MCPB	Thistrol	4-(4-chloro-o-tolyloxy)butyric acid	4.80	44	
Silvex	Kuron	2-(2,4,5-trichlorophenoxy)propionic acid	~3.0	140	low
Chloramben	Amiben	3-amino-2,5-dichlorobenzoic acid	3.40	700	very low
Dicamba	Banvel D	3,6-dichloro-o-anisic acid	1.93	4500	
Tricamba	Banvel T	3,5,6-trichloro-o-anisic acid	~1.5	sl.sol.	
2,3,6-TBA	Benzac	2,3,6-trichlorobenzoic acid	~1.5	8400	negligible
TIBA	Floraltone	2,3,5-triiodobenzoic acid	~1.5	sl.sol.	
Fenac	Fenac	2,3,6-trichlorophenylacetic acid	3.70	sl.sol.	
Benzadox	Topicide	Benzamidooxyacetic acid	~4.7	19,000	
Picloram	Tordon	4-amino-3,5,6-trichloropicolinic acid	1.90	430	0.62
Endothall	Endothall	7-oxabicyclo[2.2.1]heptane-2,3-dicarboxylic acid	~4.0	10,000	negligible
Naptalam	Alanap	N-1-naphthylphthalmic acid	~4.0	200	low
Dalapon	Dowpon	2,2-dichloropropionic acid	1.84	450,000	0
TCA	TCA	trichloroacetic acid	0.63	1,300,000	0
DMSA	Alar	N-dimethylamino succinamic acid	~4.0	sl.sol.	
MH	MH-30	1,2-dihydro-3,6-pyridazinedione	~4.0	6000	~0
Dinoseb	Premerge	2-sec-butyl-4,6-dinitrophenol	4.40	52	>1.0
DNOC	Sinox	4,6-dinitro-o-cresol	4.35	130	1.0-52.0
Ioxynil	Certrol	4-hydroxy-3,5-diiodobenzonitrile	3.96	1.8	negligible
Bromoxynil	Brominil	3,5-dibromo-4-hydroxybenzonitrile	4.08	130	negligible

With suspension pH values of 6, 5, 4 and 3, it was estimated that the pH at the clay surface was approximately 4.5, 4.0, 3.2 and 2.5 respectively. Water, therefore, is more acidic at clay surfaces than in the bulk solution. This phenomenon may play a role in the adsorption of acidic compounds so that the extent of adsorption to clay surface is higher than is expected from the pH of the bulk solution.

However, anion adsorption can occur in soil systems to other soil components. The reaction of an organic anion with iron and aluminum oxides can be illustrated as follows:

$$Anion^- + Al(OH)_3^+ = A\text{-}Al(OH)_3 \quad (3\text{-}25)$$

Arsenates and phosphates can also react as follows:

$$CH_3\text{-}As\text{-}OHOO^- + Fe(OH)^{2+} = CH_3\text{-}As\text{-}OHOO^- \text{ --- } Fe(OH)_2 \quad (3\text{-}26)$$

Therefore, adsorption of organic anions does occur, and usually does not involve ion exchange, as with organic cations such as paraquat and diquat.

Acidic organic molecules, including pesticides, ionize in soil solution to form anionic species. The pH at which ionization occurs depends upon the pK_a value for each organic acid. Under normal pH values for soil solutions, the anionic forms predominate. Since soil inorganic and organic fractions are generally negatively charged at normal pH values for soil, organic acids tend to be relatively mobile. Thus, highly water soluble, acidic organic molecules comprise one of the most mobile classes of organic chemicals in soil systems.

Nonionic Compounds

According to Weber (1972), nonionic pesticides consist of the following families of compounds: 1) chlorinated hydrocarbon; 2) organophosphates; 3) substituted anilines; 4) phenyl carbamates; 5) phenylureas; 6) substituted anilides; 7) phenylamides; 8) thio carbamates, carbotioates, and acetamides; 9) Benzonitrites; and 10) esters. Nonionic compounds do not ionize significantly in aqueous or soil systems. These families of compounds are highly variable in their properties, not only among the ten families of compounds mentioned, but also among the individual compounds within each family of compounds (Table 3-42).

The nonionic nonpolar compounds in general adsorb at the hydrophobic regions of the soil. In other words, organic carbon content of soil is a very important factor in adsorption of nonionic nonpolar compounds (Lichtenstein et al. 1960; Graham-Bryce 1967; Kay and Elrick 1967; Lambert 1967; Whitney 1967; Harris et al. 1969; Shin et al. 1970; Peterson et al. 1971; Felsot and Dahm 1979; Karickhoff et al. 1979; Wahid and Sethunathan 1978, 1979, 1980; Kozak and Weber 1983; Mkedi-Kizza et al. 1983). For polar compounds, adsorption occurs in competition with water molecules on adsorbent surface (Guenzi and Bread 1967). The adsorption is negatively correlated to the solubility of compounds, while desorption was proportional to the solubility (Sharom et al. 1980).

TABLE 3-42. SELECTED PROPERTIES OF SOME NONIONIC PESTICIDES (WEBER 1972)
Used by permission, see Copyright Notice

Common Name or Designation	Trade Name	Chemical Name	Water Solubility 20-25°C,ppm	Vapor Pressure mm Hg 20-25 °C (x10-6)	Parachor
Organophosphates					
Dimethoate	Cygon	0,0-dimethyl-S-(N-methylcarbamoyl-methyl)-phosphorodithioate	20,000	100	484
Methyl parathion	Metron	0,0-dimethyl-0-p-nitrophenyl phosphorothioate	50		529
Phorate	Thimet	0,0-diethyl-S-(ethylthio)-methyl phosphorodithioate	80-85	2300	573
Demeton	Systox	0,0-diethyl-0(and S)-[2-ethylthio)-ethyl]phosphorothioates	100	1000	553-583
Parathion	Phoskil	0,0-diethyl-0-p-nitrophenyl phosphorothioate	24	37.8	609
Disulfoton	Disyston	0,0-diethyl-S-[2-(ethylthio)-ethyl] phosphorodithioate	60-66	300	613
Dursban	Dursban	0,0-diethyl-0-(3,5,6-trichloro-2-pyridyl) phosphorothioate	2	18.7	659
Diazinon	Diazol	0,0-diethyl-0-(2-isopropyl-6-methyl-4-pyrimidinyl) phosphorothioate	40	140	693
Malathion	Cythion	0,0-dimethyl-S-(1,2-dicarbethoxy-ethyl) phosphorodithioate	145	40	699
Carbophenothion	Trithion	0,0-diethyl-S-(p-chlorophenylthio-methyl) phosphorodithioate	1-2		707
Ethion	Nialate	0,0,0',0'-tetraethyl-S,S'-methylene biophosphorodithioate	1		815
Schraden	OMPA	Octamethylpyrophosphoramide	miscible	1000	689

TABLE 3-42. (CONTINUED)

Common Name or Designation	Trade Name	Chemical Name	Water Solubility 20-25°C,ppm	Vapor Pressure mm Hg 20-25 °C (x10-6)	Parachor
Chlorinated hydrocarbons					
DDT	Gesapon	1,1,1-trichlor-2,2-bis(p-chloro-phenyl)-ethane	0.001-0.04	0.15	658
Methoxychlor	Marlate	2,2-bis(p-methoxyphenyl)-1,1,1-trichloroethane	0.1-0.25		699
Endrin	Endrin	1,2,3,4,10,10-hexachloro-6,7-epoxy-1,4-4a,5,6,7,8,8a-octa-hydro-1,4-endo-endo-5,8-di-methanonaphthalene	0.23	0.20	494
Dieldrin	Octalox	1,2,3,4,10,10-hexachloro-6,7-epoxy-1,4-4a,5,6,7,8,8a-octahydro-1,4-endo-exo-5,8-dimethanonaphtha-lene	0.1-0.25	0.18	494
Aldrin	Drinox	1,2,3,4,10,10-hexachloro-1,4,4a,5,8,8a-hexahydro-1,4,-endo-exo-5,8-dimethanonaphthalene	0.01-0.2	6.0	493
Toxaphene	Phenacide	mixture made by chlorinating cam-phene to 69% chlorine	0.4	1.0	mixture
Lindane	Gamexane	1,2,3,4,5,6-hexachlorocyclohexane	7.3-10.0	9.4-45.0	478
Chlordane	Octa-Klor	1,2,3,5,6,7,8,8-octachloro-2,3,3a,4,7-7a-hexahydro-4,7-methanoin-dene	very low	10.0	647
Heptachlor	Drinox H-34	1,4,5,6,7,8,8a-heptachloro-3a,4,7a-tetrahydro-4,7-methanoindene	very low	300	596
Substituted anilines					
Nitralin	Planavin	4-(methylsulfonyl)-2,6-dinitro-N,N-dipropylaniline	0.6	1.0	768
Benefin	Balan	N-butyl-N-ethyl-α,α,α-trifluro-2,6-dinitro-p-toluidine	0.5	38.9	671
Trifluralin	Treflan	α,α,α-trifluro-2,6-dinitro-N,N-dipropyl-p-toluidine	0.05	114	671

TABLE 3-42. (CONTINUED)

Common Name or Designation	Trade Name	Chemical Name	Water Solubility 20-25°C,ppm	Vapor Pressure mm Hg 20-25 °C (x10-6)	Parachor
Phenylcarbamates (and carbanilates)					
Propham	Chem-hoe	isopropyl carbanilate	250-254		426
Dichlormate	Rowmate	3,4-dichlorobenzyl methylcarbamate	170		466
Chlorpropham	Chloro IPC	isopropyl m-chlorocarbanilate	88-102	10	466
Carbaryl	Sevin	1-naphthyl-N-methylcarbamate	40-99		453
Barban	Carbyne	4-chloro-2-butynyl-m-chlorocarbanilate	11-12		522
Terbutol	Azak	2,6-di-tert-butyl-p-tolyl methylcarbamate	6-7		706
Acetamide					
CDAA	Randox	2-chloro-N,N-diallylacetamide	20,000	9.4-9.6	408
Benzonitrile					
Dichlobenil	Casoron	2,6-dichlorobenzonitrile	18-25	0.5	334
Esters					
Methyl ester of chloramben	Amiben methylester	methyl ester of 3-amino-2,5-dichlorobenzoic acid	120		426
Isopropyl ester of 2,4-D	2,4-D ester	isopropyl ester of 2,4-dichloro-phenoxyacetic acid		10-15	533
DCPA	Dacthal	tetrachloroterephthalic acid, dimethyl ester	0.5	<10	580
Phenylureas					
Fenuron	Dybar	1,1-dimethyl-3-phenylurea	2900-3850		399
Monuron	Telvan	3-(p-chlorophenyl)-1,1-dimethylurea	230	0.5	439
Monolinuron	Aresin	3-(p-chlorophenyl)-1-methoxy-1-methylurea	580	150	459

TABLE 3-42. (CONTINUED)

Common Name or Designation	Trade Name	Chemical Name	Water Solubility 20-25°C,ppm	Vapor Pressure mm Hg 20-25 °C (x10-6)	Parachor
Fluometuron	Cotoran	1,1-dimethyl-3-(α,α,α-trifluro-m-tolyl)-urea	90	0.5	470
Metobromuron	Patoran	3-(p-bromophenyl)-1-methoxy-1-methylurea	330	3.0	473
Diuron	Karmex	3-(3,4-dichlorophenyl)-1,1-dimethylurea	42		479
Linuron	Lorox	3-(3,4-dichlorophenyl)-1-methoxy-1-methylurea	75	15	499
C-6313	Maloran	3-(4-bromo-3-chlorophenyl)-1-methoxy-1-methylurea	50	0.4	513
Norea	Herban	3-(hexahydro-4,7-methanoindan-5-yl)-1,1-dimethylurea	150		539
Siduron	Tupersan	1-(2-methylcyclohexyl)-3-phenyl-urea	18		569
Neburon	Kloben	1-butyl-3-(3,4-dichlorophenyl)-1-methylurea	4.8		599
Fluorodifen	Preforan	p-nitrophenyl-α,α,α-trifluro-2-nitro-p-tolyl ether	<2.0	0.07	592
Chloroxuron	Tenoran	3-[p-(p'-chlorophenoxy)phenyl]-1,1-dimethylurea	2.7-3.7	0.02	634
Substituted anilides					
Propachlor	Ramrod	2-chlor-N-isopropylacetanilide	700		486
Propanil	Rogue	3',4'(-dichloropropionanilide	500		446
Alachlor	Lasso	2-chlor-2',6'-diethyl-N-(methoxy-methyl) acetanilide	148		626
Dicryl	Dicryl	3',4'-dichloro-2-methylacrylanilide	8-9		474
Solan	Solan	N-(3-chloro-4-methylphenyl)-2-methyl-pentanamide	8-9		566

TABLE 3-42. (CONTINUED)

Common Name or Designation	Trade Name	Chemical Name	Water Solubility 20-25°C,ppm	Vapor Pressure mm Hg 20-25 °C ($x10^{-6}$)	Parachor
Phenylamide					
Diphenamide	Enide	N,N-dimethyl-2,2-diphenylacetamide	260		581
Thiocarbamates					
EPTC	Eptam	S-ethyl dipropylthiocarbamate	370-375	20-34	482
CDEC	Vegadex	2-chloroallyl diethyl-dithiocarbamate	92	2.2	500
Pebulate	Tillam	S-propyl butylethylthiocarbamate	90-92	4.3-4.8	522
Vernolate	Vernam	S-propyl dipropylthiocarbamate	90-109	5.4-10.4	522
Cycloate	Ro-Neet	S-ethyl-N-ethylthiocyclohexanecarbamate	85-90	2.0-6.2	532
Butylate	Sutan	S-ethyl diisobutylthiocarbamate	45	13	562
Carbothioate					
Molinate	Ordram	S-ethyl hexahydro-1-H-azepine-1-carbothioate	800-912	5.6	455

Weber (1972) discusses each family of compounds in detail. Chlorinated hydrocarbons are insoluble in water (< 1 ppm) except for lindane. DDT is the most studied compound in this family; it is the least soluble, and it is highly immobile in soils (Weed and Weber 1974, Grau and Peterle 1979). DDT is also very persistent in soils (half-life is 10-20 years). The other compounds are less persistent than DDT, but still relatively immobile (Harris et al. 1969).

Organophosphates are more soluble and more volatile (higher vapor pressure) than chlorinated hydrocarbon. These compounds are more degradable than chlorinated hydrocarbons. The mobility of organophosphate is directly related to solubility (King and McCarty 1968, Fuhremann and Lichtenstein 1980). Organophosphate still is considered to be immobile in soil systems (Harris 1967, Jenkins et al. 1978).

Quantitative Description of Adsorption

Adsorption can be described in two ways: 1) adsorption equilibrium, and 2) adsorption kinetics. The first approach is based on the assumption that the adsorption process is instantaneous, i.e., equilibrium is achieved in relatively very short times. The second approach is based on the assumption that adsorption is a time dependent process. The quantitative description of adsorption, when coupled with a transport equation, is important in assessing the fate of pollutants in soil systems as discussed previously. Travis and Etnier (1981) have reviewed available adsorption models, which are summarized in Table 3-43. Some of the models will be described in this section.

Adsorption Equilibrium--

Equilibrium adsorption is described by an adsorption isotherm, which is the relationship between the amount of solute adsorbed and the equilibrium concentration of solute in the soil solution. Isotherms can be divided into four general types according to Giles et al. (1960) and are presented in Figure 3-39. The L-type curve, also known as a Langmuir type curve, occurs when there is no strong competition from the solvent for sorption sited on the adsorbent surface. The S-type curve shows an initial slow rate of adsorption at low concentration. The rate of adsorption increases as the concentration increases. The C-curve is applicable where constant partitioning occurs between solution and adsorbent phases. Finally, the H-curve is applicable where there is a very high affinity of the adsorbate for the adsorbent. Chemisorption, for example, usually produces an H-curve adsorption isotherm. H-type isotherm has been observed for diquat (6, 7-dihydrobipyrid) (1.2-a:2-, 1-c) pyrazidinium dibromide) and paraquat (1,1- dimethyl-4, 4-bipyridinium dichloride) on Na-montmorillonite and Na-kaolinite clays and for promotone on H-montmorillonite (Weed and Weber 1974). Examples of H-isotherms for some pesticides obtained from Weber (1972) are shown in Figure 3-40.

Specific adsorption isotherms that are used to describe immobilization of organic constituents in soils include the following: 1) Langmuir isotherm, 2) Freundlich isotherm, and 3) Brunauer, Emmet and Teller (BET) isotherm.

TABLE 3-43. EQUILIBRIUM AND NONEQUILIBRIUM MODELS (TRAVIS AND ETNIER 1981)
Used by permission, see Copyright Notice

1.0 Equilibrium sorption models

1.1 Langmuir isotherm

$$S = \frac{K_1 K_2 C}{1 + K_2 C}$$

1.2 Freundlich isotherm

$S = KC^N$ (nonlinear)

$S = K_dC$ (linear)

1.3 BET isotherm

$$S = \frac{BQ^o}{(C_s - C)\,[1 + (B - 1)(C/C_s)]}$$

1.4 University phosphate adsorption

$S = ac + X_o$ $\quad C \geq C_c$

$S = ac + KC^N$ $\quad C \leq C_c$

where $C_c = (X_o/K)^{1/N}$

1.5 Langmuir two surface isotherm

$$S = \frac{k_1 b_1 C}{1 + K_1 C} + \frac{k_2 b_2 C}{1 + k_2 C}$$

1.6 Competitive Langmuir

$(C_1/C_2)/S = (b_2/k_1 b_1) + (C_1/b_2 C_2)$

2.0 Nonequilibrium sorption models

2.1 Reversible linear

$$\frac{dS}{dt} = \frac{k_1 \Theta}{\rho} C - k_2 S$$

2.2 Reversible nonlinear

$$\frac{dS}{dt} = \frac{k_1 \Theta}{\rho} C^n - k_2 S$$

TABLE 3-43. CONTINUED.

2.3 Kinetic product model

$$\frac{dS}{dt} = a\ C^{\gamma} S^{d}$$

2.4 Bilinear adsorption model

$$\frac{ds}{dt} = k_1 C(b - s) - k_2 s$$

2.5 Mass transfer model

$$\frac{ds}{dt} = K(C - C^{*})$$

2.6 Elovich model

$$\frac{dq}{dt} = A_1 \exp(-B_2\ a)$$

q = fraction of adsorption sites in soil matrix covered by solute

2.7 Fava and Eyring model

$$\frac{d\phi}{dt} = 2k_1\phi \quad \sinh(b\phi)$$

$$\phi(t) = \frac{S(t) - S(\infty)}{S(0) - S(\infty)}$$

S(0) and S(∞) are the initial amount sorbed and amount sorbed at equilibrium respectively.

2.8 Two site kinetic model

$$\frac{ds}{dt} = \left(\frac{k_1}{k_2}\frac{\Theta}{P} M\ C^{M-1}\frac{\partial C}{\partial t}\right) + \left[k_3 \frac{\Theta}{P} C^{N} - k_4(1 - f)C\right]$$

f denotes fraction of sorption sites occupied which are type 1.

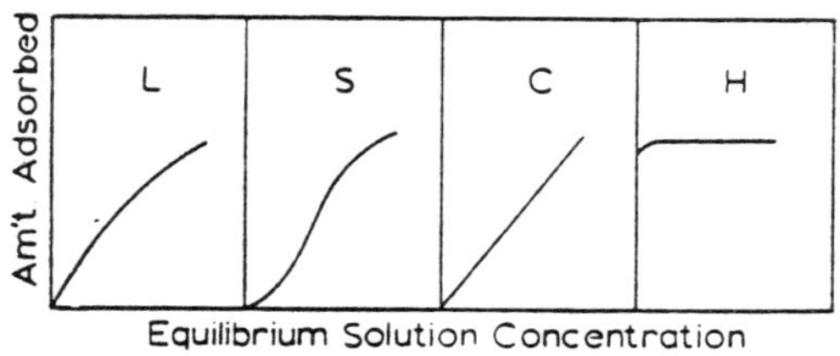

Figure 3-39. Classification of adsorption isotherms (Giles et al. 1960. Used by permission, see Copyright Notice.

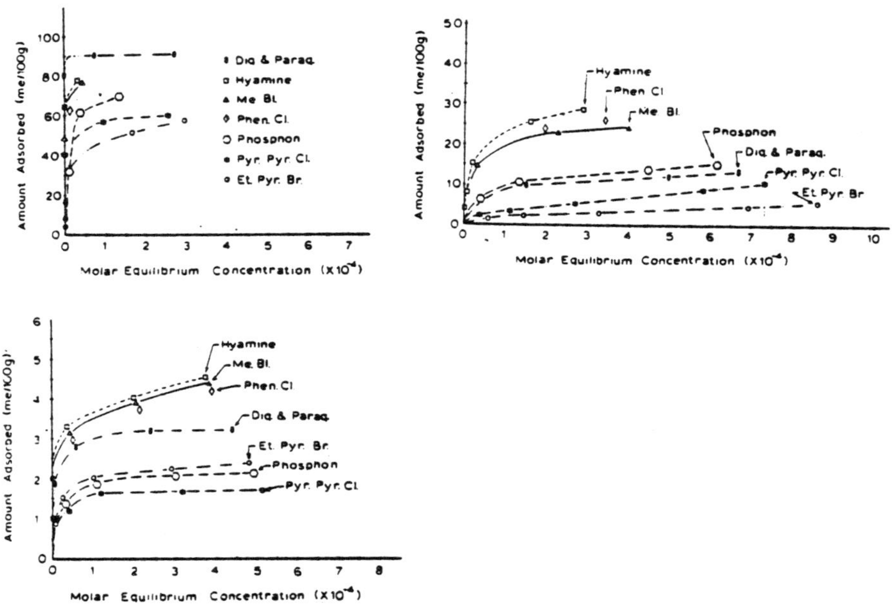

Figure 3-40. Isotherms for adsorption of several cationic pesticides on (top) Na-montmorillonite, (center) Na-kaolinite, (bottom) soil organic matter (Weber 1972). Used by permission, see Copyright Notice.

1) Langmuir isotherm: originally this isotherm was developed for adsorption of gases by solids. The derivation was based on three assumptions: 1) constant energy of adsorption, i.e., independent of the extent of surface covered, 2) no interaction between adsorbed molecules, and 3) only a monolayer adsorption is formed on the solid surface. The Langmuir adsorption equation may be expressed as:

$$S = \frac{K_1 K_2 C}{1 + K_2 C} \tag{3-27}$$

where S is the amount of adsorbed solute per unit mass adsorbent, K_1 represents the maximum amount of solute that can be adsorbed by the soil matrix, K_2 is a measure of the bond strength holding the sorbed solute on soil surface, and C is the equilibrium concentration in the soil solution. The Langmuir isotherm has been used extensively (e.g., Cavallaro and McBride 1978) for organic and inorganic constituents.

2) Freundlich isotherm: The Freundlich isotherm is expressed as:

$$S = KC^N \tag{3-28}$$

where K and N are constants. This isotherm was an empirical formulation which was used to describe gas adsorption on solid phase. However, the Freundlich isotherm was derived by Halsey and Taylor (1947) based on the assumption that the heat of adsorption is a logarithmic function of the surface covered. The Freundlich isotherm has two important properties. The first is that the Freundlich isotherm is a very flexible equation. Considering the flexibility of two parameters, the isotherm can fit a wide range of data. The second property is that the Freundlich isotherm does not have a maximum limit for the amount of substance sorbed. Due to the flexibility of this Freundlich isotherm it is used extensively and has been proposed to be used with a transport equation to describ the movement of synthetic organic chemicals (pesticides) in soils (e.g., Van Genuchten et al. 1974).

The linear isotherm is an important isotherm which can be obtained from the Freundlich isotherm when N = 1. The linear isotherm has been used for low concentration of pesticides because of the available analytical solutions when coupled with a transport equation (Kay and Elrick 1967, Green et al. 1972, Davidson et al. 1968, Davidson and Chang 1972, Hugenberger et al. 1972, and Davidson and McDougal 1973). The linear isotherm may be expressed as:

$$S = k_d C \tag{3-29}$$

where K_d is called the distribution coefficient. This isotherm can also be obtained from the Langmuir isotherm (Equation 3-15) if it is assumed that $K_2C \ll 1$, in which case $K_d = K_1K_2$.

Under normal soil moisture conditions, with sorption constants greater than one (e.g., usually for nonionic compounds), most of the chemical is present in the soil system in an adsorbed state. The percent sorbed can be expressed in terms of K_d and soil moisture content Θ as follows:

where Θ = soil moisture content (weight basis, assuming density of water as 1.0 g/cc). Figure 3.41 illustrates the extent of sorption (percent sorbed) as a function of distribution coefficient K_d for different values of Θ. Hamaker and Thompson (1972) point out that this explains the observation that chemicals tend to resist leaching and remain in the upper soil layers longer than would be expected based on chemical solubilities.

These observations have important implications with respect to management/ treatment of hazardous waste contaminated soil. Careful control of soil water content will determine, to a large extent, the relative immobilization of a given set of chemical constituents identified at a remedial site. Optimization of cost effective and efficient treatment may require a compromise between optimum soil moisture content for biodegradation versus adsorption. Implications for specific treatment methods are discussed in detail in Chapter 2, Volume I.

The K or K_d value is dependent on the type of soil, even for the same chemical. K_{oc}, the normalized coefficient of adsorption with respect to organic carbon content (OC), is defined as:

$$K_{oc} = K_d/OC\% \times 100 \quad \text{for linear isotherm} \tag{3.31}$$

$$K_{oc} = K/OC\% \times 100 \quad \text{for nonlinear Freundlich isotherm} \tag{3.32}$$

The K_{oc} parameter is less variable than the original coefficient of adsorption and is nearly independent of soil type (Hamaker and Thompson 1972, Rao and Davidson 1980, Kenaga and Goring 1980, and Karickhoff 1981). Thus the implicit assumption is that the organic matter content plays the most important role in adsorption of organics by soils. If this assumption is not true, the K_{oc} will be dependent on the mineral content of soils as well. Also, the argument does not hold for soils with very low and very high organic carbon contents (Hamaker and Thompson 1972, Rao and Jessup 1982). K_{oc} values for a multitude of organic compounds are listed in several publications (Hamaker and Thompson 1972, Kenaga and Goring 1980, and Karickhoff 1981). Recently it has been reported that Koc is also independent of soil fractions i.e., particle size separates (Karickhoff et al. 1979, Rao et al. 1982, and Nkedi-Kizza et al. 1983).

Due to the large number of toxic organic substances contaminating soil systems, it is impractical to measure adsorption constants for all substances. Hence, indirect measurement of adsorption constants by relating them to properties of the specific compound appears to be efficient and economical. Lambert (1967) has derived a relationship between adsorption by soil and chemical structure of certain classes of chemicals. The relationship is based on extrathermodynamic linear free energy approximations and uses "parachor" (p) as a measure of the molar volume (V_m) and surface tension (r) of the chemical under investigation: $p = V_m r^{1/4}$. Two assumptions were made in this model: 1) the soil organic matter is the dominant sorbing medium, and 2)

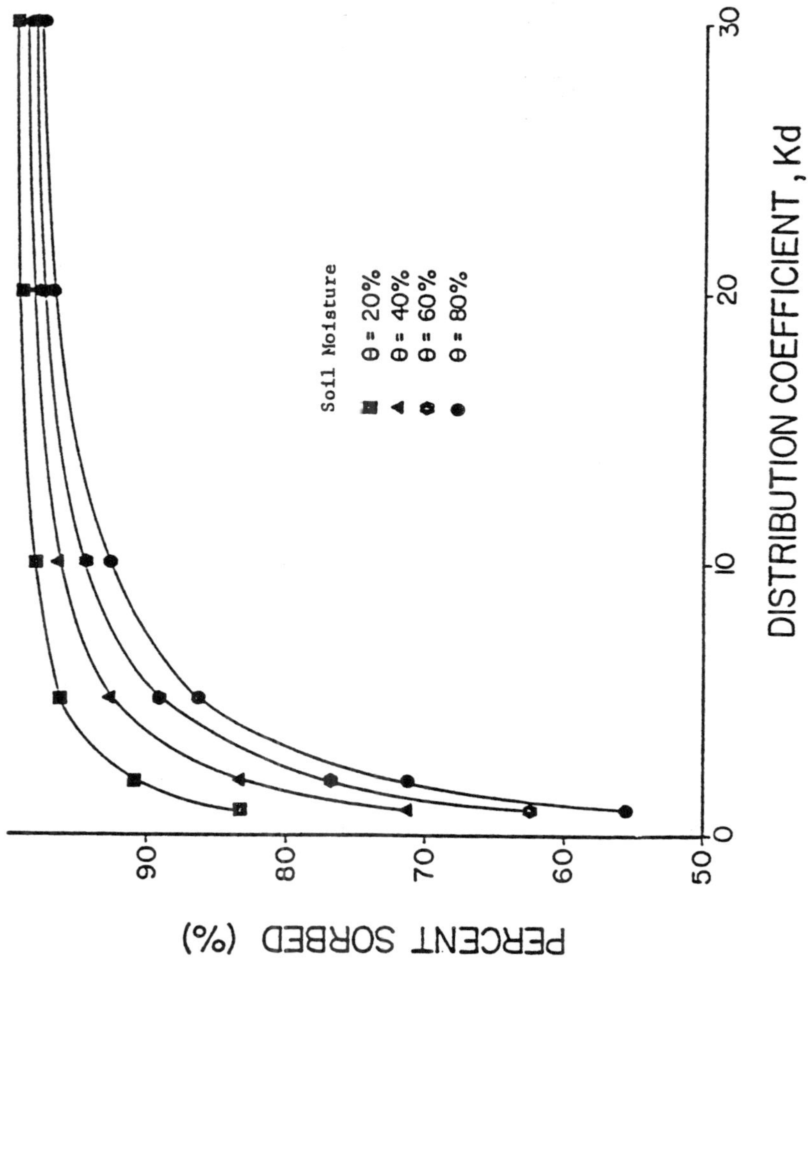

Figure 3-41. Extent of sorption as a function of soil moisture θ and K_d.

the adsorbate molecule is an uncharged molecule. Because of difficulties in measuring parachor for pesticides, this concept has not been thoroughly investigated (Rao and Davidson 1980). Recently there have been some attempts to correlate K_{OC} with solubility and octanol-water partition coefficient (K_{OW}) if a compound (Karickhoff et al. 1979; Karickhoff 1981; Kenaga and Goring 1980; Schwarzenbach and Westall 1981; Chiou et al. 1982). Briggs (1981) has demonstrated both theoretical and experimental justification for such relationships. Table 3-44 summarizes these relationships.

As shown by the equations in Table 3-44, K_{OC} values are correlated with octanol-water partition coefficients (K_{OW}). The K_{OW} values were generally measured by direct liquid-liquid partitioning. However, recently Veith et al. (1979) and Ellgehausen et al. (1981) have used the retention time of constituents in high performance liquid chromatography (HPLC) for estimating K_{OW}. Veith et al. (1979) obtained a calibration curve using a mixture of standard chemicals. The average variation of the calculated K_{OW}'s for 18 other chemicals was 22.5 percent, which is comparable to other methods of estimating K_{OW}. Laboratory measurement of K_{OW} is often highly variable (Mingelgrin and Gerstl 1983, and Baes and Sharp 1983).

Sabljić and Protić (1982) have used molecular connectivity indices to predict K_{OC}. Molecular connectivity indices are numerical characteristics of a molecule depending on the number and types of atoms and bonds and the adjacent environment. Polycyclic aromatic hydrocarbon data was used in this study, which was developed by Karickhoff et al. (1979).

3. Brunauer, Emmet, and Teller (BET) adsorption isotherm: The BET adsorption theory is actually an extension of Langmuir monolayer adsorption theory. The BET model assumes that the Langmuir equation applies to each layer adsorbed. Also, it assumes that the first layer need not be completed before another layer can start. The BET equation may be expressed as:

$$S = \frac{B_c Q^o}{(c_s - e)\,[1 + (B-1)(c - c_s)]} \tag{3-33}$$

where c_s is the saturation concentration of the solute, c is the concentration of the solute in liquid phase, Q^o is the number of moles adsorbed per unit weight of adsorbent in forming a complete surface layer, q_e is the number of moles of solutes adsorbed per unit weight at concentration c, and B is a constant. The BET isotherm has been used for adsorption of pesticides (Jurinak 1957).

The isotherms discussed above can be presented graphically as straight lines using convenient transformations (Figure 3-42). More models are available and are listed in Table 3-43.

Nonequilibrium Adsorption--

The summary of kinetics models available are shown in Table 3-43. A detailed discussion of the kinetic equations is given by Hamaker and Thompson (1972) and Travis and Etnier (1981). The relevant kinetic models will be discussed in conjunction with transport models in another section of this report.

TABLE 3-44. K_{OC} RELATIONSHIPS WITH SOLUBILITY AND OCTANOL-WATER PARTITION COEFFICIENT

1. Karickhoff et al. (1979), for the polycyclic aromatic class of chemicals

   $\log K_{OC} = 1.0 \log k_{ow} - 0.21$ (1.1)

   $\log K_{OC} = -0.54 \log X_{sol} + 0.44$ (1.2)

   where X_{sol} is water solubility expressed as a mole fraction

2. Karickhoff (1981) obtained from data for five polycyclic hydrophobic compounds (benzene, napthalene, phenonthrene, anthracene, and pyrene)

   $\log K_{OC} = -0.921 \log X_{sol} - 0.00953\ (MP-25) - 1.045$ (2.1)

   $\log K_{OC} = -0.594 \log X_{sol} - 0.197$ (2.2)

   $\log K_{OC} = 0.987 \log K_{OW} - 0.336$ (2.3)

   where MP is melting point (°C), if liquid at 25°C MP = 25

3. Chiou et al. (1979)

   $\log K_{OM} = 4.04 - 0.557 \log WS$ (3.1)

   where WS is solubility in μmole/l

4. Chiou et al. (1982)

   $\log K_{OM} = -0.813 \log (S\bar{V}) - 0.993$ (4.1)

   $\log K_{OM} = -0.729 \log (S)$ (4.2)

   where S is solubility in mole/l and $\bar{V}$ is the molar volume (l/mole).

5. Kenaga and Goring (1980)

   $\log K_{OC} = 3.64 - 0.55 \log WS$ (5.1)

   where WS is the solubility in mg/l

6. Schwarzenbach and Westall (1981)

   $\log K_{OC} = 0.72 \log K_{OW} + 0.49$ (6.1)

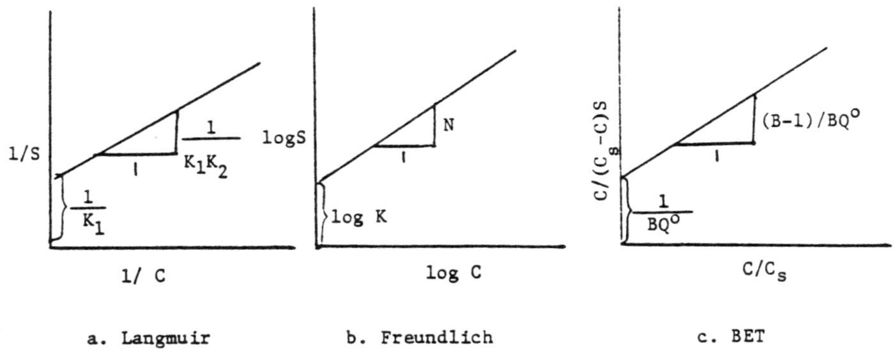

Figure 3-42. Linear transforms of adsorption isotherms.

Factors Affecting Sorption

It was stated previously that adsorption is a complex process involving the interaction of a solvent, solute, and adsorbent system. Hence, it is expected that the properties of the solvent, solute, and asorbent will influence the adsorption process. Bailey and White (1970, and Hamaker and Thompson (1972) have discussed in detail the various factors that influence adsorption. In this section the factors important for predicting sorption of hazardous constituents on soil are highlighted.

Characteristics of the adsorbate that affect adsorption are: 1) chemical character, 2) dissociation constant, 3) water solubility, 4) charge distribution of organic cation, 5) molecular size, 6) polarity, and 7) polarizability. These factors relate specifically to the following aspects of sorption:

1. The mechanism by which the adsorbate will be adsorbed (e.g., hydrogen bonding vs. ion exchange).

2. The effect of pH on adsorption and reversibility of of adsorption.

3. Location of adsorption, in other words exterior surface vs. interlayer surface of clay mineral.

4. Orientation of molecule with respect to adsorbent surface.

Characteristics of a chemical such as parachor, solubility, and connectivity index have been used as mentioned previously to determine adsorption constants (Lambert 1967, Karickhoff et al. 1979, Chiou et al. 1979, Karickhoff 1981, Chiou et al. 1982, and Sabljić and Protić 1982).

The physicochemical characteristics of adsorbents that affect adsorption are mainly related to surface characteristics. That is the surface area, shape, and charge (magnitude and distribution) of the adsorbent surface. The colloidal part of soil is important due to its high surface area. The colloidal part consists of clay minerals, organic matter and hydrated metal oxides.

Clay minerals are minerals with sheet silicate structures. A layer silicate may be a combination of two types of structural units. The first unit is a tetrahedral sheet which is an Si^{4+} atom joined with four oxygen atoms forming a tetrahedron. The second structural unit is an octahedral sheet with an Al^{3+} (typically) in octahedral coordination with six oxygen atoms or hydroxyl anions. The relative proportions of the two sheets describes the clay type. A 1:1 clay (e.g., kaolinite) has low specific area and low cation exchange capacity (CEC). The adsorption in 1:1 clay occurs only on the external surface area. A 2:1 expanding clay (e.g., montmorillonite and vermiculite) has high specific area and high CEC. Adsorption may occur between the layers as well as or the external surface. The organic molecule should have some polarity in order to penetrate the basal plane (between the layers). Organic molecules adsorbed between the layers might not be available for biodegradation (Weber and Scott 1966). Other 2:1 clays that are not expandable due to strong bonds between the layers (e.g., illite and chlorite), have lower specific area and lower CEC. Table 3-45 has a summary of properties of different clay minerals obtained from Bailey and White (1970). Clay minerals have a net negative charge which is the result of isomorphic substitution (Bohn et al. 1979). The charge induced by isomorphic substitution is a permanent charge and is independent of pH. The magnitude and location of this charge is important in attracting cations to clay surface.

Soil organic matter (OM) is a complex mixture of organic compounds in soil. Qualitatively OM consists of two fractions, humic and nonhumic. Humic organic matter is the transformed (microbiologically or chemically) component of plants, animals, and microorganisms. Humic organic matter is divided into three components: 1) fulvic acid, 2) humic acid, and 3) humins. These are similar in structure but rather different in molecular weight and functional groups. Nonhumic organic matter is represented by the unaltered remains of plants, animals, and microorganisms such as cellulose, starch, proteins, chitin, and fats (Bohn et al. 1979, Morrill et al. 1982). Soil OM generally has high CEC, however, the CEC is pH dependent (Bohn et al. 1979). Many reseachers consider soil OM the most important parameter for pesticide adsorption (Sheets et al. 1962, Roberts and Wilson 1965, Hamaker et al. 1966, Savage and Wauchope 1974, and Nkedi-Kizza et al. 1983). Also it has been reported that desorption of organic compounds from soil OM is slower than that from clay minerals (McGlamery and Slife 1966; Helling et al. 1971; Jamet and Piedallu 1975). Desorption is not complete if soil is allowed to dry after adsorption (Hamaker and Thompson 1972). Walker and Crawford (1968) showed an absolute increase in adsorption of four triazine herbicides when straw was added to soil. Their study was performed on 36 different soils. Also, they found a high correlation between adsorption and present organic carbon for soils with percent OC>5. Thus the implications of enhanced adsorption of hazardous constituents by adding a source of organic matter to the soil such as straw is important in terms of "immobilizing" the contaminants of the soil.

TABLE 3-45. SELECTED PHYSICAL PROPERTIES OF SOIL CONSTITUENTS (BAILEY AND WHITE 1970)
Used by permission, see Copyright Notice

Soil constituent	Physical Property	
	Cation exchange capacity (meq/100 g)	Surface area (sq. m/g)
Organic matter	200 to 400	500 to 800
Vermiculite	100 to 150	600 to 800
Montmorillonite	80 to 150	600 to 800
Dioctahedral vermiculite	10 to 150	50 to 800
Illite	10 to 40	65 to 100
Chlorite	10 to 40	25 to 40
Kaolinite	3 to 15	7 to 30
Oxides and hydroxides	2 to 6	100 to 800

Amorphous minerals are non-crystalline metal oxides and hydroxides (mainly Al and Fe). These minerals represent the transition stage between unweathered parent materials and well crystallized secondary soil minerals. The CEC of amorphous mineral is pH dependent. The charge for freshly deposited mineral is positive which becomes negative with aging. The positive charge plays an important role in anion adsorption (Bohn et al. 1979).

Nonsingularity of Adsorption-Desorption--

The adsorption isotherms described are obtained by implicitly assuming that the adsorption is a reversible process. However, in some cases adsorption is not reversible. That is, the desorption path will be different from that of the adsorption path. Nonsingularity (hysteresis) is defined when there is a residual retained on soil at a given equilibrium concentrate when desorption occurs (Rao and Davidson 1980). Desorption has been observed frequently (Hornsby and Davidson 1973, Swanson and Dutt 1973, Farmer and Aochi 1974, Savage and Wauchope 1974). Hornsby and Davidson (1973) found that the desorption distribution coefficient is a function of the amount adsorbed prior to desorption. Desorption generally follows the Freundlich isotherm equation but with a different coefficient. The desorption exponent (N_a) was equal to the adsorption coefficient divided by 2.3 (Swanson and Dutt 1973). Rao and Davidson (1980) gave three reasons for non-singularity: 1) artifacts created due to some aspects of the desorption method used, 2) failure to establish complete equilibrium during the adsorption phase, and 3) chemical and/or biological transformations of pesticides during the experiment. Koskinen et al. (1979) have studied the desorption of 2,4,5-T. They noticed that nonsingularity increases as the equilibrium time allowed for adsorption increases. Also, they found that when concentrations were corrected for biodegradation, which they monitored by CO_2 emission, the hysteresis (reaction direction dependence) was decreased considerably. Sharom et al. (1980), studied desorption of 12 insecticides, and found desorption was proportional to the solubility of the compound. They also noticed that insecticides desorbed in decreasing order from sand, sandy loam, and sediments, while the relative order of desorption of the 12 insecticides remained the same in each soil.

Rao and Davidson (1980) have shown the variation of error in estimating adsorbed material when desorption is neglected and the following argument is taken from them. As indicated previously, both adsorption and desorption phases follow Freundlich isotherm but with different coefficients:

$$S_a = K_a C^{N_a} \tag{3-34}$$

$$S_d = K_d C^{N_d} \tag{3-35}$$

where subscripts a, d denote adsorption and desorption respectively. Since the degree of non-singularity is dependent on the maximum amount adsorbed (S_m) before desorption (e.g., Van Genuchten et al. 1974), therefore:

$$K_d = K_a^{\beta} S_m^{1-\beta} \tag{3-36}$$

where $\beta = N_d/N_a$. Since S_m can be expressed using Equation 3-35 as:

$$S_m = K_a C_m^{N_a} \tag{3-37}$$

substituting S_m from Equation 3-37 into Equation 3-36 we get:

$$K_d = K_a C_m^{N_a - N_d} \tag{3-38}$$

Substitution of Equation 3-38 into Equation 3-35 for K_d yields:

$$S_d = K_a C_m^{N_a - N_d} C^{N_d} \tag{3-39}$$

Calculating the ratio S_d/S_a by dividing Equation 3-39 by Equation 3-35:

$$\frac{S_d}{S_2} = (C_m^{N_a - N_d})(C^{N_d - N_c}) \tag{3-40}$$

Using $N_a = 2\text{-}3N_d$ (Swanson and Dutt 1973), and assuming that the largest $C_m = 10$ mg/l, Equation 3-40 is simplified to

$$\frac{S_d}{S_3} = (10^{0.5652\ N_a})(C^{-0.5652\ N_a}) \tag{3-41}$$

This equation is plotted for different N_a values in Figure 3-43. From Figure 3-43 it is apparent that the error in estimating the adsorbed amount becomes larger as the concentration decreases, and as adsorption becomes a linear function, i.e., as N_a approaches 1.

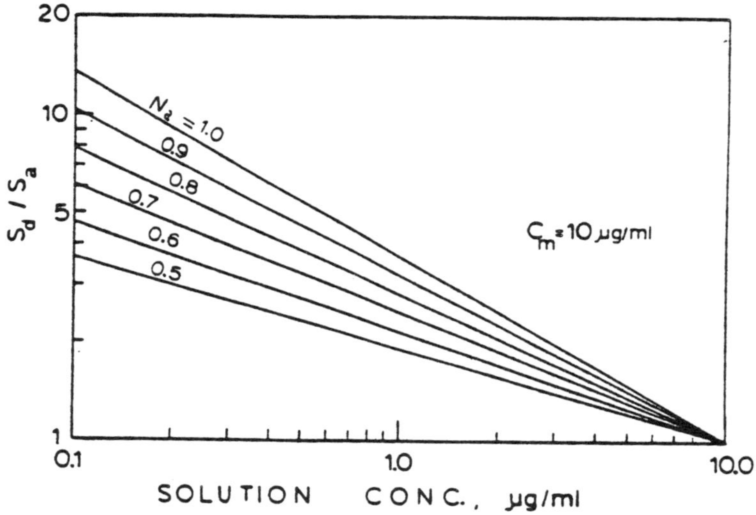

Figure 3-43. Errors introduced by the assumption that adsorption-desorption isotherms are singular when they are nonsingular (Rao and Davidson 1980). Used by permission, see Copyright Notice.

The information presented here concerning processes and factors influencing adsorption of solute species by soil systems can be used to assess the relative mobility and potential for immobilization in a specific waste-soil system and aid in determining specific in situ treatment schemes. Examples and methodologies concerning the use of such information are presented in the modeling section of this chapter and in Volume I of this manual.

SOIL MICROBIOLOGICAL FACTORS RELATED TO IN SITU TREATMENT

The Soil Microbial Ecosystem

Physical and Chemical Properties--

The soil environment is one of heterogeneity with scales of particulate size ranging from meters to nanometers. The environment having direct effect on the biological activity of a single bacterium or yeast, or microcolony of bacteria in the soil is probably no larger than a cubic millimeter. The hyphae of filamentous fungi may, however, interact with a range of soil microenvironments extending over a much larger volume. Variations in water availability, soil atmosphere composition, inorganic and organic nutrient availability, soil texture, and soil structure all make up the environment of soil microbes and affect their metabolism and growth. In a practical sense, we are forced to deal with management of the soil microbial environment on a relatively large scale, relying on the average condition at microsites in the soil to be conducive to the microbial function we are trying to encourage. We are, however, more likely to be successful in our approach if we keep in mind the nature of the soil microbial environment and the microsites of which it is made. It is, of course, beyond the scope of this manual to deal extensively with soil microbiology or microbial ecology. Good textbooks are available (Alexander 1977, Atlas and Bartha 1981) if more complete information on these topics is needed. Gray (1978) has discussed microbial aspects of the soil relating to pesticide transformations. We will also briefly review principles of soil microbiology which pertain to the broader topic of toxic and hazardous waste transformations, with emphasis on environmental factors and microbial populations or microbial community structure which might be manipulated to accomplish treatment of contaminated soil.

Soil microbes are greatly affected by soil water. Water is of course, necessary for microbial life and the water potential against which microorganisms must extract water from the soil regulates their activity. The Soil Science Society of America (1981) has published proceedings of a symposium on the relation of soil water potential to soil microbiology. Many microorganisms are capable of metabolic activity at water potentials lower than -15 bar. The lower limit for all bacterial activity is probably abouth -80 bar, but some organisms cease activities at -5 bar. Fungi appear to be more tolerant of low soil water potential, than are the bacteria (Gray 1978, Harris 1981), and microbial decomposition of organic materials in dryer soils is probably due primarily to fungi. When the soil becomes too dry many microorganism form spores, cysts, or other resistant forms, while many others are killed by desiccation. Soil water also serves as the transport medium through which many nutrients diffuse to the microbial cell, and through which waste products are removed. The decomposition of pesticides and other xenobiotic organic compounds as well as natural organic matter in the soil depends on soil water potential and its influence on microbial activity (Sommers 1981).

The degree to which the soil pore space is filled with water also affects the exchange of gasses through the soil. Microbial respiration, plant root respiration, and the respiration of other organisms removes oxygen from the soil atmosphere and enriches it with carbon dioxide. Gasses diffuse into the soil from the air above it, and gasses in the soil atmosphere diffuse into the air, but the oxygen concentration in ordinary soil may be only half that in air while carbon dioxide concentrations may be many times that of air (Brady 1974). Even so, a large fraction of the microbial population within the soil depends on oxygen as the terminal election acceptor in metabolism. Bacteria of the genus Pseudomonas, members of which are often linked to the transformation of xenobiotic compounds in the soil, are of this type, i.e. strict aerobs. When soil pores become filled with water, the diffusion of gasses through the soil is severely restricted, oxygen is consumed faster than it can be replaced, and the soil becomes anaerobic. This loss of oxygen as a metabolic electron acceptor induces a drastic change in the make up of the soil microflora. Facultative anaerobic organisms, organisms which use alternative election acceptors such as nitrate (denitrifiers) or sulfate (sulfate reducers), and strict anaerobic organisms become the dominant species. Generally, microbial metabolism shifts from oxidative to fermentative, and becomes less efficient in terms of biosynthetic energy production. Soil structure and/or texture primarily determine the size of soil pores, and hence the water content at which gas diffusion is significantly limited in a given soil, and the rate at which anaerobiosis sets in.

Soil texture and clay mineralogy are also important factors in soil microbial ecology (Stotzky 1972, 1980). Clays with a 1:1 crystal lattice (e.g. kaolinite) are non-swelling and have lower cation exchange capacity, while 1:2 crystal lattice clays (e.g. montmorillonite) swell enormously trapping water and perhaps other compounds between the lattices. The high cation exchange of clays like montmorillonite greatly increases the buffer capacity at microsites within the soil, reducing the impact of protons released into the environment as a product of microbial metabolism. Differential sorption of organic compounds and inorganic ions, including hazardous waste components by different clays, also affects the availability of substrates and micronutrients to microogranisms. Stotzky (1980) has investigated the biodegradabilty of proteins, peptides, and amino acids adsorbed by montmorillonite clay and found that they were much less susceptible to degradation by microorganisms readily able to use them in solution. However, since all clay in soil is apparently not complexed with organic matter, some mechanism (mechanical, chemical, or biological) must remove or prevent complexing of clay with organic matter in situ. Perhaps organics in situ are not bound directly to the clay but to polymeric hydous oxides of Fe, Al, and Mn associated with the clay. Fluctuations in soil (microsite) pH would disrupt these inorganic polymers, releasing the organic matter to solution. It could then either be degraded or rebind to soil clays (Stotzky 1980).

Soil organic matter is extremely important to the microbial ecology and activity of the soil. Its high cation exchange capacity and high density of reactive functional groups play important roles in the retention of bacteria in the soil as well as binding both organic and inorganic compounds which may be added to the soil. These adsorbed or chemically bound compounds or ions may be available for microbial attack and transformation. This microbial activity may detach the compound or its metabolite from the soil organic matter, increasing its mobility in the soil (Bartha 1980). Although generally considered to be recalcitrant, soil organic matter is in a state of flux. Mineralization of soil organic matter may provide important source nutrients to the soil from organically bound nitrogen, phosphorus, and sulfur (Woodmansee et al. 1978).

Due to their high surface to volume ratio, soil microogranisms are well adapted to take up inorganic nutrients from the soil. If biodegradable organic materials are added to the soil so as to raise the carbon to nitrogen (C:N) ratio higher than about 20:1, mineral nitrogen in the soil will be immobilized into microbial biomass, and the decomposition process will be slowed considerably. Similar immobilization of phosphorous can occur when carbon is in excess (Alexander 1977). If during the treatment of hazardous waste contaminated soils, the soil must be managed to decompose organic matter, nitrogen and phosphorus may be required to bring the C:N:P ratio to approximately 120:10:1, the approximate ratio found in bacterial biomass (Alexander 1977, Kowalenko 1978).

Soils contaminated by waste materials may contain elevated concentrations of salts. In addition, treatment approaches such as the use of fertilizers or chemical agents, may add salinity to the soil. Increased salinity increases the osmotic potential of the soil water, and soil microbes may be restricted in their activity due to osmotic stress. Monitoring of soil salinity prior to and during treatment is important to the success of any biological treatment process.

The soil reaction, or pH, affects the activity of soil microbes. Fungi are typically more tolerant of acidic soil conditions (below pH 5) than are bacteria in aqueous media, but the differences are less clear in soil where the buffer capacity of clay and humic materials affect the concentration of protons at microsites. (Gray 1978). Soils may need to be limed to raise pH or treated with sulfur or other acid forming materials to lower pH so that the soil pH ranges between 5.5 and 8.5 to encourage microbial activity. Phosphorus solubility is maximized at pH 6.5; this may be the ideal soil pH.

Adsorption of organic molecules on negatively charged clay colloids strongly depends on the solution pH. Critical pH values are dependent on the pK_a of the adsorbate and the magnitude of charge on the adsorbent. Adsorption of pesticides of widely varying molecular structure on clays increases as pH decreases (Frissel and Bolt 1962, Harris and Warren 1964, Talbert and Fletchall 1965). The pH effect is apparently less important for peat and some other organic soils which suggests that a different sorptive mechanism is occurring for those types of soils.

The Bacteria--

The metabolic diversity of the procaryotic organisms in the soil focuses a great deal of attention on them when biodegration of xenobiotic or hazardous materials is of concern. In general, bacteria may be classified metabolically as heterotrophic; that is, they derive their energy and carbon for survival and growth from the decomposition of organic materials, or as autotrophic, meaning that they fix the carbon they need for growth from carbon dioxide and (usually) obtain their energy from light (photosynthetic) or the oxidation of inorganic compounds (lithotrophic). The heterotrophic bacteria are most important in the transformation of organic hazardous compunds, and soil treatment schemes may be directed toward enhancing their activity. However the availabilty of nutrients (especially nitrogen) to the heterotrophic microorganisms of the soil may depend on the transformation of inorganic materials by lithotrophic organisms (Focht and Verstraete 1977, Painter 1977). Transformations of toxic metals or metalloids and their compounds may directly or indirectly involve the activities of lithotrophic or photosynthetic bacteria.

The Fungi--

The fungi are eucaryotic microorganisms which lack photosynthetic apparatus and depend on heterotrophic metabolism. They may be either filamentous or uncellular (yeasts or amoeboid), while some types aggregate to form macroscopic forms such as a plasmodium or fruit-bodies (mushrooms). Fungi constitute a large fraction of microbial biomass in soil, especially in acidic soil, and their enzymatic activity is important to most decomposition processes. When levels of available substrate are low or water availability is low, a major portion of the fungal biomass in soil is either dormant or dead. Fungal spores or other resistant structures can survive in soil under adverse conditions for long periods of time, and then quickly germinate and grow when conditions become favorable. Clay mineralogy, temperature, and other soil environmental conditions affect the speciation and diversity of fungi in the soil. Most filamentous fungi are aerobic, and yeasts are often facultatively anaerobic. Most fungi are mesophilic, and even thermophilic fungi do not grow above about 65°C. Many fungi grow at temperatures below 10°C.

Algae--

The soil surface usually contains appreciable populations of eucaryotic algae and cyanobacteria. Since the zone of light penetration into the soil is severely restricted, the algal biomass in soil is usually low. Algae may be important, however, in enhancing photodecomposition of hazardous organic compounds at the soil surface.

Higher Life Forms--

The soil also contains many other higher organisms which graze the bacterial or fungal populations, or feed on detrital matter and its associated microflora. Protozoa, nematodes, insects, and worms are important in this regard. These organisms affect the decomposition process by controlling bacterial or fungal population size through grazing (Bryant et al. 1982), by harboring microbes within their intestinal tract which may decompose a compound of interest, by comminuting plant materials (insects), or by mixing the soil and contributing to its aeration.

Microbial Interactions--

Interactions between microorganisms in the soil are complex, and undoubtedly play an important role in the activity of microorganisms important to the transformation or decomposition of hazardous waste compounds. Some of these interactions have already been suggested in earlier discussion. Competition for growth requirements among microorganisms is intense. Many microorganisms are metabolically specialized (e.g., the autotrophic nitrifiers) and are less reliant on preformed organic substrates and/or growth factors. However the nutritional requirements of many soil microbes overlap to some degree, and those organisms which are to survive in the soil must be able to effectively compete with their neighbors for these required nutrients.

To increase their opportunities for survival in the competitive soil environment, many microorganisms have aquired antagonistic abilities with which they may limit the growth of other microorganisms. The antibiotics produced by soil microorganisms are probably the best known microbial antagonistic agents, but other kinds of inhibitors, including acids, bases and other inorgranic compounds are used (Alexander 1971, Atlas and Bartha 1981). Suppressive influences are often expressed between microorganisms and higher life forms in the soil as well (Rice 1974).

Soil microbes may also be closely associated with one another physically and/or metabolically. True mutualistic or symbiotic relationships exist, where the organisms relationships are necessary to their survival in the niche they occupy. Less restrictive relationships are perhaps more common. It is not uncommon for the degradation of a xenobiotic compound to involve sequential metabolism by two or more microorganisms (Beam and Perry 1974); a relationship which may benefit only one partner (commensalism) or both (protocooperation or synergism) (Atlas and Bartha 1981).

Life on the Verge of Starvation--

Readily assimilable substrates are short lived in most soil environments. The majority of time, carbon and energy supplies are limited and microbial growth rates are restricted. Soil organic matter, including many hazardous waste components, is recalcitrant to microbial attack and supplies carbon and energy slowly to the soil microbial ecosystem. Organisms able to survive in such a growth limited environment must have evolved strategies which compensate for this state of affairs. Two major kinds of survival strategies are apparent in the microbial ecosystem. Winogradsky (1949) observed, in about 1924, that soil harbored a population of bacteria which increased their numbers rapidly when a readily "fermentable" substrate was introduced to the soil. These "zymogenous" organisms would dominate the population until the added substrate was depleted, and then would die away. Another population of "autochthonous" bacteria grew much more slowly, and their population did not change significantly when a fermentable substrate was added. Population ecologists often classify organisms which depend on high growth rates for competitive advantage as "r" strategists. This type of organism would correspond to Winogradsky's zymogenous bacteria. Organisms which reproduce more slowly and have many competitive adaptations are classified as "k" strategists, implying that their growth is related to the carrying capacity of the ecosystem. These organisms correspond to Winogradsky's autochthonous bacteria (Atlas and Bartha 1981).

The inoculation of hazardous waste contaminated soils with microorganisms selected for their ability to degrade or transform hazardous materials is a very attractive treatment concept. However, the treatment manager must be aware that the restrictive environment and complex ecology of the soil microbial system may limit the ability of introduced microbes to become self-perpetuating and perform their specialized functions for more than a short period of time. Reinoculation may need to be repeated several times before satisfactory levels of treatment are achieved. Liang et al. (1982) conducted experiments on the survival of antibiotic resistant microorganisms after inoculation into sterile and nonsterile water, sewage, and soil. They suggest that microorganisms which are tolerant of multiple kinds of stresses (e.g., abiotic stress, starvation, and antagonism) have a higher potential for survival in the soil after genetic manipulation (engineering) in the laboratory, than do microorganisms with less tolerance versatility. Much of the current genetic engineering work directed at producing bacteria capable of complete degradation of xenobiotic compounds is being done with bacteria (e.g., Pseudomonas) which are zymogenous (Chakrabarty 1982). It is encouraging that some recent isolations of xenobiotic degrading bacteria have turned out to be Arthrobacter spp. These organisms were isolated using chemostat techniques which allow time for more slow growing, autotrophic bacteria to become dominant and their degradative ability to be recognized (Stanlake and Finn 1982). Recently reported experiments have shown the potential for use of these more autochthonous bacteria to treat hazardous waste contaminated soils (Edgehill and Finn 1983).

Biogeochemistry of Toxic Metals and Metalloids

Arsenic, barium, cadmium, chromium, lead, mercury, nickel, selenium, silver, thallium, and vanadium and/or one or more of their compounds are listed as hazardous waste constituents by the U. S. Environmental Protection Agency (40 CFR 261). The toxic nature of these elements or their compounds adversly affects some fraction of the microbial population when the soil is contaminated, while other organisms or groups of organisms may be able oxidize, reduce, methylate, demethylate or otherwise transform these elements so that their solubility, sorption, or volatility in the soil is greatly affected.

Arsenic--

Since arsenic exists in the -3, +3, or +5 valence states it is subject to oxidation and reduction by microbial activity. The positive trivalent state is generally the most toxic (Zajic 1969). There is no evidence that organisms which oxidize arsenite to arsenate use the energy available from the reaction for growth. Apparently, the oxidation is done by heterotrophic organisms. Many heterotrophs can also reduce arsenate (Alexander 1977, Zajic 1969).

Arsenic can be lost from soils as volatile arsine or as methylarsines. Cheng and Focht (1979) studied the volatilization of arsenicals from flooded soils amended with glucose and urea in aerobic flasks. They found that soil treated with arsenate or arsenite gave off arsine, while soil amended with disodium methylarsonate gave off arsine and monomethylarsine, and soil amended with dimethylarsonate (cacodylic adic) yielded arsine and dimethylarsine. No evidence was found for methylation of any of the arsenical substrates in the three soils tested. Apparently, demethylation was more important under these conditions. Other studies (Cox and Alexander 1973, Woolson and Kearney 1973, Kaufman 1974, and Woolson 1977) have found methylation of arsenicals as an important process in soils, and have also found that trimethylarsine is an important gaseous product. These results suggest that soils contaminated with arsenic must be managed to minimize volatilization through microbial reduction. Hassler (1982) found that essentially only soluble arsenic was volatile in soil and processed oil shale.

Barium--

Although barium is distributed in biological systems in comparable amounts with strontium, very little is known about its interactions with soil microorganisms (Zajic 1969).

Cadmium--

The growth of soil bacteria is retarded by cadmium, and the community structure is affected. Cells of Escherichia coli exposed to Cd suffer single stranded breakage in their deoxyribonucleic acid (Mitra and Bernstein 1978). In experiments with bacteria isolated from Cd contaminated soils in Great Britain, Shearer and Olson (1983) found that the bacteria would grow in media containing 10ppm Cd (regardless of chemical species tested), but that growth rate was inversely proportional to Cd concentration (0, 0.1 and 10 ppm) in a defined growth medium. Chemical species of Cd tested were toxic to the bacteria in the order: Cd^{2+} > Cd aspartate > $Cd_3(PO_4)_2$ > $Cd_3(PO_4)_2s$. Tripp et al. (1983) isolated 50 strains of bacteria on media containing 0, 10, and

100 ppm Cd from six sewage sludge amended sites in England. Gram negative bacteria were most common on the 100 ppm medium, and the number of genera of bacteria decreased with increasing Cd concentration. The number of genera isolated on cadmium medium was significantly lower from sites not contaminated with cadmium than from contaminated sites. Apparently, the bacterial population had adapted to the heavy metal contamination. Khazael and Mitra (1981) found that cells of E. coli which were adapted to growth in 0.34 ppm Cd in a minimal salts medium produced a Cd binding high-molecular-weight cytoplasmic component which may sequester Cd^{2+} in the cells and allow normal metabolic function. It is not clear if this mode of adaptation to Cd in the environment is common to other microorganisms.

Chaney et al. (1978) studied respiration rates of black oak forest soil and litter which had been both "naturally" and experimentally contaminated with Cd and Zn. After 23 days incubation, soil and litter microcosms treated with 10 ppm Cd and no zinc had significantly lower respiration rates than the control. The most severely reduced respiration rate occurred when 10 ppm Cd plus 1000 ppm Zn was applied. But even this highest combined treatment rate was not significantly lower than the control. Apparently the litter-soil system used in this study was quite resilient to contamination by these metals.

The soil microflora may have significant impact on Cd availability. Work by Kurek et al. (1982) has shown that soil microbial biomass may contribute significantly to the soil Cd binding capacity. Dead bacterial cells sorbed more Cd from liquid medium than did live cells.

Since cadmium exists in nature only in the valence state of +2, microbial oxidation or reduction of this element is unlikely.

Chromium--

Although chromium has common valence states of +2, +3, and +6, and should be amenable to oxidation or reduction by microbes, no information is available on the biochemical transformation of chromium (Zajic 1969). Hexavalent chromium has been shown to be toxic and mutagenic to Salmonella typhimurium, while trivalent chromium was neither toxic or mutagenic in the same tests (Petrilli and De Flora 1977). Ross et al. (1981) tested the toxicity of Cr^{+6} and Cr^{+3} to liquid cultures of soil bacteria and concluded that Cr^{+6} was more toxic than Cr^{+3}, and that gram negative bacteria were more sensitive to Cr^{+6} than were gram positive bacteria. When they treated a loam and fine sandy loam soil with Cr^{+3} or Cr^{+6} under laboratory conditions, they observed decreased soil respiration (CO_2 evolution) in both treatments. Small amounts of extractable Cr^{+6} were observed within 6 days of treatment with 100 ppm Cr^{+3}, but no Cr^{+6} could be extracted from this treatment after 13 days incubation. Extractable levels of Cr^{+6} decreased to less than 5 ppm after 13 days incubation of the soils treated with 100 ppm Cr^{+6}, but no increase in soil respiration was observed in any of Cr^{+6} or Cr^{+3} soils in up to 22 days incubation. Apparently, it should not be assumed that Cr^{+3} is harmless to the soil microflora. Working with laboratory media alone, Babich et al. (1982) demonstrated that Cr^{+6} was significantly more toxic than Cr^{+3} in terms of spore germination, growth, and sporulation to six species of soil fungi. However, at 100 ppm (the lowest treatment reported) of either Cr^{+3} or Cr^{+6}, the mycelial growth rate of Penicillum vermiculatum was not significantly different from the control culture.

Lead--

Inorganic lead has been shown to be toxic to a broad range of microorganisms including cyanobacteria, marine algae, fungi, and protozoa. Lead and its compounds have also been shown to affect microbial activities in soil, including inhibition of nitrogen mineralization, stimulation of nitrification, and the synthesis of soil enzymes. Species diversity of microbes in lead contaminated soils has also been found to be lower. Babich and Stotzky (1979) found different sensitivities to lead among eight species of fungi. Rhizoctonia soloni ceased growth at 500 µg Pb/ml of medium, while Aspergillus niger was not appreciably inhibited by 1000 µg Pb/ml of medium. In the same study, Ph, phosphate and carbonate ions, clay minerals, particulate humic acid, and soluble organics affected the toxicity of lead. Low pH of 5 or 6 increased the toxicity of lead, probably through the increased dominance of Pb^{2+} ion, while higher pH and the other abiotic factors tested reduced toxicity. Apparently, hydroxylated, sorbed, or complexed forms of Pb are less toxic to fungi, and probably to other soil microorganisms.

Mercury--

Microbiological processes are responsible for the oxidation, reduction, methylation, and demethylation of mercury. Mercurial pesticides, phenylmercury acetate (PMA), ethylmercuric phosphate, and methyl mercuric chloride are all biodegradable (Kaufman 1974). Metallic mercury is often the reduced product of the decomposition of such compounds (Alexander 1977). Monomethyl or dimethyl mercury may also be formed biologically in the soil from Hg^{2+} or from organomercurials. Landa (1978) monitored ^{203}Hg loss from five Montana soils of varying texture and classification which were ammended with 1ppm Hg as $^{203}HgNO_3$, and incubated for 7 weeks at room temperature at 1/3-bar moisture content. Mercury loss ranged from 5 to 30 percent with the higher initial rates of loss observed in soil ammended with glucose. Losses of Hg from autoclaved, but not necessarily sterile, soil of each type was much lower. After 2-24 days, depending on treatment and soil, very little if any Hg loss was observed. Inoculation of the autoclaved soils or reamendment of the treated soil after 38 days did not initiate further Hg loss. Apparently, the remaining Hg was stabilized in the soil and unavailable for microbial processing. Shariat et al. (1979) screened 40 common soil bacteria for their ability to tolerate and degrade methylmercury chloride. Twenty-seven (67.5 percent) could tolerate CH_3HgCl at 0.37 to 2.5 mg/l, and 21 (52.5 percent) could degrade it. The disappearance of CH_3HgCl from the medium was accompanied by a loss of total Hg, probably to volatilization of Hg° produced from reduction of CH_3HgCl. Mason et al. (1979) investigated the kinetics of methylmercury chloride degradation by the bacteria Enterobacter aerogenes and Serratia marcesens. They found that the initial rate depended on CH_3HgCl concentration and pH, and that the kinetic pattern was characteristic of enzymatic reactions. E. aerogenes exhibited uniform kinetics over the pH range tested, while S. marcescens had the highest rate of degradation at pH 8. Callister and Winfrey (1983) found that both aerobic and anaerobic heterotrophic bacteria were resistant to 14 ppm Hg^{2+}. Mercury methylation was highest in anaerobically incubated surface sediments. Nearly all (99 percent) of the Hg added as $^{203}Hg\ (NO_3)^2$ bound to the sediment within one hour, but more than 7 percent of the added Hg was methylated during a 7-day incubation. This suggests that bound inorganic mercury was available for methylation. The processes of mercury methylation and demethylation have usually been ascribed to different bacteria strains. However, Pan-Hou et al. (1980) found that a strain of the mercury methylating, anaerobic bacterium Clostridium cochlearium T-2 acquired the ability to decompose methylmercury and other organomercurials. Apparently this ability was coded for on transferable genetic material, probably a plasmid. Silver and Kinscherf (1982) have recently reviewed the genetics and biochemistry of microbial transformations of mercury and its compounds. They point out that mercury resistance is an inducible trait of organisms possessing this ability, and that the genetic material coding for this ability is transposable.

Nickel--

Nitrification, carbon mineralization, and the activities of acid and alkaline phosphatase and arylsulfatase are all inhibited in soil by nickel. Babich and Stotzky (1982, 1983a, 1983b) have studied the toxicity of nickel to microorganisms in pure culture and in soil. In pH 4.9 soil without montmorillonite clay minerals, incipient Ni toxicity to fungi ranged from 50 ppm for Aspergillus clavatus to 750 ppm for Trichoderma viride. Only one species of fungus (Gliocladium sp.) showed any growth at 1000 ppm Ni. Addition of kaolinite or montmorillonite clay reduced the toxicity of Ni to fungi with montmorillonite being most effective. Increasing the pH to approximately 7.0 also reduced toxicity. At 1000 ppm Ni, adjusting the soil pH to approximately 7.0 with $CaCO_3$ essentially doubled the fungal growth rate. Incipient decreases in survival of unicellular microbes in the same pH 4.9, Ni treated soil began at 250 ppm for Agrobacterium radiobacter, Bacillus megateriurm (both bacteria), Cryptococcus terreus and Torulopsis glabrata (both yeasts). Proteus vulgaris, Bacillus cereus (both bacteria), Nocardia rhodochrous (an actinomycete), Cryrtococcus terrus, Rhodotorula rubra, and Torulopsis glabrata (all yeasts) were viable after 7 days exposure to 1000 ppm Ni in the pH 4.9 soil. In a pH 7.7 soil, 1000 ppm Ni did not affect the viability of the six bacteria and yeasts tested. The pH 7.7 soil also contained some montmorillonite clay, and significantly higher amounts of magnesium than the pH 4.9 soil. Apparently, alkaline pH, type and amount of clay minerals, and other ions (e.g., Mg) greatly affect the microbial toxicity of nickel. Babich and Stotzky (1983b) have also reported that Mg^{2+} or Zn^{2+} cations, and S^{2-} and PO_4^{3-} anions also reduce Ni toxicity. However, K^+, Na^+, Ca^{2+}, Fe^{3+}, amino acids, tryptone, casamino acids, yeast extract, and chelating agents (citrate, EDTA, DPA, NTA) did not reduce Ni toxicity.

Selenium--

The soil microflora, as a whole, are capable of several transformations of selenium. Since organic selenium compounds do not accumulate in the soil in areas where organic selenium compounds, including volatile dimethylselenide, are common in plant tissue, these compounds must be mineralized and the selenium released as selenite and/or selenate. Apparently, the oxidation of elemental selenium in soil is at least partly mediated by microorganisms, but the mechanisms and organisms participating in the process have yet to be discovered. The reduction of selenite and selenate by soil microorganisms is a relatively common attribute, but bacteria do not seem to use selenate or selenite as metabolic electron acceptors. Selenium may be methylated to dimethylselenide by a variety of soil fungi and bacteria, and rates of methylation are greatly accelerated when a readily available carbon source, such as glucose, is added to the soil (Alexander 1977).

Silver--

The bacteriocidal property of silver is well known (Chambers et al. 1962, Woodward 1963). Sokol and Klein (1975) and Klein and Molise (1975) studied the effects of silver from cloud seeding agents on soil and sediment microbial activity. In laboratory culture, silver ions from $AgNO_3$ or from saturated AgI exhibited about the same level of toxicity to Arthrobacter sp. Anaerobic cellulose degradation was inhibited in muds amended with 100 ppm AgI, and apparently, organic matter decomposition was inhibited in soil contaminated with AgI from a cloud seeding devise. Silver concentration in these contaminated forest soils reached nearly 1500 ppm (organic matter free basis). From this and studies in a semiarid grassland soil, these investigators concluded that the threshold level of impact of silver to soil decomposer activity was 1-2 ppm Ag. Approximately 11 percent of the bacteria in the grassland soil which was not treated with Ag could reduce Ag^+ ions to metalic silver. Experimental plots treated with approximately 100 ppm Ag to a depth of 2 cm had significantly higher numbers (22 percent of total) of Ag^+ reducers.

Vanadium--

Ammonium vanadate and vanadium pentoxide may be components of hazardous waste (40 CFR 261). Vanadate may be reduced by microorganisms under appropriate conditions to vanadium oxide. Ferric ion produced by the oxidation of Fe^{2+} by the Thiobacilli has been shown to oxidize V^{+3} to V^{+5}. Vanadium is chelated by EDTA and presumably by humic acids and other natural ligands (Zajic 1969).

Decomposition of Xenobiotic Organic Compounds

The Role of Microorganisms--

Most of the compounds that can render a waste hazardous are organic, and many of these hazardous organic compounds are synthetic and without close structural analogs in nature. The molecular structure of some of these compounds was purposefully designed to resist microbial attack and hence prolong the persistence of the compound in the environment. However, many of these compounds are susceptible to some kind of transformation or complete mineralization by microorganisms in the environment. Where biodegradation is possible, augmentation of the soil microbiota with cultured organisms and/or manipulation of the soil environment through chemical or physical means to optimize the desired activity may accelerate the biotransformation or mineralization process. Alexander (1981) has pointed out that few abiotic processes completely mineralize complex organic compounds in nature and complete degradation of these compounds depends on microbial activity. However, physical/chemical transformation processes may act synergistically with biochemical decomposition in the decomposition of compounds of environmental concern. For example, simultaneous photo and microbial decomposition of 2,4,5-trichloroanaline in river water was nearly twice as fast as photodegradation alone, and more than 10 times as fast as microbial mineralization in the dark (H.-M. Hwang and R. E. Hodson, personal communication, May 1983, University of Georgia, Athens).

Genetic Engineering--

The resistance of many organic chemicals to degradation in the environment has prompted a search for techniques to improve the capabilities of microorganisms in degrading these compounds. The use of genetic manipulation of microorganisms to enhance the production of biological materials of commercial importance has made significant advancements in recent years (Hopwood 1981), and many of the principles used can be and have been employed to improve the degradative capabilities of microorganisms. Many degradative capabilities in bacteria are encoded on extrachromosomal elements called plasmids. Under appropriate conditions, plasmids can be transferred from one organism to another, and once inside the cell can impart the plasmid coded trait to the new host. Using plasmid manipulation techniques it has been possible to "engineer" organisms that have extended degradative capability, i.e., organisms that can degrade more than one xenobiotic substrate or that can completely mineralize highly recalcitrant molecules such as 2,4,5-trichlorophenoxyacetic acid (2,4,5-T) (Kamp and Chakrabarty 1979, Chakrabarty 1980, Chatterjee et al. 1981, Johnston and Robinson 1982, Pierce 1982). Continuing efforts to genetically manipulate microorganisms to be able to degrade recalcitrant hazardous compounds hold out great hope for future treatment of contaminated soil and other environments.

Concerns have been expressed over the ability of genetically engineered microorganisms to survive and function in the complex soil or aquatic environment (Stotsky and Krasovsky 1981). Since the permutations of interactions of factors in the environment which influence microbial activity and growth in the soil are essentially unlimited, it is difficult to predict the fate of organisms with new or extra genetic material (Stotsky and Krasovsky 1981). Unless the new genetic information increases an organism's ability to detoxify its environment, or increases its ability to use a broader substrate range, maintaining the genetic material would seem energetically disadvantageous for the organism, and the organism or its new genes would be selected against. However, Liang et al. (1982) have suggested that mutant organisms that tolerate abiotic stresses, resist starvation, and can coexist with antagonists may persist in the environment for extended periods. Kilbane et al. (1983) observed that a genetically engineered Pseudomonas cepacia, which could use 2,4,5-T as its sole source of carbon, maintained high populations in soil as long as appreciable amounts of 2,4,5-T were present, but decreased to unmeasurable levels when 2,4,5-T was exhausted from the soil.

There is some indirect evidence that plasmid born genes may be transferred among bacteria in the soil, and it has been suggested that some degradative plasmids have evolved through the natural combinations of plasmid DNA from different organisms (Pemberton et al. 1979, Chatterjee et al. 1981). No information is available on the ability of introduced microbes to transfer genes to indigenous organisms in natural habitats (Stotsky and Krasovsky 1981).

Effect of Structure--

Many factors, some of them only poorly understood, determine the degradability of an organic compound. Molecular structure and degree of substitution may be important. Apparently, for polychlorinated biphenyls (PCBs) recalcitrance is related to the degree of chlorination and the position of the chlorine moieties on the ring structure of the molecule (Furukawa 1982). The number of fused rings in polynuclear aromatics (PNAs) is important in determining the rate of their decomposition (Sims and Overcash 1983). Paris et al. (1983) found an inverse relationship and a high correlation between microbial transformation rates and van der Waal's radius of 8 phenols and suggest that this molecular property may be useful in predicting degradability of xenobiotics. More basically, the biodegradability of complex organic molecules depends on the existence or inducibility of enzymes in the microbial population which are capable of catalyzing the reactions needed. Alexander (1981) has compiled lists of type reactions for transformation of chemicals of environmental concern, but points out that the list is far from complete, and that more research is needed to provide this kind of information.

Cometabolism--

Xenobiotic organic compounds are usually transformed or degraded by microorganisms in either a metabolic sequence which provides energy and nutrients (i.e., C, N, P, etc.) for growth or maintenance of the organism, or by a biochemically mediated reaction which provides neither energy or nutrients to the cell. The first process usually results in the complete biodegradation of the organic molecule to mineral products (CO_2, H_2O, NH_4^+, PO_4^{3-}, etc.) and is called mineralization. The second process usually results in only a minor transformation of the organic molecule and is called cometabolism or cooxidation (Alexander 1973, 1977, 1982). Two or more substrates are required for cometabolism; one is the nongrowth substrate which is neither essential for, nor sufficient to, support replication of the microorganism (Hulbert and Krawiec 1977, Perry 1979), while the other compound(s) does support growth. The nongrowth substrate is only incompletely oxidized (or otherwise transformed) by the microorganism involved, although other microorganisms sometimes can utilize by-products of the cometabolic process (de Klerk and van der Linden 1974, Perry 1979).

It is generally recognized that cometabolism results when a non-specific enzyme attacks a recalcitrant molecule resulting in metabolism of the compound (Horvath and Alexander 1970, de Klerk and van der Linden 1974). This could be called a "metabolic mishap" since the organism supplying the enzyme gains no benefit from the metabolic transformation. The enzyme is made by the microorganism to metabolize some other organic compound in the environment which, when fully metabolized, could provide energy to the microorganisms (Perry 1979, McKenna 1977). Oxygenases are often involved in cometabolism because they can be induced by, and can attack, a large set of substrates (Perry 1979, McKenna 1977).

Cometabolism may be a prerequisite for the mineralization of many recalcitrant molecules in the environment. For example, de Klerk and van der Linden (1974) found the oxidation of cyclohexane to involve two distinct steps. First, the conversion of cyclohexane to cyclohexanol by a Pseudomonad bacterium which was using n-heptane as its energy source. Second, cyclohexanol was readily utilized by a second strain of Pseudomonad. Beam and Perry (1973, 1974) reported the same general results for the oxidation of unsubstituted cycloparaffinic hydrocarbons. Over 100 strains of bacteria were tested and none could use the hydrocarbons as their sole carbon and energy source. However, many strains could partially oxidize the hydrocarbons when a suitable energy source was present (e.g., n-alkane). The cycloalkanones resulting from the partial oxidation were readily oxidized and used as an energy source by other strains of bacteria (Beam and Perry 1974). Unsubstituted cycloparaffinic hydrocarbons are readily mineralized in natural soil systems, presumably by a process including cometabolism (Beam and Perry 1973, 1974).

Cometabolism probably occurs frequently in natural soil systems since numerous genera of bacteria, fungi, and actinomycetes have been shown to participate in the process (Horvath and Alexander 1970, Alexander 1977). Compounds incompletely degraded by cometabolism are diverse and include saturated hydrocarbons, halogenated hydrocarbons, numerous pesticides, and single ringed polycyclic aromatic hydrocarbons (Horvath 1972, Raymond et al. 1969, Horvath and Alexander 1970, Alexander 1977, Merkel and Perry 1977, McKenna 1977, Perry 1979, Sims and Overcash 1983). Since cometabolism is a generally occurring process and has the potential for transforming very refractory compounds to compounds which are more easily degraded, it may be important in the biodegradation of complex organics in hazardous waste contaminated soils. Treatment to encourage cometabolism may include adding a more easily degraded compound which is a chemical analog to the hazardous compound that must be decomposed in the soil. Figures 3-44 and 3-45 illustrate the effect of several different growth substrates on the rate at which CO_2 was released from malathion. Other growth substrates, including glucose, glycerol, and glycerophosphate, did not have an appreciable effect on the rate of [^{14}C] carbon dioxide evolved from malathion (Merkel and Perry 1977). Table 3.46 shows the relative effects of naphthalene and phenanthrene as growth substrates for the cometabolism of four polynuclear aromatic hydrocarbons in water (McKenna 1977). Sims and Overcash (1981) showed that the rate of cometabolism of benz(a)pyrene was significantly increased when the soil was enriched with phenanthrene as an analog.

Degradation of Organic Pesticides--

Seventy of the organic based compounds listed by the U.S. EPA as constituents of hazardous waste have been or are used as pesticides. These compounds are listed by broad classes of compounds in Table 3.47. Insecticides, acaricides, rodenticides, and herbicides are all represented. Concern over the environmental fate and public health effects of pesticides has focused a great deal of research on the biodecomposition or transformation of many of these compounds, and much can be learned about hazardous organic compound degradation from these studies. Here we will concentrate on the classes and specific pesticide compounds included as constituents of hazardous waste.

Decomposition of the chlorinated hydrocarbon pesticides has been the frequent topic of research. The literstuve pertaining to this topic has been reviewed by Kaufman (1974), and Matsumura and Benezet (1978). Reductive dechlorination under anaerobic conditions and cometabolism seem to be central mechanistic principles for these compounds (Horowitz et al. 1983, Sulflita et al. 1983). However, greater metabolic diversity of microorganisms degrading these compounds is being discovered (Stanlake and Finn 1982, Edgehill and Finn 1983, Nielson et al. 1983), and the importance of sequential metabolism by microbial consortia is being recognized. For example, a strain

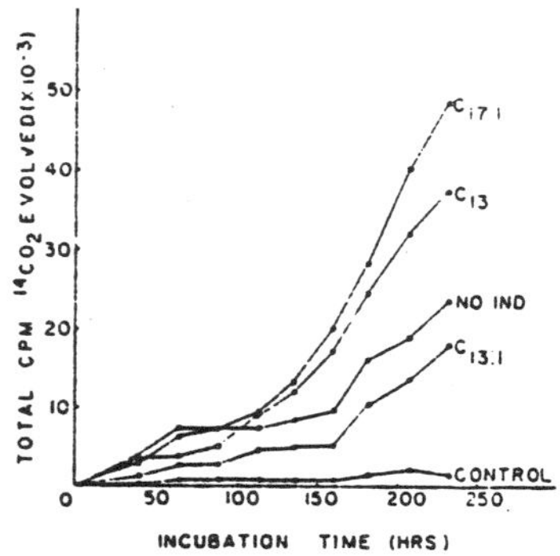

Figure 3-44. Evolution of $^{14}CO_2$ from Core Creek soil suspensions amended with [$_{14}$C]malathion. (0.1 μCi of [^{14}C]malathion added to each flask at beginning of the experiment. NO IND = no inducer added; $C_{17:1}$ = 1-heptadecene; C_{13} = n-tridecane; $C_{13:1}$ = 1-tridecene. Control was 1 g of soil/50 ml of basal medium autoclaved for 50 min on three consecutive days.) (From Merkel and Perry 1977.) Used by permission, see Copyright Notice.

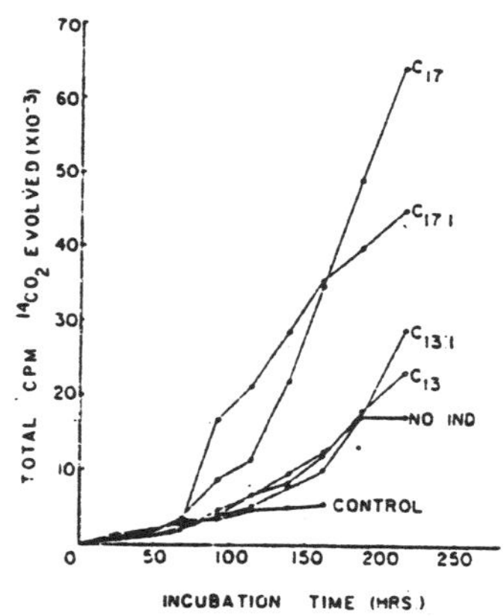

Figure 3-45. Evolution $^{14}CO_2$ from tobacco field soil suspensions. Conditions same as Figure 3-44 with C_{17} = n-heptadecane also added (from Merkel and Perry 1977). Used by permission see Copyright Notice.

TABLE 3-46. PERSISTENCE OF POLYNUCLEAR AROMATIC HYDROCARBONS IN NATURAL WATERS (FROM McKENNA 1977)
Used by permission, see Copyright Notice

Non-Growth Substrate	Growth Substrate	Amount Non-Growth Substrate Remaining After Four Weeks
Pyrene	Naphthalene	36.7%
	Phenanthrene	47.2
3,4-Benzpyrene	Naphthalene	83.5
	Phenanthrene	38.3
1,2-Benzanthracene	Naphthalene	58.3
	Phenanthrene	33.8
1,3,5,6-Dibenzanthracene	Naphthalene	92.7
	Phenanthrene	32.9

of Hydrogenomonas in consortium with the fungus Fusarium can mineralize DDM (bis-(p-chlorophenyl)methane), a metabolite of DDT (Focht 1972). Chlorinated hydrocarbon pesticides tend to be persistent in soil, but several of them are short lived. Hexachlorocyclohexane (Lindane) for example disappears from soil quite rapidly through a combination of chemical and biological degradation mechanisms. The persistence of chlorinated hydrocarbons and their metabolic products in soil may be affected by their incorporation into soil organic matter (Bartha 1980, Bollag and Liu 1983). Some chlorinated hydrocarbon pesticides have been shown to be toxic to soil microbial activity (Chendrayan and Sethunathan 1980, Subba-Rao and Alexander 1980, Tam and Trevors 1981, Trevors 1982).

The microbial degradation of organophosphate pesticides has been reviewed by Kaufman (1974), Matsumura and Benezet (1978), and Munnecke et al. (1982). Organophosphate pesticides exhibit relatively low persistence in soil, and this property influences their widespread use despite their mammalian toxicity (Hsu and Bartha 1979). Microbial decomposition is the major degradative route for organophosphorus insecticides (Miles et al. 1979). Decomposition usually proceeds through oxidative pathways, but reductive mechanisms have been reported (i.e., parathion → aminoparathion) (Matsumura and Benezet 1978). Hsu and Bartha (1979) point out that microbial metabolism of diazinon and parathion appears to be cometabolic at least in the first step. The enriched environment of the plant rhicosphere apparently stimulates microbial activity and accelerates microbial mineralization of organophosphate insecticides (Hsu and Bartha 1979, Reddy and Sethunathan 1983). The diffusion of parathion through the soil is appreciably restricted by microbial activity after a lag

TABLE 3-47. PESTICIDES INCLUDED AS CONSTITUENTS OF HAZARDOUS WASTE (40 CFR 261)

Chlorinated Hydrocarbons

Aldrin	Heptachlor (oxidation product-Heptachorepoxide)
Chlordane	Hexachlorbenzene
Chlorbenzilate	Hexachlorocyclohexane (Lindane)
DDD	Hexachloroethane
DDE	Hexachlophene
DDT	Methoxychlor
1,2-Dichlorobenzene	Pentachloronitrobenzene (PCNB)
1,4-Dichlorobenzene	Pentachlorophenol (PCP)
1,2-Dichloroethane	1,2,4-Trichlorobenzene
Dichloromethane	1,1,1-Trichloroethane
1,2-Dichloropropane	Trichloroethylene
1,3-Dichloropropene	Trichlorophenol
Dieldrin	Toxaphene
Endosulfan	
Endrin (and metabolites)	

Organophosphates

O,O-Diethyl S- 2-(ethylthiolethyl phosphorodithioate (Disulfoton)
O,O-Diethyl O-2-pyrazinyl phosphorothioate (Zinophos)
Methyl parathion
Parathion
Phosphorodithioric acid (Phorate)
Phosphorothioric acid (Famphur)
Tetraethylpyrophosphate (TEPP)

Thiocarbamates

Diallate (Avadex)

Dithiocarbamates

Ethylenebisdithiocarbamic acid, salts and esters (Nabam, Maneb, Zineb)

Phenoxyalkanoates

2,4-Dichlorophenoxyacetic acid (2,4-D)
2,4,5-Trichlorophenoxyacetic acid (2,4,5-T)
2,4,5-Trichlorophenoxypropionic acid (Silvex)

Aldehydes

Acrolein (Aqualin)
Formaldehyde

TABLE 3-47. CONTINUED

Alkyl halides
Bromomethane
1.2-Dibromomethane
Tetrachloromethane
Cyclic ketones
Kepone
Maleic hydrazide
Dinitrophenols
2-sec-Butyl-4,6-Dinitrophenol (Dinoseb)
2-Cyclohexyl-4,6-Dinitrophenol (DN-111)
4,6-Dinitro-o-cresol (DNOC)
2,4-Dinitrophenol
Miscellaneous
Acrylonitrile (Acritet)
Allyl alcohol
4-Aminopyridine
Amitrole
Aramite
Bis (2-chloroethyl) ether
Carbon Disulfide
Creosote
Cyclophosphamide (Endoxin)
Decamox
Dimethoate
Dimethyl phthalate
Diphenylamide
Ethylene oxide
Fluoroacetamide
Naphthalene
1-Naphthyl-2-thiourea (Antu)
Nicotine
Pronamide (Kerb)
Strychnine
Warfarin

period or period of adaptation of the microbial population (Gerstl et al. 1979). Phorate has been shown to be inhibitory to nitrification in soil for up to 4 weeks, after which activity returned. In the same study, soil respiration (O_2 consumption) increased when phorate was added (Tu 1980). Forrest et al. (1981) found that soil that had been adapted to diazinon through several annual treatments degraded diazinon rapidly. Successive treatments of the soil with diazinon further increased degradative ability. Freezing and thawing the soil did not adversely affect its degradation ability. A Flavobacterium isolated from the soil hydrolysed both parathion and diazinon.

Kaufman (1974), Matsumura and Benezet (1978), and Munnecke et al. (1982) have reviewed the microbial decomposition of carbamate pesticides. The only carbamate pesticides listed as hazardous waste constituents are members of the thiocarbamate and dithiocarbamate subgroups. The thiocarbamate diallate is an herbicide which controls growth by interfering with cell division. The half-life of diallate in soil has been reported in one study as 12 days (Banting 1967) and 4 weeks in another (Anderson and Domsch 1976). Mineralization of ^{14}C-labeled diallate has been observed and a fungus (Tridioderma harzianium) isolated which can use diallate as a carbon source. Other fungi could transform diallate cometabolically (Anderson and Domsch 1976).

The salts and esters of ethylene bisdithrocarbamic acid, which includes the fungicides nabam, maneb, and zineb are the only dithiocarbamates included as constituent compounds of hazardous waste (40 CFR 261). These compounds degrade primarily by chemical means, although limited microbial degradation has been observed. One decomposition product is ethylenethiourea (ETU), a known carcinogen. Apparently, ETU degrades rapidly with ethyleneurea as an intermediate to other compounds including CO_2 (Kaufman 1974).

Phenoxyalkonate herbicide degradation by microbes has been reviewed by Kaufman (1974) Cripps and Roberts (1978), and Munnecke et al. (1982). Recent concern over the mutagenicity of these compounds and the dioxin contamination of 2,4,5-T has prompted a renewed interest in the degradation of these compounds. Many strains of bacteria and fungi can degrade these herbicides. Complete mineralization of 2,4-D by soil microorganisms has been demonstrated, but 2,4,5-T is more resistant to degradation. Until recently, cometabolism has been the only degradative mechanism described. Recently, reductive dehalogenation of 2,4,5-T to 2,4-D by microbial activity in anaerobic environments has been described (Horowitz et al. 1983, Sulflita et al. 1983). Kilbane et al. (1982) have produced, through "plasmid assisted molecular breeding" (genetic engineering) a strain of Pseudomonas cepacia which will grow on 2,4,5-T as a sole carbon source. This organism has been used to detoxify 2,4,5-T in contaminated soil (Kilbane et al. 1983).

Degradation of Polychlorinated Biphenyls (PCBs)--

The fate of PCBs in soil-plant systems has been reviewed recently by Pal et al. (1980). They conclude that all isomers of mono- and di-chlorobiphenyls are degraded by various aerobic soil microbes, but that a few tri- and tetrachlorinated PCBs are more resistant. Only a few species with five or six chlorines are known to biodegrade. Biodegradation of the more highly chlorinated PCBs has not been studied. Furukawa (1982) has also reviewed the microbial degradation of PCBs. This review points out that the position of chlorine atoms on the PCB molecule is important to degradability as well as the number of chlorines. Transformations of highly chlorinated PCBS are probably cometabolic, but the mixed microbial population of the soil may perform extensive decomposition. Apparently, highly chlorinated PCBs are quite photolabile, and Furukawa (1982) suggests that if photolysis and microbial degradation of PCBs were combined more efficient and rapid decomposition of PCBs might be expected. Liu (1982) also concluded from fermentor studies with a PCB degrading *Pseudomonas* sp. that the position of chlorine substitution on the biphenyl molecule helps determine a PCB isomer's biodegradability, and that higher chlorinated PCB formulations was much slower. A genetically engineered pseudomonad, with the capability to completely degrade mono- or dichlorobenzoates, and an *Acinetobacter* sp. or *Arthrobacter* sp. capable of growing on 4-chlorobiphenyl, in combined culture were able to utilize more than 98 percent of mono- and dichlorobiphenyls, with liberation of equivalent amounts of chloride ions (Furukawa and Chakabarty 1982).

Degradation of Aromatic Hydrocarbons--

Sims and Overcash (1983) have recently reviewed the microbial decomposition of polynuclear aromatic compounds (PNAs) in the soil environment. The toxicity of PNAs to microorganisms is related to their water solubility. Cometabolism appears to be the biotransformation process for PNAs with more than three rings, and acclimation of soils to PNAs enhances their ability to degrade these compounds.

Degradation of Phthalic Acid Esters--

A broad spectrum of soil bacteria and fungi have been shown to degrade phthalic acid esters (PAEs). Mixed populations of microorganisms degrade PAEs more rapidly than pure cultures (Overcash et al. 1982).

Degradation of Substituted Benzenes--

The microbial decomposition of chlorophenols and phenol in soil has been studied by Baker and Mayfield (1980). They found that phenol, *o*, and *p*-chlorophenol, 2,4- and 2,6-dichlorphenol, and 2,4,6-trichlorophenol were rapidly degraded in aerobically incubated soil, but *m*-chlorophenol, 3,4-dichlorophenol, 2,4,5-trichlorophenol and pentachlorophenol were degraded very slowly under aerobic conditions. None of the compounds tested were degraded in anaerobic soil. Suflita et al. (1982) have demonstrated the potential for dehalogenation of haloaromatic substances by anaerobic microbial communities using halogenated benzoates.

QUANTITATIVE DESCRIPTION OF ORGANIC DECOMPOSITION

Microorganisms are highly effective in removing certain hazardous organic constituents from soil systems. Principles of biological degradation of organic substances derived from the study of decomposition in aqueous systems provide the starting point for predicting attenuation in soil systems. These principles also provide the fundamentals for explaining the effects of toxic constituents and solid surfaces on quantitative decomposition in soil systems. Organic chemical decomposition in soil systems is a very complex process which depends upon organic substance properties, soil properties, and environmental factors (Hamaker 1972, Goring et al. 1975). The addition of exogenous toxic organic wastes to soil systems greatly increases the complexity of soil metabolic processes. This is partly because organisms are exposed to at least three different types of chemicals: 1) readily consumable substrates, 2) toxic constituents which may inhibit metabolic processes, and 3) non-readily consumable organics which may not be toxic, but which do not serve as sources of carbon and energy for cellular metabolism (cometabolites). The effects of these functional classes of organics on the rate and extent of metabolism of each class and the effects of solid surfaces within the soil system is very complex and very difficult to describe quantitatively. Nevertheless, studies concerning herbicide and insecticide degradation in soil systems have provided a quantitative basis for extrapolating from aqueous system fundamentals to soil systems which can be utilized as a tool for describing the biodegradation of hazardous waste constituents.

In addition to biological decomposition, the degradation of a chemical or class of chemicals in soil can also be influenced by chemical degradation, volatilization, photodecomposition, and plant uptake and metabolism. The latter process will not be addressed in this manual.

The goal in this section is to describe models for evaluating and predicting the disappearance of certain compounds or classes of compounds in soils with time. Mathematically, the rate of decomposition represents a sink term in organic transport models which are needed to predict potential groundwater contamination with respect to magnitude and type of contamination and the time factor for contamination (rate of transport). Thus, a description of quantitative organic chemical decomposition in soil systems is necessary to determine potential contamination of groundwater, and to determine the chemicals or classes of chemicals that require management, through control of mass transport, and treatment for destruction.

Useful degradation rate models of organics in soils have been described by Hamaker (1966, 1972) and Goring et al. (1975), and Rao and Jessup (1982). Two basic models which may be used to model the fate of toxic organic substances in soil systems were described by Hamaker (1972). The "power rate model" can be expressed as:

$$dc/dt = -KC^N \tag{3-42}$$

where, C is the concentration remaining in soil solution at time t, k is a rate constant, and N is the order of reaction. The second model, the "hyperbolic rate model" can be expressed as:

$$dc/dt = -K_1 C/K_2 + C \tag{3-43}$$

where, K_1, K_2 are constants. The constant K represents the maximum rate of degradation that is approached as the concentration increases.

The "power rate model" is applicable to chemical reactions in homogenous solutions and reaction occurs in proportion to the concentration in solution. The "hyperbolic rate model," which is similar to Michaelis-Menten enzyme kinetics, simulates a catalytic process. That is, degradation is catalyzed by either microorganisms, adsorption, or complex formation with soils.

A first order rate model can be obtained from both models. When N = 1 in Equation 3-42, the rate law becomes:

$$dc/dt = - KC \quad (3\text{-}44)$$

Also, when $K_2 >> C$, Equation 3-43 reduces to Equation 3-44 with $K = K_1/K_2$. The first order rate model is widely used in modeling degradation of organic compounds in soil systems (Goring et al. 1975) because of its simplicity. The first order rate model when coupled with a transport equation, provides a simple model with an analytical solution for transport of pollutants in soil systems. Analytical solutions are generally preferred over numerical solutions that require computing facilities and large amounts of computer time. Also, the half-life of a pollutant, a concept associated with the first order rate, is independent of the initial pollutant concentration. The degradation rate constant in Equation 3-44 actually represents a lumped parameter for several complicated processes responsible for dissipation of organics taking place in soil. Rao and Jessup (1982) recently pointed out that it is satisfactory to assume a constant value for "global" degradation rate coefficients. However, the degradation rate could be represented as a function of soil temperature and soil-water potential, two very important parameters influencing the rate of degradation in soil systems (Walker 1976a, b).

Effect of sorption on the rate and extent of biodegradation is difficult to assess for the wide range of organics and complex mixtures involved. Hamaker and Goring (1976) developed a model for decomposition of pesticides to explain the low degradation rates at low residual concentrations observed in experiments that were monitored for relatively long periods. The model organic constituent (pesticide) is assumed to be divided into available and unavailable fractions. The available fraction is considered to be mobile, labile, and is subject to degradation and movement. The unavailable fraction is characterized as immobile and non-labile with no degradation allowed. However, a first order transfer rate is allowed to describe movement to and from the unavailable fraction. The decomposition is also assumed to be first order (Figure 3-46). The fresh chemical in soil is assumed to be available initially, but as time proceeds, the unavailable fraction increases. Hence, based on this model, it is expected that desorption will be more difficult for aged samples, as observed by Saha et al. (1969).

The "two compartment" model is formulated as follows:

$$dc/dt = - (K + K_1)\, C_1 + K_{-1}C_2 \quad (3\text{-}45)$$

$$dc/dt = K_1C_1 - K_{-1}C_2 \quad (3\text{-}46)$$

where C_1 and C_2 are concentrations in the available and unavailable fractions, respectively; K is the first order decomposition rate constant; and K_1 and K_{-1} are first order transfer rate constants between the two fractions (Figure 3-46). Solution of the two equations and an estimation of parameters are presented by Hamaker and Goring (1976).

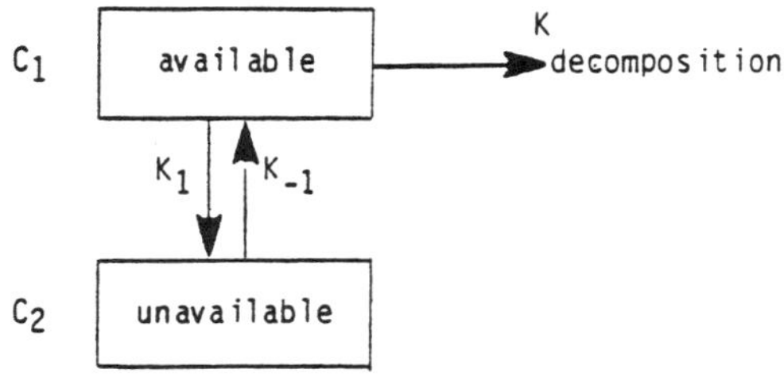

Figure 3-46. Schematic description of "two compartment" model (after Hamaker and Goring 1976). Used by permission, see Copyright Notice.

The ratio (K_1/K_{-1}) is a measure of the equilibrium between the sorbed (unavailable) and desorbed (available) fractions. It has an effect in the overall degradation rate (Figure 3-47). The degradation rate deviation from first order increases as the ratio K_1/K_{-1} increases, i.e., the rate of degradation decreases as the residual concentration decreases (Figure 3-47).

The rate law models discussed thus far do not consider the effect of environmental factors on the degradation rate. Three of these factors will be discussed, including: 1) initial concentration of the chemical, 2) temperature, and 3) moisture content. These factors have been reviewed by Hamaker (1972) and recently by Hurle and Walker (1980).

Initial Concentration

Hamaker (1972) presented data showing that the 50 percent disappearance time (DT-50%) increases as the initial concentration increases, i.e., the persistence increases as the initial concentration increases. Walker (1976a) has also observed the same phenomenon. The reduced rate is explained either by the limited active sites available (Hance and McKone 1971), or by a toxic effect on microorganisms or enzyme inhibition (Hurle and Walker 1980).

Sims and Overcash (1983) compiled data from the literature and conducted experiments at North Carolina State University concerning polynuclear aromatic hydrocarbon (PNAs) initial degradation rates and initial concentrations (Figure 3-48). The data indicate an increasing trend of initial rate of degradation as the initial concentration increases. The data were normalized to 20°C temperature, using Arrhenius equation, with $\Theta = 1.013$ (Table 3-48). No apparent toxic effects of PNA on microbial activity were evident at the concentrations studied.

Temperature

Generally, an increase in temperature increases the rate of degradation of organic compounds in soil. This phenomenon is attributed to a decrease in

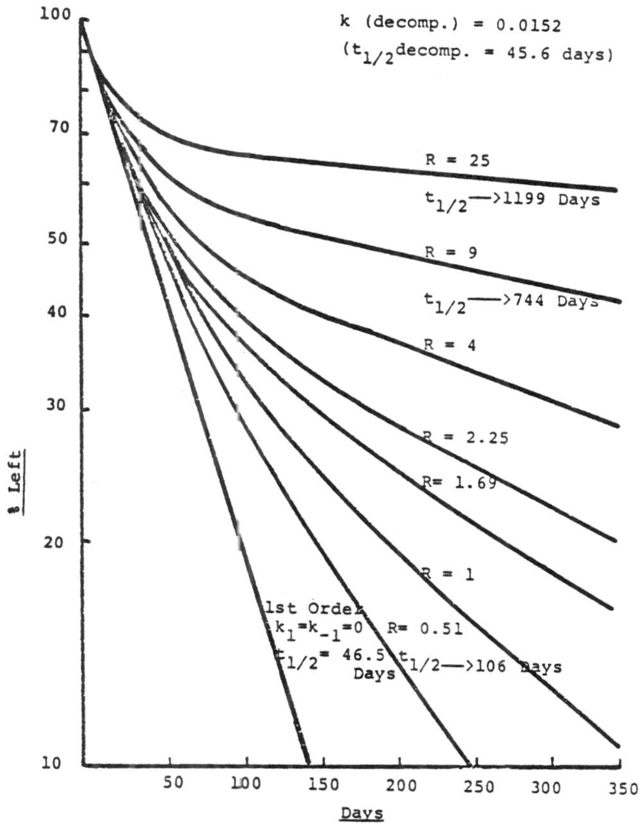

Figure 3-47. Disappearance of total chemical for different sizes of bound residue reservoir, i.e., k_1 (binding)/k_{-1} (unbinding) = R (Hamaker and Goring 1976). Used by permission, see Copyright Notice.

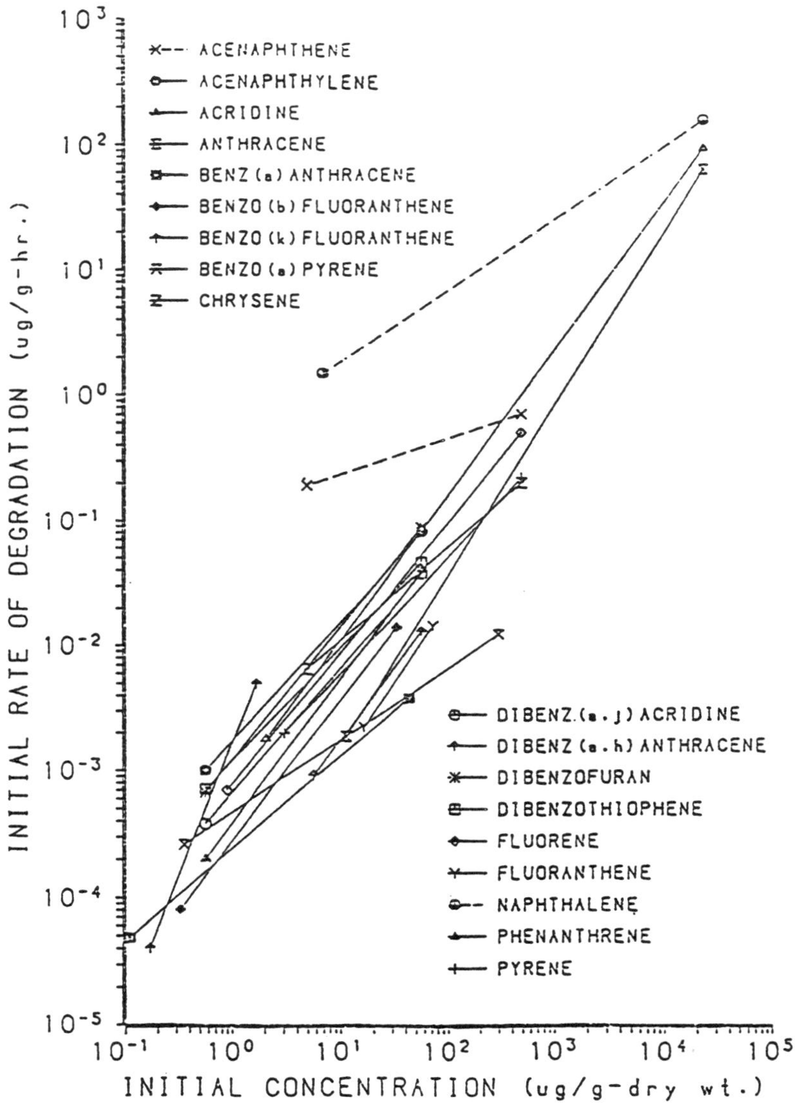

Figure 3-48. Rates of transformation of PNA compounds in soil as a function of initial soil concentration.

TABLE 3-48. KINETIC PARAMETERS DESCRIBING RATES OF DEGRADATION OF AROMATIC COMPOUNDS IN SOIL SYSTEM (SIMS AND OVERCASH 1983)

PNA	Initial Concentration (μg/g soil)	k (day^{-1})	Rate of transformation (μg/g-day)	$t_{1/2}$[a] (days)	Reference
Pyrocatechol	500	3.47	1,735	0.2 m	Medvedev & Davidov (1972)
Phenol	500	0.693	364.5	1.0 m	Medvedev & Davidov (1972)
Phenol	500	0.315	157.5	2.2 l	Medvedev & Davidov (1972)
Fluorene	0.9	0.018	0.016	39 m	Groenewegen & Stolp (1976)
Fluorene	500	0.347	173.3	2 m	Medvedev & Davidov (1972)
Indole	500	0.693	364.5	1.0 m	Medvedev & Davidov (1972)
Indole	500	0.315	157.5	2.2 l	Medvedev & Davidov (1972)
Naphthol	500	0.770	385	0.9 m	Medvedev & Davidov (1972)
Naphthalene	7.0	5.78	40.4	0.12 m	Herbes & Schwall (1978)
Naphthalene	7.0	0.005	0.035	125 l	Herbes & Schwall (1978)
Naphthalene	25,000	0.173	4,331	4 h	Sisler & Zobell (1947)
1,4-Naphthoquinone	500	0.578	288.8	1.2 m	Medvedev & Davidov (1972)
Acenaphthene	500	0.173	86.6	4 m	Medvedev & Davidov (1972)
Acenaphthene	5	2.81	22.6	0.3 m	Medvedev & Davidov (1972)
Anthracene	3.4	0.21	0.714	3.3 l	Gardner et al. (1979)
Anthracene	13.7	0.004	0.054	175 m	Gardner et al. (1979)
Anthracene	10.3	0.005	0.050	143 m	Gardner et al. (1979)
Anthracene	11.4	0.006	0.073	108 m	Gardner et al. (1979)
Anthracene	40.0	0.005	0.208	138 m	Gardner et al. (1979)
Anthracene	36.4	0.005	0.196	129 m	Gardner et al. (1979)
Anthracene	25,000	0.198	4,950	3.5 h	Sisler & Zobell (1947)
Phenanthrene	2.1	0.027	0.056	26 m	Groenewegen & Stolp (1976)
Phenanthrene	25,000	0.277	6,930	2.5 h	Sisler & Zobell (1947)
Carbazole	500	0.067	33	10.5 m	Medvedev & Davidov (1972)
Carbazole	5	0.231	1.16	3 m	Medvedev & Davidov (1972)
Acridine	500	0.075	37.67	9.2 m	Medvedev & Davidov (1972)
Acridine	5	0.281	1.16	3 m	Medvedev & Davidov (1972)
Benz(a)anthracene	0.12	0.046	0.005	15.2 l	Herbes & Schwall (1978)
Benz(a)anthracene	0.12	0.0001	0.00001	6250 m	Herbes & Schwall (1978)
Benz(a)anthracene	3.5	0.007	0.024	102 m	Groenewegen & Stolp (1976)

TABLE 3-48. (CONTINUED)

PNA	Initial Concentration (μg/g soil)	k (day-1)	Rate of transformation (μg/g-day)	$t_{1/2}$[a] (days)		Reference
Benz(a)anthracene	20.8	0.003	0.062	231	m	Gardner et al. (1979)
Benz(a)anthracene	25.8	0.005	0.134	133	m	Gardner et al. (1979)
Benz(a)anthracene	17.2	0.008	0.060	199	m	Gardner et al. (1979)
Benz(a)anthracene	22.1	0.006	0.130	118	m	Gardner et al. (1979)
Benz(a)anthracene	42.6	0.003	0.118	252	m	Gardner et al. (1979)
Benz(a)anthracene	72.8	0.004	0.257	196	m	Gardner et al. (1979)
Benz(a)anthracene	25,000	0.173	4,331	4	h	Sisler & Zobell (1947)
Fluoranthene	3.9	0.016	0.061	44	m	Groenewegen & Stolp (1976)
Fluoranthene	18.8	0.004	0.072	182	m	Gardner et al. (1979)
Fluoranthene	23.0	0.007	0.152	105	m	Gardner et al. (1979)
Fluoranthene	16.5	0.005	0.080	143	m	Gardner et al. (1979)
Fluoranthene	20.9	0.006	0.125	109	m	Gardner et al. (1979)
Fluoranthene	44.5	0.004	0.176	175	m	Gardner et al. (1979)
Fluoranthene	72.8	0.005	0.379	133	m	Gardner et al. (1979)
Pyrene	3.1	0.020	0.061	35	m	Groenewegen & Stolp (1976)
Pyrene	500	0.067	33	10.5	m	Medvedev & Davidov (1972)
Pyrene	5	0.231	1.16	3	m	Medvedev & Davidov (1972)
Chrysene	4.4	0	0	-		Groenewegen & Stolp (1976)
Chrysene	500	0.067	33	10.5	m	Medvedev & Davidov (1972)
Chrysene	5	0.126	0.63	5.5	m	Medvedev & Davidov (1972)
Benz(a)pyrene	0.048	0.014	0.007	50	l	Herbes & Schwall (1978)
Benz(a)pyrene	0.01	0.001	0.00001	694	l	Herbes & Schwall (1978)
Benz(a)pyrene	3.4	0.012	0.041	57	m	Groenewegen & Stolp (1976)
Benz(a)pyrene	9.5	0.002	0.022	294	m	Gardner et al. (1979)
Benz(a)pyrene	12.3	0.005	0.058	147	m	Gardner et al. (1979)
Benz(a)pyrene	7.6	0.003	0.020	264	m	Gardner et al. (1979)
Benz(a)pyrene	18.5	0.023	0.312	30	m	Gardner et al. (1979)
Benz(a)pyrene	17.0	0.002	0.028	420	m	Gardner et al. (1979)
Benz(a)pyrene	32.6	0.004	0.129	175	m	Gardner et al. (1979)
Benz(a)pyrene	1.0	0.347	0.347	2	h	Shabad et al. (1971)

TABLE 3-48. (CONTINUED)

PNA	Initial Concentration (µg/g soil)	k (day^{-1})	Rate of transformation (µg/g-day)	$t_{1/2}$[a] (days)		Reference
Benz(a)pyrene	0.515	0.347	0.179	2	h	Shabad et al. (1971)
Benz(a)pyrene	0.00135	0.139	0.0002	5	h	Shabad et al. (1971)
Benz(a)pyrene	0.0094	0.002	0.00002	406	l	Shabad et al. (1971)
Benz(a)pyrene	0.545	0.011	0.006	66	l	Shabad et al. (1971)
Benz(a)pyrene	28.5	0.019	0.533	37	l	Shabad et al. (1971)
Benz(a)pyrene	29.2	0	0			Shabad et al. (1971)
Benz(a)pyrene	9,100	0.018	161.7	39	h	Lijinsky & Quastel (1956)
Benz(a)pyrene	19.5	0.099	1.93	7	h	Poglazova et al. (1967b)
Benz(a)pyrene	19.5	0.139	2.70	5	h	Poglazova et al. (1967b)
Benz(a)pyrene	19.5	0.231	4.50	3	h	Poglazova et al. (1967b)
Benz(a)pyrene	130.6	0.173	22.63	4	h	Poglazova et al. (1967b)
Benz(a)pyrene	130.6	0.116	15.08	6	h	Poglazova et al. (1967b)
Dibenz(a,h)anthracene	9,700	0.033	320.1	21	h	Lijinsky & Quastel (1956)
Dibenz(a,h)anthracene	25,000	0.039	962.5	18	h	Sisler & Zobell (1947)

[a]l = low temperature range (< 15°C), m = medium temperature range (15-25°C), and h = high temperature range (> 25°C).

adsorption with increasing temperature which makes more organics available, or to an increase in biological activity, or both. The second factor has an optimum temperature beyond which biological activity decreases.

The effect of temperature can be described quantitatively by the change in the degradation constant which can be determined using the Arrhenius equation:

$$K = A_0 e{-}(E_a/T) \qquad (3\text{-}47)$$

where A_0 is a constant, R is the gas constant, T is the absolute temperature, and E_a is the activation energy. Equation 3-47 implies that the rate law is known. Usually, the rate law is assumed to be first order (e.g., Walker 1978). The half-lives at different temperatures can be estimated by:

$$\log[t_{1/2}(T_1)] - \log[t_{1/2}(T_2)] = Ea/4.575\ (1/T_1 - 1/T_2) \qquad (3\text{-}48)$$

where $t_{1/2}(T_1)$ and $t_{1/2}(T_2)$ are the half-lives at absolute temperatures T_1 and T_2.

Temperature effect on degradation rate is important in terms of assessment of the seasonal and geographical variation of degradation rates. Knowledge of temperature effects allows an estimation of assimilative capacities for organic wastes in soil systems for different regions of the United States. The effect of controlling soil temperature on the rate of degradation of organics can also be quantified and evaluated. Overcash and Pal (1979) quantified the seasonal and geographical variation of oil waste decomposition (Figure 3-49). The annual rate of degradation is higher in high temperature regions than in low temperature regions. Field measurements also show increased disappearance rates with seasonal high temperatures (Walker 1970, Schweizer 1976).

Moisture Content

Hamaker (1972) presented data for several pesticides indicating an increase in the degradation rate with increasing soil moisture content. It was observed that the degradation rate increased as the moisture content increased up to field capacity (Walker, 1976 a,b,c; 1978). Hamaker (1972) suggested the following empirical equation to quantitatively describe the rate dependence on moisture content for organic constituents in soil systems:

$$\text{rate} = r + KW^n \qquad (3\text{-}49)$$

where W is the moisture content less than saturation, n is a constant less than one, K is a constant, and r is the rate of degradation when N = 0. Note that when using Equation 3-49, it is assumed all other factors are constant.

Harris et al. (1969) conducted field measurements of persistence of atrazine and chlorfenanc at 12 locations. For both chemicals there was more persistence in cooler northern and drier western states than in the semi-tropical southern states.

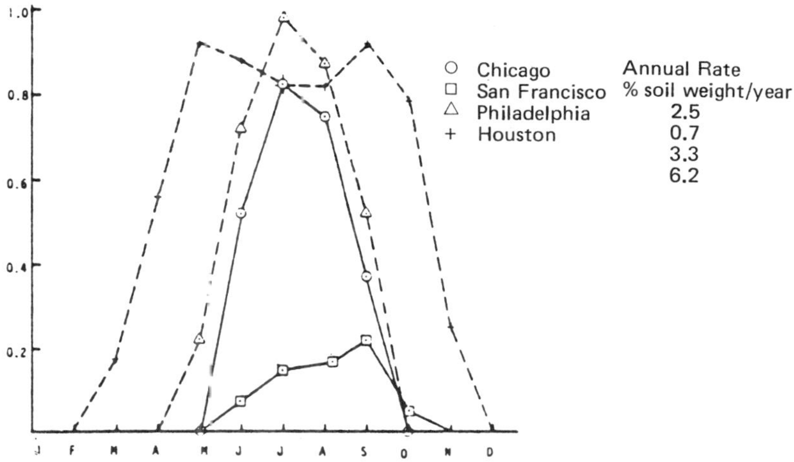

Figure 3-49. Effect of climatic conditions at major refinery locations on the annual pattern of oil decomposition (Overcash and Pal 1979). Used by permission, see Copyright Notice.

The quantitative approach to determining organic constituent degradation is utilized in Volume I of the manual. The models presented provide a rational framework for determining relative degradation rates of constituents in a soil system as a function of temperature and soil moisture. When models for degradation are combined with simple transport models, the dynamic profile in a soil system can be determined and used to identify those constituents that require treatment first. Thus a prioritization of treatment steps and processes can be developed based on the fate of contaminants, specifically with respect to the potential danger to public health.

CHEMICAL REACTIONS IN THE SOIL MATRIX

Introduction

For a site characterized by large quantities of hazardous waste contaminated soils, consideration of in situ treatment requires certain information related specifically to chemical reactions in the soil-waste matrix. These chemical reactions can be classified into two functional groups for simplicity: 1) chemical reactions involving the soil medium, and 2) chemical reactions involving chemicals in the soil or soil solution. The latter group of reactions may occur among chemicals already present at the site, or may occur as a result of addition of a treatment reagent to the soil-waste matrix. For in situ treatment it is anticipated that all immediate chemical reactions that affect public health (explosion, heat generation, flammability, etc.) will have occurred prior to site assessment for treatment. Therefore chemical reactions as a result of addition of treatment agent will be emphasized.

This section is concerned with the first group of chemical reactions identified above, i.e., chemical reactions involving the soil medium. The effects of gross chemical structure of inorganic and organic chemicals added to soil systems are considered, including acids, bases, salts, nonionic organics, surfactants, and aqueous-solvent mixtures. Effects on soil include: 1) dissolution, 2) permeability 3) sorption, 4) polymerization, and 5) microbial activity. These effects on soil relate specifically to soil treatment process, i e., immobilization and degradation/transformation.

Reactions among chemicals in the soil or soil solution are considered with respect to health hazard (heat generation, fire, explosion, formation of toxic fumes, formation of flammable gases, solubilization of substances, and volatilization of toxic substances) in Volume I for specific waste constituent-treatment agent combinations.

Effects of Waste Type on Soil Properties

Acidic Wastes--

Table 3-49 summarizes several industrial categories that generate acidic waste constituents. Organic acids in soil systems are very likely to undergo biodecomposition with evolution of CO_2, stimulation of microbial populations, and consumption of O_2 resources. If there are carbonates in the system, such as in calcareous and sodic soils, CO_2 may evolve through reactions of the acid with carbonate and bicarbonate species (Pal et al. 1977). Under acidic conditions in soils, soubilities of complexed cations such as Cu and Zn increase, and Fe, Mn, and Cu are easily reduced to more soluble forms. Acidic wastes may also be used as treatment process for saline-sodic soils.

The decrease in soil pH due to the presence of hazardous acidic wastes will also change the microorganism distribution within the soil systems. Under acidic conditions ($pH < 7$) fungi predominate and thrive in soil systems. Biochemical reactions resulting from fungal metabolism of certain hazardous chemicals may make soil pH an important determinant, not only of the type of microorganism, but also of the metabolic pathway of degradation.

Hydrocarbon-utilizing organisms possess a class of enzymes, known as oxygenases, which incorporates atmospheric oxygen into the inert ring-structure as the initial step in aromatic hydrocarbon transformation and degradation. The substrate specificities of bacterial hydroxylation mechanisms, however, may differ from those of microsomal enzyme systems present in fungal tissues. Fungal systems may transform aromatic hydrocarbons by means of monooxygenases into arene oxides, the mutagenic form of PNAs (Cerniglia et al. 1979), while bacterial systems carry out the dioxygenation of the aromatic nucleus to form a cis-glycol as the first stable intermediate, not the arene oxide.

These differences in the mechanism of aromatic hydrocarbon metabolism by microogranisms have important implications for engineering techniques for controlling and possibly detoxifying PNAs in hazardous waste contaminated soils. Because of the major differences with respect to microbial oxidation of aromatic hydrocarbons between bacteria and fungi, it may be especially

TABLE 3-49. INDUSTRIAL CATEGORIES GENERATING ACIDIC WASTE CONSTITUENTS (PAL ET AL. 1977)
Used by permission, see Copyright Notice

Organic	Inorganic
Winery Stillage (3.5-5)	Mineral Acid Manufacturing (3-4)
Paper and Pulp (4-6)	Dredge Mines (2-4)
Organic Acid Manufacture (3-6)	Coal Processing (3-5)
Leather Processing (5-6)	Steel Cleaning (3-5)
	Metal Plating (2-4)
	Petrochemical Manufacturing (4-6)

important in soil systems to encourage bacterial growth and competition vis-a-vis fungi. Since soil pH has a significant effect on the soil bacterial/fungal proportions, pH may be an important engineering tool to direct the pathway of PNA degradation.

Dissolution of clay minerals in the presence of organic acids has also been observed. Those elements comprising the framework cations of clay minerals, i.e., silicon, aluminum, magnesium, and iron, greatly increased in solutions phases containing organic acids (Huang and Keller 1971). The organic acids studied included aspartic, citric, salicyclic, taratric, and tannic. The most rapid rate of dissolution occurs during the first 24 hours of exposure of clay minerals to organic acids. After 45 days of exposure, the solution concentration reaches a constant value.

The ability of organic acids to solubilize soil minerals has also been observed with humic and fulvic acid fractions. Fulvic acids (FA) and low-molecular weight humic acids (HA) can attack and degrade soil minerals to form water-soluble and water-insoluble metal complexes, depending on how much metal per unit weight of HA or FA is dissoved. The ability of these organic acid fractions to dissolve clay minerals is believed to be due to their ability to complex di- and trivalent metal ions. Fulvic acid can dissolve a proportion of the dominant cation in the clay, forming a soluble complex and replacing the removed cation with H+. If this process continues the FA will degrade the clay structure (Schnitzer and Kodama 1977).

Table 3-50 presents results summarizing the increase in solubility of framework cations of clay minerals, including kaolinite, illite, and montmorillonite in the presence of the organic acids listed previously. Aluminum dissolution from clay minerals reached 60 ppm with illite. This concentration is very high compared to less than 1 ppm solubility typical in aqueous solutions. Iron disolution was observed to be similar to aluminum in ratios and concentrations. In strongly complexing acids, silica solubility rose by a factor of 2 to 6 for the kaolinites, and by larger factors for the 2:1

TABLE 3-50. EFFECT OF ORGANIC ACIDS ON CLAY SOLUBILITY (HUANG AND KELLER 1971

Element	Increase in solubility (factor) over water		
	Kalonite	Illite	Montmorillonite
Si	2-6	15-30	12-25
Al	10-50	10-100	10-50
Fe	10-50	10-100	10-50

layer clays. Aspartic acid was especially active in dissolving calcium and magnesium. Clay mineral framework cation concentrations are much lower in the presence of water.

Implications of these observations include: 1) susceptibility of weathering of silicate minerals in water is not followed in reactions with organic acids; 2) dissolution of clay minerals by organic acid constitutents of hazardous wastes may directly impact mass transport and permeability aspects of the soil-waste matrix; 3) toxic metal ions at hazardous wastes sites may also become mobile through chelation with natural and exogenous organic acid fractions.

Basic Wastes--

Basic waste addition to acidic soil has been reported to increase the pH of the surface layer (4" -18") but not the subsoil (Brown 1975). Basic wastes would be expected to react with acidic groups of the soil system to neutralize the buffer capacity of the soil, and increase the exchangeable sodium percentage (ESP), percent base saturation, and soil pH. Basic waste added to soil may result in physical damage to the soil system with resultant low permeability, decreased hydraulic conductivity, and high runoff potential. However, organic bases added to soil systems may increase soil buffer capacity and exchange capacity as the bases are degraded. Generally an increase in soil pH is associated with an increase in cation exchange capacity, which increases the buffering action of a soil (Pal et al. 1977).

Salts--

Salt affected soils may be classified based on chemical conditions of the soil-waste system identified in Table 3-51.

Soils contaminated by salty wastes become friable, flocculated, and permeable with a high osmotic concentration of the soil solution. With sodium as the dominant ion, and the SAR of the waste above 30, sodium dominates on the exchange sites. Subsequent leaching of salts by precipitation results in dispersed, deflocculated, and impermeable soils. In the absence of salts,

TABLE 3-51. CLASSIFICATION OF SALT-AFFECTED SOILS

Soil Group	EC of saturation extract (mmhos/cm at 25° C)	ESP (%)	pH
Saline	>4	<15	<8.5
Saline-sodic	>4	>15	=8.5
Sodic	<4	>15	>8.5

sodium soils become extremely dispersed and impermeable to water. The effect of sodium on hydraulic conductivity is indicated in Figure 3-50.

In regions of base unsaturated soils (Southeastern U. S.), addition of salt wastes may improve base saturation, decrease exchange acidity, and increase nutrient supply (Pal et al. 1977). Sodic soils, however, can only be treated in place by chemical treatment with calcium salts (gypsum) and leaching of excessive sodium with solutions low in sodium and high in other salts (Ca and Mg). Critical SAR values for soil are listed in Table 3-52.

Overcash and Pal (1979) point out that treatment of a soil that has been salt-damaged is a difficult and slow process. If the damage is restricted to the shallow soil surface layer, restoration is easier since amendments can be mechanically incorporated into the soil. Restoration may be accomplished within months. However, if structure deterioration has progressed to the upper B horizon, calcium salts must be leached into the B horizon zone through the severely restrictive deflocculated zone. This process for soil restoration may require several years.

Solvents--

Very little information has been generated concerning the behavior and effects of the solvent class of organic constituents on terrestrial systems and on transport through terrestrial systems. However, assessment of hazardous waste sites has identified numerous organic solvents present in soil as well as groundwater samples.

Volatile solvents are known to exert an initial soil sterilization effect on soil systems as the critical dose level is achieved. Soil recovery is indicated by an increase in soil microbial activity and degradation of the solvent. Solvents of low volatility produce very little soil sterilization when applied in moderate to high levels. Table 3-53 lists solvents, the critical soil levels for soil microbial populations, and the time period for recovery at the critical dose.

Solvents also have an effect on the physical properties of soil. Several studies have indicated that organic solvents, in general, will react with clay

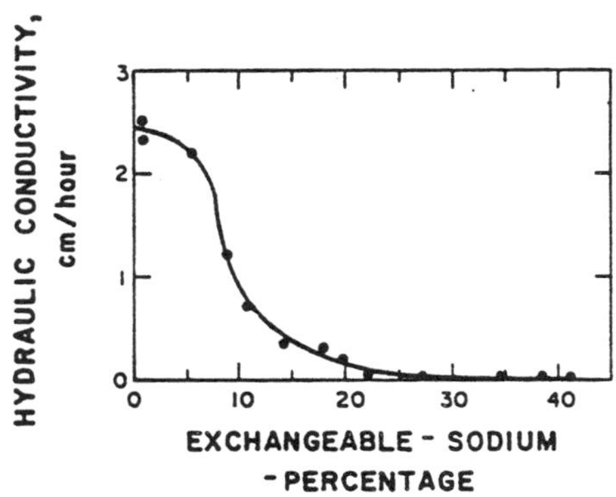

Figure 3-50. Influence of exchangeable sodium percentage on the hydraulic conductivity of a clay loam (Overcash and Pal 1979). Used by permission, see Copyright Notice.

TABLE 3-52. CRITICAL SAR VALUES FOR SOIL (OVERCASH AND PAL 1979)
Used by permission, see Copyright Notice

Soil	SAR
Swelling clay (bentonite)	8-10
Nonswelling clay	20
Pure sand	750
Loam or finer textures (>10% clay)	5-15

TABLE 3-53. RESPONSE OF SOIL MICROBIAL POPULATIONS TO APPLICATION OF SOLVENTS (OVERCASH AND PAL 1979)
Used by permission, see Copyright Notice

Solvent	Critical soil level (ppm)	Time period for recovery at critical dose (days)
Cyclohexane	840	within 37 days
Hexane	430	within 19 days
Heptane	10,000	within 24-63 days
Pentane	7,200	within 30-53 days
Formaldehyde	150-300	within 22 days
Chloroform	590	within 12 days
Ether	7,400	within 14 days
Acetone	58,000	within 12 days
Pyridine	7,900	within 16-30 days

minerals in soil systems resulting in an increase in permeability of the clay fraction (Buchanan 1964, White 1976, Anderson 1981, Anderson and Brown 1981, Schram 1981, Anderson et al. 1982, Brown and Anderson 1983, and Brown et al. 1983). A change in permeability is related to a change in the relative volume that a given clay occupies as a function of the adsorbing liquid. A solvent property that can be used to describe the magnitude of the result of the clay-solvent interaction, or change in permeability, is the dielectric constant of a liquid. A clay surface in contact with a high dielectric constant liquid undergoes swelling. Conversely, a low dielectric constant liquid in contact with a clay surface will cause very little swelling. In the situation where a soil system containing a high dielectric constuant fluid (e.g., water) receives a concentrated waste containing a low dielectric constant liquid (solvent), an increase in soil permeability may be anticipated as the relative clay-liquid volume decreases.

Table 3-54 contains dielectric constants for several solvents identified at hazardous waste sites. It is also apparent that organic solvents generally become less water soluble with decreasing dielectric constant.

The effect of specific solvents on soil permeability is summarized in Table 3-55 (Anderson and Brown 1981). General classes of organics, including acidic, basic, neutral polar, and neutral nonpolar are included. From Table 3-55 it is evident that most solvent organics increase permeability of clay minerals in soil systems.

While a decrease in water solubility is generally associated with an increase in sorption for the solvents listed, this is not true for several solvents, including carbon tetrachloride, ethylene dibromide, and trichloroethylene (Rogers and McFarlane 1981). These chlorinated hydrocarbons had relatively low adsorption in batch isotherm tests to several surfaces, including two silty clay loam soils, aluminum-saturated montmorillonite, and calcium-saturated montmorillonite. Experimental results are indicated in Table 3-56. For these chemicals the K values indicate that only minimal sorption would occur in soil or clay. The sorption of chemials did not exceed 6 percent, except for 10 percent sorption of TCE by Al-mortmorillonite. It is important to point out that the elevated volatility of these solvents may result in atmospheric mass transfer as the primary pathway in the soil-atmospheric system.

Surfactants--

Surfactants are organic molecules widely used in industry for cleaning purposes. Cationic surfactants are usually quaternary ammonium salts, while nonionic surfactants are characterized by polymers of oxyethylene (C_2H_4O) with both polymer ends attached to alcohol and are nondissociated in water. Anionic surfactants comprise the largest class of detergents (80-85 percent) and consist of a sodium cation and the active organic anionic surfactant (Overcash and Pal 1979). The anion part is usually a sulfate or sulfonate.

Application of surfactants to the soil will affect physical and biological properties. For a total soil concentration of approximately 100 ppm, no substantial adverse microbial response has been noted. A large amount of information exists concerning field movement of surfactants to subsurface

TABLE 3-54. DIELECTRIC CONSTANTS, DENSITIES AND WATER SOLUBILITIES OF VARIOUS HALOGENATED AND NONHALOGENATED SOLVENTS (ANDERSON ET AL. 1983, MELLAN 1977)

Name	Dielectric Constant	Density (g cm-3)	Water Solubility @ 25°C
Water	78.5	1.00	-
Methanol	32.7	0.79	Miscible
Ethanol	24.6	0.79	Miscible
Benzyl Chloride	23.0	1.10	Moderately Miscible
Acetone	20.7	0.79	Miscible
1-Propanol	20.3	0.80	Miscible
1-Butanol	17.5	0.81	Miscible
1-Pentanol	13.5	0.81	Miscible
Pyridine	12.4	0.97	Miscible
Phenol	9.8	1.05	Miscible
Dichloromethane	8.9	1.31	1.32%
1-Bromopropane	8.1	1.34	NA
1-Chloropropane	7.7	0.89	NA
1,1,1-Trichloroethane	7.5	1.34	Slightly Soluble
Aniline	6.9	1.02	Soluble
Chloroform	4.8	1.48	0.82%
Bromoform	4.4	2.89	0.10%
1,1,2-Trichloroethylene	3.4	1.48	0.10%
Toluene	2.4	0.87	Slightly Soluble
Benzene	2.3	0.88	Slightly Soluble
Carbon tetrachloride	2.2	1.60	0.08%
Cyclohexane	2.0	0.78	45 ppm
Hexane	1.9	0.65	NA

NA - not available.

TABLE 3-55. EFFECT OF ORGANIC SOLVENT IN CLAY PERMEABILITY (ANDERSON AND BROWN 1981)

Chemical Class	Solvent	Effect on Permeability	Comment
Acid	Acetic acid	decrease	dissolution of clay
Base	Aniline	increase	100 fold for Lufkin[a] 10 fold for H.B.[a]
Neutral Polar	Acetone	increase	1000 fold for Lufkin 50 fold for H.B.
	Ethylene Glycol	increase	100 fold for Lufkin 4 fold for H. B.
Neutral Nonpolar	Xylene	increase	1000 fold for Lufkin 100 fold for H.B.
	Heptane	increase	similar to Xylene•

[a]Smectite clay.

TABLE 3-56. SORPTION OF HALOGENATED ORGANICS ON SOIL AND CLAY (ROGERS AND MCFARLANE 1981)

Adsorbent	Clay (%)	Organic carbon (%)	Freundlich constants CT[a]		EDB[b]		TCE[c]	
			K	1/n	K	1/n	K	1/n
Silty clay loam	31	2.6	0.62	1.08	1.31	0.93	3.89	0.82
Silty clay loam	34	1.8	1.18	1.04	1.49	0.92	1.57	0.93
Al-saturated montmorillonite	100	0	1.75	0.88	0.47	1.04	6.96	0.92
Ca-saturated montmorillonite	100	0	NS[d]	NS	0.05	1.03	NS	NS

[a]CT = carbon tetrachloride.
[b]EDB = ethylene dibromide.
[c]TCE = trichloroethylene.
[d]NS = Not sufficiently sorbed to determine constants.

waters (Sebastiani et al. 1971). These situations involve very concentrated solutions in land fill operations. Overcash and Pal (1979) point out that the adsorption process for surfactants in most soils cannot be considered instantaneous, but rather occurs at a finite rate. Therefore saturated flow conditions should be avoided. Table 3-57 lists results of the adsorption of a cationic and a nonionic surfactant by montmorillonite clay as a function of surfactant concentration.

While surfactants may be toxic to soil microoganisms at high concentrations (> 10,000 ppm), the most severe effects of surfactants in soil systems would most likely involve mass transport of toxic and hazardous constituents through soil systems. Very little information is available concerning actual mass transport of toxic substances in soil systems due.to surfactant addition.

Desorption

A very important area where initial soil-waste interactions may affect subsequent constitutent mobility is desorption. Desorption of chemicals from soil systems may result from precipitation events following a contamination event, or may occur as a result of addition of amendments for land treatment. The chemical matrix of the waste appears to have an important effect on the subsequent behavior of waste constituents in soil systems.

TABLE 3-57. ADSORPTION OF SURFACTANTS ON MONTMORILLONITE (HOWER 1970)
Used by permission, see Copyright Notice

Surfactant Added mg/g montmorillonite	Surfactant adsorbed (mg/g montmorillonite)	
	Cationic	Nonionic
100	97	99
200	188	192
400	351	321
600	451	412
1,000	510	512
1,500	528	540

Three matrices were utilized by Bowman and Sans (1982) to place pesticides on soils. The pesticide matrices included: 1) aqueous solutions, 2) hexane-acetone (95:S volume/volume) solvent, and 3) pure acetone. Two levels of spiking were used with the hexane-acetone solvent, one at 5.0 ug/g (low spike) and one at the level obtained with aqueous equilibrium isotherm studies. With the acetone matrix, spiking was done at the level indicated by the aqueous equilibrium isotherm method. Four desorption cycles were used with two pesticides of different water solubilities, fensulfothion (water solubility = 2000 mg/l) and its related sulfide (f. sulfide) (water solubility = 1.7 mg/l). Three mineral soils and a muck soil were used to determine the influence of organic matter on desorption (Table 3-58).

Figures 3-51 and 3-52 illustrate desorption of the pesticides as functions of application matrix and soil type. For both pesticides desorption was greater for solvent applied pesticide (hexane-acetone, and acetone) than for aqueous applied pesticide in the mineral soils. Also an obvious effect of solvent matrix on desorption is evident for f. sulfide. Less f. sulfide was desorbed in the acetone treated soil than in the hexaneacetone (high spike) treated soil. This was not observed for fensulfothion. Thus the type of solvent also appears to play an important role in desorption behavior of organic wastes.

The muck soil did not show differences in desorption to the extent exhibited by the mineral soils. This was true for both pesticides studied. However, while the maximum desorption of f. sulfide was 10 percent after four desorption cycles (of the high spike hexane-acetone method), 60 to 70 percent fensulfothion was desorbed from the muck soil. Desorption was also consistently greater for fensulfothion from the mineral soils. This would be expected for the more water soluble fensulfothion.

Thus the chemical matrix in which organic constituents are dissolved may have a very important effect on the subsequent mobilization of the

TABLE 3-58. PROPERTIES OF SOIL ADSORBENTS (BOWMAN AND SANS 1982)
Used by permission, see Copyright Notice

Adsorbent	% Sand	% Silt	% Clay	% Organic Matter	pH
Sandy	91.5	1.5	7	0.7	6.9
Creek sediment	71	22	7	2.3	6.5
Sandy loam	77	15	8	3.9	6.9
Muck soil	52	34	14	36.7	6.3

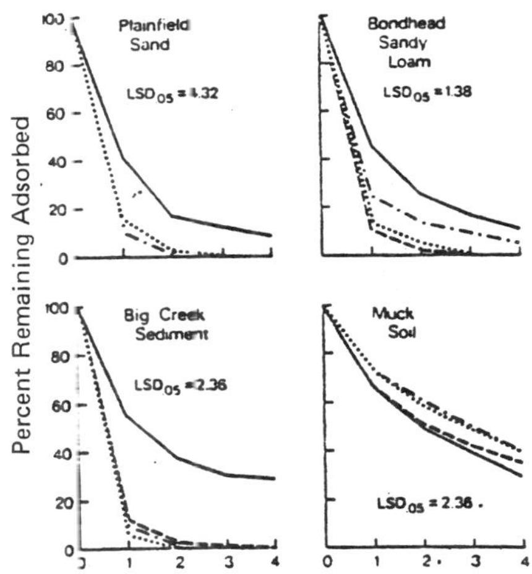

Figure 3-51. Desorption of fensulfothion from four soils. ( ) Low spike; ( ) high-spike acetone; (---) high-spike hexane; (——) equilibrium (Bowman and Sans 1982). Used by permission, see Copyright Notice.

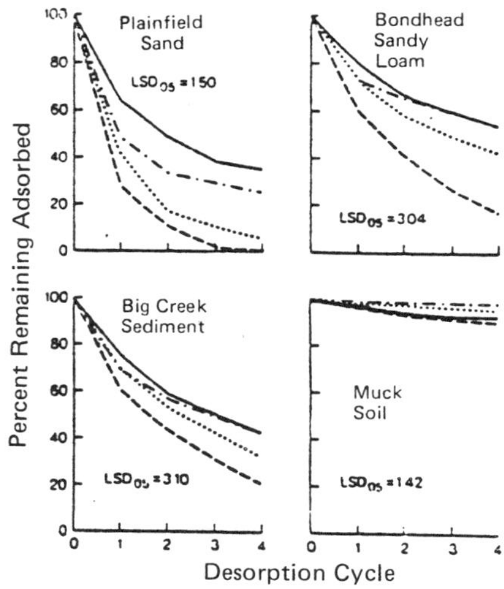

Figure 3-52 Desorption of fensulfothion sulfide from four soils. ( ) Low spike; ( ) high-spike acetone; (---) high-spike hexane; (——) equilibrium (Bowman and Sans 1982). Used by permission, see Copyright Notice.

constituents. For typical mineral soils, which are likely to be typical of hazardous waste sites, insecticides adsorbed from aqueous solutions would be desorbed to a lesser extent than insecticides adsorbed from organic solvents. Also, less water soluble constitutents may desorb to a less extent than more water soluble constituents.

The work of Brown et al. (1983) and Bowman and Sans (1982) may have direct application to hazardous waste sites. If solvent-action generally results in clay shrinkage, the number of sites available for sorption of chemicals will be reduced. Less total sorption may be anticipated as a result. If the application of an aqueous solution follows, as with a rain event or amendment addition, rapid desorption of poorly sorbed chemicals is anticipated. Also, the mass transport of chemicals in the soil solution is excerbated due to the increased permeability of the soil as a result of soil shrinkage. Thus the effects of solvents in hazardous waste contaminated soils may include two factors: 1) decrease in total sorption to soils, and 2) increase in leaching potential through changes in soil structure.

<u>Soil Catalysis</u>

Many chemical reactions that occur in soil systems may occur independently of the soil, or may be soil-catalyzed. For example, there is a rapid release from soil of toxic methyl isothiocyanate after application of dazomet or metham to moist soil. Catalysis of dazomet by clay surfaces resulted in the formation of isothiocyanate within 15-20 minutes of application to soil compost (Brady 1971). The reaction generally increases with decrease in soil moisture. This is due to the increased availability of O_2 in soil with decreasing moisture content, and the O_2 dependence of the reaction:

$$\text{metham} \xrightarrow[O_2]{} \text{methyl isothiocyanate}$$

The ability of clays to catalyze chemical reactions is attributed to their strongly acidic nature. Heptachlor is rapidly degraded in the presence of the acid clay attapulgite ($pK_a < 1$). Significant decomposition of several substances, including: 1) chlordane, 2) toxaphene, 3) heptachlor, 4) DDT, 5) dieldrin, and 6) endrin can be achieved with kaolinite and attapulgite (Fawkes et al. 1960). The chemical 2, 3-dichloropropene has been shown to be hydrolyzed up to three times faster in moist soil than in solution.

Hydrolysis reactions are important for attenuation of chemicals in soils. Sorption-catalyzed reactions are especially important for two classes of chemicals, the chloro-s-triazines, and the organophosphates. Hydrolysis of atrazine, simazine, and propazine to their hydroxy derivatives in soils incubated at 30°C for 8 weeks was reported by Harris (1967).

The mechanism of chemical degradation of atrazine and other chloro-s-triazines is believed to be explained by the reaction of triazines with carboxyl groups (COOH) associated with soil organic matter. Sorption of triazines occurs between protonated carboxyl groups and triazine ring N atoms. Hydrogen bonding of the ring N and a soil organic matter carboxyl group causes electron withdrawal, in addition to that caused by electronegative chlorine and nitrogen atoms, from the electron deficient carbon atom enabling soil water to replace the chlorine groups.

Chemical degradation of organophosphates involves primarily soil-sorption-catalyzed hydrolysis of ester linkages. Although hydrolysis leads to degradation, the hydrolysis products may exhibit pesticidal properties (Cowart et al. 1971). The most important factors controlling rate and product formation include pH (Faust and Suffet 1966) and sorption. Degradation of diazinon, ciodrin, and malathion occur by chemical hydrolysis. While malathion degradation may be chemical or biological, chemical hydrolysis is complete before the biological lag phase is complete (Konrad et al. 1969). Hydrolysis in soil systems is sorption catalyzed. Specifically, sorption of organophosphate insecticides through complexation by soil-bound cations is the suggested mechanism for sorption-catalyzed hydrolysis.

Malathion degradation half-lives were 6-8 hr in soil systems at pH 7. First order kinetics of degradation were observed. Results of Hindin (1963) suggest that in alkaline soils (pH 8), hydrolysis of malathion may be base rather than sorption catalyzed.

It has been suggested that the extent of sorption-catalyzed hydrolysis of chemicals may be related to susceptiblity to undergo acidic or basic hydrolysis in soil-free systems (Guenzi 1971).

Polymerization

Organic compounds added to soil systems often react with soil oragnic matter to form complexes. The covalent binding of the pesticide class of organic chmicals to soil humic material is discussed by Bartha (1980). For example, 3, 4-dichloroaniline, a principal residue of several phenylamide herbicides, exhibited a half-life in soil at 500 ppm as the humic acid-dichloroaniline (HA-DCA) complex of 693 days. The addition of aniline at 1000 ppm to the HA-DCA containing soil decreased the half-life, based on first order kinetics, to 178 days (You et al. 1982). Incorporation of substituted aniline compounds into soil humic material may occur through the attachment of substituted anilines to the quinoidal subunits of soil humus (You et al. 1982, Hsu and Bartha 1976, and Parris 1980). Thus although the recalcitrance of compounds may increase in the soil system through polymerization reactions, addition of analogs to the system may significantly reduce the half-life of constituents and increase degradation. Results of chemical incorporation of constituents into soil humic material through polymerization and subsequent effects on biodegradation are sometimes confused with the effects of physical sorption of chemical constituents on biodegradation rate and extent.

MODELING THE BEHAVIOR OF WASTE CONSTITUENTS IN SOIL SYSTEMS

Transport Models

Transport models are important prediction tools for assessing the fate of organics in the soil environment and the potential for groundwater pollution. Transport models in principle consist of two parts: 1) a water flow model, and 2) a solute transport model (Selim and Iskandar 1981, Rao and Jessup 1982). In this section one dimensional transport models will be described and plant uptake will be assumed to be zero. The water flow model may be expressed as:

$$\frac{\partial \Theta}{\partial x} = \frac{\partial}{\partial x} \left(-K(\Theta) \frac{\partial H}{\partial x}\right) \tag{3-50}$$

where Θ is the volumetric water-soil content, K is the hydraulic conductivity, x is the soil depth, and H is the hydraulic head. Equation 3-50 can also be expressed as:

$$\frac{\partial \Theta}{\partial x} = \frac{\partial}{\partial x} \left(\alpha(\Theta) \frac{\partial \Theta}{\partial x}\right) - \frac{\partial}{\partial x} \left(K(\Theta)\right) \tag{3-51}$$

by substituting $H = h + x$, where h is the pressure head and $D(\Theta)$ is the soil water diffusivity, and $\alpha(\Theta) = -k(\Theta)\frac{dh}{d\Theta}$. The term $\frac{dh}{d\Theta}$ is found from the soil-water characteristic, which is a functional relationship between soil water content (Θ) and the pressure head (h). However, the soil-water characteristic is not a unique relationship, and it exhibits hysteresis (Nielsen et al. 1981a). That is, the response is different depending on whether the soil is in a wetting or drying cycle (Figure 3-53). The solute transport model can be derived for a "representative elementary value" (REV) of a porous medium. In this REV level it is assumed that the variability in medium properties is minimal (Bear 1972). The solute transport for REV is formulated as

$$\frac{\partial(\Theta_c)}{\partial t} = \frac{\partial}{\partial x} (\Theta D_h \frac{\partial c}{\partial x} (v\Theta c) - \frac{\partial}{\partial t} (ps) + \sum_{i=1}^{n} Qi \qquad (3\text{-}52)$$

where D_h is the hydrodynamic dispersion coefficient, c is the solution concentration of solute, S is the amount of solute adsorbed per unit weight of soil, and Qi represents source and sink terms. Sink terms include degradation (chemical and/or microbial), volatilization, decay, and any other removal mechanisms. The models for describing degradation and volatilization are discussed in other sections of this report. The hydrodynamic dispersion coefficient is composed of two components: 1) mechanical dispersion and 2) molecular diffusion. Mechanical dispersion is caused by: 1) velocity distribution within each pore, i.e., the velocity of fluid is zero at the soil particle surface and increases away from the particle surface, 2) velocity change due to pore size change, and 3) continuous fluctuation in velocity direction with respect to the mean direction of flow. Molecular diffusion is caused by the concentration gradient, i.e., Fick's law (the concentrations tend to diffuse from high to low concentration regions) (Fried 1977). The hydrodynamic dispersion coefficient is the sum of the mechanical dispersion and molecular diffusion (Dm) (Fried and Combarnous 1971):

$$Dx = av^n + Dm \qquad (3\text{-}53)$$

where the second term is the mechanical dispersion, v is interstitial velocity (Darcy's flux divided by the water content), and a and n are constants. When the soil is unsaturated a, n, and Dm are functions of soil-water content (Yule and Gardner 1978, Kirda et al. 1973). For saturated conditions n is assumed to be 1, and a is defined as dispersivity (Freeze and Cherry 1979). Hydrodynamic dispersion relationships with interstitial velocity were discussed by Fried and Combarnous (1971), Fried (1977), Biggar and Nielsen (1980), Nielsen et al. (1981b), and Gelham and Cherry (1982). It should be recognized that the hydrodynamic dispersion coefficient and the interstitial velocity are highly variable in the field (Nielsen et al. 1981b). Figures 3-54 and 3-55 show the skewed distribution of the interstitial velocity and the hydrodynamic dispersion coefficient for 360 field observations (Biggar and Nielsen 1976). The practical significance will be discussed later in this section.

The solution of the solute transport equation is possible if the spatial and temporal distribution of water content (Θ) and interstitial velocity (v) are known with appropriate boundary conditions (Table 3-59). Under transient flow conditions the distribution of Θ and v can be determined. Finite difference methods have been used for solving transport equations under unsaturated conditions (Bresler and Hanks 1969, Bresler 1973, Kirda et al. 1973, Wood and Davidson 1975, Tilloston et al. 1980). Selim (1978) has used a finite difference method for multilayered soil under unsaturated flow conditions with reactive solutes. Gurechian et al. (1979) have also solved the transport equation under transient unsaturated flow conditions for a layered soil.

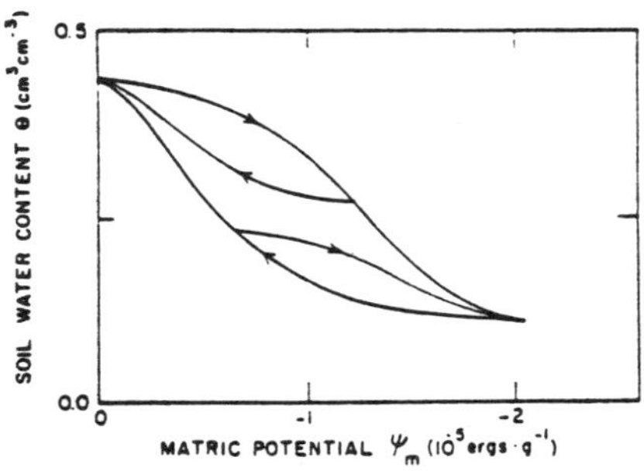

Figure 3-53. Soil-water characteristic relating the volumetric water content Θ to the matric potential ψ_m. Hysteresis properties are illustrated by the directional arrows (Nielsen et al. 1981a). Used by permission, see Copyright Notice.

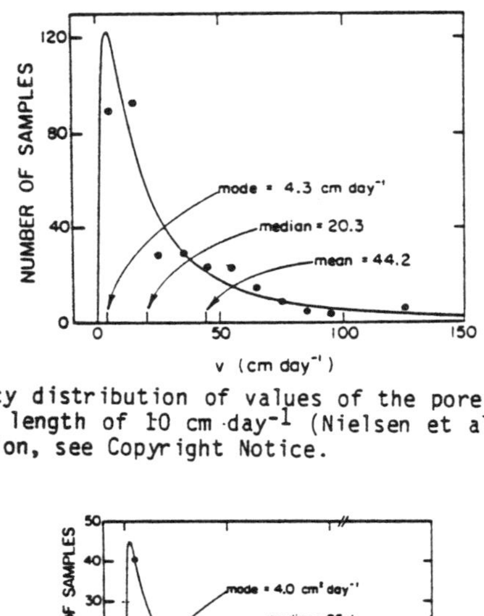

Figure 3-54. Frequency distribution of values of the pore-water velocity for a class length of 10 cm·day^{-1} (Nielsen et al. 1981b). Used by permission, see Copyright Notice.

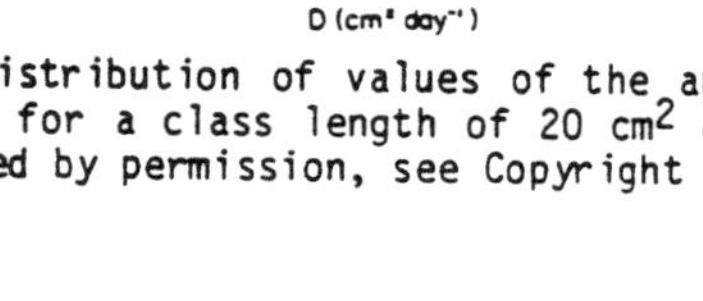

Figure 3-55. Frequency distribution of values of the apparent diffusion coefficient D for a class length of 20 cm^2 day^{-1} (Nielsen et al. 1981b). Used by permission, see Copyright Notice.

TABLE 3-59. BOUNDARY CONDITIONS FOR THE TRANSPORT EQUATION

1.0 Water Flow Boundary Conditions:

1.1 Soil surface boundary condition

1.1.1 Water head boundary condition

$$h = h_o(t) \text{ at } z = 0$$

1.1.2 Water flux boundary condition

$$q(t) = -D(\Theta)\frac{\partial \Theta}{\partial z} - K(\Theta) \text{ at } z=0$$

where $q(t)$ = water flux - it can be either constant or time dependent.

1.2 Bottom boundary condition

1.2.1 Impervious barrier
The flux is zero at the impermeable surface
(e.g. heavy clays)

$$-D(\Theta)\frac{\partial \Theta}{\partial z} + K(\Theta) = 0 \quad \text{at } Z = L$$

1.2.2 Soil profile extends to a great depth (semi-infinite medium)

$$\frac{\partial \Theta}{\partial z} = 0 \quad , \quad Z \to \infty$$

This boundary condition can be used in a well drained soil.

1.2.3 Groundwater table

$$\Theta = \Theta s \qquad Z = L \qquad t > 0$$
$$h = 0 \qquad Z = L \qquad t \geq 0$$

Θ_s = saturation water content

2.0 Mass Transport Boundary Conditions

2.1 Application of waste

$$vC_s = -\Theta D\frac{\partial c}{\partial z} + vc \qquad Z = 0 \qquad t < T$$

where C_s = concentration of applied wastewater
v = flux of wastewater application (cm hr-1)
T = duration of application

The same boundary condition can be used for a rainfall event

2.2 Boundary condition at the bottom of soil profile

$$\frac{\partial c}{\partial z} = 0 \qquad Z = L \qquad t \geq 0$$

Finite element methods have also been used for the transport equation (e.g. Duguid and Reeves 1974, Segol 1977). Sykes et al. (1982) have used Galerkin-finite element methods to solve the transport equation under steady state flow conditions. Adsorption was neglected and Michaelis-Menten kinetics (hyperbolic rate law) were used to model the biodegradation of organics in leachate from a landfill. Recently Van Genuchten (1982) has compared different numerical schemes that can be used in solving one dimensional saturated-unsaturated flow and solute transport equations.

Under steady state flow conditions Equation 3-52 becomes:

$$\frac{\partial c}{\partial t} = D_h \frac{\partial^2 c}{\partial x^2} - v \frac{\partial c}{\partial x} - \frac{\rho}{\Theta} \frac{\partial S}{\partial t} \qquad (3\text{-}54)$$

assuming a constant dispersion coefficient, and neglecting source and sink terms. Equation 3-54 can be solved if an appropriate relationship between S and c is defined in addition to appropriate boundary and initial conditions. Two types of adsorption equations are used in conjunction with the transport equation: 1) the equilibrium equation and 2) the nonequilibrium equation (Rao and Jessup 1982). Equilibrium equations are simply one of the adsorption isotherms described previously in this report. The rationale in using adsorption isotherm in the transport equation is that, in most cases, adsorption reaches equilibrium instantaneously. In other words, the sorption rate is very high compared to the pore water velocity. The Freundlich isotherm has been most often used with the transport equation. When the Freundlich isotherm is assumed for adsorption Equation 3-54 becomes:

$$R \frac{\partial c}{\partial t} = D \frac{\partial^2 c}{\partial x^2} - v \frac{\partial c}{\partial x} \qquad (3\text{-}55)$$

where R is the retardation factor, which is defined as

$$R = 1 + \frac{\rho\ KNC^{N-1}}{\Theta} \qquad (3\text{-}56)$$

The retardation factor represents the ratio between average pore water velocity and average pollutant front velocity. R becomes independent of concentration when N=1, i.e., when the isotherm is linear. The use of a linear isotherm with the transport equation gives rise to an analytic solution of the transport equation (Lapidus and Amundson 1952). The linear isotherms have been used with the transport equation very frequently (Kay and Elrick 1967, Davidson et al. 1968, Davidson and Chang 1972, Davidson and McDougal 1973). Van Genuchten and Alves (1981) have compiled analytical solutions for the diffusive convective transport equation with linear isotherms for different combinations of boundary conditions.

The assumption of a linear isotherm appears to be reasonable at low concentration ranges for organic constituents in soil systems. However, at high concentration ranges, i.e., close to solubility limits (such conditions may exist at hazardous waste sites), the exponent N of the Freundlich isotherm is usually less than one (Hamaker and Thompson 1972, Davidson et al. 1980b, Rao and Davidson 1980). In fact the assumption of a linear isotherm at high concentration seems to underestimate the leachable concentration, which may have dangerous consequences with respect to management of hazardous wastes sites (Davidson et al. 1978, Rao and Davidson 1979, Davidson et al. 1980b,c). Also, deviations between calculated and observed breakthrough curves (BTC's) often occur when linear isotherms are used. Van Genuchten and Wierenga (1976a) listed the experimental conditions under which asymmetry (tailing) is observed in BTC: 1) unsaturated flow, 2) aggregated media, and 3) pore-water velocity. The inadequate prediction of transport with equilibrium models led some investigators to use a kinetic (nonequilibrium) adsorption equation. Two approaches were used in developing nonequilibrium models: 1) the use of chemical-process models and 2) the use of physical-process models. In the chemical-process models the assumption is that adsorption is a time dependent reaction and that sorption equilibrium is not attained instantaneously. Travis and Etnier (1981) have reviewed adsorption kinetic models (see Table 3-29) In the second approach the assumption is that sorption is instantaneous. However, the rate of sorption is controlled by the rate of diffusion of the solute to the soil surface (Rao et al. 1980, Rao and Jessup 1982).

Chemical-process models can be divided into: 1) one-site models and 2) two-site models. In one-site models the sorption is assumed to be uniform throughout the soil surface. Lapidus and Amundson (1952) have presented an analytical solution for a transport equation with first order reversible kinetic adsorption. Oddson et al. (1970) have also presented an analytical solution for a transport equation with first order kinetic adsorption. However, the authors neglected the diffusion term in the transport equation. Lindstrom and Boersma (1978) presented an analytical solution for their model, developed earlier (Lindstrom and Boersma 1971), coupled with first order kinetic sorption. Hornsby and Davidson (1973) and Mansell et al. (1977) have also used the kinetic model that was developed by Lindstrom and Boersma (1973). The one-site kinetic models showed much improvement of results compared with equilibrium adsorption models. However, the improvements were only limited to low pore-water velocities (Davidson and McDougal 1973, Hornsby and Davidson 1973, Van Genuchten et al. 1974). In two-site models the sorption sites are divided into two types. The first site type is where sorption occurs instantaneously; with the second type of site the sorption is time dependent (Cameron and Klute 1977). The advantage of a two-site model over a one-site model is that two-site model can better simulate BTCs obtained from miscible displacement studies (Rao and Jessup 1982). The adsorption rate is defined as:

$$\frac{\partial S_1}{\partial t} = NKC^{N-1}\frac{\partial d}{\partial t} \tag{3-57}$$

and

$$\frac{\partial S_2}{\partial t} = \alpha[K_2C^N - S_2] \tag{3-58}$$

where the subscripts 1 and 2 denote the sites type 1 and type 2, respectively. The total amount adsorbed, S, is

$$S = S_1 + S_2 \tag{3-59}$$

The rate of change of S is found by substituting Equations 3-57 and 3-58 into the rate of change of Equation 3-59. The expression can be simplified further by setting N=1, and assuming that F is the fraction of sorption site 1 of the total sorption sites. The rate of amount sorbed becomes:

$$\frac{\partial S}{\partial t} = F\,K\,\frac{\partial c}{\partial t} + \alpha[(1 - F)K\,C - S_2] \tag{3-60}$$

The dimensionless form of the transport equation is (Van Genuchten 1981):

$$R_1\,\frac{\partial C_1}{\partial T} + (R_T - R_1)\,\frac{\partial C_2}{\partial T} = \frac{1}{P}\,\frac{\partial^2 C_1}{\partial Z^2} - \frac{\partial C_1}{\partial Z} \tag{3-61}$$

where the transformed dimensionless variables are defined as:

$$R_1 = 1 + \frac{\rho F K}{\Theta} \quad ; \quad R_T = 1 + \frac{\rho K}{\Theta} \tag{3-62}$$

$$T = vt/L \quad ; \quad Z = x/L \tag{3-63}$$

$$w = kL/v \tag{3-64}$$

$$C_1 = \frac{C - C_i}{C_o - C_i} \quad ; \quad C_2 = \frac{S_2 - (1 - F)KC_i}{(1 - F)K(C_o - C_i)} \tag{3-65}$$

T is dimensionless time, L is column length, P is Peclet number (a measure of convection relative to dispersion); C_1 and C_2 are dimensionless concentrations; and C_i and C_o are the initial and applied solution concentrations, respectively. Van Genuchten and Cleary (1979) and Rao et al. 1982 have stated that for $w \geq 5$ equilibrium conditions can be assumed.

In physical process models the liquid phase in the porous medium is divided into "mobile" and "immobile" regions. Convective-dispersive solute transport occurs only in the mobile region. Adsorption is assumed to occur instantaneously. However, the sorption rate is controlled by the rate of transfer between the mobile and immobile regions. The rate of transfer between the two regions is driven by the difference in the concentrations (Van Genuchten and Wierenga 1976, 1977). This model is formulated as follows:

$$[\Theta_m + f\rho k]\,\frac{\partial C_m}{\partial t} + [\Theta_{im} + (1-f)\rho k]\,\frac{\partial C_{im}}{\partial t} = \Theta_m D_m \frac{\partial^2 C_m}{\partial x^2} - V_m \Theta_m \frac{\partial C_m}{\partial x} \tag{3-66}$$

$$[\Theta_m + (1-f)\rho k] \frac{\partial C_{im}}{\partial t} = \alpha(C_m - C_{im}) \tag{3-67}$$

where subscripts m and im denote the variables for mobile and immobile regions, respectively, and f is the fraction of sorption sites in contact with the mobile region. The dimensionless form of Equations 3-66 and 3-67 are:

$$R_1 \frac{\partial C_m}{\partial t} + (R_T - R_1) \frac{\partial C_{im}}{\partial T} = \frac{1}{P} \frac{\partial^2 C_m}{\partial Z^2} - \frac{\partial C_m}{\partial Z} \tag{3-68}$$

$$(R_T - R_1) \frac{\partial C_{im}}{\partial T} = \beta(C_m - C_{im}) \tag{3-69}$$

where the transformed dimensionless variables are defined as follows:

$$T = vmt\ \phi/L \quad ; \quad z = x/L \tag{3-70}$$

$$\Theta_s = (\Theta_m + \Theta_{im}) \quad ; \quad \phi = (\Theta_m/\Theta_s) \tag{3-71}$$

$$R_1 = (\phi + \frac{f\rho k}{\Theta_s}) \quad ; \quad R_T = (1 - \frac{\rho k}{\Theta_s}) \tag{3-72}$$

$$P = (VnL/Dm) \tag{3-73}$$

$$\beta = (\frac{\alpha L}{V_m \Theta_m}) \tag{3-74}$$

$$C_1 = \frac{C_m - C_i}{C_o - C_i} \quad ; \quad C_2 = \frac{C_{im} - C_i}{C_o - C_i} \tag{3-75}$$

ϕ represents the fraction of moisture content in the mobile region. Although the chemical process model and physical process model are conceptually different, mathematically they are equivalent for given boundary and initial conditions. The validity of any of these models can be proven experimentally if independent measurements of the parameters of the models are available. Unfortunately this is not usually the case. Nonlinear least squares have been used to optimize the parameter values in order to make the difference between the model and observed points minimum. Rao et al. (1980b) used the analytic expression developed to calculate the transfer coefficient as an independent measure of the parameter. The results showed good agreement with the BTC's observed for $^{38}Cl^-$ and ^3H_2O over a wide range of pore-water velocities. Based on theoretical and experimental analysis, Rao et al. (1980a) have found that the transfer coefficient is a function of system variables.

Limitations

The parameters for the models described above were estimated using nonlinear least squares. The parameters were optimized so that the difference between observed data and computed data is minimum. Hence, the validity of the models is questionable since there is no independent measurements of parameters for the models. In addition to the unavailability of independent parameter estimates, the field applicability of these models is further limited by the heteorogeneity of most soils. As demonstrated in Figures 3-54 and 3-55 pore water velocities and diffusion coefficients in field measurements are highly variable. The pore water velocity and the diffusion coefficient field measurements were shown to be log normally distributed (Biggar and Nielsen 1976, Van De Pol et al. 1977). These uncertainties in the parameters will be reflected in decreased accuracy of predictions. Dagan and Bresler (1979), using mixed stochastic and deterministic differential equations, attributed the variability of concentration profile in the field mainly to the uncertainty of hydraulic properties of the porous medium. In their analysis they neglected molecular diffusion and any interaction between the solute and soil. Recently, Amoozegar-Fard et al. (1982) used a Monte Carlo simulation and generated 2000 values of C/C_0 using the solute transport equation. They assumed normal distribution for ln D, ln v, and the water filled porosity, and no solute interaction with the soil. The results showed wide differences between solute profiles when deterministic values of v and D were used. The variability of D was much less important than the variability of v.

Therefore, according to the above discussions, conceptual models have been developed but not much attention is paid to parameter evaluation. Davidson et al. (1980) have shown that differences between the physical nonequilibrium model Equations 3-68 and 3-69, and the analytic solution of Equation 3-55 for nonreacting solutes is negligible (Figure 3-56). Hence, the simple model is preferred over the more complicated in many situations. For the purpose of site assessment for in situ treatment, a simple model has been selected. The model chosen is described in detail by Enfield et al. (1982). The model includes a first order degradation term and assumes a linear adsorption isotherm. The model is a one dimensional advective-dispersive transport model with a linear adsorption isotherm, and a first order degradation rate. This model can be expressed as follows:

$$\left(1 + \frac{\rho K_d}{\Theta}\right) \frac{\partial C}{\partial t} = D_h \frac{\partial^2 C}{\partial x^2} - v \frac{\partial C}{\partial x} - k_1 C \qquad (3\text{-}76)$$

where K_1 is the first order degradation rate constant and the rest of the terms are as defined previously. Steady state unsaturated flow condition was assumed in this model. The water content is estimated such that mass conservation holds. Assuming a unit hydraulic gradient, the water recharge rate (Darcy's flux) is equal to the hydraulic conductivity. The hydraulic conductivity is related to water content by the following empirical equation developed by Clapp and Hornberger (1978):

$$k = (\Theta/\Theta_s)^{2b+3} \qquad (3\text{-}77)$$

where k is the hydraulic conductivity, Θ_s is the saturation water content, Θ is the water content, and b is a constant. Clapp and Hornberger (1978) have estimated Θ_s and b for a range of soil textures (Table 3-60). Hence, Equation 3-77 can be used to estimate the water content, knowing Θ_s, b and the recharge

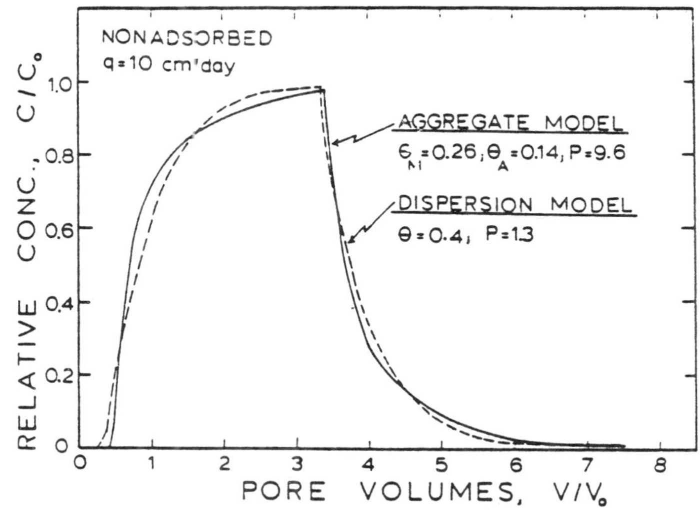

Figure 3-56. Calculated relative effluent concentrations for a nonadsorbed solute. Solid line obtained using analytic solution of Equations 3-68, 3-69; dashed line obtained with analytic solution of Equation 3-55 (Davidson et al. 1980). Used by permission, see Copyright Notice.

TABLE 3-60. REPRESENTATIVE VALUES OF HYDRAULIC PARAMETERS (STANDARD DEVIATION IN PARENTHESES) (CLAPP AND HORNBERGER 1978)
Used by permission, see Copyright Notice.

Soil texture	No. of soils	b	s cm	s cm^3/cm^3
Sand	13	4.05 (1.78)	12.1 (14.3)	0.395 (0.056)
Loamy sand	30	4.38 (1.47)	9.0 (12.4)	0.410 (0.068)
Sandy loam	204	4.90 (1.75)	21.8 (31.0)	0.435 (0.086)
Silt loam	384	5.30 (1.87)	78.6 (51.2)	0.485 (0.059)
Loam	125	5.39 (1.87)	47.8 (51.2)	0.451 (0.078)
Sandy clay loam	80	7.12 (2.43)	29.9 (37.8)	0.420 (0.059)
Silty clay loam	147	7.75 (2.77)	35.6 (37.8)	0.477 (0.057)
Clay loam	262	8.52 (3.44)	63.0 (51.0)	0.476 (0.053)
Sandy clay	19	10.4 (1.64)	15.3 (17.3)	0.426 (0.057)
Silty clay	441	10.4 (4.45)	49.0 (62.1)	0.492 (0.064)
Clay	140	11.4 (3.70)	40.5 (39.7)	0.482 (0.050)

rate. The model will be useful, not in terms of the absolute estimate, but rather for establishing priorities with respect to chemicals and chemical classes that require treatment. An example of the application of the model and data requirements are presented in Volume I of this report.

ATMOSPHERIC ASPECTS OF IN SITU TREATMENT: VOLATILIZATION AND PHOTODEGRADATION

Volatilization of Organics

The primary emphasis in monitoring at sites of hazardous waste contaminated soils has been related to the impact of these contaminants on the water environment. Studies have concentrated on the movement of metals, salts and hazardous organic compounds under uncontrolled hazardous waste sites and through soil systems via groundwater and leachates. Information obtained from ambient air measurements of the Love Canal area and from the Hudson River Basin indicate that significant concentrations of toxic materials can be released from landfills and dump sites, often much greater than emissions via water transport (Shen and Tofflemire 1980). A number of hazardous compounds, ranging from benzene to vanadium pentoxide, have been identified at uncontrolled hazardous waste sites in varying concentrations as shown in Table 3-61, indicating that air releases of many hazardous compounds from these waste sites may be significant.

Volatilization of Organic Contaminants--

Although a paucity of information exists relating to the modeling of organic contaminant emissions from hazardous waste disposal sites, much information exists concerning the volatilization of organics, primarily pesticides, from soil surfaces. General definitions of volatilization include the loss of chemicals from surfaces in the vapor phase, indicating that volatilization requires the vaporization and movement of chemicals from a surface into the atmosphere above the surface. The rate of contaminant volatilization is a complex function of the properties of the contaminant and its surrounding environment. The driving force for volatilization is derived from the vaporized chemical in the soil pore spaces, and the volatilization rate of the material will be affected greatly by the adsorption onto soil particles and the absorption or solubility of the compound within the soil organic matter and within the soil water.

For organics in soil systems Spencer and Cliath (1977) indicate that the factors affecting volatilization include:

1. Contaminant vapor pressure

2. Contaminant concentration

3. Soil/chemical adsorption reactions

4. Contaminant solubility in soil water

5. Contaminant solubility in soil organic matter

TABLE 3-61. HAZARDOUS CHEMICAL VAPORS DETECTED AT UNCONTROLLED HAZARDOUS WASTE SITES

Chemical	Number of Sites
Benzene	5
Hexane	1
Toluene	6
Dichloroethane	1
Acetone	1
Tetrachloroethylene	1
Xylene	3
Methane	6
Vinyl chloride monomer	3
PCB	3
Other VHO's and VNHO's	2
Dioxin	1
Ozone	1
Ionizable vapors	1
Vanadium pentoxide	1
TCE	1
THC	1
Methyl furan	1
Benzaldehyde	1
Bis(2-chloroethyl) ether	1
Thorium	1
Radium	1
Methylene chloride	1
Trichloro ethane	1
Phenol	2

6. Soil temperature, water content, organic content, porosity, and bulk density

The major contaminant property affecting volatilization is its vapor pressure, while the major environmental factors affecting the contaminant's vapor pressure are the soil/water and air/water partition coefficients that exist for the various soil/water/air environments within the soil system. Additional complexity results if the contaminant is added along with an additional adsorbing fluid such as oil in refinery wastes, where partitioning of the contaminant between the oil/soil, oil/ water, and oil/air phases would also be expected to affect the volatilization or vapor pressure of the volatile compounds.

Equilibrium Chemical Partitioning--

The tendency of a compound to migrate from one environmental medium to another is a function of the driving force to reach equilibrium between media, and is related to the compound's chemical potential energy or its fugacity. MacKay (1979) originally proposed the fugacity approach for compound partitioning as an indication of the potential fate of compounds released in multicomponent environments. Further development of the equilibrium partition approach has been presented by McCall et al. (1983) for estimations of chemical partitioning based on the:

1. Soil sorption constant - Koc = $\frac{\mu g \text{ chemical/g organic carbon}}{\mu g \text{ chemical/g water}}$

2. Water/air partition constant - KW = $\frac{C_{water}}{C_{air}} = \frac{T(WS)}{16.04\ PM} = \frac{1}{H}$

3. Bioconcentration factor - BCF = $\frac{\mu g \text{ chemical/g fish}}{\mu g \text{ chemical/g water}}$

where

T = temperature, °K

C_{air}, C_{water} = concentration of the chemical in the air and water, respectively, mg/l

WS = water solubility, mg/l

P = vapor pressure of pure chemical (mm Hg)

M = chemical molecular weight, and

H = Henry's law constant

Correlations between partitioning coefficients have been presented by a number of authors and have been tabulated in Table according to chemical class. These correlation equations are valid for nonionic compounds and describe partition coefficients in terms of a compound's vapor pressure, water solubility, and melting point. Knowing the compound's partitioning among the various media allows the estimation of its equilibrium distribution throughout the environment, and indicates the relative importance of each medium as the contaminant concentration moves toward an equilibrium level.

A compound's potential for volatilization from aqueous solution is indicated by its air to water partition coefficient, $1/K_w = H$. From the expression for K_w given above, volatility is seen to increase as water solubility decreases or vapor pressure increases. Volatilization from soils is also affected by soil sorption allowing an expression for volatilization from wet soils to be developed through a modification of the water volatilization term (Swann et al. 1983):

$$\text{Volatility rate} \propto \frac{1}{K_w K_{oc}} \qquad (3\text{-}78)$$

$$\frac{PM}{(WS)TK_{oc}} \qquad (3\text{-}79)$$

An analysis of a compound's physical properties in the above manner will indicate its potential for volatility as compared to other compounds of interest. McCall et al. (1983) showed that tetrachlorobiphenyl (K_w = 2; K_{oc} = 32,500; P = 4.9x10^{-4} mm Hg) has a large tendency to volatilize from aqueous solutions due to its low K_w value, even with a vapor pressure five orders of magnitude less than 1,3-dichloropropene (K_w = 18, K_{oc} = 68, P = 25 mm Hg), yet its volatilization from soils is less than that of 1,3-dichloropropene because of tetrachlorobiphenyl's large K_{oc} value.

Because of the importance of partition coefficients in estimating contaminant movement, and the large number of compounds of concern, efforts have been made to estimate partition coefficient values based on rapidly determinable surrogate parameters. Swann et al. (1983) have presented equations relating various partition coefficients to retention times obtained by reverse phase high-performance liquid chromotagraphy (RP-HPLC). A Waters Associates HPLC system was utilized with an 85:15 v/v mobile phase of methanol-water pumped at room temperature at 1.0 ml/min. Expressions for the various partition coefficients of interest based on RP-HPLC retention time (Rt) and compound melting point (MP) were as follows:

$$\ln K_{oc} = 3.446 \ln Rt + 1.029 \qquad (3\text{-}80)$$

$$\ln K_{ow} = 5.505 \ln Rt - 3.780 \qquad (3\text{-}81)$$

$$\ln BCF = 5.147 \ln Rt - 6.977 \qquad (3\text{-}82)$$

$$\ln WS = 7.618 \ln Rt - 0.01 (MP-25) + 18.328 \qquad (3\text{-}83)$$

Correlation coefficients for the above equations ranged from 0.95 to 0.98, indicating that a rapid estimation of compound partitioning is possible based on RP-HPLC data. The equations were presented for nine pesticides; benzene;

bromobenzene; bibenzyl; biphenyl; 2,2',4,5,5'-PCB; and anthracene and cannot be considered general purpose relationships. They do suggest, however, the validity of using RP-HPLC retention for partition coefficient estimates if correlation equations for a specific compound class are developed using a particular RP-HPLC operation.

Dynamic Vapor Movement Models--

Evaluation of chemical characteristics using equilibrium concepts as described above indicate gross chemical movement potential but do not provide information concerning the rate of movement toward equilibrium nor the variation of rates with time. Analysis of contaminant vapor movement using diffusion concepts provides this time variant analysis for movement through soil systems.

Volatilization from an Adsorbing Surface--

Adsorption of a compound onto an adsorbing surface reduces its chemical activity, or fugacity, resulting in a reduction in its vapor pressure (Spencer and Cliath 1977). This reduction in vapor pressure decreases significantly the vaporization rate of the compound. This fact invalidates volatilization equations based on compound aqueous vapor pressure measurements and on the rate of volatilization of a model compound assuming nonadsorbing surface volatilization as presented by Hartley (1969):

$$F_b = \frac{P_b\ (M_b)^{1/2}}{P_a\ (M_a)^{1/2}} \cdot F_a \qquad (3\text{-}84)$$

where

F = vapor flux rate

P = vapor pressure

M = molecular weight

a,b = model compound and volatilizing compound, respectively

A relationship similar to Equation 3-84 could be used to estimate volatilization from soil systems if vapor pressure measurements of the compound in the soil are determined. Spencer and Cliath (1969, 1983) describe a modification to a gas saturation method they have developed which utilizes a pesticide-treated soil, in place of a pesticide treated quartz sand, in a test saturator along with a humidity control device for maintenance of soil moisture content during testing. Diffusion time variation information for the volatilizing compound can then be obtained from Equation 3-84 if diffusion variation of the model compound is known.

Further complications result when the compound is soil incorporated. Under soil incorporation conditions, volatilization of the compound involves the desorption of the compound from liquid layers that coat the soil particles, diffusion through the air filled pore spaces within the soil column to the air/soil interface, followed finally by the diffusion from the soil surface to the overlying atmosphere (Thibodeaux 1979). Vaporization under soil incorporation conditions occurs at a much slower rate as compared to vaporization from soil surfaces due to reductions in the vapor pressure of the compound and the slow rate of diffusion through the soil column to the air/soil interface. As volatilization occurs, a concentration gradient develops between equilibrium and actual concentration levels in all phases resulting in a driving force for continued diffusion. The rate of diffusion declines with time, however, as the concentration gradient is reduced due to an ever increasing diffusion path length to the air/soil surface (Hamaker 1972). Simplification of this complex problem, by assuming a compound concentration at the soil surface equal to zero and a soil column of infinite depth, has resulted in relationships for mass flux rate with time based on Fick's second law of diffusion in the general form as presented by Mayer et al. (1974):

$$F = \frac{D_s \; C_{Ao}}{(\Pi \; Dt)^{\frac{1}{2}}} \qquad (3\text{-}85)$$

where

F = component mass flux rate through the soil surface

D_S = soil diffusion coefficient

C_{Ao} = initial component concentration

t = time

Contaminant Advection--

An additional source of contaminant volatilization from soil systems is an advection process, labeled the "wick effect" by Hartley (1969), that describes the net contaminant transport via a large upward diffusion of water toward the soil surface due to evaporation of water from soil surfaces. The impact of this advection term will vary from compound to compound and is a function of the compound's soil adsorption characteristics, water solubility, and partition coefficients in the air, soil, and water phases. A simple relationship for this flux term was presented by Spencer and Cliath (1973):

$$F = F_W \times c \qquad (3\text{-}86)$$

where

F_W = water mass flux rate

c = component concentration in soil water

A complete accounting for the mass flux of a volatile component from a soil system can then be written using the summation of Equations 3-85 and 3-86 to account for flux due to diffusion and due to mass transport via advection with evaporated soil moisture.

Vapor Transport from Hazardous Waste Sites--

The models developed as described above are limited in that they lack the ability to include intimate soil and waste interactions as occurs in hazardous waste sites where drums of chemicals leak into soil systems and around soil particles. To accurately model volatile organic emissions from hazardous waste sites, both the soil pore diffusion and soil surface diffusion phenomenon must be considered, and means must be provided to predict diffusion as a function of soil and diffusion length characteristics.

Modeling of the evaporation and diffusion of chemicals within the pore spaces of soil systems using the concept of a "dried-out" zone was presented by Thibodeaux (1979) to describe soil contamination from liquid spills. In his model, soil contamination to a soil depth of h was assumed, with compound evaporation from soil surfaces, vapor diffusion into soil air spaces, and movement of the vapor up and out of the air/soil interface. A "dried-out" zone develops at the air/soil surface which is relatively free of adsorbed contaminant but through which vapors from the lower level must travel. With time, this "dried-out" zone increases in depth, correspondingly reducing the contaminated zone to an ever decreasing thickness, y. The soil column is assumed to be isothermal, capillary action is considered minimal, and soil adsorption of vapor through the dried-out zone along with biodegradation are also considered negligible.

Vapor diffusion through soil pores in the "dried-out" zone is considered limiting, resulting in the following expression for compound mass flux rate from the contaminated zone through the dry surface zone:

$$F_A = \frac{D_A}{h - y}(C_A^* - C_{A_i}) \qquad (3\text{-}87)$$

where

D_A = component diffusion coefficient in soil air spaces

h = initial depth of soil contamination

y = variable thickness of soil contamination after onset of diffusion

C_A^* = equilibrium concentration of component in pore spaces at the evaporating plane

C_{A_i} = concentration of the compound at the air/soil interface

The time for all of the liquid to vaporize from the contaminated zone was given as:

$$t = \frac{h\, m_A}{2D_A(C_A^* - C_{A_i})A} \qquad (3\text{-}88)$$

where

A = surface area of contaminated region

t = time for all liquid to vaporize.

m_A = mass of component originally in the contaminated zone

Upon complete vaporization within the contaminated zone, diffusion can be modeled as the diffusion of a chemical from vapor filled pore spaces that are saturated to a depth of h. Analysis of the multicomponent continuity equation with appropriate boundary conditions results in an expression for the average concentration in the contaminated zone at time t (Thibodeaux 1979):

$$\bar{C}_A = C_{A_i} + (C_A{}^* - C_{A_i}) \frac{8}{\Pi^2} \sum_{n=0}^{\infty} \frac{1}{(2n+1)^2} \exp\left[-\frac{D_A(2n+1)^2\Pi^2 t}{4h^2}\right] \qquad (3\text{-}89)$$

where

$\bar{C}_A$ = average compound concentration in the pore spaces, and other terms are as defined above.

Thibodeaux (1979) presented a graphical representation of the fraction of chemical remaining, $F_A = \bar{C}_A/C_A{}^*$, versus dimensionless time, log (D_At/h^2) (Figure 3-57), allowing the determination of compound lifetime for pore diffusion. Total decontamination time is thus the sum of results from Equation 3-88 for vaporization time and Equation 3-89 for vapor diffusion time.

Refinement of the "dried-out" zone approach to air emissions from landfarming of petroleum wastes has been carried out by Thibodeaux and Hwang (1982) and represents the state-of-the-art description for the volatilization of organics from land treatment operations. This model is applicable to the estimation of air emissions from hazardous waste sites with slight modifications as discussed below. The model as developed assumes an isothermal soil column, no capillary action through the soil layer, no adsorption in the soil pore space, and no biodegradation of applied organics within the soil column. The description of vapor movement through the soil/waste matrix is valid for surface or subsurface waste applications through the use of surface injection depth, h_s, and depth of penetration or plow slice depth, h_p. As applied to hazardous waste sites, h_s=0 and h_p are equated to the estimated or measured depth of the contaminated soil layer. Under steady-state conditions, the time for the intial mass applied to completely volatilize is determined through the analysis of the component vaporization rate, assuming a diffusion length

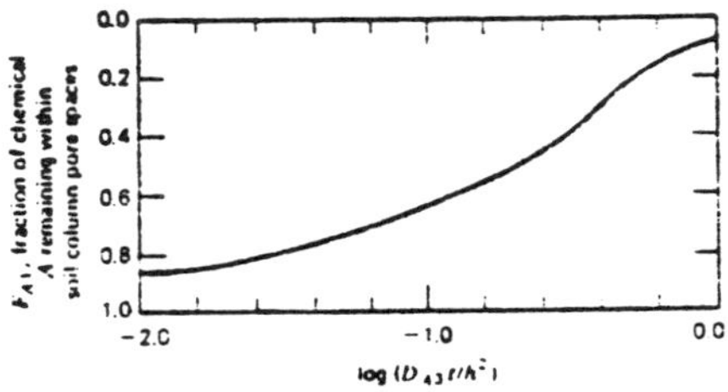

Figure 3-57. Fraction of contaminant remaining versus dimensionless time from Thibodeaux 1979). Used by permission, see Copyright Notice.

varying from the surface incorporation depth, h_s, at time t=0 to a value of the penetration depth, h_p, at complete volatilization. The average diffusion length is:

$$\frac{h_p - h_s}{2} + h_s = \frac{h_p + h_s}{2} \tag{3-90}$$

the flux rate into the pore spaces from Fick's first law becomes:

$$F_A = \frac{D_A \, (C_A - C_{Ai})}{\left(\frac{h_p + h_s}{2}\right)} \tag{3-91}$$

Equating F to its principal components and solving for the lifetime of the diffusion reaction assuming a mass concentration at the air/soil interface equal to 0 yields an expression for the lifetime, t, of the applied component:

$$t = \frac{M_A \, (h_p + h_s)}{2 \; A \; D_A \; C_A} \tag{3-92}$$

where

C_A = component concentration within the pore air spaces

M_A = total mass component volatilized

A = applied surface area or area of soil contamination

It should be recognized that M_A/A represents the waste application rate on the landfarming area if all mass is volatilized. In waste spills analysis,

M_A represents the estimated mass of component A originally in the contaminated zone, again assuming all mass is volatilized.

Through mass balance relationships the mass flux rate of component A through the wet zone-dry zone interface with appropriate boundary conditions can be presented as:

$$F = \frac{D_A C_A}{\left(\frac{h_s^2 + 2 D_A t A (h_p - h_s) C_A}{M_A}\right)^{\frac{1}{2}}} \qquad (3\text{-}93)$$

where

t = time after waste application

The component pore-space air concentration is related to the component concentration within the applied oil by equating the rate of transport through the oil phase to that through the dry soil column. The transfer rate equality takes the form:

$$a_s A\bar{y} \left(\frac{D_o}{Z_o}\right) (C_{io} - C_L) = \left(\frac{D_A A}{h_p - \bar{y}}\right) (C_A - 0) \qquad (3\text{-}94)$$

where

a_s = interfacial area per unit volume of soil

D_o = component diffusion in the oil phase

Z_o = oil-layer diffusion length

C_{io} = component concentration on the oil side of the air/oil interface

$\bar{y}$ = average thickness of the wet zone

The concentration of the component in the air and oil phases within the soil pore space is related by Henry's law constant to yield:

$$C_A = H_C C_L \qquad (3\text{-}95)$$

where

H_C = Henry's law constant in molar concentration form

Substitution of Equation 3-94 into Equation 3-95 allows for the expression of the concentration of the component in the soil vapor phase in terms of its initial concentration within the oil as:

$$C_A = \left[\frac{H_C}{1 + H_C \left(\frac{D_A Z_o}{D_o a_s \bar{y}(h_p - \bar{y})} \right)} \right] C_{io} \qquad (3\text{-}96)$$

The expression $\bar{y}(h_p-\bar{y})$ indicates the impact of a lengthening dry zone and is expressed as:

$$\bar{y}(h_p - \bar{y}) = \frac{h_p^2 + h_p h_s - 2h_s^2}{6} \qquad (3\text{-}97)$$

resulting in

$$C_A = \left[\frac{H_c}{1 + H_c \left(\frac{6 D_A Z_o}{D_o a_s (h_p^2 + h_p h_s - 2h_s^2)} \right)} \right] C_{io} \qquad (3\text{-}98)$$

The relative importance of the oil layer diffusion rate is highly dependent upon the oil-layer diffusion length, Z_o, and the interfacial area, a_s, which should be intimately tied to the waste application rate and the nature of the soil in the soil system. Thibodeaux and Hwang (1982) presented equations for Z_o and a_s for oil/soil interactions that result in either "film" forms or "lump" forms within the soil column. Oil interactions resulting in a thin coating around hypothetical spherical soil particles results in "film" forms with the following characteristics:

$$Z_o = \frac{d_p \bar{\rho}_p f}{6\bar{\rho}_o} \qquad (3\text{-}99)$$

$$a_s = \frac{6}{d_p} \qquad (3\text{-}100)$$

where

d_p = particle diameter

$\bar{\rho}_p$ = soil particle density

$\bar{\rho}$ = oil density

f = fraction of oil in film form

Soil aggregation results in the clumping of soil particles and causes the entrapment of oil or contaminant "lumps" within the soil particle aggregate. Diffusion film thickness and interfacial area expressions for soil clumps assuming an orthogonal arrangement of particles surrounding an oil "lump" yield:

$$Z_o = \frac{d_p}{2} \qquad (3\text{-}101)$$

$$a_s = \frac{2.70}{d_p} \qquad (3\text{-}102)$$

The fraction of pore spaces that is air filled is assumed to be 50 percent, yielding an estimated f value of 0.5.

With a thin oil diffusion length, on the order of soil particle size thickness, Equation 3-96 simplifies to:

$$C_A = H_c\, C_{io} \qquad (3\text{-}103)$$

Transport between the oil and air pore spaces is at equilibrium under these conditions and diffusion in the air spaces limits the overall mass transfer rate from the landfarming area.

The diffusivity of the component in the air filled pore spaces is related to that in air by taking into account the reduction in molecular diffusion in a soil medium due to cross-sectional area occupied by soil particles (porosity correction) and to twists and turns in the diffusion path due to irregularities of the pore spaces within the soil column (tortuosity correction). Mathematically this correction takes the form:

$$D_A = \frac{D_{A_i} \cdot \varepsilon}{\tau} \qquad (3\text{-}104)$$

where

D_{A_i} = component diffusion in air

ε = soil porosity = $1 - \frac{\text{soil bulk density}}{\text{particle density}}$

τ = soil tortuosity

With the use of Equations 3-92, 3-93, 3-98, 3-99, 3-102, and 3-104, the time for decontamination and the rate of organic emissions from hazardous waste sites can be determined using the Thibodeaux-Hwang model, once the following parameters are measured: soil parameters including bulk density, particle diameter and particle density; compound parameters including air and oil molecular diffusivity and Henry's law constant; and operational parameters including surface injection and penetration or plow splice depth, surface area of application, mass application, and time.

This model represents the state-of-the-art for landfarm air emission rate (AERR) modeling and for a general case of soil contamination from spills or leaking drums, volatilization from contaminated soil without the oil film diffusion term can be modeled utilizing partitioning coefficients described earlier. The partitioning between soil organic carbon and soil water would be controlled by K_{oc} and between the soil water and the soil air pore spaces by K_w. Assuming equilibrium conditions occur between phases before volatilization begins results in an expression for soil air pore concentration:

$$C_A = C_w/K_w \tag{3-105}$$

where

C_w = the component concentration on the water side of the air/water interface

Following the development of an expression for C_A as carried out above assuming the rate of diffusion through that the water phase and dry soil column can be equated and the water film diffusion length is small yields the following equation:

$$C_A = C_{iw}/K_w \tag{3-106}$$

where

C_{iw} = the initial component concentration measured in the soil water

C_{iw} will represent the measured value established based both on partitioning between the soil/water and water/air phases if equilibrium conditions are reached.

In the development of the above models, the air/soil surface concentration of the contaminant was assumed to be equal to zero due to surface air transport of vaporized components away from the site. Movement in both the vertical direction, due to vertical turbulent diffusion as modeled by the Thornthwaite-Holzman equation (Thibodeaux 1979), and the horizontal direction due to advection, result in contaminant transport away from the diffusing surface. The integration of surface flux rate models for transport through soil systems, with surface transport models, enables the evaluation of the impact of micrometeorological conditions on the emission and transport of toxic organics from hazardous waste sites and allows an evaluation to be made of ambient air quality level impacts from hazardous waste sites.

Factors Controlling Contaminant Volatilization

In reviewing the models presented for prediction of contaminant volatilization from soil systems, the following soil and contaminant characteristics can be identified as significantly impacting compound volatibility:

1. Soil moisture content, for compounds with low K_{OC} and K_W values when soil moisture content is low, causing an increase in diffusion due to water competition for soil adsorption sites (Hamaker 1972, Spencer and Cliath 1977, Thibodeaux 1979). The model parameter reflecting this effect would be Henry's law constant, H_C, which varies with vapor pressure from the relationship:

$$H_C = H\ Cg\ (\frac{10^6}{P}) \qquad (3\text{-}107)$$

where

H_C = Henry's law constant in molar concentration terms

H = Henry's law constant in partial pressure/molar concentration terms

P = total pressure

Cg = molar gas density = P^o/RT

Po = component partial pressure

R = gas law constant

T = absolute temperature

An increase in component partial pressure would increase C_g, which in turn increases H_C, which in turn increases C_A, resulting in an expected increase in the mass flux rate F until greater than one monolayer exists. With greater than a single water monolayer within the soil (high moisture content), reduced porosity affects becomes controlling.

2. Temperature, causing an increase in vapor pressure, water advection rates, soil/water/air partition coefficients, and biodegradation rates (Hamaker 1972, Thibodeaux 1979).

3. Soil organic content, causing the partitioning between the soil/vapor phase and a reduction in compound diffusion coefficient due to a reduction in its vapor pressure (Hamaker 1972, Thibodeaux 1979).

4. Surface air flow rate, causing an enhancement of flux rates due to maintenance of minimal organic and water vapor concentrations above the air/soil interface, maintaining the driving force for diffusion. This effect is reported to be of major concern for surface applied wastes due to the pore diffusion limitations that exist for volatilization from subsurface applications (Spencer and Cliath 1977, Thibodeaux 1979).

5. Biodegradability, causing the removal of the compound via uptake and metabolism thus reducing its volatility. Degradation products have been reported to be more volatile than parent compounds however (Spencer

and Cliath 1977), and may potentially cause an increase in component flux rates.

Other contaminant characteristics such as K_{OW}, K_{OC}, K_W, BCF, MP, and vapor pressure affect volatility due to their interaction with soil/ water conditions within the environment and are not generally controllable. Their relationship with one another for a given compound will indicate the potential effect of organic amendment addition and water addition on volatility, however, and can serve as tools for contaminant management as will be discussed in the treatment section of this report.

Compound Photoreactivity

Photooxidation is the use of incident solar radiation to carry out photoreaction processes. It utilizes cost-free solar energy as the driving force for reaction but suffers from the variable supply of incident solar energy and the attenuation of UV light as it travels through the atmosphere.

Photodegradation of organic compounds may occur by two processes: direct photodegradation and sensitized photooxidation. In the direct photoreaction process, each particle of light, or quanta, excites one substrate molecule. Although many organic compounds absorb light, their adsorption is primarily in the ultraviolet region, rendering direct photodegradation an inefficient treatment process (Thorington 1980).

In the sensitized photooxidation process, sensitizing molecules absorb light in the visible region where there is a wealth of energy reaching the earth's surface. The electronically excited sensitizing molecule, or triplet sensitizer, then returns to ground state by transferring its excess energy to molecular oxygen, resulting in the formation of singlet oxygen. Sensitized rates of photolysis are often orders of magnitude greater than those of direct photolysis (Acher and Saltzman 1980) and are also characteristically photooxidations, whereas direct photolysis may proceed by a variety of mechanisms including isomerism, dehalogenation and dissociation.

Singlet oxygen is a highly reactive species of oxygen with a lifetime of 3 µsec (Kearns 1971). It readily attacks organic substrates, yielding oxidized photoproducts. Spikes and Straight (1967) stated that many classes of organic molecules are susceptible to sensitized photooxidation. Among these are alcohols, nitrogen heterocycles, organic acids, phenols, and polycyclic aromatic hydrocarbons. Sargent and Sanks (1974) concluded that many compounds found in industrial wastes, such as phenols, cresols, trinitrotoluene, and unsaturated nitriles should also be susceptible to photooxidation.

The rate of photolysis is influenced by a number of factors including the intensity and wavelength distribution of light reaching the reaction media, their diurnal and seasonal variations, the absorption spectra of the contaminant or photosensitizer, the concentration of reacting compounds, and the energy yield produced during photon absorption. Additionally, the nature of the media in which the reaction takes place and the interactions between the contaminant and its surroundings has a major effect on photochemical reactions as will be discussed briefly below.

Photolysis in Soils--

The evaluation of pesticide reactivity has been of primary concern to past investigators due to the loss of effectiveness of pesticides undergoing photooxidation, and both direct and sensitized photolysis have been observed for pesticides in aqueous, soil, and vapor phases (Miller and Zepp 1983). Although the occurrence of the soil photoreaction of adsorbed chemicals has been identified, the importance of this reaction as compared to aqueous or vapor photoactivity and the reaction pathways through which the reactions occur, have not been identified to date. General soil characteristics including soil particle size, organic content, mineral composition, light absorption characteristics, and moisture content have been identified as affecting the nature of photodecomposition products and the rate and extent of the photoreaction of some pesticides (Miller and Zepp 1983).

The relative importance of the photooxidation reaction of a chemical on or within a soil will depend to a large extent upon its partitioning between the air/water/soil media within the soil system. The soil/air distribution coefficient, K_{SA}, can be calculation from other constants as follows (Lemaire et al. 1982):

$$K_{SA} = \frac{C_{soil}}{C_{air}} = \frac{C_{water}}{C_{air}} \cdot \left(\frac{1}{r} + K_D\right) = K_W \cdot \left(\frac{1}{r} + K_D\right) \qquad (3\text{-}108)$$

where

C_{soil} = concentration of chemical in wet soil on a dry basis, μg/g

C_{air} = concentration of chemical in air, $\mu g/cm^3$)

r = weight of soil/unit weight of water

K_D = soil/water adsorption coefficient, μg/g

Once partitioning is determined, determination of the importance of volatilization or mobility to the groundwater may be made. Soil photodecomposition will be of concern if the model compound remains relatively stationary within the contaminated soil. Soil strongly absorbs incident light energy and continuous soil turn over may be necessary to expose adsorbed chemicals to solar radiation.

Soil organic content would be expected to affect the partitioning of chemicals between the soil organic material and the soil water. Partitioning within soil organic matter reduces the availability of the compound for photodecomposition reactions as shown by the reduction in the photoconversion of parathion to paraoxon as the organic matter in soil increased from 0.1 percent to 4.2 percent (Spencer et al. 1980). Additional soil characteristics such as transition metal content (Nilles and Zabik 1975) and soil pigment content (Hautala 1978) have also been indicated as affecting the photochemical reactions taking place within a soil system.

Moisture content is an additional parameter affecting the relative distribution of a compound within the soil/water/air matrix within a soil system and would be expected to affect contaminant photolysis accordingly. Burkhard and Guth (1979) reported that photolysis rates of profenfos and diazinon increased slightly as dry soils were amended to 12 percent moisture content. Hautala (1978) found a significant decrease in the photolysis of carboyl in wet soils as compared to dry soils, while the photolysis of parathion was independent of moisture content.

Photoreactions of halogenated aromatic compounds have also recently been of interest because of a 2,3,7,8-tetrachlorodibenz-p-dioxin (TCDD) release near Seveso, Italy, in 1976 (Choudhry and Hutzinger 1982). Photolysis of TCDD on soil surfaces was reported by Crosby et al. (1971) in the presence of suitable hydrogen sources in the form of polar solvents, and Plimmer and Klingeble (1973) indicated that methanol used as a solvent for TCDD photooxidation experimentation also acted as a hydrogen donor in the photolysis reaction. Investigations of the use of alternative hydrogen donors for in situ treatment of contaminated surfaces in the Seveso area was reported by Wipf et al. (1978). Solutions of 80 percent olive oil and 20 percent cyclohexane at 350 l/ha and 40 percent aqueous emulsion with 4 percent biodegradable emulsifying agent at 400 l/ha were found to produce a thin film on vegetation and other smooth surfaces to provide a maximum reaction surface for TCDD photolysis. Under laboratory conditions, the oil and emulsion solutions reduced the half life of TCDD by a factor of 25 when irradiated with light of approximate solar wave-length distribution. Liberti et al. (1978) reported that a 1:1 solution of ethyl oleate and xylene used as hydrogen donors also resulted in complete TCDD degradation on building surfaces in approximately 1 hour at 2m W/cm^2 and 72 hours at 20 $\mu W/cm^2$ light intensity.

Despite the isolated observations of soil photodecomposition, little substantive information exists to allow accurate predictions of the importance of soil surfaces on general chemical photodecomposition. As Helling et al. (1971) indicated in 1971 and Miller and Zepp (1983) repeated 12 years later, the effects of soil sensitization, reaction quenching, radical formation, and light screening and the soil characteristics such as particle size, organic content, and temperature which alter these effects remain poorly understood. A great deal of additional basic data is needed if soil photolysis rate and magnitude predictions are to be possible in the future.

Photolysis in Air--

Because of the relatively high volatility of pesticides and their transport by the air medium, vapor phase photochemistry of these compounds has been studied extensively. Reviews of general photooxidation mechanisms for pesticides have been presented (Crosby 1971, Plimmer 1971) and indicate that oxidation is the most widespread reaction mechanism of pesticides under normal environmental conditions. Vapor phase studies have been hampered, however, by a lack of general experimental protocol for reactor design, reaction kinetics, light source selections, and photoproduct collection and analysis.

Seiber et al. (1975) described methods for evaluating vapor phase photodecomposition of trifluralin on a lab and field scale, identifying four photodecomposition products of the parent compound. The decomposition products were also found within the soil matrix, preventing an absolute identification of them as vapor phase products or volatilized soil phase products. Definite vapor phase production of photodegradation products of DDT (Crosby and Moilanen 1977), and trifluralin and parathion (Woodrow et al. 1978) have been observed in the laboratory. Woodrow et al. (1983) have recently reviewed pesticide vapor-phase photochemistry and presented information concerning observed photochemical vapor phase reactions under laboratory and field conditions for a number of additional pesticides as shown in Table 3-62. Some compounds require ozone, an hydroxyl radical, or hydrocarbon free radicals to photoreact while others display a reaction rate enhancement in the presence of an oxidant (Table 3-62). Pesticides not amenable to photodecomposition included toxaphene, methylbromide, and sulfuryl fluoride (Vikane).

Polynuclear aromatic hydrocarbons (PNA) are known to readily decompose in the air medium upon reaction with ozone, NO_x and SO_x (NAS 1972). PNAs are not readily volatile as a class due to their partitioning into the soil medium, making vapor phase photodecomposition of these materials in soils under ambient conditions an insignificant pathway for their degradation.

Information regarding the photooxidation of other organic compounds within the vapor phase is generally lacking but information from the pesticide investigations indicates that major photoreactions within the atmosphere are secondary reactions, i.e., reaction of the model compound with photochemically produced reactive species such as radicals, ozone, singlet oxygen, etc. The major secondary reactions that occur are those between these organic vapors and the OH radical or ozone, of which the OH radical is the photoactive species of greatest reactivity (Lamaire et al. 1982). The international standard hydroxide radical environmental concentration of 4×10^5 molecules/cm^3 (Lamaire et al. 1982) provides a basis for estimating lifetimes of compounds within the atmosphere if reaction rates of the compound, or its homologs, with the hydroxide radical are known.

The half life of the chemical in air may be determined as follows:

1. The reaction rate expression takes the form

$$dC/dt = -k_{OH^\bullet}[OH^\bullet]_{env}\,[C] \qquad (3\text{-}109)$$

where

- dC/dt = rate of compound decomposition, mole/sec
- $k_{OH^\bullet}$ = bimolecular hydroxide radical rate constant, 1/mole·sec
- $[OH^\bullet]_{env}$ = environmental hydroxide radical concentration, 4×10^5 molecules/cm^3 = 6645×10^{-19} moles/cm^3
- $[C]$ = component concentration, mole/cm^3

Integration of Equation 3-109 yields:

TABLE 3-62. PESTICIDE PHOTOCHEMICAL REACTIONS IN THE VAPOR-PHASE (SUMMARY) (WOODROW ET AL. 1983)
Used by permission, see Copyright Notice

Chemical	Reaction observed in lab	Reaction observed in field	Effect of oxidant	Major pathway
Trifluralin	yes	yes	rate enhanced	>-NR_2 → >-NH(R)
Parathion	yes	yes	rate enhanced	P → S → P → O
Folex	-	yes	-	P(III) → P(V)O
DEF	yes	no	required	P(SR)O → P(OH)O
Molinate	yes	no	required	-CH_2N- → -C(O)-N- >N-C(O)-S- → >N-H
Chloropicrin	yes	yes	-	Cl_3C-NO_2 → Cl-C(O)-Cl
Aldrin	yes	-	-	>C = C< → >C(-O-)C<
Dieldrin	yes	-	-	→ photodieldrin
DDT	yes	-	-	>CH-CCl_3 → >C=CCl_2
DDE	yes	-	-	>C=CCl_2 → >C=O
Telone	yes	-	required	>C=CHCl → >CH-C(O)-Cl
Tetrachloroethylene	yes	-	-	>C=CCL_2 → >CCl-C(O)-Cl+Cl-C(O)-Cl

$$\ln C/C_o = -k_{OH^\bullet}[OH^\bullet]_{env}\ t \quad (3\text{-}110)$$

2. Component half life occurs when $c/c_o = 1/2$ therefore

$$\ln 1/2 = -0.693 = -k_{OH^\bullet}[OH^\bullet]_{env}\ t_{1/2} \quad (3\text{-}111)$$

$$t_{1/2} = 0.693/(k_{OH^\bullet}[OH^\bullet]_{env}) \quad (3\text{-}112)$$

Table 3-63 contains a number of $k_{OH^\bullet}$ values for various compounds in air as presented by Lamaire et al. (1982) and Cupitt (1980). Additional $k_{OH^\bullet}$ values will be required to allow the evaluation of atmospheric residence time of the full range of materials anticipated to be generated from hazardous waste sites.

Biological Interactions--

Biological interactions in terms of enhancement of volatility of biological metabolic end products, and the impacts of photodecomposition products on biological activity become important in the evaluation of the significance and applicability of vapor phase reactions for the treatment of materials at hazardous waste sites.

The major evaluation of photodecomposition products has been related to atmospheric reaction of ozone, NO_x and SO_x with the numerous hydrocarbons found in urban air. These analyses are concerned with the impact of the oxidizer/hydrocarbon reactions on photochemical smog production (National Academy of Sciences 1976). Of concern, from a hazard mitigation standpoint, is the production of hazardous compounds as photodecomposition products of parent compounds of less hazard. Such occurrences have been documented, e.g., aldrin photooxidation to the more toxic doeldrin and paraoxon formation from parathion (Crosby 1971) along with the detection of phosgene from the photooxidation of chloropicrin, and the formation of PCBs from the photoreaction of DDT (Woodrow et al. 1983), and the potential for such occurrences with additional parent compounds would be expected to be high. Full evaluation of the potential for hazardous photodecomposition product formation is not possible at this time due to a lack of general information related to the nature of photodecomposition products formed, and further research is needed in this area.

Studies of the enhancement of compound volatility due to biological activity are few and a general lack of information exists. Studies concerning the enhanced volatility of mercury from soils due to biological activity have recently been reported (Landa 1978, Rogers and McFarlane 1979) and only through more research related to microbial degradation/vapor phase interaction will an understanding of this process be obtained.

The use of photochemical reactions for the enhancement of compound biodegradation is an additional area of interest for hazard mitigation from hazardous waste sites. The major photolysis process in the vapor phase is one of compound oxidation and it would be expected to aid in microbial degradation through the oxidation of resistant complex structures (Crosby 1971, Sims and Overcash 1983). Information regarding this aspect of the vapor phase photolysis process is also lacking and research related to aqueous phase photolysis microbial degradation enhancement presently underway (Dupont and Sims 1983) may shed light upon its usefulness as a treatment option in soil systems.

TABLE 3-63. RATE CONSTANTS FOR THE HYDROXIDE RADICAL REACTION IN AIR WITH VARIOUS ORGANIC SUBSTANCES, $K_{OH^{\circ}}$ IN UNITS OF /MOLE-SEC (ADOPTED FROM LEMAIRE ET AL. (1980) AND CUPITT (1980))

Substance	log $k^{OH^{\circ}}_{air}$
Acetaldehyde	9.98
Acrolein	10.42
Acrylonitrile	9.08
Allyl chloride	10.23
Benzene	8.95
Benzyl chloride	9.26
Bis(chloromethyl)ether	9.38
Carbon tetrachloride	<5.78
Chlorobenzene	8.38
Chloroform	7.78
Chloromethyl methyl ether	9.26
Chloroprene	10.44
o-,m-,p-cresol*	10.52
p-cresol	10.49
Dichlorobromobenzene*	8.26
Diethyl ether	9.73
Dimethyl nitrosamine	10.37
Dioxane	9.26
Epichlorohydrin	9.08
1,2-epoxybutane	9.16
Epoxypropane	8.89
Ethanol	9.28
Ethyl acetate	9.06
Ethyl propionate	9.03
Ethylene dibromide	8.18
Ethylene dichloride	8.12
Ethylene oxide	9.08
Formaldehyde	9.78
Hexachlorocyclopentadiene	10.55
Maelic anhydride	10.56
Methanol	8.78
Methyl acetate	8.04
Methyl chloroform	6.86
Methyl ethyl ketone	9.32
Methylene chloride	7.93
Methyl iodide	6.38
Methyl propionate	8.23
Nitrobenzene	7.56
Nitromethane	8.81
2-nitropropane	10.52
n-nitrosodiethylamine	10.19
Nitrosoethylurea	9.89

TABLE 3-63. CONTINUED

Substance	$\log k_{air}^{OH^o}$
n-propylacetate	9.41
Perchloroethylene	8.01
Phenol	10.01
Phosgene	nonreactive
Polychlorinated biphenyls	<8.78
Propanol	9.51
Propylene oxide	8.89
Tetrahydrofuran	9.95
Toluene	9.52, 9.56
Trichloroethylene	9.12
Vinylidene chloride	9.38
o-,m-,p-xylene*	9.98

REFERENCES

Acher, A. J., and S. Saltzman. 1980. Dye-sensitized photooxidation of bromacil in water. Jour. Environ. Qual. 9(2):190-194.

Adriano, D. C., and H. E. Doner. 1982. Bromine, chlorine, and fluorine. p. 479-483. In A. L. Page, R. H. Miller, and D. R. Keeney (eds.) Methods of soil analysis. Part 2. Chemical and microbiological properties. Second edition. Agronomy Monograph No. 9(2), American Society of Agronomy, Inc., Madison, WI.

Agricultural Research Service. 1974. Factors involved in land application of agricultural and municipal wastes. ARS-USDA, Soil, Water and Air Sciences, Beltsville, MD.

Ahlrichs, J. L. 1972. The soil environment. p. 3-26. In C. A. I. Goring and J. W. Hamaker (eds.) Organic chemicals in the soil environment. Volume 1. Marcel Dekker, Inc., New York, NY.

Alexander, M. 1971. Microbial ecology. John Wiley and Sons, New York, NY. 511 p.

Alexander, M. 1973. Nonbiodegradable and other recalcitrant molecules. Biotechnology and Bioengineering 15:611-647.

Alexander, M. 1977. Introduction to soil microbiology. Second edition. John Wiley and Sons, New York, NY. 467 p.

Alexander, M. 1981. Biodegradation of chemicals of environmental concern. Science 211:132-138.

Amoozegar-Fard, A., D. R. Nielsen, and A. W. Warrick. 1982. Soil solute concentration distributions for spatially varying pore-water velocities and apparent diffusion coefficients. Soil Sci. Soc. Am. J. 46:3-9.

Anderson, D. C. 1981. Organic leachate effects on the permeability of clay soils. M. S. Thesis. Texas A & M University, College Station, Texas.

Anderson, D. C., and K. W. Brown. 1981. Organic leachate effects on the permeability of clay liners. p. 119-130. In D. W. Schultz (ed.) Proc. Seventh Ann. Research Symp., EPA-600/9-81-002B.

Anderson, D. C., K. W. Brown, and J. Green. 1982. Organic solvent effects on the permeability of clay soils. p. 179-190. In D. W. Schultz (ed.) Land disposal of hazardous waste. EPA-600/9-82-002.

Anderson, D. C., C. Smith, and K. W. Brown. 1983. Fate of constituents in the soil environment (Chapter 6). In K. W. Brown, Gordon Evans, and B. D. Frentrup (eds.) Hazardous waste land treatment. Ann Arbor Science, Ann Arbor, MI. 692 p.

Anderson, J. P. E., and K. H. Domsch. 1976. Microbial degradation of the thio-carbamate herbicide, diallate, in soils and by pure cultures of soil microorganisms. Arch. Environ. Contam. Toxicology 4:1.

Anderson, W. P. 1977. Weed science: Principles. West Publishing Co., St. Paul, MN.

Armstrong, D. E., G. Chesters, and R. F. Harris. 1967. Atrazine hydrolysis in soil. Soil Sci. Soc. Am. Proc. 31:61-66.

Artiole, J. and W. H. Fuller. 1979. Effect of crushed limestone barriers on chromium attenuation in soils. J. Environ. Qual. 8:503-510.

Atlas, R. M., and R. Bartha. 1981. Microbial ecology; fundamentals and applications. Addison-Wesley Publishing Company, Reading, MA. 560 p.

Babich, H., and G. Stotzky. 1979. Abiotic factors affecting the toxicity of lead to fungi. Appl. Environ. Microbiol. 38:506-513.

Babich, H., and G. Stotzky. 1982. Toxicity of nickel to microorganisms in soil: Influence of some physicochemical characteristics. Environ. Pollut. (Ser. A) 29:303-315.

Babich, H., and G. Stotzky. 1983a. Nickel toxicity to microbes and a bacteriophage in soil and aquatic ecosystems: Mediation by environmental characteristics. p. 261. In Abstracts of the Annual Meeting of the ASM, American Society for Microbiology, Washington, DC.

Babich H., and G. Stotzky. 1983b. Abiotic factors affecting nickel toxicity: Pure culture studies. p. 261. In Abstracts of the Annual Meeting of the ASM, American Society for Microbiology, Washington, DC.

Babich, H., M. Schiffenbauer, and G. Stotzky. 1982. Comparative toxicity of trivalent and hexavalent chromium to fungi. Bull. Environ. Contam. Toxicol. 28:452-459.

Baes, C. F., and R. D. Sharp. 1983. A proposal for estimation of soil leaching and leaching constants for use in assessment models. J. Environ. Qual. 12:17-28.

Baham, J., N. B. Ball, and G. Sposito. 1978. Gel filtration studies of trace metal-fulvic acid solutions extracted from sewage sludge. J. Environ. Qual. 7:181-188.

Baham, J., and G. Sposito. 1983. Chemistry of water-soluble metal complexing ligands extracted from anaerobically digested sewage sludge. J. Environ. Qual. 12:96-100.

Bailey, G. W., and J. L. White. 1970. Factors influencing the adsorption, desorption, and movement of pesticides in soils. Residue Reviews 32:29-93.

Baker, M. D., and C. I. Mayfield. 1980. Microbial and nonbiological decomposition of chlorophenols and phenols in soil. Water Air Soil Poll. 13:411-424.

Banting, J. D. 1967. Factors affecting the activity of diallate and triallate. Weed Res. 7:302.

Bartha, R. 1980. Pesticide residues in humus. ASM News 46:356-360.

Bartlett, R. S., and J. M. Kimble. 1976a. Behavior of chromium in soils: I. Trivalent forms. J. Environ. Qual. 5:379-383.

Bartlett, R. J., and J. M. Kimble. 1976b. Behavior of chromium in soils: II. Hexavalent forms. J. Environ. Qual. 5:383-386.

Bartlett, R. J., and B. James. 1979. Behavior of chromium in soils: III. Oxidation. J. Environ. Qual. 8:31-35.

Beam, H. W., and J. J. Perry. 1973. Co-metabolism as a factor in microbial degradation of cycloparaffinic hydrocarbons. Arch. Mikrobiol. 91:87-90.

Beam, H. W., and J. J. Perry. 1974. Microbial degradation of cyloparaffinic hydrocarbons via cometabolism and commensalism. J. Gen. Microbiol. 82:163-169.

Bear, J. 1972. Dynamics of fluids in porous media. Elsevier, New York, NY.

Becker, C. D., J. E. Rogers, D. J. Silviera, D. A. Neitzel, and C. E. Cushing. 1982. EPA field guide for biological treatment of hazardous wastes in the environment. PNL-4601, Prepared by Pacific Northwest Laboratory for U.S. Environmental Protection Agency, Corvallis, OR.

Beckman Instruments, Inc. 1982. Beckman Microtox™ system operation manual.

Behel, D., D. W. Nelson, and L. E. Sommers. 1983. Assessment of heavy metals equilibria in sewage sludge-treated soil. J. Environ. Qual. 12:181-186.

Benjamin, M. M., and J. O. Leckie. 1982. Effects of complexation by Cl, SO_4, and S_2O_3 on adsorption behavior of Cd on oxide surfaces. Environ. Sci. Technol. 16:162-170.

Biggar, J. W., U. Mingelgrin, and M. W. Cheung. 1978. Equilibrium and kinetics of adsorption of picloram and parathion with soils. J. Agr. Food Chem. 26:1306-1312.

Biggar, J. W., and D. R. Nielsen. 1976. Spatial variability of the leaching characteristics of a field soil. Water Resour. Res. 12:78-84.

Biggar, J. W., and D. R. Nielsen. 1980. Mechanisms of chemical movement in soils. In A. Banin and U. Kafkafi (eds.) Agrochemicals in soils. Pergamon Press, London.

Blackman, W. C., Jr., B. E. Benson, K. E. Fischer, and V. Z. Hardy. 1980. Enforcement and safety procedures for evaluation of hazardous waste disposal sites. p. 91-95. In Management of uncontrolled hazardous waste sites. Hazardous Materials Control Research Institute, Silver Spring, MD.

Bloomfield, C., and G. Pruden. 1980. The behavior of Cr(VI) in soil under aerobic and anaerobic conditions. Environ. Pollut. Ser. A:103-114.

Bohn, H. L., B. L. McNeal, and G. A. O'Conner. 1979. Soil chemistry. John Wiley and Sons, New York.

Bollag, J.-M., and S.-Y. Liu. 1983. Incorporation of pentachlorophenol into humic acid polymers by a fungal laccase. p. 273. In Abstracts of the Annual Meeting of the ASM, American Society for Microbiology, Washington, DC.

Bowman, B. T. 1978. Effect of fulvic acid on adsorption of methyl parathion and parathion by Ca and Fe montmorillonite suspensions. Soil Sci. Soc. Am. J. 42:441-446.

Bowman, B. T., and W. W. Sans. 1982. Influence of methods of pesticide application on subsequent desorption from soils. J. Agr. Food Chem. 30:147-150.

Bowman, R. S., M. E. Essington, and G. A. O'Connor. 1981. Soil sorption of nickel: Influence of solution composition. Soil Sci. Soc. Am. J. 45:860-865.

Boyd, S. A., L. E. Sommers, and D. W. Nelson. 1979. Infrared spectra of sewage sludge fractions: Evidence for an amide binding site. Soil Sci. Soc. Am. J. 43:893-899.

Brady, N. C. 1971. Advances in Agronomy. Vol. 23. Academic Press, Inc., New York, NY.

Brady, N. C. 1974. The nature and properties of soils. Eighth edition. Macmillan Publishing Co., Inc., New York, NY.

Bremner, J. M., and C. S. Mulvaney. 1982. Nitrogen - total. p. 595-624. In A. L. Page, R. H. Miller, and D. R. Keeney (eds.) Methods of soil analysis. Part 2. Chemical and microbiological properties. Second edition. Agronomy Monograph No. 9(2), American Society of Agronomy, Inc., Madison, WI.

Bresler, E. 1973. Simultaneous transport of solute and water under transient unsaturated flow conditions. Water Resour. Res. 9:975-986.

Bresler, E., and R. J. Hanks. 1969. Numerical methods for estimating simultaneous flow of water and salt in unsaturated soils. Soil Sci. Soc. Am. Proc. 33:827-832.

Briggs, G. G. 1981. Theoretical and experimental relationships between soil adsorption, octanol-water partition coefficients, water solubilities, bioconcentration factors, and the parachor. J. Agr. Food Chem. 29:1050-1059.

Brown, K. W. 1975. An investigation of the feasibility of soil disposal of waste water for the Jefferson Chemical Co., Monroe, Texas. Final

Report Phase I, Texas A & M Research Foundation. P. O. Box H. College Station, TX.

Brown, K. W., and D. C. Anderson, 1983. The influence of selected organics liquid on the permeability of soils. EPA-600/2-83-016.

Brown, K. W. and Associates Inc. 1980. Hazardous waste land treatment. Draft edition. SW-874. U.S. Environmental Protection Agency, Cincinnati, OH.

Brown, K. W., J. Green, and J. Thomas. 1983. The influence of selected organics liquid on the permeability of clay liners. In D. W. Schultz (ed.) Land disposal, incineration, and treatment of hazardous waste (In Press).

Bryant, R. J., L. E. Woods, D. C. Coleman, B. C. Fairbanks, J. F. McClellan, and C. V. Cole. 1982. Interactions of bacterial and amoebal populations in soil microcosms with fluctuating moisture content. Appl. Environ. Microbiol. 43:747-752.

Buchanan, P. N. 1964. Effect of temperature and adsorbed water on permeability and consolidation characteristics of sodium and calcium montmorillonite. Ph.D. Dissertation. Texas A & M University, College Station, TX.

Burkhard, N., and J. A. Guth. 1979. Photolysis of organophosphorous insecticides on soil surfaces. Pest. Sci. 10, 313.

Caldwell, S., K. W. Barrett, and S. S. Chang. 1981. Ranking system for releases of hazardous substances. p. 14-20. In Management of uncontrolled hazardous waste sites. Hazardous Materials Control Research Institute. Silver Spring, MD.

Callister, S. M., and M. R. Winfrey. 1983. Microbial mercury resistance and methylation rates in the upper Wisconsin River. p. 261. In Abstracts of the Annual Meeting of the American Society for Microbiology, Washington, DC.

Calvet, R. 1980. Adsorption-desorption phenomena. In Interaction between herbicides and the soil Academic Press, NY.

Cameron, D. A., and A. Klute. 1977. Convective-dispersive solute transport with a combined equilibrium and kinetic adsorption model. Water Resour. Res. 13:183-188.

Cavallaro, N., and M. B. McBride. 1978. Copper and cadmium adsorption characteristics of selected acid and calcareous soils. Soil Sci. Soc. Am. J. 42:550-556.

Cavallaro, N., and M. B. McBride. 1980. Activities of Cu^{2+} and Cd^{2+} in soil solutions as affected by pH. Soil Sci. Soc. Am. J. 44:729-732.

Cerniglia, C. E., R. L. Herbert, R. H. Dodge, P. J. Szaniszlo, and D. T. Gibson. 1979. Some approaches to studies on the degradation of aromatic hydrocarbons by fungi. p. 360. In A. L. Bourquin and H. Pritshard (eds.) Microbial degradation of pollutants in marine environments. EPA-600/9-79-012.

Chakrabarty, A. M. 1980. Plasmids and dissimilation of synthetic environmental pollutants. p. 21-30. In C. Stuttard and K. R. Rozee (eds.) Plasmids and transposons, environmental effects and maintenance mechanisms. Academic Press, New York, NY.

Chakrabarty, A. M. 1982. Genetic mechanisms in the dissimilation of chlorinated compounds. p. 127-139. In Biodegradation and detoxification of environmental pollutants. CRC Press, Inc., Boca Ratov, FL.

Chambers, C. W., C. M. Porter, and P. W. Kabler. 1962. Bactericidal effect of low concentrations of silver. J. Am. Water Works Assoc. 54:208-216.

Chaney, W. R., J. M. Kelly, and R. C. Strickland. 1978. Influence of cadmium and zinc on carbon dioxide evolution from litter and soil from a black oak forest. J. Environ. Qual. 7:115-119.

Chatterjee, D. K., S. T. Kellogg, D. R. Watkins, and A. M. Chakrabarty. 1981. Plasmids in the biodegradation of chlorinated aromatic compounds p. 519-528. In S. B. Levy, R. C. Clowes, and E. L. Koenig (eds.) Molecular biology, pathogenicity and ecology of bacterial plasmids. Plenum Press, New York, NY.

Chendrayan, K., and N. Sethunathan. 1980. Effects of HCH, carbaryl, benomyl, and atrazine on dehydrogenase activity in flooded soil. Bull. Environ. Contam. Toxicol. 24:379-382.

Chepil, W. S., and N. P. Woodruff. 1963. The physics of wind erosion and its control. Advan. Agron. 15:211-302, Academic Press, New York, NY.

Cheng, C.-N., and D. D. Focht. 1979. Production of arsine and methylarsines in soil and in culture. Appl. Environ. Microbiol. 38:494-498.

Cheung, M. W., U. Mingelgrin, and J. W. Biggar. 1979. Equilibrium and kinetics of desorption of picloram and parathion in soils. J. Agr. Food Chem. 27:1201-1206.

Chiou, C. T., L. J. Peters, and D. W. Schmedding. 1979. A physical concept of soil-water equilibria for nonionic organic compounds. Science 206: 831-832.

Chiou, C. T., P. E. Porter, and D. W. Schmedding. 1983. Partition equilibria of nonionic organic compounds between soil organic matter and water. Environ. Sci. Technol. 17:227-231.

Chiou, C. T., D. W. Schmedding, and M. Manes. 1982. Partitioning of organic compounds in octanol-water systems. Environ. Sci. Technol. 16:4-10.

Choudhry, G. G., and O. Hutzingher. 1982. Photochemical formation and degradation of polychlorinated dibenzofurans and dibenzo-p-dioxins. Residue Reviews 84:111-161.

Christensen, D. C., and W. C. Weimer. 1979. Enhanced degradation of persistent halogenated organic materials. 34th Annual Purdue Industrial Waste Conference Proceedings, p. 160-166.

Clapp, R. B., and G. M. Hornberger. 1978. Empirical equations for some soil hydraulic properties. Water Resour. Res. 14:601-604.

Clark, F. E. 1967. Bacteria in soil. p. 15-49. In A. Burges and F. Raw (eds.) Soil biology. Academic Press, New York, NY.

Cook, M. G., and G. M. Pace. 1978. Concepts in soil science. North Carolina State University, Raleigh, NC.

Cook, R. J., and R. I. Papendick. 1970. Soil water potential as a factor in the ecology of Fusarium roseum f. sp. cerealis "colmorum." Plant Soil 32:131-145.

Corbin, M. 1980. General considerations for hazardous waste management facilities. p. 158-215. In A. A. Metry (ed.) The handbook of hazardous waste management. Technomic Publishing Col., Inc., Chicago, IL.

Cowart, R. P., F. L. Bonner, and E. A. Epps, Jr. 1971. Rate of hydrolysis of seven organophosphate pesticides. Bull. Environ. Contam. Toxicol. 6:231-234.

Cox, D. P., and M. Alexander. 1973. Production of trimethyl-arsine gas from various arsenic compounds by three sewage fungi. Bull. Environ. Contam. Toxicol. 9:84-88.

Cripps, R. E., and T. R. Roberts. 1978. Microbial degradation of herbicides. p. 669-730. In I. R. Hill and S. J. L. Wright (eds.) Pesticide microbiology. Academic Press, New York, NY.

Crosby, D. G. 1971. Environmental photooxidation of pesticides in degradation of synthetic organic molecules in the biosphere. National Academy of Sciences, Washington, D.C. p. 260-290.

Crosby, D. G., and K. W. Moilanen. 1977. Vapor-phase photodecomposition of DDT. Chemosphere p. 167.

Crosby, D. G., A. S. Wong, J. R. Plimmer, and E. A. Woolson. 1971. Photodecomposition of chlorinated dibenzo-p-dioxins. Science 173:748.

Cupitt, L. T. 1980. Fate of toxic and hazardous materials in the air environment. USEPA, Athens, Georgia, EPA-600/53-80-084.

Dagan, G., and E. Bresler. 1979. Solute dispersion in unsaturated heteorogeneous soil at field scale: 1. Theory. Soil Sci. Soc. Am. J. 43:461-467.

Davidson, J. M., and R. K. Chang. 1972. Transport of picloram in relation to soil physical conditions and pure-water velocity. Soil Sci. Soc. Am. Proc. 36:257-261.

Davidson, J. M., and J. R. McDougal. 1973. Experimental and predicted movement of three herbicides in a water-saturated soil. J. Environ. Qual. 2:428-433.

Davidson, J. M., L. T. Ou, and P. S. C. Rao. 1978. Adsorption, movement, and biological degradation of high concentrations of selected pesticides in soils. EPA-600/2-78-016.

Davidson, M. M., P. S. C. Rao, R. E. Green, and H. M. Selim. 1980a. Evaluation of conceptual process models for solute behavior in soil-water systems. In A. Banin and U. Kafkafi (eds.) Agrochemicals in soils. Pergamon Press, London.

Davidson, J. M., P. S. C. Rao, L. T. Ou, W. B. Wheeler, and D. F. Rothwell. 1980b. Adsorption, movement, and biological degradation of large concentrations of selected pesticides in soils. EPA-600/2-80-124.

Davidson, J. M., P. S. C. Rao, and L. T. Ou. 1980c. Movement and biological degradation of large concentrations of selected pesticides in soils. Proc. of Sixth Annual Res. Symp. Disposal of Hazardous Waste. EPA-600/9-80-010.

Davidson, J. M., C. M. Rieck, and P. W. Santelmann. 1968. Influence of water flux and porous material on the movement of selected herbicides. Soil Sci. Soc. Am. Proc. 32:629-633.

Davies, C. W. 1962. Ion association. Butterworths, London.

Davies, J. T., and E. K. Rideal. 1963. Interfacial phenomena. Academic Press, New York.

Dawson, G. W., and S. M. Brown. 1981. New assessment methods to aid site restoration. p. 79-83. In Management of uncontrolled hazardous waste sites. Hazardous Materials Control Research Institute, Silver Spring, MD.

Dawson, G. W., B. W. Mercer, and C. H. Thompson. 1980. Strategy for the treatment of spills on land. National Conference on Control of Hazardous Materials Spills, May.

de Klerk, H., and A. C. van der Linden. 1974. Bacterial degradation of cyclohexane. Participation of a co-oxidation reaction. Antonie van Leeuwenhoek 40:7-15.

Doner, H. E. 1978. Chloride as a factor in mobilities of Ni (II), Cu(II) and Cd(II) in soil. Soil Sci. Soc. Am. J. 42:882-885.

Doner, H. E., A. Pukite, and E. Yang. 1982. Mobility through soils of certain heavy metals in geothermal brine water. J. Environ. Qual. 11:389-394.

Duguid, J. O., and M. Reeves. 1976. Material transport through porous media: A finite element Galerkin model. Oak Ridge National Laboratory. ORNL-4928.

Dupont, R. R., and R. C. Sims. 1983. Engineering treatment of oil shale wastewater for microbial degradation enhancement, reuse, and ultimate disposal. Utah Water Research Laboratory, Research in Progress. WR-83.

Edgehill, R. U., and R. K. Finn. 1983. Microbial treatment of soil to remove pentachlorophenol. Appl. Environ. Microbiol. 45:1122-1125.

Ekwall, P. K. Garth, and V. Runnstrom-Reio. 1963. Surface chemistry. Academic Press, New York.

Ellgehausen, H. C. D'Hondt, and R. Fuerer. 1981. Reversed-phased chromatography as a general method for determining octan-1-ol/water partition coefficient. Pestic. Sci. 12:219-227.

Elliott, H. A., and C. M. Denneny. 1982. Soil adsorption of cadmium from solutions containing organic ligands. J. Environ. Qual. 11:658-663.

Elprince, A. M., and G. Sposito. 1981. Thermodynamic derivation of equations of Langmuir-type for ion equilibrium in soils. Soil Sci. Soc. Am. J. 45:277-282.

Elrashidi, M. A., and G. A. O'Connor. 1982. Influence of solution composition on sorption of zinc by soil. Soil Sci. Soc. Am. J. 46:1153-1158.

Enfield, C. G., R. F. Carsel, S. Z. Cohen, T. Phan, and D. M. Walters. 1982. Approximating pollutant transport to ground water. Ground Water 20:711-722.

Esser, J., and N. El Bassam. 1981. On the mobility of cadmium under aerobic soil conditions. Environ. Pollut. Ser. A. 15-31.

Everett, L. G., E. W. Hoylman, L. G. McMillion, and L. G. Wilson. 1982. Vadose zone monitoring concepts at landfills, impoundments, and land treatment disposal areas. p. 100-105. In Management of uncontrolled hazardous waste sites. Hazardous Materials Control Research Institute, Silver Spring, MD.

Farmer, W. J., and Y. Aochi. 1974. Picloram sorption by soils. Soil Sci. Soc. Am. Proc. 38:418-423.

Faust, S. D., and I. H. Suffet. 1966. Recovery, separation and identification of organic pesticides from natural and potable waters. Residue Rev. 15:44-116.

Federal Register. 1982. Regulations for hazardous waste land treatment. 47(143):32325-32331.

Felsot, A., and P. A. Dahm. 1979. Sorption of organophosphorous and carbamate insecticides by soil. J. Agr. Food Chem. 27:557-563.

Ferguson, A. H. 1976. Acceptability of wastewater effluents by soils. p. 85-100. In R. L. Sanks and T. Asano (eds.) Land treatment and disposal of municipal and industrial wastewater. Ann Arbor Science, Ann Arbor, MI.

Flaig, W., H. Beutelspacher, and E. Rietz. 1975. Chemical composition and physical properties of humic substances. p. 1-211. In J. E. Gieseking (ed.) Soil components, Vol. 1. Springer-Verlag, New York, NY.

Fluker, B. J. 1958. Soil temperatures. Soil Sci. 86:35-46.

Focht, D. D. 1972. Microbial degradation of DDT metabolites to carbon dioxide, water and chloride. Bull. Environ. Contam. Toxicol. 7:52-56.

Focht, D. D., and W. Verstraete 1977. Biochemical ecology of nitrification and denitrification. Adv. Microbial Ecol. 1:135-214.

Forrest, M., K. A. Lord, and N. Walker. 1981. The influence of soil treatments on the bacterial degradation of diazinon and other organophosphorus insecticides. Environ. Pollut. (Ser. A) 27:93-104.

Foster, G. R. 1981. Soil erosion modeling: Special considerations for nonpoint pollution evaluation of field-sized areas. p. 213-240. In M. R. Overcash and J. M. Davidson (eds.) Environmental impact of nonpoint source pollution. Ann Arbor Science Publishers, Inc., Ann Arbor, MI.

Foth, H. D. 1978. Fundamentals of soil science. Sixth edition. John Wiley and Sons, New York, NY.

Fowkes, F. M., H. S. Benesi, L. B. Ryland, W. M. Sawyer, K. D. Detling, E. S. Loeffler, F. B. Folckremer, M. R. Johnson, and Y. F. Sun. 1960. Clay-catalyzed decomposition of insecticides. J. Agr. Food Chem. 8:203-210.

Freeze, R. A., and J. A. Cherry. 1979. Groundwater. Prentice Hall, Inc., New York, NY.

Fried, J. J. 1977. Groundwater pollution. Theory, methodology, modeling and practical rules. Elsevier, Amsterdam.

Fried, J. J., and M. A. Combarnous. 1971. Dispersion in porous media. Adv. Hydrosci. 7:169-282.

Frissel, M. J. 1961. The adsorption of some organic compounds especially herbicides on clay minerals. Versl. Landbouwk. Onderz. N. R. 67.3. Wageningen.

Frissel, M. J., and G. H. Bolt. 1962. Interactions between certain ionizable organic compounds (herbicides) and clay minerals. Soil Sci. 94:284.

Fuhremann, T. W., and E. P. Lichtenstein. 1980. A comparative study of the persistence, movement, and metabolism of six carbon-14 insecticides in soils and plants. J. Agr. Food Chem. 28:446-452.

Fuller, W. H. 1978. Investigations of landfill leachate pollutant attenuation by soils. EPA-600/2-78-158, U.S. Environmental Protection Agency, Cincinnati, OH.

Furukawa, K. 1982. Microbial degradation of polychlorinated biphenyls (PCBs). p. 33-57. In A. M. Chakrabarty (ed.) Biodegradation and detoxification of environmental pollutants. CRC Press, Inc., Boca Raton, FL.

Furukawa, K., and A. M. Chakrabarty. 1982. Involvement of plasmids in total degradation of chlorinated biphenyls. Appl. Environ. Microbiol. 44:619-626.

Gamble, D. S. 1970. Titration curves of fulvic acid: the analytical chemistry of a weak acid polyelectrolyte. Can. J. Chem. 48:2662-2669.

Gardner, W. S., R. F. Lee, K. R. Tenore, and L. W. Smith. 1979. Degradation of selected polycyclic aromatic hydrocarbons in coastal sediments: importance of microbes and polychaete worms. Water, Air, and Soil Pollution 2:339.

Geering, H. R., E. E. Cory, L. H. P. Jones, and W. H. Alloway. 1968. Solubility and redox criteria for the possible forms of selenium in soils. Soil. Sci. Soc. Am. Proc. 32:35-40.

Gelham, R. W., and J. A. Cherry. 1982. Contaminant migration in saturate unconsolidated geologic deposits. Geol. Soc. Am. Special Paper 189:31-62.

Gerritse, R. G., R. Vriesema, J. W. Dalenberg, and H. P. DeRoos. 1982. Effect of sewage sludge on trace element mobility in soils. J. Environ. Qual. 11:359-364.

Gerstl, Z., P. H. Nye, and B. Yaron. 1979. Diffusion of a biodegradable pesticide: II as affected by microbial decomposition. Soil Sci. Soc. Am. J. 43:843-848.

Gibson, D.T., and W. K. Yeh. 1973. The microbial degradation of aromatic hydrocarbons. In The microbial degradation of oil pollutants. LSU-SG-73-01. Center for Wetland Resources, Louisiana State Univ., Baton Rouge, LA.

Giles, H., T. H. MacEwan, S. N. Nakhwa, and D. Smith. 1960. Studies in adsorption. Part XI. A system of classification of solution adsorption isotherms and its use in diagnosis of adsorption mechanisms and in measurement of specific surface areas of solids. J. Chem. Soc. 111:3973-3993.

Gilmour, J. T., and N. T. Coleman. 1971. Buffer curves and acid strength of humic acid and two resins. Soil Sci. Soc. Am. Proc. 35:710-714.

Goring, C. A. I., D. A. Laskowski, J. W. Hamaker, and R. W. Miekle. 1975. Principles of pesticides degradation in soil. In R. Haque and V. H. Freed (eds.) Environmental dynamics of pesticides. Plenum, NY.

Graham-Bryce, I. G. 1967. Adsorption of disulfoton by soil. J. Sci. Food Agr. 18:72-77.

Grau, B. L, and T. J. Peterle. 1979. Compartmentalization of ^{14}C-DDT in an experimental old-field plot. Bull. Environ. Contam. Toxicol. 22:846-853.

Gray, T. R. G. 1978. Microbiological aspects of the soil, plant, aquatic, air, and animal environments: The soil and plant environments. p. 19-38. In I. R. Hill and S. J. L. Wright (eds.) Pesticide Microbiology. Academic Press, New York, NY.

Green, R. E., P. S. C. Rao, and J. C. Corey. 1972. Solute transport in aggregated soils: tracer zone shape in relation to pore-velocity distribution and adsorption. Proc. 2nd Sym. Fundamentals of Transport Phenomena in Porous Media. IAHR ISSS. Guelph. 7-11 August, 1972. Vol. 2:732-752.

Greenland, D. J. 1965a. Interaction between clays and organic compounds in soils. Part I. Mechanisms of interaction between clays and defined organic compounds. Soils and Fert. 28:415-425.

Greenland, D. J. 1965b. Interaction between clays and organic compounds in soils. Part II. Adsorption of soil organic compounds and its effect on soil properties. Soils and Fert. 28:521.

Griffin, R. A., and A. K. Au. 1977. Lead adsorption by montmorillonite using a competitive Langmuir equation. Soil Sci. Soc. Am. J. 41-880-882.

Griffin, R. A., and N. F. Shimp. 1978. Attenuation of pollutants in municipal landfill leachate by clay minerals. EPA-600/2-78-157.

Groenewegen, D., and H. Stolp. 1976. Microbial breakdown of polycyclic aromatic hydrocarbons. Tbl Bakt. Hyg. I. Abt. Orig. B162. 225.

Guenther, W. B. 1975. Chemical equilibrium: A practical introduction for the physical and life sciences. Plenum Press, New York, NY.

Guenzi, W. D. 1974. Pesticides in soil and water. Soil Sci. Soc. of Amer., Inc., Madison, WI.

Guenzi, W. D., and W. E. Bread. 1967. Anaerobic biodegradation of DDT to DDD in soil. Science 156:1116-1117.

Guenzi, W. D., and W. E. Bread. 1967. Movement and persistence of DDT and lindane in soil columns. Soil Sci. Soc. Am. Proc. 31:644-647.

Gureghian, A. B., D. S. Ward, and R. W. Cleary. 1979. Simultaneous transport of water and reacting solutes through multilayered soil under transient unsaturated conditions. J. Hydrol. 41:253-278.

Halsey, G., and H. S. Taylor. 1947. The adsorption of hydrogen on tungsten powders. J. Chem. Phys. 15:624-630.

Hamaker, J. W. 1966. Mathematical prediction of cumulative levels of pesticides in soil. Adv. Chem. Series. 60:122-131.

Hamaker, J. W. 1972. Decomposition: quantitative aspects. In C. A. I. Goring and J. W. Hamaker (eds.) Organic chemicals in soil environment. Marcel Dekker, Inc., New York.

Hamaker, J. W. 1972. Diffusion and volatilization of organic chemicals in the soil environment. C. A. I. Loring and J. W. Hamaker (eds.) Organic chemicals in the soil environment. Marcel Decker, Inc., New York.

Hamaker, J. W., and C. A. I. Goring. 1976. Turnover of pesticide residues in soil. In D. D. Kaufman et al. (eds.) Bound and conjugate pesticide residues. ACS Symp. Series 29. Washington, D. C.

Hamaker, J. W., C. A. I. Goring, and C. R. Youngson. 1966. Sorption and leaching of 4-amino-3,5,6-trichloropicolinic acids in soils. Organic Pesticides in the Environment. Adv. in Chem. 60:23-37.

Hamaker, J. W., and J. M. Thompson. 1972. Adsorption. In C. A. I. Goring and J. W. Hamaker (eds.) Organic chemicals in the soil environment. Marcel Dekker, Inc, New York.

Hance, R. J., and C. E. McKone. 1971. Effect of concentration on the decomposition rates in soil of atrazine, linuron and picloram. Pestic. Sci. 3:31-34.

Harris, C. I. 1967a. Fate of two s-triazine herbicides in soil. J. Agr. Food Chem. 15:157-162.

Harris, C. I. 1967b. Movement of herbicides in soil. Weed. 15:214-216.

Harris, C. I., and G. F. Warren. 1964. Adsorption and desorption of herbicides in soil. Weeds 12:120-126.

Harris, C. I., E. A. Woolson, and B. E. Hummer. 1969. Dissipation of herbicides at three soil depths. Weed Sci. 17:27-31.

Harris, R. F. 1981. Effect of water potential on microbial growth and activity. p. 23-95. In Water potential relations in soil microbiology. SSSA special publication number 9. Soil Sci. Soc. Am., Madison, WI.

Harter, R. D. 1979. Adsorption of copper and lead by Ap and B2 horizons of several Northeastern United States soils. Soil Sci. Soc. Am. J. 43:679-683.

Harter, R. D. 1983. Effect of soil pH on adsorption of lead, copper, zinc and nickel. Soil Sci. Soc. Am. J. 47:47-51.

Harter, R. D., and J. L. Ahlrichs. 1967. Determination of clay surface acidity by infrared spectroscopy. Soil Sci. Soc. Am. Proc. 31:30.

Hartley, G. S. 1969. Pesticidal formulations research, physical and colloidal aspects. Adv. Chem. Ser., 86:115.

Hassler, R. A. 1982. Microbial contributions to soluble and volatile arsenic dynamics in retorted oil shale. M.S. Thesis, Colorado State University, Fort Collins, CO. 73 p.

Hassett, J. J., J. C. Means, W. L. Banwart, and S. G. Wood. 1980. Sorption properties of sediments and energy related pollutants. EPA-600/3-80-041.

Hautala, R. 1978. Surfactant effects on pesticide photochemistry in water and soil. U. S. Environmental Protection Agency, Athens, Georgia, p. 175. EPA-600/3-78-060.

Hayes, M. H. B. 1970. Adsorption of triazine herbicides on soil organic matter. Including a review on soil organic matter chemistry. Residue Rev. 32:131-174.

Helling, C. S., P. C. Kearney, and M. Alexander. 1971. Behavior of pesticides in soils. p. 147-240. In C. N. Brady (ed.) Advances in agronomy. Vol. 23. Academic Press, New York.

Herbes, S. E., and L. R. Schwall. 1978. Microbial transformations of polycyclic aromatic hydrocarbons in pristine and petroleum-contaminated sediments. Applied Environ. Microbiol. 35:306.

Hindin, E. 1963. Analysis of organic pesticides by gas chromatography. Washington State University, Institute of Technology. Final report to P. H. S. Grant No. PW-00215.

Hogg, T. J., J. W. B. Steward, and J. R. Bettany. 1978. Influence of the chemical forces of mercury on its adsorption and ability to leach through soils. J. Environ. Qual. 7:440-444.

Holtzclaw, K. M., G. Sposito, and G. R. Bradford. 1976. Analytical properties of the soluble, metal-complexing fractions in sludge-soil mixtures: I. Extraction and purification of fulvic acid. Soil Sci. Soc. Am. J. 40:254.

Hopwood, D. A. 1981. The genetic programming of industrial microorganisms. Scientific American 245(3):91-102.

Hornsby, A. G., and J. M. Davidson. 1973. Solution and adsorbed fluometuron concentration distribution in a water-saturated soil: experimental and predicted evaluation. Soil Sci. Soc. Am. Proc. 37:823-828.

Horowitz, A., J. M. Sulflita, and J. M. Tiedje. 1983. Reductive dehalogenations of halobenzoates by anaerobic lake sediment microorganisms. Appl. Environ. Microbiol. 45:1459-1465.

Horrobin, S. 1963. The hydrolysis of some chloro-1,3,5-triazines. Mechanism-structure and reactivity. J. Chem. Soc. Part IV. 4130-4135.

Horvath, R. S. 1972. Microbial co-metabolism and the degradation of organic compounds in nature. Bacteriol. Rev. 36:146-155.

Horvath, R. S., and M. Alexander. 1970. Cometabolism of m-chlorobenzoate by an Arthorobacter. Appl. Microbiol. 20:254-258.

Howar, W. F. 1970. Adsorption of surfactants on montmorillonite. Clays and Clay Minerals. 18:97-105.

Hsu, T. S., and R. Bartha. J. Agr. Food Chem. 24:118-122.

Hsu, T.-S., and R. Bartha. 1979. Accelerated mineralization of two organophosphate insecticides in the rhizosphere. Appl. Environ. Microbiol. 37:36-41.

Huang, W. H., and W. D. Keller. 1971. Dissolution of clay minerals in dilute organic acids at room temperature. The Am. Mineralogist. 56:1082-1095.

Huber, W. C., and J. P. Heaney. 1981. Operational models for stormwater quality management. p. 397-444. In M. R. Overcash and J. M. Davidson (eds.) Environmental impact of nonpoint source pollution. Ann Arbor Science Publishers, Inc., Ann Arbor, MI.

Hugenberger, F., J. Letey and W. J. Farmer. 1972. Observed and calculated distribution of Lindane in soil columns as influenced by water movement. Soil Sci. Soc. Am. Proc. 36:544-548.

Hulbert, M. H., and S. Krawiec. 1977. Cometabolism: A critique. J. Theor. Biol. 69:287-291.

Hurle, K., and A. Walker. 1980. Persistence and its prediction. In R. J. Hance (ed.) Interaction between herbicides and the soil. Academic Press, London.

James, B. R., and R. J. Bartlett. 1983. Behavior of chromium in soils: V. Fate of organically complexed Cr(III) added to soil. J. Environ. Qual. 12:169-172.

James R. O., and T. W. Healy. 1972. Adsorption of hydrolyzable metal ions of the oxide-water interface. III. A thermodynamic model of adsorption. J. Colloid and Interface Sci. 40:65-81.

Jamet, P., and M. A. Piedallu. 1975. Adsorption and desorption of pyrazon (5-amino-4-chloro-2-phenyl-3-C2H1-pyridazinone) by different types of soils. Weed Res. 15:113-121.

Jenkins, D., S. A. Klein, M. S. Young, R. J. Wagenet, and J. W. Biggar. 1978. The accumulation, translocation and degradation of biocides at land wastewater treatment disposal sites: The fate of Malathion, Carbaryl, diazinon, and 2,4-D butoxethyl ester. Water Res. 12:713-723.

Jenne, E. A. 1968. Controls on Mn, Fe, Co, Ni, Cu, and Zn concentrations in soils and water the significant role of hydrous Mn and Fe oxides. Adv. in Chem 7:337-387.

Johnston, J. B., and S. G. Robinson. 1982. Opportunities for development of new detoxication processes through genetic engineering. p. 301-314. In J. H. Exner (ed.) Detoxication of hazardous waste. Ann Arbor Science, Ann Arbor, MI.

Jurinak, J. J. 1957. Adsorption of 1,2-Dibromo-3-Chloropropane vapor by soils. J. Agr. Food Chem. 5:598-601.

Kamp, P. F., and A. M. Chakrabarty. 1979. Plasmids specifying p-chlorobiphenyl degradation in enteric bacteria. p. 275-285. In K. N. Timmis and A. Puhler (eds.) Plasmids of medical, environmental and commercial importance. Elsevier/North Holland, Amsterdam, The Netherlands.

Karickhoff, S. W. 1981. Semi-empirical estimation of sorption of hydrophobic pollutants on natural sediments. Chemosphere 10:833-846.

Karickhoff, S. W., D. S. Brown, and T. A. Scott. 1979. Sorption of hydrophobic pollutants on natural sediments. Water Res. 13:241-248.

Kaufman, D. D. 1974. Degradation of pesticides by soil microorganisms. p. 133-202. In W. D. Guenzi (ed.) Pesticides in soil and water. Soil Sci. Soc. Am., Madison, WI.

Kay, B. D., and D. E. Elrick. 1967. Adsorption and movement of lindane in soils. Soil Sci. 104:314-322.

Kearns, D. R. 1971. Physical and chemical reactions of singlet oxygen. Chem. Review 71:395-425.

Keeney, D. R., and D. W. Nelson. 1982. Nitrogen-inorganic forms. p. 643-698. In A. L. Page, R. H. Miller, and D. R. Keeney (eds.) Methods of soil analysis. Part 2. Chemical and microbiological properties. Second edition. Agronomy Monograph No. 9(2), American Society of Agronomy, Inc., Madison, WI.

Kenaga, E. E., and C. A. I. Goring. 1980. Relationship between water solubility, soil sorption, octanol-water partitioning, and concentration of chemicals in biota. p. 78-115. In J. C. Eaton, P. R. Parrish, and A. C. Hendricks (eds.) Aquatic toxicology. Special publication No. 707. Am. Soc. Testing & Materials, Washington, D.C.

Khan, S., D. Nonden, and N. N. Khan. 1982. The mobility of some heavy metals through Indian red soil. Environ. Pollut. Ser. B. 119-125.

Khazael, M. B., and R. S. Mitra. 1981. Cadmium-binding component in *Eschuricia coli* during accommodation to low levels of this ion. Appl. Environ. Microbiol. 41:46-50.

Kilbane, J. J., D. K. Chatterjee, and A. M. Chakrabarty. 1983. Detoxification of 2,4,5-trichlorophenoxyacetic acid from contaminated soil by *Pseudomonas cepacia*. Appl. Environ. Microbiol. 45:1697-1700.

Kilbane, J. J., D. K. Chatterjee, J. S. Karns, S. T. Kellogg, and A. M. Chakrabarty. 1982. Biodegradation of 2,4,5-trichlorophenoxyacetic acid by a pure culture of Pseudomona cepacia. Appl. Environ. Microbiol. 44:72-78.

King, P. H., and P. L. McCarty. 1968. A chromatographic model for predicting pesticide migration in soils. Soil Sci. 106:248-261.

Kinniburgh, D. G., and M. L. Jackson. 1978. Adsorption of mercury (II) by iron hydrous oxide gel. Soil Sci. Soc. Am. J. 42:45-47.

Kinniburgh, D. G., and M. L. Jackson. 1982. Concentration and pH dependence of calcium and zinc adsorption by iron hydrous oxide gels. Soil Sci. Soc. Am. J. 46:56-61.

Kipling, J. J. 1965. Adsorption from solutions of non-electrolytes. Academic Press, New York.

Kirda, C., D. R. Nielsen, and J. W. Biggar. 1973. Simultaneous transport of chloride and water during infiltration. Soil Sci. Soc. Am. Proc. 37:339-345.

Klein, D. A., and E. M. Molise. 1975. Ecological ramifications of silver iodide nucleating agent accumulation in a semi-arid grassland environment. J. Applied Meteorology 14:673-680.

Klopffer, W. 1980. Rapid test for simulation of photo-oxidative degradation in the gas phase. Int. Workshop on Test Methods and Assessing Procedures for the Determination of the Photochemical Degradation Behavior of Chem. Substances, Berlin, Dec. 2-4.

Konrad, J. G., G. Chesters, and D. E. Armstrong. 1969. Soil degradation of malathion, a phosphorodithioate. Soil Sci. Soc. Amer. Proc. 33:259-262.

Koskinen, W. C., G. A. O'Connor, and H. H. Cheng. 1979. Characterization of hysteresis in the desorption of 2,4,5-T from soils. Soil Sci. Soc. Am. J. 43:871-874.

Kowalenko, C. G. 1978. Organic nitrogen, phosphorus and sulfur in soils. p. 95-136. In M. Schnitzer and S. U. Kahn (eds.) Soil organic matter. Elsevier Scientific Publishing Co., New York, NY.

Kozak, J., and J. B. Weber. 1983. Adsorption of five phenylurea by selected soils of Czechoslovakia. Weed Sci. 31:368-372.

Kuo, S., and A. S. Baker. 1980. Sorption of copper, zinc, and cadmium by some acid soils. Soil Sci. Soc. Am. J. 44:969-974.

Kurek, E., J. Czaban, and J.-M. Bollag. 1982. Sorption of cadmium by microorganisms in competition with other soil constituents. Appl. Environ. Microbiol. 43:1011-1015.

Lambert, S. M. 1967. Functional relationship between sorption in soil and chemical structure. Soil Sci. 106:248-261.

Landa, E. R. 1978. Microbial aspects of the volatile loss of applied mercury (II) from soils. J. Environ. Qual. 7(1):84-86.

Lapidus, L., and N. R. Amundson. 1952. Mathematics of adsorption in beds. VI. The effects of longitudinal diffusion in ion exchange and chromatographic columns. J. Phys. Chem. 56:984-988.

Lavy, T. L. 1968. Micromovement mechanisms of s-triazines in soil. Soil Sci. Soc. Am. Proc. 32:377.

Lemaire, J., I. Campbell, H. Hulpke, J. A. Guth, W. Merz, L. Philip, and C. Von Waldow. 1982. An assessment of test methods for photodegradation of chemicals in the environment. Chemosphere 11(2):119-164.

Leo, A., C. Hansch, and D. Ekins. 1971 Partition coefficients and their uses. Chemical Reviews 71:525-616.

Liang, L. M., J. L. Sinclair, L. M. Mallory, and M. Alexander. 1982. Fate in model ecosystems of microbial species of potential use in genetic engineering. Appl. Environ. Microbiol. 44:708-714.

Liberti, A., D. Brocco, I. Allegrini, and G. Bertoni. 1978. Field photodegradation of TCDD by ultra-violet radiations. p. 195-200. In F. Cattabeni, A. Cavallaro, and G. Galli (eds.) Dioxin: toxicological and chemical aspects. Spectrum Publications, Inc., New York.

Lichtenstein, E. P., L. J. DePew, E. L. Eshbaugh, and J. P. Sleesman. 1960. Persistence of DDT, aldrin, and lindane in some soils. J. Econ. Entomol. 53:136-142.

Lijinsky, W., and J. H. Quastel. 1956. Metabolism of carcinogenic hydrocarbons by soil microorgnaisms. Arch. Biochem. Biophys. 63:160.

Lindsay, W. L. 1979. Chemical equilibria in soils. John Wiley and Sons, New York.

Lindstrom, F. T., and L. Boersma. 1971. A theory on the mass transport of previously distributed chemicals in a water saturated sorbing porous medium. Soil Sci. 111:192-199.

Lindstrom, F. T., and L. Boersma. 1973. A theory on the mass transport of previously distributed chemicals in water saturated sorbing porous medium: III. Exact solution for first-order kinetic sorbtion. Soil Sci. 115:5-10.

Lindstrom, F. T., L. Boersma, and D. Stockard. 1971. A theory on the mass transport of previously distributed chemicals in a water saturated sorbing porous medium: isothermal cases. Soil Sci. 112:291-300.

Liu, D. 1982. Assessment of continuous biodegradation of commercial PCB formulations. Bull. Environ. Contam. Toxicol. 29:200-207.

MacKay, D. 1979. Finding fugacity feasible. Environ. Sci. Tech. 13, 1218.

Maes, A., and A. Cremers. 1975. Cation-exchange hysteresis in montmorillonite a pH dependent effect. Soil Sci. 119:198-202.

Mansell, R. S., H. M. Selim, P. Kanchanasut, J. M. Davidson, and J. G. A. Fiskell. 1977. Experimental and simulated transport of phosphorus through sandy soils. Water Resour. Res 13:189-194.

Marshall, C. E. 1964. The physical chemistry and mineralogy of soils. Volume 1. Wiley, New York, NY.

Martell, A. W., and R. N. Smith. 1977. Critical stability constants. Vol. 3. Plenum Press, New York.

Mason, J. W., A. C. Anderson, and M. Shariat. 1979. Role of demethylation of methymercuric chloride by Enterobacter aerogenes and Serratia marcesceus. Bull. Environ. Contam. Toxicol. 29:262-268.

Matsumura, F., and H. J. Benezet. 1978. Microbial degradation of insecticides. p. 623-667. In I. R. Hill and S. J. L. Wright (eds.) Pesticide Microbiology. Academic Press, New York, NY.

Mattigod, S. W., G. Sposito, and A. L. Page. 1981. Factors affecting the solubilities of trace metals in soils. In Chemistry in the soil environment. ASA Special Public 40.

Mattigod, S. W., and G. Sposito. 1979. Chemical modeling of trace metals equilibrium in contaminated soil solutions using the computer program GEOCHEM. In E. A. Jenne (ed.) Chemical modeling in aqueous systems. ACS No. 93 Am. Chem. Soc., Washington. D. C.

Mayer, R., W. J. Farmer, and J. Letey. 1974. Models for predicting volatilization of soil-incorporated pesticides. Soil Sci. Soc. Amer. Proc., 38(4): 563-568.

McBride, M. B., and J. J. Bhasiak. 1979. Zinc and copper solubility as a function of pH in an acid soil. Soil Sci. Soc. Am. J. 43:866-870.

McCall, P. J., D. A. Laskowski, R. L. Swann, and H. L. Dishburger. 1983. Estimation of environmental partitioning of organic chemicals in model ecosystems. Residue Reviews 85:231-244.

McDuff, R. E., and F. M. M. Morel. 1974. Description and use of the chemical equilibrium program REDEQL2. Tech. Dept. EQ-73-02 W. M. Keck Lab. (at Inst. Tech, Pasadena).

McGlamery, M. D., and F. W. Slife. 1966. The adsorption and desorption of Atrazine as affected by pH, temperature and and concentration. Weeds 14:237-239.

McKenna, E. J. 1977. Biodegradation of polynuclear aromatic hydrocarbon pollutants by soil and water microorganisms. Presented at the 70th Annual Meeting of the Am. Inst. Chem. Eng., New York, NY.

McLean, E. O. 1982. Soil pH and lime requirement. p. 199-224. In A. L. Page, R. H. Miller, and D. R. Keeney (eds.) Methods of soil analysis. Part 2. Chemical and microbiological properties. Second edition. Agronomy Monograph No. 9(2), American Society of Agronomy, Madison, WI.

Medvedev, V. A., and V. D. Davidov. 1972. Transformation of individual organic products of the coke industry in chernozemic soils. Pochvovedenie 11:22.

Mellan, I. 1977. Industrial solvents handbook. Second edition. Noyes Data Corporatión. Park Ridge, NJ. 567 p.

Merkel, G. J., and J. J. Perry. 1977. Increased cooxidative biodegradation of malathion in soil via cosubstrate enrichment. J. Agr. Food Chem. 25:1011-1012.

Milberg, R. P., D. L. Brower, and J. V. Lagerwerff. 1978. Exchange adsorption of trace quantities of cadmium in soils treated with calcium and sodium: A reappraisal. Soil Sci. Soc. Am. J. 42:892-894.

Miles, J. R. W., C. M. Tu, and C. R. Harris. 1979. Persistence of eight organophosphorus insecticides in sterile and non-sterile mineral and organic soils. Bull. Environ. Contam. Toxicol. 22:312-318.

Miller, G. C., and R. G. Zepp. 1983. Extrapolating photolysis rates from the laboratory to the environment. Residue Reviews 85:89-110.

Mingelgrin, U., and Z. Gerstl. 1983. Reevaluation of partitioning as a mechanism of nonionic chemicals adsorption in soils. J. Environ. Qual. 12:1-11.

Mitra, R. S., and I. A. Bernstein. 1978. Single-strand breakage in DNA of Eschericia coli exposed to Cd^{2+}. Bacteriol. 133:75-80.

Morrill, L. G., B. C. Mahilum, and S. H. Mohinddin. 1982. Organic compounds in soil: Sorption, degradation and persistence. Ann Arbor Science Publishing Inc., Michigan.

Munnecke, D. M., L. M. Johnson, H. W. Talbot, and S. Barik. 1982. Microbial metabolism and enzymology of selected pesticides. p. 1-32. In A. M. Chakrabarty (ed.) Biodegradation and detoxification of environmental pollutants. CRC Press, Inc., Boca Raton, FL.

Murray, D. S. P. W. Santelman, and J. M. Davidson. 1975. Comparative adsorption, desorption, and mobility of dipropetryn and prometryn in soil. J. Agr. Food Chem. 23:578-582.

National Academy of Sciences, National Research Council. 1972. Particulate polycyclic organic matter. Washington, D.C. National Academy of Sciences.

National Academy of Sciences, National Research Council. 1976. Vapor-phase organic pollutants, volatile hydrocarbons and oxidation products. Washington, DC.

Neilson, A. H., A.-S. Allard, P.-A. Hynning, M. Remberger, and L. Landner. 1983. Bacterial nethylation of chlorinated phenols and quaiacols: Formation of veratroles from quaiacols and high molecular weight liguin. Appl. Environ. Microbiol. 45:774-783.

Nelson, D. W., and L. E. Sommers. 1982. Total carbon, organic carbon, and organic matter. p. 539-579. In A. L. Page, R. H. Miller, and D. R. Keeney (eds.) Methods of soil analysis. Part 2. Chemical and micro-biological properties. Second edition. Agronomy Monograph No. 9(2), American Society of Agronomy, Inc., Madison, WI.

Nielsen, D. R., J. W. Biggar, and C. S. Simmons. 1981a. Mechanisms of solute transport in soils. In I. K. Iskandar (ed.) Modeling wastewater renovation land treatment. Wiley-Interscience, New York.

Nielsen, D. R., J. Metthey, and J. W. Biggar. 1981b. Soil hydraulic properties, spatial variability, and soil-water movement. In I. K. Iskandar (ed.) Modeling wastewater renovation land treatment. Wiley-Interscience, New York.

Nkedi-Kizza, P. P. S. C. Rao, and J. W. Johnson. 1983. Adsorption of diruon and 2,4,5-T on soil particle separates. J. Environ. Qual. (In press).

Nriagu, J. O. 1974. Lead orthophosphates. IV. Formation and stability in the environment. Geochim. Cosmochim. Acta. 38:887-898.

Nilles, G. P., and M. J. Zabik. 1975. Photochemistry of bioactive compounds. Multiple photodegradation and mass spectral analysis of basagran. J. Agr. Food Chem. 23:410.

Oddson, J. K., J. Letey, and L. V. Weeks. 1970. Predicted distribution of organic chemicals in solution and adsorbed as a function of position and time for various chemical and soil properties. Soil Sci. Soc. Am. Proc. 34:412-417.

Osgerby, J. M. 1970. Sorption of unionized pesticides by soils. In Sorption and transport processes in soil. Soc. Chem. Ind. Monograph 37:63-78.

Otis, R. J. 1983. Site evaluation for onsite treatment and disposal systems. p. C:1-39. In Design workshop: Small alternative wastewater systems. U. S. Environmental Protection Agency, West Virginia State University and EPA Small Wastewater Flows Clearinghouse. Presented in Seattle, WA, March 21-24.

Overcash, M. R., and D. Pal. 1979. Design of land treatment systems for industrial wastes--theory and practice. Ann Arbor Science Publishers, Ann Arbor, MI.

Overcash, M. R., J. B. Weber, and M. L. Miles. 1982. Behavior of organic priority pollutants in the terrestrial system. Di-N-butyl phthalate ester, toluene, and 2,4-dinitrophenol. Report No. 171. Water Resources Research Institute of the University of North Carolina, North Carolina State University, Raleigh, NC. 94 p.

Page, A. L., R. H. Miller, and D. R. Keeney (eds.). 1982. Methods of soil analysis. Part 2. Chemical and microbiological properties. Second edition. Agronomy Monograph No. 9(2), American Society of Agronomy, Inc., Madison, WI.

Painter, H. A. 1977. Microbial transformation of inorganic nitrogen. Prog. Wat. Technol. 8:3-29.

Pal, D., M. R. Overcash, and P. W. Westerman. 1977. Land disposal of acidic basic and salty wastes from industries. p. 151-159. In Proc. 1977 National Conference on Treatment and Disposal of Industrial Wastewaters and Residues.

Pal, D., J. B. Weber, and M. R. Overcash. 1980. Fate of polychlorinated biphenyls (PCBs) in soil-plant systems. Residue Reviews 74:45-98.

Pan-Hou, H. S. K., M. Hosono, and N. Imura. 1980. Plasmid-controlled mercury biotransformation by Clostridium cochleariun T-2. Appl. Environ. Microbiol. 40:1007-1011.

Paris, D. F., N. L. Wolfe, W. C. Steen, and G. L. Baughman. 1983. Effect of phenol molecular structure on bacterial transformation rate constants in pond and river samples. Appl. Environ. Microbiol. 45:1153-1155.

Parr, J. F., and S. Smith. 1973. Degradation of trifluralin under laboratory conditions and soil anaerobiosis. Soil Sci. 115:55-63.

Parr, J. F., and S. Smith. 1976. Degradation of toxaphene in selected anaerobic soil environments. Soil Sci. 121:52-57.

Parris, G. E. 1980. Covalent bonding of aromatic amines to humates. 1. Reactions with carbonyls. Environ. Sci. Technol. 14:1099-1106.

Pemberton, J. M., B. Corney, and R. H. Don. 1979. Evaluation and spread of pesticide degrading ability among soil microorganisms. p. 287-299. In K. N. Timmis and A. Puhler (eds.) Plasmids of medicial, enviornmental, and commercial importance. Elsevier/North Holland Biomedical Press, Amsterdam, The Netherlands.

Perry, J. J. 1979. Microbial cooxidation involving hydrocarbons. Microbiological Reviews 43:59-72.

Peterson, J. R., R. S. Adams, Jr., and L. K. Cutkomp. 1971. Soil properties influencing DDT bioactivity. Soil Sci. Soc. Am. Proc. 35:72-78.

Petrilli, F. L., and S. De Flora. 1977. Toxicity and mutagenicity of hexavalent chromium on Salmonella typhimurium. Appl. Environ. Microbiol. 33:805-809.

Phung, T., L. Barker, D. Ross, and D. Bauer. 1978. Land cultivation of industrial wastes and municipal solid wastes: State-of-the-art study. Volume I: Technical summary and literature review. EPA-600/2-78-140a. U.S. Environmental Protection Agency, Cincinnati, OH.

Pierce, G. E. 1982. Potential role of genetically engineered microorganisms to degrade toxic chlorinated hydrocarbons. p. 315-322. In J. H. Exner (ed.) Detoxication of hazardous waste. Ann Arbor Science, Ann Arbor, MI.

Plimmer, J. R. 1971. Principles of photodecomposition of pesticides. p. 279-290. In Degradation of synthetic organic molecules in the biosphere. National Academy of Sciences, Washington, DC.

Plimmer, J. R., and U. I. Klingebiel. 1973. Photochemistry of dibenzo-p-dioxins. p. 44-54. In E. H. Blair (ed.) Chlorodioxins--origin and fate. American Chemical Society, Washington, DC.

Poglazova, M. N., G. E. Fedoseeva, and A. J. Khesina. 1967. Further investigations of the decomposition of benz(a)pyrene by soil bacteria. Doklady Akademii Nauk SSSR. 176:1165.

Poglazova, M. N., G. E. Fedoseeva, and A. J. Khesina. 1968. The oxidation of benz(a)pyrene by microorganisms in relation to its concentration in the medium. Doklay Akademii Nauk SSSR. 179:1460.

Porter, L. K., and W. E. Bread. 1968. Retention and volatilization of lindane and DDT in the presence of organic colloids isolated from soils and leonardite. J. Agr. Food Chem. 16:344-347.

Rao, P. S. C., and J. M. Davidson. 1979. Adsorption and movement of selected pesticides at high concentrations in soils. Water Res. 13:375-380.

Rao, P. S. C., and J. M. Davidson. 1980. Estimation of pesticide retention and transformation parameters required in nonpoint source pollution models. p. 23-67. In M. R. Overcash and J. M. Davidson (eds.) Environmental impact of nonpoint source pollution. Ann Arbor Sci. Pub. Inc., Ann Arbor, MI.

Rao, P. S. C., J. M. Davidson, R. E. Jessup, and H. M. Selim. 1979. Evaluation of conceptual models of describing nonequilibrium adsorption-desorption of pesticides during steady-flow in soils. Soil Sci. Soc. Am. J. 43:22-28.

Rao, P. S. C., and R. E. Jessup. 1982. Development of verification of simulation models for describing pesticides dynamic in soils. Ecological Modeling. 16:67-75.

Rao, P. S. C., and R. E. Jessup. 1982. Sorption and movement of pesticides and other toxic organic substances in soils. In D. W. Nelson, K. Tanji, and D. E. Elrick (eds.) Chemical mobility and reactivity in soil systems. American Society of Agronomy Special Publication (in press).

Rao, P. S. C., R. E. Jessup, D. E. Rolston, J. M. Davidson, and D. P. Kilcrease. 1980a. Experimental and mathematical desorption of nonadsorbed solute transfer by diffusion in spherical aggregates. Soil Sci. Soc. Am. J. 44:684-688.

Rao, P. S. C., P. Nkedi-Kizza, and J. M. Davidson. 1982. Retention and transformation of pesticides in relation to nonpoint source pollution from croplands. In G. W. Bailey and F. Schaller (eds.) Agricultural management and water quality. Iowa State University Press, Iowa, in press.

Rao, P. S. C., D. E. Rolston, R. E. Jessup, and J. M. Davidson. 1980b. Solute transport in aggregated porous media: theoretical and experimental evaluation. Soil Sci. Soc. Am. J. 44:1139-1146.

Raymond, R. L., V. W. Jamison, J. O. Hudson. 1969. Microbial hydrocarbon resins. Applied Microbiology 17:512-515.

Reddy, B. R., and N. Sethunathan. 1983. Mineralization of parathion in the rice rhizosphere. Appl. Environ. Microbiol. 45:826-829.

Reeve, R. C., and N. R. Fausey. 1974. Drainage and timeliness of farming operations. p. 55-66. In J. V. Schilfgaarde (ed.) Drainage for agriculture. Agronomy Monograph No. 17, American Society of Agronomy, Madison, WI.

Rhoades, J. D. 1982. Soluble salts. p. 167-179. In A. L. Page, R. H. Miller, and D. R. Keeney (eds.) Methods of soil analysis. Part 2. Chemical and microbiological properties. Second edition. Agronomy Monograph No. 9(2), American Society of Agronomy, Inc., Madison, WI.

Rice, E. L. 1974. Allelopathy. Academic Press, New York, NY. 353 p.

Roberts, H. A. and B. J. Wilson. 1965. Adsorption of chlorprophan by different soils. Weed Res. 5:348-350.

Rogers, R. D. 1979. Volatility of mercury from soils amended with various mercury compounds. Soil Sci. Soc. Am. J. 43:289-291.

Rogers, R. D., and J. C. McFarlane. 1979. Factors influencing the volatilization of mercury from soil. Jour. Environ. Qual. 8(2):255-260.

Rogers, R. D., and J. McFarlane. 1981. Sorption of carbon tetrachloride, ethylene dibromide, and trichloroethylene on soil and clay. Env. Monitoring and Assessment: I:155-162.

Ross, D. S., R. E. Sjogren, and R. J. Bartlett. 1981. Behavior of chromium in soils: IV. Toxicity to microorganisms. J. Environ. Qual. 10:145-148.

Ryan, J. R., and R. C. Loehr. 1981. Site selection methodology for the land treatment of wastewater. Special Report 81-28, United States Army Corps of Engineers, Cold Regions Research and Engineering Laboratory, Hanover, NH.

Sabljic, A., and M. Protic. 1982. Relationship between molecular connectivity indices and soil sorption coefficients of polycyclic aromatic hydrocarbons. Bull. Environ. Contam. Toxicol. 28:162-165.

Saha, J. G., B. Bhavaraju, and Y. W. Lee. 1969. Validity of using soil fortification with dieldrin to measure solvent extraction efficiency. J. Agr. Food Chem. 17:874-876.

Santillan-Medrano, J., and J. J. Jurinak. 1975. The chemistry of lead and cadmium in soil: Solid phase formation. Soil Sci. Soc. Am. Proc. 29:851-856.

Sargent, J. W., and R. L. Sanks. 1974. Light-energized oxidation of organic wastes. J. Water Pollution Control Federation 46:2554-2574.

Savage, K. E., and R. D. Wauchope. 1974. Floumeturon adsorption equilibrium in soils. Weed Sci. 21:106-110.

Schindler, P. W. 1967. Heterogeneous equilibria involving oxides, hydroxides, carbonates and hydroxide carbonates. In R. F. Gould (ed.) Equilibrium concepts in natural water systems. Adv. in Chem. Ser. No. 67., Am. Chem. Soc., Washington, D. C.

Schnitzer, M. 1975. Chemical, spectroscopic, and thermal methods for the classification and characterization of humic substances. p. 293-310. In D. Povoledo and H. L. Golterman (eds.) Humic substances, their structure and function in the biosphere. Center for Agricultural Publishing and Documentation, Wageningen.

Schnitzer, M. 1978. Humic substances: chemistry and reactions. p. 1-64. In M. Schnitzer and S. U. Khan (eds.) Soil organic matter. Elsevier North-Holland, Inc., New York, NY.

Schnitzer, M. 1982. Organic matter characterization. p. 581-594. In A. L. Page, R. H. Miller, and D. R. Keeney (eds.) Methods of soil analysis. Part 2. Chemical and microbiological properties. Second edition. Agronomy Monograph No. 9(2), American Society of Agronomy, Inc., Madison, WI.

Schnitzer, M., and H. Kodama. 1977. Reactions of minerals with soil humic substances. p. 741-770. In J. B. Dixon and S. B. Weed (eds.) Minerals in soil environments. Soil Science Society of America, Madison, WI.

Schram, M. 1981. Permeability of soils to four organic solvents and water. M. S. Thesis. University of Arizona, Tucson, Arizona.

Schwab, G. O., R. K. Frevert, T. W. Edminster, and K. K. Barnes. 1966. Soil and water conservation engineering. Second Edition. John Wiley and Sons, Inc., New York, NY.

Schwarzenbach, R. P., and J. Westall. 1981. Transport of nonpolar organic compounds from surface water to groundwater. Laboratory sorption studies. Environ. Sci. Technol. 15:1360-1367.

Schweizer, E. E. 1976. Persistence and movement of ethofumesate in soil. Weed Sci. 14:22-26.

Sebastiani, A., A. D. Simonetti, and A. Borgioli. 1971. Behavior of synthetic detergents in soil. Nuovi Ann. d'Igiene Microbiolozia V. 22(1): 11-27: V. 22(2):81-95; V. 22(4):229-242.

Segol, G. 1977. A three-dimensional Galerkin-finite element model for the analysis of contaminant transport in saturated-unsaturated porous media. In W. G. Gray et al. (eds.) Finite elements in water resources. Pentech Press, London.

Seiber, J. N., J. E. Woodrow, T. M. Shofik, and H. F. Enos. 1975. Determination of pesticides and their transformation products in air. In R. Haque and V. H. Freed (eds.) Environmental dynamics of pesticides. Plenum Press, New York.

Selim, H. M. 1978. Transport of reactive solute during transient, unsaturated water flow in multilayered soils. Soil Sci. 126:127-135.

Selim, H. M., and I. K. Iskandar. 1981. A model for predicting nitrogen behaviour in slow and rapid infiltration system. In I. K. Iskandar (ed.) Modeling wastewater renovation land treatment. Wiley-Interscience, New York.

Shabad, L. M., Y. L. Cohan, A. Ya Khesina, H. P. Schuback, and G. A. Smirnov. 1969. The carcinogenic hydrocarbon benzo(a)pyrene in the soil. J. Nat. Cancer Inst. 47:1179.

Shariat, M., A. C. Anderson, and J. W. Mason. 1979. Screening of common bacteria capable of dimethylation of methylmercurie chloride. Bull. Environ. Contam. Toxicol. 21:255-261.

Sharom, M. S., J. R. W. Miles, C. R. Harris, and F. L. McEwen. 1980. Behaviour of 12 insecticides in soil and aqueous suspensions of soil and sediments. Water Res. 14:1095-1100.

Shearer, D. F., and B. H. Olson. 1983. The role of chemical speciation on bacterial resistance to cadmium. p. 260. In Abstracts of the Annual Meeting of the ASM. American Society for Microbiology, Washington, DC.

Sheets, T. J., A. S. Crafts, and H. R. Drever. 1962. Influence of soil properties on phytotoxicity of S-triazines. J. Agr. Food Chem. 10:458-462.

Shen, T. T., and T. J. Tofflemire. 1980. Air pollution aspects of land disposal of toxic wastes. J. Environmental Engineering Division, ASCE, EE1, p. 211-226.

Shin, Y. O., J. J. Chodan, and A. R. Wolcott. 1970. Adsorption of DDT by soils, soil fractions, and biological materials. J. Agr. Food Chem. 18:1129-1133.

Silver, S., and T. G. Kinscherf. 1982. Genetic and biochemical basis for microbial transformations and detoxification of mercury and mercurial compounds. p. 85-103. In A. M. Chakrabarty (ed.) Biodegradation and detoxification of environmental pollutants. CRC Press, Inc, Boca Raton, FL.

Sims, J. L., and W. H. Patrick., Jr. 1978. The distribution of micronutrient cations in soil under conditions of varing redox potential and pH. Soil Sci. Soc. Am. J. 42:258-262.

Sims, R. C., and M. R. Overcash. 1981. Land treatment of coal conversion wastewaters. p. 218-230. In Environmental aspects of fuel conversion technology, VI. A Symposium on Coal Based Synfuels. EPA-600/9-82-017 U. S. Environmental Protection Agency, Washington, DC.

Sims, R. C., and M. R. Overcash. 1983. Fate of polynuclear aromatic compounds (PNAs) in soil-plant systems. Residue Reviews 88:1-68.

Sisler, F. D., and C. E. Zobell. 1947. Microbial utilization of carcinogenic hydrocarbons. Science 106:521.

Soil Science Society of America. 1981. Water potential relations in soil microbiology. SSSA special publication number 9. Soil Science Society of America, Madison, WI. 151 p.

Soil Survey Staff. 1975. Soil taxonomy. Agriculture Handbook No. 436. Soil Conservation Service. U. S. Department of Agriculture, Washington, DC. 754 p.

Sokol, R. A., and D. A. Klein. 1975. The responses of soils and soil microorganisms to silver iodide weather modification agents. J. Environ. Qual. 4:211-214.

Sommers, L. E., C. M. Gilmour, R. E. Wildung, and S. M. Beck. 1981. The effect of water potential on decomposition processes in soils. p. 97-117. In Water potential relations in soil microbiology. SSSA special publication number 9. Soil Science Society of America, Madison, WI.

Spencer, W. F., J. D. Adams, T. D. Shoup, and R. C. Spear. 1980. Conversion of parathion to paraoxon on soil dusts and clay minerals as affected by ozone and UV light. J. Agr. Food Chem. 28:369.

Spencer, W. F., and M. M. Cliath. 1969. Vapor density of dieldrin. Environ. Sci. Tech. 3:670.

Spencer, W. F., and M. M. Cliath. 1973. Pesticide volatilization as related to water loss from soil. Environ. Qual. 2(2):284-289.

Spencer W. F., and M. M. Cliath. 1977. The solid-air interface: transfer of organic pollutants between the solid-air interface. In I. H. Suffet (ed.) Fate of pollutants in air and water environments, Part 1. Advances in environmental science and technology. John Wiley and Sons, New York.

Spencer, W. F., and M. M. Cliath. 1983. Measurement of pesticide vapors. Residue Reviews 85:57-71.

Spikes J. D., and R. Straight. 1967. Sensitized photochemical processes in biological systems. Annual Rev. Phys. Chem. 18:409-440.

Sposito, G. 1979. Derivation of the Langmuir equation for ion exchange reactions in soils. Soil Sci. Soc. Am. J. 43:197-198.

Sposito, G., F. T. Bingham, S. S. Yadav, and C. A. Inwye. 1982. Trace metal complexation by fulvic acid extracted from sewage study. II. Development of chemical models. Soil Sci. Soc. Am. J. 46:51-56.

Sposito, G., and K. M. Holtzclaw. 1977. Titration studies on the polynuclear, polyacidic nature of fulvic acid extract from sewage sludge-soil mixtures. Soil Sci. Soc. Am. J. 41:330-336.

Sposito, G., K. M. Holtzclaw, and D. A. Kecch. 1977. Proton binding in fulvic acid extracted from sewage sludge-soil mixtures. Soil Sci. Soc. Am. J. 41:1119-1125.

Sposito, G., K. M. Holtzclaw, and C. S. LeVesque-Madore. 1979. Cupric ion complexation by fulvic acid extracted from sewage sludge-soil mixtures. Soil Sci. Soc. Am. J. 43:1148-1155.

Sposito, G., K. M. Holtzclaw, and C. S. LeVesque-Madore. 1981. Trace metal complexation by fulvic acid extracted from sewage sludge: I. Determination of stability constants and linear correlation analysis. Soil Sci. Soc. Am. J. 45:465-468.

Stanlake, G. J., and R. K. Finn. 1982. Isolation and characterization of a pentachlorophenol degrading bacterium. Appl. Environ. Microbiol. 44:1421-1427.

Steenhuis, T. S., and M. F. Walter. 1979. Definitions and qualitative evaluation of soil and water conservation practices. p. 14-38. In D. A. Haith and R. C. Loehr (eds.) Effectiveness of soil and water conservation practices for pollution control. EPA-600/3-79-106. U. S. Environmental Protection Agency, Athens, GA.

Stevenson, F. J. 1982. Nitrogen-organic forms. p. 625-641. In A. L. Page, R. H. Miller, and D. R. Keeney (eds.) Methods of soil analysis. Part 2. Chemical and microbiological properties. Second edition. Agronomy Monograph No. 9(2), American Society of Agronomy, Inc., Madison, WI.

Stotzky, G. 1972. Activity, ecology, and population dynamics of microorganisms in soil. CRC Critical Reviews and Microbiology 2:59-137.

Stotzky, G. 1980. Surface interactions between clay minerals and microbes, viruses and soluble organics, and the probable importance of these interactions to the ecology of microbes in soil. p. 231-247. In R. C. W. Berkeley, J. M. Lindi, J. Melling, P. R. Rutter, and B. Vincent (eds.) Microbial adhesion to surfaces. John Wiley and Sons, New York, NY.

Stotsky, G., and V. N. Krasovsky. 1981. Ecological factors that affect the survival, establishment, growth, and genetic recombination of microbes in natural habitats. p. 31-42. In S. B. Levy, R. C. Clowes, and E. L. Koenig (eds). Molecular biology, pathogenicity, and ecology of bacterial plasmids. Plenum Press, New York, NY.

Stumm, W. and J. J. Morgan. 1981. Aquatic chemistry. Second Edition. John Wiley and Sons, New York.

Subba-Rao, R. V., and M. Alexander. 1980. Effect of DDT metabolites on soil respiration and on an aquatic alga. Bull. Environ. Cont. Toxicol. 25:215-220.

Sulflita, J. M., A. Horowitz, D. R. Shelton, and J. M. Tiedje. 1982. Dehalogenation: A novel pathway for the anaerobic biodegradation of haloaromatic compounds. Science 218:1115-1117.

Sulflita, J. M., J. A. Robinson, and J. M. Tiedje. 1983. Kinetics of microbial dehalogenation of haloaromatic substrates in methanogenic environments. Appl. Environ. Microbiol. 45:1466-1473.

Swann, R. L., D. A. Laskowski, P. J. McCall, and K. Vander Kuy. 1983. A rapid method for the estimation of environmental parameters K_{OW}, K_{OC}, and water solubility. Residue Reviews 85:15-28.

Swanson, R. A., and G. R. Dutt. 1973. Chemical and physical processes that affect Atrazine and distribution in soil systems. Soil Sci. Soc. Am. Proc. 37:892-876.

Sykes, J. F., S. Soyupak, and G.J. Farquhar. 1982. Modeling of leachate organic migration and attenuation in groundwaters below sanitary landfills. Water Resour. Res. 18:135-145.

Takai, Y., and T. Kamura. 1966. The mechanism of reduction in waterlogged paddy soil. Folia Microbiol. (Prague) 11:304-313.

Talbert, R. E., and O. H. Fletchall. 1965. The adsorption of some s-triazines in soils. Weeds 13:46-52.

Talibudeen, O. 1955. Complex formation between montmorillonoid calys and amino acids and proteins. Trans. Farady Soc. 51:582.

Tam, T.-Y., and J. T. Trevors. 1981. Toxicity of pentachlorophenol to Azotobacter vinlanail. Bull. Environ. Contam. Toxicol. 29:727-730.

Taylor, J. M., J. F. Parr, L. J. Sikora, and G. B. Willson. 1980. Considerations in the land treatment of hazardous wastes: Principles and practices. In Proceedings of the Second Oil and Hazardous Material Spills Conference and Exhibition, Hazardous Materials Control Research Institute, Silver Spring, MD.

Taylor, S. A., and G. L. Ashcroft. 1972. Physical edaphology. W. H. Freeman and Co., San Francisco, CA.

Thibodeaux, L. J. 1979. Chemodynamics, environmental movement of chemicals in air, water, and soil. John Wiley and Sons, New York. 501 p.

Thibodeaux, L. J., and S. T. Hwang. 1982. Landfarming of petroleum wastes -- modeling the air emission problem. Env. Progress 1(1):42-46.

Thorington, L. 1980. Actinic effects of light and biological implications. Photochem. Photobiol. 32:117-129.

Tilloston, W. R., C. W. Robbins, R. J. Wagenet, and R. J. Hanks. 1980. Soil water, solute and plant growth simulation. Utah Agricultural Experiment Station Bulletin 502. Utah State University, Logan, Utah.

Tirsch, F. S., J. H. Baker, and F. A. Diglano. 1979. Copper and cadmium reactions with soils in land applications. J. Water Pollut. Control Fed. 51:2649-2660.

Tisdale, S. L., and W. L. Nelson. 1975. Soil fertility and fertilizers. Macmillan Publishing Co., Inc., New York, NY.

Travis, C. C., and E. L. Etnier. 1981. A survey of sorption relationships for reactive solutes in soils. J. Environ. Qual. 10:8-17.

Trevors, J. T. 1982. Effect of pentachlorophenol on electron transport system activity in soil. Bull. Environ. Contam. Toxicol. 29:727-730.

Tripp, S., T. Barkay, and B. H. Olson. 1983. The effect of cadmium on the community structure of soil bacteria. p. 260. <u>In</u> Abstracts of the Annual Meeting of the ASM, American Society for Microbiology, Washington, DC.

Tu, C. M. 1980. Influence of pesticides and some of the oxidized analogues on microbial populations, nitrification and respiration activities in soils. Bull. Environ. Contam. Toxicol. 24:13-19.

U. S. Army Engineer Waterways Experiment Station, Corps of Engineers. 1956. Trafficability of soils, a summary of trafficability studies through 1955. Tech. Memo. No. 3-240, 14th Supplement. 63 p.

U. S. Environmental Protection Agency. 1976. Rationale for recommended list of priority pollutants. Memo to Director of Effluent Guideline Division. November 19.

U. S. Environmental Protection Agency. 1977. Process design manual for land treatment of municipal wastewater. EPA 625/1-77-008. U. S. Environmental Protection Agency, Cincinnati, OH.

U.S. Environmental Protection Agency. 1979. Methods for chemical analysis of water and wastes. EPA-600/4-79-020. U.S. Environmental Protection Agency, Cincinnati, OH.

U. S. Environmental Protection Agency 1980. Design manual: Onsite wastewater treatment and disposal systems. U. S. Environmental Protection Agency, Washington, DC.

U.S. Environmental Protection Agency. 1982a. EPA field guide for scientific support activities associated with Superfund emergency response. EPA-600/18-25-025. Office of Emergency and Remedial Response, U.S. Environmental Protection Agency, Washington, DC.

U.S. Environmental Protection Agency. 1982b. Test methods for evaluating solid waste: physical/chemical methods. Second edition. SW-846. U.S. Environmental Protection Agency.

U.S. Environmental Protection Agency. 1983. Hazardous waste land treatment. Revised edition. SW-874. U.S. Environmental Protection Agency, Cincinnati, OH.

Van De Pol, R. M., P. J. Wierenga, and D. R. Nielsen. 1977. Solute movement in a field soil. Soil Sci. Soc. Am. J. 41:10-13.

Van Genuchten, M. Th. 1980. Determining transport parameters from solute displacement experiments. USDA Res. Report No. 118.

Van Genuchten, M. Th. 1981. Non-equilibrium transport parameters from miscible displacement experiments. USDA Res. Report No. 199.

Van Genuchten, M. Th. 1982. A comparison of numerical solutions of the one-dimensional unsaturated-saturated flow and mass transport equations. Adv. Water Resour. 5:47-55.

Van Genuchten, M. Th., and W. J. Alves. 1981. A compendium of available analytical solutions of the one-dimensional convective-dispersive transport equation. U. S. Salinity Laboratory, Riverside, Ca.

Van Genuchten, M. Th., and R. W. Cleary. 1979. Movement of solutes in soil: computer simulated and laboratory results. In G. H. Bolt and M. G. M. Bruggewet (eds.) Soil chemistry. Elsevier, Amesterdam.

Van Genuchten, M. Th., J. M. Davidson, and P. J. Wierenga. 1974. An evaluation of kinetic and equilibrium equations for the prediction of pesticide movement through porous media. Soil Sci. Soc. Am. Proc. 38:29-35.

Van Genuchten, M. Th., and P. J. Wierenga. 1976. Mass transfer studies in sorbing porous media: I. Analytic solutions. Soil Sci. Soc. Am. J. 40:473-480.

Van Genuchten, M. Th., and P. J. Wierenga. 1977. Mass transfer studies in sorbing porous media: II. Experimental evaluation with tritium (^3H_2O). Soil Sci. Soc. Am. J. 41:272-277.

Van Genuchten, M. Th., P. J. Wierenga, and G. A. O'Connor. 1977. Mass transfer studies in sorbing porous media. III. Experimental evaluation with 2,4,5-T. Soil Sci. Soc. Am. J. 41:278-285.

Van Olphen, H. 1963. Clay colloid chemistry. Wiley Publication, New York.

Veith, G. D., N. M. Austin, and R. T. Morris. 1979. A rapid change for estimating log P for organic chemicals. Water Res. 13:43-49.

Veith, J. A., and G. Sposito. 1977. On the use of the Langmuir equation in the interpretation of adsorption phenomena. Soil Sci. Soc. Am.

Wahid, P. A., and P. A. Dahm. 1978. Sorption-desorption of parathion in soils. J. Agr. Food Chem. 26:101-105.

Wahid, P. A., and N. Sethunathan. 1979. Sorption-desorption of , , isomers of hexachlorocyclohexine in soils. J. Agr. Food Chem. 27:1050-1053.

Wahid, P. A., and N. Sethunathan. 1980. Sorption-desorption of lindane by anaerobic and aerobic soils. J. Agr. Food Chem. 28:623-625.

Walker, A. 1970. Persistence of pronamide in soil. Pestic. Sci. 1:237-239.

Walker, A. 1976a. Simulation of herbicide persistence in soil. I. Simazine and prometryne. Pestic. Sci. 7:41-49.

Walker, A. 1976b. Simulation of herbicide persistence in soil. II. Simazine and Linuron in longterm experiments. Pestic. Sci. 7:50-58.

Walker, A. 1976c. Simulation of herbicide persistence in soil. III. Propyzamide in different soil types. Pestic. Sci. 7:59-64.

Walker, A. 1978. Simulation of the persistence of eight soil-applied herbicides. Weed Res. 18:305-313.

Walker, A., and D. V. Crawford. 1968. The role of organic matter in adsorption of the triazine herbicides by soils. p. 91-108. In Isotopes and radiation in soil organic matter studies. Proc. 2nd Symp. Int. Atomic Energy Agency, Vienna.

Walters, L. T. and T. J. Wolery. 1974. Transfer of heavy metals pollutants from Lake Erie bottom-sediment to the overlying water. Ohio State University, Columbus, Ohio.

Warner, D. L. 1976. Hydrogeologic aspects of land application of wastewater. p. 133-146. In R. L. Sanks and T. Asano (eds.) Land treatment and disposal of wastewaters. Ann Arbor Science Publishers, Ann Arbor, MI.

Warrick, A. W., J. W. Biggar, and D. R. Nielsen. 1971. Simultaneous solute and water transport for an unsaturated soil. Water Resour. Res. 7:1216-1225.

Wasik, S. P., M. M. Miller, Y. B. Tewari, W. E. May, W. J. Sonnefeld, H. DeVoe, and W. H. Zoller. 1983. Determination of the vapor pressure, aqueous solubility, and octanol/water partition coefficient of hydrophobic substances by coupled generator column/liquid chromatographic methods. Residue Reviews 85:27-42.

Weber, J. B. 1966. Molecular structure and pH effects on the adsorption of 13 s-triazine compounds on montmorillonite clay. Am. Minerologist 51:1657.

Weber, J. B. 1970. Mechanisms of adsorption of s-triazines by clay colloids and factors affecting plant availability. Residue Rev. 32:93-130.

Weber, J. B. 1972. Interaction of organic pesticides with particular matter in aquatic and soil systems. In R. F. Gould (ed.) Fate of organic pesticides in the aquatic environment. Am. Chem. Soc., Washington, DC.

Weber, J. B., T. J. Monaco, and A. D. Worsham. 1973. What happens to herbicides in the environment? Weeds Today 4(1):16-17,22.

Weber, J. B., and D. C. Scott. 1966. Availability of cationic herbicide adsorbed on clay minerals to encumber seedlings.

Weber, J. B., S. B. Weed, and T. M. Ward. 1969. Adsorption of s-triazines on soil organic matter. Weed Sci. 17:417-421.

Weed, S. B., and J. B. Weber. 1968. The effect of adsorbent charge on the competitive adsorption of divalent organic cations by layer-silicate minerals. Am. Mineralogist 53:478-490.

Weed, S. B., and J. B. Weber. 1974. Pesticides-organic matter interactions. In W. D. Guenzi (ed.) Pesticides in soil and water. Soil Sci. Soc. Am, WI.

White, R. 1976. Remolded soil samples from proposed waste landfill site north of Three Rivers, TX. Trinity Engineering Testing Corporation Report #76791. Corpus Christi, TX.

Whitney, W. K. 1967. Laboratory tests with dursban and other insecticides in soil. J. Econ. Entomol. 60:68-74.

Wilson, J. T., C. G. Enfield, W. J. Dunlap, R. L. Cosby, D. A. Fosler, and L. B. Baskin. 1981. Transport of selected organic pollutants in a sandy soil. J. Environ. Qual. 10:501-506.

Winogradsky, S. 1949. Microbiologie du sol: Problemes et methodes. Mason et Cie, Paris, France.

Wipf, H. K., E. Homberger, N. Neuner, and F. Schenker. 1978. Field trials on photodegradation of TCDD on vegetation after spraying with vegetable oil. p. 201-217. In F. Cattabeni, A. Cavallero, and G. Galli (eds.) Dioxin: toxicologial and chemical aspects, Spectrum Publications, Inc., New York.

Wood, A. L., and J. M. Davidson. 1975. Fluorneutron and water content distributions during infiltration: Measured and calculated. Soil Sci. Soc. Am. Proc. 38:820-825.

Woodmansee, R. G., J. L. Dodd, R. A. Bowman, F. E. Clark, and C. E. Dickson. 1978. Nitrogen budget of a short-grass prairie ecosystem. Occolegia (Berl) 34:363-376.

Woodrow, J. E., D. G. Crosby, T. Mast, K. M. Moilanen, and J. N. Seiber. 1978. Rates of transformation of trifluoraline and parathion vapors in air. J. Agr. Food Chem. 26:1312.

Woodrow, J. E., D. G. Crosby, and J. N. Seiber. 1983. Vapor-phase photochemistry of pesticides. Residue Reviews 85:111-125.

Woodward, R. L. 1963. Review of the bactericidal effectiveness of silver. J. Amer. Water Works Assn. 55:881-888.

Woolson, E. A. 1977a. Fate of arsenicals in different environmental substrates. Environ. Health Perspect. 19:73-81.

Woolson, E. A. 1977b. Generation of allsylarsines from soil. Weed Sci. 25:412-416.

Woolson, E. A., J. H. Axley, and P. C. Kearney. 1971. The chemistry and phytotoxicity of arsenic in soils. I. contaminated field soils. Soil Sci. Soc. Amer. Proc. 35:938-943.

Woolson, E. A., and P. C. Kearney. 1973. Persistence and reactions of ^{14}C-cacodylic acid in soils. Environ. Sci. Technol. 7:42-50.

Yalkowsky, S. H., S. C. Valvani, and D. MacKay. 1983. Estimation of the aqueous solubility of some aromatic compounds. Residue Reviews 85:41-55.

Yang, J. T., and W. E. Bye. 1979. A guidance for protection of ground water resources from the effects of accidental spills of hydrocarbons and other hazardous substances. EPA-570/ 9-79-017, U. S. Environmental Protection Agency, Washington, DC.

You, In-Soon, R. A. Jones, and R. Bartha. 1982. Evaluation of a chemically defined model for the attachment of 3,4-dichloro-aniline to humus. Bull. Environm. Contam. Toxicol. 29:476-482.

Yule, D. F., and W. R. Gardner. 1978. Longitudinal and transverse dispersion coefficients in unsaturated plainfield sand. Water Resour. Res. 14:582-588.

Zajic, J. E. 1969. Microbial biogeochemistry. Academic Press, New York, NY. 345 p.

Appendix A: Parameters for Assessing Soil/Waste Interaction

Compound Name: Acenaphthene

Compound Properties: Structure

M.W.	154.2	Water solubility	3.47 mg/l
M.P.	96.2	Log K_{ow}	4.33
B.P.		Chemical class	
Sp. gr.		Chemical reactivity	
Vapor pressure	2.0×10^{-2} torr (20°C)		

Adsorption in Soils

Adsorption Parameters			Soil Properties					
					Percent (%)			
K	N	K_{oc}	pH	CEC	OC	Sand	Silt	Clay

Degradation in Soils

Degradation Parameters				Soil Properties							
			Initial	Temp.	Mois-ture			Percent (%)			
$t_{1/2}$	K_s	n	conc.	°C		pH	CEC	OC	Sand	Silt	Clay
4	0.178		500	15-25							
0.3	2.81		5	15-25							

Volatilization in Soils

Volatilization Parameters		Soil Properties				
K_w	K_wK_D	OC (%)	Moisture	Bulk Density	Temp. °C	Porosity

Compound Name: Aldicarb [2-methyl-2-(methylthio)-propionaldehyde O-(methylcarbamoyl)oxime]

Compound Properties: Structure

M.W.	162.2	Water solubility	40000 ppm
M.P.		Log K_{ow}	0.85
B.P.		Chemical class	
Sp. gr.		Chemical reactivity	
Vapor pressure			

Adsorption in Soils

Adsorption Parameters			Soil Properties					
					Percent (%)			
K	N	K_{oc}	pH	CEC	OC	Sand	Silt	Clay
0.17	0.93	33.3	7.30	5.71	0.51	77	15	8
0.20	0.95	18.7	6.83	6.10	1.07	83	9	8
0.62	0.86	23.4	5.00	21.02	2.64	37	42	21
0.88	0.85	23.2	7.30	37.84	3.80	21	55	24
3.47	0.89	18.9	6.98	77.34	18.36	42	39	19

Degradation in Soils

Degradation Parameters				Soil Properties							
								Percent (%)			
$t_{1/2}$	K_s	n	Initial conc.	Temp. °C	Moisture θ	pH	CEC	OC	Sand	Silt	Clay
	0.00273					5.4		0.58			
	0.0087					7.8		1.16			
	0.0420					7.5		0.15			
	0.0322					7.5		0.15			
	<0.0032										

Volatilization in Soils

Volatilization Parameters		Soil Properties				
K_w	K_wK_D	OC (%)	Moisture θ	Bulk Density	Temp. °C	Porosity

Compound Name: Aldrin [1,2,3,4,10,10-Hexachloro-1,4,4a,5,8,8a-hexachloro-1,4-endoexo-5, 8-dimethanonaphthalene]

Compound Properties: Structure

M.W.	364.9	Water solubility	0.025 ppm
M.P.	104	Log K_{ow}	-0.14
B.P.		Chemical class	
Sp. gr.		Chemical reactivity	
Vapor pressure	$2.31x1-^{-5}$ torr (20°C)		

Adsorption in Soils

Adsorption Parameters			Soil Properties					
					Percent (%)			
K	N	K_{oc}	pH	CEC	OC	Sand	Silt	Clay
1.29[a]	0.778[a]	253[a]						

Degradation in Soils

Degradation Parameters				Soil Properties							
								Percent (%)			
$t_{1/2}$	K_s	n	Initial conc.	Temp. °C	Moisture θ	pH	CEC	OC	Sand	Silt	Clay
	0.0149					7.8		0.93			
	0.0165					8.6		0.15			
	0.0061										
	0.0096										
	0.0038										

Volatilization in Soils

Volatilization Parameters		Soil Properties				
K_w	K_wK_D	OC (%)	Moisture θ	Bulk Density	Temp. °C	Porosity

[a]These are average values from three soils.

Compound Name: 2-Ammoanthracene

Compound Properties:

Structure

M.W.	193.2	Water solubility	1.3 μg/l
M.P.		Log K_{OW}	4.13
B.P.		Chemical class	
Sp. gr.		Chemical reactivity	
Vapor pressure			

Adsorption in Soils

Adsorption Parameters			Soil Properties		Percent (%)			
K	N	K_{oc}	pH	CEC	OC	Sand	Silt	Clay
321.6		26580	6.35	3.72	1.21	67.5	13.9	18.6
329.2		15904	7.79	23.72	2.07	3.0	41.8	55.2
304.1		13338	7.74	19.00	7.28	33.6	35.4	31.0
259.5		36039	7.83	33.01	0.72	0.2	31.2	68.6
79.0		52659	8.32	3.72	0.15	82.4	10.7	6.8
103.7		94276	8.34[a]	12.40	0.11	7.1	75.6	17.4
145.1		30225	4.54[a]	18.86	0.48	2.1	34.4	63.6
191.9		41248	7.79	11.30	0.95	15.6	48.7	35.7
283.0		42878	7.76	15.43	0.66	34.6	25.8	39.5
458.7		35287	5.50[a]	8.50	1.30	0.0	71.4	28.6
531.9		28292	7.60	8.33	1.88	50.2	42.7	7.1
502.1		30069	7.55	8.53	1.67	26.1	52.7	21.2
875.2		36772	6.70	31.15	2.38	17.3	13.6	69.1
688.7		48537	7.75	20.86	1.48	1.6	55.4	42.9

Degradation in Soils

Degradation Parameters				Soil Properties				Percent (%)			
$t_{1/2}$	K_s	n	Initial conc.	Temp. °C	Moisture Θ	pH	CEC	OC	Sand	Silt	Clay

Volatilization in Soils

Volatilization Parameters		Soil Properties				
K_w	K_wK_D	OC (%)	Moisture Θ	Bulk Density	Temp. °C	Porosity

[a]Soils; all others are sediments.

Compound Name: 6-aminochrysene

Compound Properties: Structure

M.W.	243.2	Water solubility	0.155 μg/l
M.P.	210-211°C	Log K_{ow}	4.99
B.P.		Chemical class	
Sp. gr.		Chemical reactivity	
Vapor pressure			

Adsorption in Soils

Adsorption Parameters			Soil Properties					
					Percent (%)			
K	N	K_{oc}	pH	CEC	OC	Sand	Silt	Clay
1735.5		143427						
3115.7		150519						
3972.5		174232						
1078.7		149817						
573.3		382185						
686.4		624022						
924.3		192553						
1292.2		136025						
1424.5		215835						
871.9		67070						
2616.0		139149						
1459.0		87363						
3923.3		164844						
1688.8		114108						

Degradation in Soils

Degradation Parameters				Soil Properties							
								Percent (%)			
$t_{1/2}$	K_s	n	Initial conc.	Temp. °C	Moisture Θ	pH	CEC	OC	Sand	Silt	Clay

Volatilization in Soils

Volatilization Parameters		Soil Properties				
K_w	K_wK_D	OC (%)	Moisture Θ	Bulk Density	Temp. °C	Porosity

Compound Name: Anthracene

Compound Properties: Structure

M.W.	178.2	Water solubility	0.075 mg/l (15°C)
M.P.	216°C	Log K_{OW}	4.45
B.P.	342°C	Chemical class	
Sp. gr.		Chemical reactivity	

Vapor pressure 196×10^{-4} torr (20°C)

Adsorption in Soils

Adsorption Parameters			Soil Properties					
					Percent (%)			
K	N	K_{oc}	pH	CEC	OC	Sand	Silt	Clay

Degradation in Soils

Degradation Parameters				Soil Properties							
								Percent (%)			
$t_{1/2}$	K_s	n	Initial conc.	Temp. °C	Moisture Θ	pH	CEC	OC	Sand	Silt	Clay
3.3	0.21		3.4	<15							
175	0.004		13.7	15-25							
143	0.005		10.3	15-25							
108	0.006		11.4	15-25							
138	0.005		40.0	15-25							
129	0.005		36.4	15-25							
3.5	0.198		25000	>25							
42	0.017		41								
36	0.019		0.41								
45	0.015		47								
43	0.016		0.49								
39	0.018		62								
43	0.016		0.60								

Volatilization in Soils

Volatilization Parameters		Soil Properties				
K_W	$K_W K_D$	OC (%)	Moisture Θ	Bulk Density	Temp. °C	Porosity

Compound Name: Anthracene-9-Carboxylic acid

Compound Properties: Structure

M.W.	222.2	Water solubility	85 μg/l
M.P.		Log K_{ow}	3.11
B.P.		Chemical class	
Sp. gr.		Chemical reactivity	
Vapor pressure			

Adsorption in Soils

Adsorption Parameters			Soil Properties						
					Percent (%)				
K	N	K_{oc}	pH	CEC	OC	Sand	Silt	Clay
5.27		463						
5.49		265						
7.96		349						
5.47		760						
1.84		1227						
2.82		2564						
10.03		2090						
2.66		280						
1.78		270						
13.27		1021						
6.45		343						
5.59		335						
9.88		415						
7.50		507						

Degradation in Soils

Degradation Parameters				Soil Properties							
								Percent (%)			
$t_{1/2}$	K_s	n	Initial conc.	Temp. °C	Moisture Θ	pH	CEC	OC	Sand	Silt	Clay

Volatilization in Soils

Volatilization Parameters		Soil Properties				
K_w	K_wK_D	OC (%)	Moisture Θ	Bulk Density	Temp. °C	Porosity

Compound Name: Atrazine [2-chloro-4-ethylamino-6-isopropylamino-1,3,5-triazine]

Compound Properties: Structure

M.W. 215.7
M.P. 173-175°C
B.P.
Sp. gr.
Vapor pressure 1.4×10^{-6} mm Hg (30°C)

Water solubility 70 mg/l (20°C
33 mg/l (27°C)
Log K_{ow} 2.68
Chemical class
Chemical reactivity

Adsorption in Soils

Adsorption Parameters			Soil Properties					
						Percent (%)		
K	N	K_{oc}	pH	CEC	OC	Sand	Silt	Clay
6.03	0.73	155.8	7.3	54.7	3.87	18.4	45.3	38.3
0.89	1.04	98.9	5.6	6.8	0.90	65.8	19.5	14.7
0.62	0.79	110.7	5.6	5.2	0.56	93.8	3.0	3.2
0.62	0.93	124.0	7.4	35.8	0.50	56.7	16.4	22.9
0.21	1	80.8	6.5	5.1	0.26[a]	77	18	5
2.61	0.85	174	6.8	20.0	1.50[a]	16	68	16
1.2	-	1200	7.0	25.4	0.1			42.8
0.4	-	100	6.6	10.2	0.4			8.2
0.7	-	140	6.2	10.6	0.5			15.8
0.8	-	89	4.9	11.0	0.9			13.9
1.6	-	178	5.7	9.2	0.9			20.3
1.2	-	100	5.9	12.4	1.2			28.5
1.9	-	146	6.1	14.0	1.3			10.3
1.5	-	100	5.8	11.0	1.5			17.9
2.5	-	167	6.7	30.1	1.5			27.8
1.5	-	100	6.2	13.9	1.5			33.5
1.5	-	94	5.8	10.3	1.6			21.0
1.1	-	69	6.9	20.2	1.6			22.0
2.0	-	118	6.4	21.5	1.7			35.2
1.1	-	78	6.7	22.3	1.8			37.1
2.9	-	107	6.6	24.8	2.7			61.6
2.8	-	90	6.5	25.9	3.1			63.4
1.8	-	53	6.9	29.2	3.4			31.9
3.5	-	92	6.5	26.9	3.8			25.9
4.1	-	89	5.7	27.4	4.6			30.2
3.7	-	57	7.6	25.7	6.5			28.1
2.9	-	37	7.2	42.0	7.8			33.8
9.5	-	101	6.8	44.6	9.4			20.1
10.3	-	104	5.3	51.3	9.9			29.5
10.6	-	84	7.1	37.6	12.6			28.4
11.8	-	87	7.1	11.2	13.6			8.1
13.8	-	99	6.9	50.4	14.0			17.6
12.3	-	87	6.5	61.3	14.1			51.2
17.4	-	113	6.5	82.1	15.4			28.7

[a]Reported as OM.

Compound Name: Atrazine (continued)

Compound Properties: Structure

M.W.	215.7	Water solubility	33 mg/l (27°C)
M.P.	173-175°C	Log K_{OW}	2.68
B.P.		Chemical class	
Sp. gr.		Chemical reactivity	
Vapor pressure 1.4×10^{-6} mm Hg (30°C)			

Adsorption in Soils

Adsorption Parameters			Soil Properties					
					Percent (%)			
K	N	K_{oc}	pH	CEC	OC	Sand	Silt	Clay
17.4	-	110	6.5	84.3	15.8			19.4
17.3	-	187	5.9	92.3	18.6			31.4
22.1	-	117	6.8	94.2	18.9			11.2
22.9	-	115	6.9	83.1	20.0			12.8
27.2	-	119	6.8	106.7	22.9			7.3
31.7	-	134	5.6	126.9	23.6			32.1
28.0	-	108	6.7	131.1	25.9			24.9
26.3	-	97	6.9	123.1	27.1			18.8

Degradation in Soils

Degradation Parameters				Soil Properties							
								Percent (%)			
$t_{1/2}$	K_s	n	Initial conc.	Temp. °C	Moisture Θ	pH	CEC	OC	Sand	Silt	Clay
	0.0131	-				4.8		0.56			
	0.0063	-				6.5		1.16			
	0.0064	-				6.5					
	0.0133	-				6.8					
	0.0149	-				6.4					

Volatilization in Soils

Volatilization Parameters		Soil Properties				
K_w	K_wK_D	OC (%)	Moisture Θ	Bulk Density	Temp. °C	Porosity

Compound Name: Benerin [N-butyl-N-ethyl-α,α,α-trifluoro-2, 6-dinitro-N, N-dipropyl-p-toluidine]

Compound Properties: Structure

M.W.	335.3	Water solubility	
M.P.		Log K_{OW}	
B.P.		Chemical class	
Sp. gr.		Chemical reactivity	
Vapor pressure			

Adsorption in Soils

Adsorption Parameters			Soil Properties					
					Percent (%)			
K	N	K_{oc}	pH	CEC	OC	Sand	Silt	Clay

Degradation in Soils

Degradation Parameters				Soil Properties							
			Initial	Temp.	Moisture			Percent (%)			
$t_{1/2}$	K_s	n	conc.	°C	Θ	pH	CEC	OC	Sand	Silt	Clay
176[a]				15		7.6	11.4	0.93[b]	64	20	14
66[a]				30		7.6	11.4	0.92[b]	64	20	14
168[a]				15		6.6	19.4	1.28[b]	35	45	20
54[a]				30		6.6	19.4	1.28[b]	35	45	20

Volatilization in Soils

Volatilization Parameters		Soil Properties				
			Moisture	Bulk	Temp.	
K_W	K_WK_D	OC (%)	Θ	Density	°C	Porosity

[a]Reported in months.
[b]Reported as OM.

Compound Name: 1,2-benzanthracene (benz(a)anthracene)

Compound Properties: Structure

M.W.	228.2	Water solubility	0.014 mg/l
M.P.	162°C	Log K_{ow}	5.61
B.P.		Chemical class	
Sp. gr.		Chemical reactivity	
Vapor pressure	5.0×10^{-9} torr (20°C)		

Adsorption in Soils

Adsorption Parameters			Soil Properties					
					Percent (%)			
K	N	K_{oc}	pH	CEC	OC	Sand	Silt	Clay

Degradation in Soils

Degradation Parameters				Soil Properties							
			Initial	Temp.	Mois-ture			Percent (%)			
$t_{1/2}$	K_s	n	conc.	°C	Θ	pH	CEC	OC	Sand	Silt	Clay
43	0.016		7			6.1		0.75			
134	0.005		0.07			6.1		0.75			
41	0.017		8.2			6.1		0.75			
142	0.005		0.1			6.1		0.75			
41	0.017		9.7			6.1		0.75			
154	0.005		0.15			6.1		0.75			

Volatilization in Soils

Volatilization Parameters		Soil Properties				
			Moisture	Bulk	Temp.	
K_w	K_wK_D	OC (%)	Θ	Density	°C	Porosity

Compound Name: Benzene

Compound Properties:

Structure

M.W.	78.1	Water solubility	1.8 g/l
M.P.		Log K_{ow}	2.13
B.P.		Chemical class	
Sp. gr.		Chemical reactivity	
Vapor pressure			

Adsorption in Soils

Adsorption Parameters			Soil Properties					
					Percent (%)			
K	N	K_{oc}	pH	CEC	OC	Sand	Silt	Clay
2.4[a]	0.89	92	5.6	17	2.6	1	68	31
1.8[a]	0.94	100	7.8	29	1.8	15	51	34
30.9[a]	1.08		4.2	80	0			100[b]
4.4[a]	0.99		6.6	80	0			100[c]
0.35		31.4		14	1.10	9	68	21

Degradation in Soils

Degradation Parameters				Soil Properties							
								Percent (%)			
$t_{1/2}$	K_s	n	Initial conc.	Temp. °C	Moisture Θ	pH	CEC	OC	Sand	Silt	Clay

Volatilization in Soils

Volatilization Parameters		Soil Properties				
K_w	K_wK_D	OC (%)	Moisture Θ	Bulk Density	Temp. °C	Porosity

[a]Reported in terms of ng to convert to g multiply by $(10^{-3})^{1-N}$
[b]Al-Montmorillonite
[c]Ca-Montmorillonite

Compound Name: Benzo(b)fluoranthene

Compound Properties: Structure

M.W.	252.3	Water solubility	0.0012 mg/l
M.P.	167	Log K_{ow}	6.57
B.P.		Chemical class	
Sp. gr.		Chemical reactivity	

Vapor pressure 5.0×10^{-7} torr (20°C)

Adsorption in Soils

Adsorption Parameters			Soil Properties					
					Percent (%)			
K	N	K_{oc}	pH	CEC	OC	Sand	Silt	Clay

Degradation in Soils

Degradation Parameters				Soil Properties							
					Mois-			Percent (%)			
			Initial	Temp.	ture						
$t_{1/2}$	K_s	n	conc.	°C	θ	pH	CEC	OC	Sand	Silt	Clay
67	0.010		33			6.1		0.75			
98	0.007		0.33			6.1		0.75			
85	0.008		46			6.1		0.75			
123	0.006		0.57			6.1		0.75			
73	0.010		53			6.1		0.75			
130	0.006		0.8			6.1		0.75			

Volatilization in Soils

Volatilization Parameters		Soil Properties				
			Moisture	Bulk	Temp.	
K_w	K_wK_D	OC (%)	θ	Density	°C	Porosity

Compound Name: Benzo(k)fluoranthene

Compound Properties: Structure

M.W.	252.3	Water solubility	0.00055
M.P.	217	Log K_{OW}	6.84
B.P.		Chemical class	
Sp. gr.		Chemical reactivity	

Vapor pressure 5.0×10^{-7} torr (20°C)

Adsorption in Soils

Adsorption Parameters			Soil Properties					
					Percent (%)			
K	N	K_{oc}	pH	CEC	OC	Sand	Silt	Clay

Degradation in Soils

Degradation Parameters				Soil Properties							
								Percent (%)			
$t_{1/2}$	K_S	n	Initial conc.	Temp. °C	Moisture Θ	pH	CEC	OC	Sand	Silt	Clay
100	0.007		1.7			6.1		0.75			
89	0.008		0.1			6.1		0.75			
87	0.008		2.6			6.1		0.75			
94	0.007		0.15			6.1		0.75			
100	0.007		2.7			6.1		0.75			
87	0.008		0.17			6.1		0.75			

Volatilization in Soils

Volatilization Parameters		Soil Properties				
K_W	$K_W K_D$	OC (%)	Moisture Θ	Bulk Density	Temp. °C	Porosity

Compound Name: 1,12-benzoperylene

Compound Properties:

Structure

M.W.	276.3	Water solubility	
M.P.		Log K_{ow}	
B.P.		Chemical class	
Sp. gr.		Chemical reactivity	
Vapor pressure			

Adsorption in Soils

Adsorption Parameters			Soil Properties					
					Percent (%)			
K	N	K_{oc}	pH	CEC	OC	Sand	Silt	Clay

Degradation in Soils

Degradation Parameters				Soil Properties							
								Percent (%)			
$t_{1/2}$	K_s	n	Initial conc.	Temp. °C	Moisture Θ	pH	CEC	OC	Sand	Silt	Clay

Volatilization in Soils

Volatilization Parameters		Soil Properties				
K_w	K_wK_D	OC (%)	Moisture Θ	Bulk Density	Temp. °C	Porosity

Compound Name: Benzo(a)pyrene [3,4-benzopyrene]

Compound Properties: Structure

M.W.	252.3	Water solubility	0.0038 mg/l
M.P.	179	Log K_{OW}	6.04
B.P.		Chemical class	
Sp. gr.		Chemical reactivity	
Vapor pressure	6.85×10^{-7} torr (20°C)		

Adsorption in Soils

Adsorption Parameters			Soil Properties					
					Percent (%)			
K	N	K_{OC}	pH	CEC	OC	Sand	Silt	Clay

Degradation in Soils

Degradation Parameters				Soil Properties							
								Percent (%)			
$t_{1/2}$	K_S	n	Initial conc.	Temp. °C	Moisture Θ	pH	CEC	OC	Sand	Silt	Clay
50	0.014		0.048	<15		6.1		0.75			
694	0.001		0.01	<15		6.1		0.75			
57	0.012		3.4	15-25		6.1		0.75			
294	0.002		9.5	15-25		6.1		0.75			
147	0.005		12.3	15-25		6.1		0.75			
264	0.003		7.6	15-25		6.1		0.75			
30	0.023		18.5	15-25							
420	0.002		17.0	15-25							
175	0.004		32.6	15-25							
2	0.347		1.0	>25							
2	0.347		0.515	>25							
5	0.139		1.00135	>25							
406	0.002		0.0094	<15							
66	0.011		0.545	<15							
37	0.019		28.5	<15							

Volatilization in Soils

Volatilization Parameters		Soil Properties				
K_W	K_WK_D	OC (%)	Moisture Θ	Bulk Density	Temp. °C	Porosity

Compound Name: Benzo(a)pyrene [3,4-benzopyrene] (continued)

Compound Properties: Structure

M.W	252.3	Water solubility	0.0038 mg/l
M.P.	179	Log K_{ow}	6.04
B.P.		Chemical class	
Sp. gr.		Chemical reactivity	
Vapor pressure	6.85×10^{-7} torr (20°C)		

Adsorption in Soils

Adsorption Parameters			Soil Properties					
					Percent (%)			
K	N	K_{oc}	pH	CEC	OC	Sand	Silt	Clay

Degradation in Soils

Degradation Parameters				Soil Properties								
$t_{1/2}$	K_s	n	Initial conc.	Temp. °C	Moisture Θ	pH	CEC	Percent (%) OC	Sand	Silt	Clay
-	0		29.2	-							
39	0.018		910.0	>25							
7	0.099		19.5	>25							
5	0.139		19.5	>25							
3	0.231		19.5	>25							
4	0.173		130.6	>25							
6	0.116		130.6	>25							
92	0.008		36			6.1		0.75			
79	0.009		0.36			6.1		0.75			
100	0.007		55			6.1		0.75			
83	0.008		0.41			6.1		0.75			
92	0.008		69			6.1		0.75			
120	0.006										

Volatilization in Soils

Volatilization Parameters		Soil Properties				
K_w	K_wK_D	OC (%)	Moisture Θ	Bulk Density	Temp. °C	Porosity

Compound Name: Bromoform

Compound Properties:

Structure

M.W. 252.8
M.P. 8.3°C
B.P.
Sp. gr.
Vapor pressure 5.6 mm (25°C)

Water solubility 3190 mg/l 30°C
Log K_{OW}
Chemical class
Chemical reactivity

Adsorption in Soils

Adsorption Parameters			Soil Properties					
					Percent (%)			
K	N	K_{OC}	pH	CEC	OC	Sand	Silt	Clay

Degradation in Soils

Degradation Parameters				Soil Properties							
$t_{1/2}$	K_S	n	Initial conc.	Temp. °C	Moisture Θ	pH	CEC	Percent (%) OC	Sand	Silt	Clay

Volatilization in Soils

Volatilization Parameters		Soil Properties				
K_W	K_WK_D	OC (%)	Moisture Θ	Bulk Density	Temp. °C	Porosity

Compound Name: Bromomethane

Compound Properties: Structure

M.W.	94.95	Water solubility	900 mg/l (20°C)
M.P.	-93°C	Log K_{ow}	1.19
B.P.		Chemical class	
Sp. gr.		Chemical reactivity	
Vapor pressure			

Adsorption in Soils

Adsorption Parameters			Soil Properties					
					Percent (%)			
K	N	K_{oc}	pH	CEC	OC	Sand	Silt	Clay

Degradation in Soils

Degradation Parameters				Soil Properties							
$t_{1/2}$	K_s	n	Initial conc.	Temp. °C	Moisture Θ	pH	CEC	Percent (%) OC	Sand	Silt	Clay

Volatilization in Soils

Volatilization Parameters		Soil Properties				
K_w	K_wK_D	OC (%)	Moisture Θ	Bulk Density	Temp. °C	Porosity

Compound Name: Carbofuran [2,3-dihydro-2, 2-dimethylbenzofuran-7-yl methylcarbonate]

Compound Properties: Structure

M.W.	219.2	Water solubility
M.P.		Log K_{ow}
B.P.		Chemical class
Sp. gr.		Chemical reactivity
Vapor pressure		

Adsorption in Soils

Adsorption Parameters			Soil Properties					
					Percent (%)			
K	N	K_{oc}	pH	CEC	OC	Sand	Silt	Clay
0.25	-	14.7		1.7	0.4			
0.74	-	61.6		7.5	1.2			
1.40	-	51.9		34.4	2.7			
1.13	-	36.5		24.8	3.1			
1.39	-	39.7		27.7	3.5			
2.22	-	29.6		55.5	7.8			
8.74	-	52.0		72.4	16.8			
92.1	1.08	210	6.1		43.7	52	34	14
1.5	0.98	93	6.6		1.62	71	22	7
4.7	1.07	324	6.8		1.45	45	30	14
0.016	0.88	39	7.0		0.41	91.5	1.5	7

Degradation in Soils

Degradation Parameters				Soil Properties							
								Percent (%)			
$t_{1/2}$	K_s	n	Initial conc.	Temp. °C	Moisture Θ	pH	CEC	OC	Sand	Silt	Clay
	0.0768										
	0.0079					8.5					

Volatilization in Soils

Volatilization Parameters		Soil Properties				
K_w	K_wK_D	OC (%)	Moisture Θ	Bulk Density	Temp. °C	Porosity

Compound Name: Carbon tetrachloride (tetra chloromethane)

Compound Properties:

Structure

M.W.	153.8	Water solubility	1160 mg/l (25°C)
M.P.	-23°C		800 mg/l (20°C)
B.P.		Log K_{OW}	2.73
Sp. gr.		Chemical class	
Vapor pressure	99 mm (20°C)	Chemical reactivity	

Adsorption in Soils

Adsorption Parameters			Soil Properties					
					Percent (%)			
K	N	K_{OC}	pH	CEC	OC	Sand	Silt	Clay

Degradation in Soils

Degradation Parameters				Soil Properties							
								Percent (%)			
$t_{1/2}$	K_s	n	Initial conc.	Temp. °C	Moisture Θ	pH	CEC	OC	Sand	Silt	Clay

Volatilization in Soils

Volatilization Parameters		Soil Properties				
K_W	K_WK_D	OC (%)	Moisture Θ	Bulk Density	Temp. °C	Porosity

Compound Name: C6A-15646 [3-(3-chloro-4-methylphenyl)-1,1-dimethylurea]

Compound Properties: Structure

M.W.	212.7	Water solubility	70 g/l
M.P.		Log K_{ow}	
B.P.		Chemical class	
Sp. gr.		Chemical reactivity	
Vapor pressure			

Adsorption in Soils

Adsorption Parameters			Soil Properties					
					Percent (%)			
K	N	Koc	pH	Texture	OC	Sand	Silt	Clay
2.18	0.79	150	8.5	S	0.06			9.8
2.49	0.81	146	8.5	S	0.58			15.0
2.89	0.88	203	7.9	S	0.81			13.0
3.61	0.81	148	7.6	S	0.87			6.8
6.80	0.90	249	7.7	S cl	0.93			31.5
4.61	0.89	246	5.1	S	1.10			10.6
4.07	0.86	333	7.8	S	0.70			18.3
16.13	0.74	346	4.6	S	2.67			4.5

Degradation in Soils

Degradation Parameters				Soil Properties							
								Percent (%)			
$t_{1/2}$	K_s	n	Initial conc.	Temp. °C	Moisture Θ	pH	CEC	OC	Sand	Silt	Clay

Volatilization in Soils

Volatilization Parameters		Soil Properties				
K_W	K_WK_D	OC (%)	Moisture Θ	Bulk Density	Temp. °C	Porosity

Compound Name: Chlorobenzene

Compound Properties: Structure

M.W.	112.6	Water solubility	490 mg/l (20-25°C)
M.P.	-45°C	Log K_{OW}	2.84
B.P.	131-132°C	Chemical class	
Sp. gr.		Chemical reactivity	
Vapor pressure			

Adsorption in Soils

Adsorption Parameters			Soil Properties					
					Percent (%)			
K	N	K_{OC}	pH	CEC	OC	Sand	Silt	Clay
0.91	-	82.7		14	1.10	9	6.8	21

Degradation in Soils

Degradation Parameters				Soil Properties							
								Percent (%)			
$t_{1/2}$	K_S	n	Initial conc.	Temp. °C	Moisture Θ	pH	CEC	OC	Sand	Silt	Clay

Volatilization in Soils

Volatilization Parameters		Soil Properties				
K_W	$K_W K_D$	OC (%)	Moisture Θ	Bulk Density	Temp. °C	Porosity

Compound Name: Chlorobromuron [3-(4-bromo-3-chlorophenyl)-1-methoxy-1-methylurea]

Compound Properties: Structure

M.W.	293.5	Water solubility	50 µg/l
M.P.		Log K_{ow}	
B.P.		Chemical class	
Sp. gr.		Chemical reactivity	
Vapor pressure			

Adsorption in Soils

Adsorption Parameters			Soil Properties					
					Percent (%)			
K	N	K_{oc}	pH	Texture	OC	Sand	Silt	Clay
1.84	0.86	1533	8.5	S	0.06			9.8
6.98	0.70	684	8.5	S	0.58			15.0
7.98	0.67	561	7.9	S	0.81			13.0
14.10	0.59	772	7.6	S	0.87			6.8
19.10	0.80	1186	7.7	S cl	0.93			31.5
28.10	0.80	1502	5.1	S	1.10			10.6
39.80	0.63	3262	7.8	S	0.70			18.3
128.00	0.50	2758	4.6	S	2.67			4.5

Degradation in Soils

Degradation Parameters				Soil Properties							
								Percent (%)			
$t_{1/2}$	K_s	n	Initial conc.	Temp. °C	Moisture Θ	pH	CEC	OC	Sand	Silt	Clay

Volatilization in Soils

Volatilization Parameters		Soil Properties				
K_W	K_WK_D	OC (%)	Moisture Θ	Bulk Density	Temp. °C	Porosity

Compound Name: Chloroethane (gas)

Compound Properties: Structure

M.W.	64.5	Water solubility	3330 mg/l (0°C)
M.P.	-138.3°C		740 mg/l (20°C)
B.P.		Log K_{ow}	1.43
Sp. gr.		Chemical class	
Vapor pressure	457 mm (0°C)	Chemical reactivity	
	700 mm (10°C)		
	1000 mm (20°C)		
	1.7 atm (30°C)		

Adsorption in Soils

Adsorption Parameters			Soil Properties					
					Percent (%)			
K	N	K_{oc}	pH	CEC	OC	Sand	Silt	Clay

Degradation in Soils

Degradation Parameters				Soil Properties							
								Percent (%)			
$t_{1/2}$	K_s	n	Initial conc.	Temp. °C	Moisture Θ	pH	CEC	OC	Sand	Silt	Clay

Volatilization in Soils

Volatilization Parameters		Soil Properties				
K_w	K_wK_D	OC (%)	Moisture Θ	Bulk Density	Temp. °C	Porosity

Compound Name: Chloroethene [Vinylchloride] (gas)

Compound Properties: Structure

M.W. 62.5
M.P. -153/-160°C
B.P.
Sp. gr.
Vapor pressure 240 mm (-40°C)
580 mm (-20°C)
2660 mm (25°C)

Water solubility 1.1 mg/l (25°C)
Log K_{ow}
Chemical class
Chemical reactivity

Adsorption in Soils

Adsorption Parameters			Soil Properties					
					Percent (%)			
K	N	K_{oc}	pH	CEC	OC	Sand	Silt	Clay

Degradation in Soils

Degradation Parameters				Soil Properties							
								Percent (%)			
$t_{1/2}$	K_s	n	Initial conc.	Temp. °C	Moisture Θ	pH	CEC	OC	Sand	Silt	Clay

Volatilization in Soils

Volatilization Parameters		Soil Properties				
K_w	K_wK_D	OC (%)	Moisture Θ	Bulk Density	Temp. °C	Porosity

Compound Name: Chloroform [trichloromethane]

Compound Properties: Structure

M.W. 119.4
M.P. -64°C
B.P.
Sp. gr.
Vapor pressure 160 mm (20°C)

Water solubility 8000 mg/l (20°C)
9300 mg/l (25°C)
10000 mg/l (15°C)
Log K_{ow} 1.96
Chemical class
Chemical reactivity

Adsorption in Soils

Adsorption Parameters			Soil Properties					
					Percent (%)			
K	N	K_{oc}	pH	CEC	OC	Sand	Silt	Clay

Degradation in Soils

Degradation Parameters				Soil Properties							
$t_{1/2}$	K_s	n	Initial conc.	Temp. °C	Moisture Θ	pH	CEC	Percent (%) OC	Sand	Silt	Clay

Volatilization in Soils

Volatilization Parameters		Soil Properties				
K_w	K_wK_D	OC (%)	Moisture Θ	Bulk Density	Temp. °C	Porosity

Compound Name: Chloromethane (gas)

Compound Properties: Structure

M.W.	51	Water solubility	4000 cm^3/l
M.P.	-97.7°C	Log K_{ow}	0.91
B.P.		Chemical class	
Sp. gr.		Chemical reactivity	
Vapor pressure	5.0 atm (20°C) 6.7 atm (30°C)		
Relative vapor density	1.8 (air = 1)		

Adsorption in Soils

Adsorption Parameters			Soil Properties					
					Percent (%)			
K	N	K_{oc}	pH	CEC	OC	Sand	Silt	Clay

Degradation in Soils

Degradation Parameters				Soil Properties							
								Percent (%)			
$t_{1/2}$	K_s	n	Initial conc.	Temp. °C	Moisture Θ	pH	CEC	OC	Sand	Silt	Clay

Volatilization in Soils

Volatilization Parameters		Soil Properties				
K_w	K_wK_D	OC (%)	Moisture Θ	Bulk Density	Temp. °C	Porosity

Compound Name: Chlorpyrifos [O,O-diethyl O-3,5,6-trichloro-2-pyridyl phosphorothioate]

Compound Properties: Structure

M.W. 350.6
M.P.
B.P.
Sp. gr.
Vapor pressure

Water solubility 1.12 ppm
Log K_{OW} 5.11
Chemical class
Chemical reactivity

Adsorption in Soils

Adsorption Parameters			Soil Properties					
					Percent (%)			
K	N	K_{OC}	pH	CEC	OC	Sand	Silt	Clay
24.51	0.86	4806	7.3	5.71	0.51	77	15	8
36.84	0.77	3443	6.83	6.10	1.07	83	9	8
147.24	0.91	5577	5.00	21.02	2.64	37	42	21
388.95	0.98	10236	7.30	37.84	3.80	21	55	24
7095	1.09	16236	6.1		43.7	52	34	14
103.3	0.98	6377	6.6		1.62	71	22	7
101.7	0.99	7014	6.8		1.45	56	30	14
13.4	0.98	3268	7.0		0.41	91.5	1.5	7

Degradation in Soils

Degradation Parameters				Soil Properties							
								Percent (%)			
$t_{1/2}$	K_S	n	Initial conc.	Temp. °C	Moisture Θ	pH	CEC	OC	Sand	Silt	Clay
120			1	20		6.2	20	1.74	-	-	51
120			0.1	20		6.2	20	1.74	-	-	51

Volatilization in Soils

Volatilization Parameters		Soil Properties				
K_W	K_WK_D	OC (%)	Moisture Θ	Bulk Density	Temp. °C	Porosity

Compound Name: Chrysene [1,2 benzophenanthrene]

Compound Properties: Structure

M.W.	228.2	Water solubility	0.002 mg/l
M.P.	254°C	Log K_{ow}	5.61
B.P.	448°C	Chemical class	
Sp. gr.		Chemical reactivity	

Vapor pressure 6.3×10^{-7} torr (20°C)

Adsorption in Soils

Adsorption Parameters			Soil Properties					
					Percent (%)			
K	N	K_{oc}	pH	CEC	OC	Sand	Silt	Clay

Degradation in Soils

Degradation Parameters				Soil Properties							
			Initial	Temp.	Moisture			Percent (%)			
$t_{1/2}$	K_s	n	conc.	°C	Θ	pH	CEC	OC	Sand	Silt	Clay
-	0		4.4								
10.5	0.067		500	15-25							
5.5	0.126		5	15-25							

Volatilization in Soils

Volatilization Parameters		Soil Properties				
			Moisture	Bulk	Temp.	
K_w	K_wK_D	CC (%)	Θ	Density	°C	Porosity

Compound Name: Cyanazine [2-[[4-Chloro-6-(ethylamino)-1,3,5-triazin-2-yl]amino]-2-methyl propionitrile]

Compound Properties: Structure

M.W.	206.2	Water solubility	
M.P.		Log K_{ow}	
B.P.		Chemical class	
Sp. gr.		Chemical reactivity	
Vapor pressure			

Adsorption in Soils

Adsorption Parameters			Soil Properties					
					Percent (%)			
K	N	K_{OC}	pH	CEC	OC	Sand	Silt	Clay
4.6	0.96	368	6.7	20.7	1.25	11	63	26
3.4	0.86	453	6.7	8.2	0.75	90	8	5

Degradation in Soils

Degradation Parameters				Soil Properties							
								Percent (%)			
$t_{1/2}$	K_s	n	Initial conc.	Temp. °C	Moisture Θ	pH	CEC	OC	Sand	Silt	Clay

Volatilization in Soils

Volatilization Parameters		Soil Properties				
K_w	K_wK_D	OC (%)	Moisture Θ	Bulk Density	Temp. °C	Porosity

Compound Name: 2,4-D [2,4-Dichlorophenoxy acetic acid]

Compound Properties:

Structure

M.W.	221.0	Water solubility	900 mg/l (25°C)
M.P.	135-138°C	Log K_{ow}	2.81
B.P.	160°C	Chemical class	
Sp. gr.		Chemical reactivity	
Vapor pressure			

Adsorption in Soils

Adsorption Parameters			Soil Properties					
					Percent (%)			
K	N	K_{oc}	pH	CEC	OC	Sand	Silt	Clay
1.30	1.00	21.4	5.9		6.08	47.5	33.2	20.3
0.78	0.98	10.8	7.7		7.20	45.5	41.2	13.5
0.47	1.01	12.5	6.5		3.75	53.3	27.5	19.2
0.13	1.00	5.4	7.7		2.41	5.3	25.3	69.5
0.31	0.97	13.1	7.8		2.36	69.3	12.3	18.5
0.36	1.01	14.5	7.8		2.48	69.0	16.0	15.0
0.09	0.79	8.7	7.5		1.03	81.6	10.4	8.0
1.90	0.90	109	5.9	19.0	1.74	12	61	27
0.21	0.86	45.7	7.7	31.1	0.46	39	30	31

Degradation in Soils

Degradation Parameters				Soil Properties								
									Percent (%)			
$t_{1/2}$	K_s	n	Initial conc.	Temp. °C	Moisture Θ	pH	CEC	OC	Sand	Silt	Clay
	0.1733	-						2.20			
	>0.0768										
	0.1386							1.91			
	0.1733							1.62			
	0.2731			10				1.86			

Volatilization in Soils

Volatilization Parameters		Soil Properties				
K_w	K_wK_D	OC (%)	Moisture Θ	Bulk Density	Temp. °C	Porosity

Compound Name: 2,4-D amine

Compound Properties:

Structure

M.W. 236.1
M.P. 85-87°C
B.P.
Sp. gr.
Vapor pressure $1x10^{-9}$ mm Hg @ 28°C

Water solubility $3x10^6$ mg/l @ 20°C
Log K_{ow}
Chemical class
Chemical reactivity

Adsorption in Soils

Adsorption Parameters			Soil Properties					
					Percent (%)			
K	N	K_{oc}	pH	CEC	OC	Sand	Silt	Clay
4.62	0.70	119.4	7.3	54.7	3.87	18.4	45.3	38.3
0.65	0.83	72.2	5.6	6.8	0.90	65.8	19.5	14.7
0.76	0.73	135.7	9.6	5.6	0.56	93.8	3.0	3.2

Degradation in Soils

Degradation Parameters				Soil Properties							
								Percent (%)			
$t_{1/2}$	K_s	n	Initial conc.	Temp. °C	Moisture Θ	pH	CEC	OC	Sand	Silt	Clay
	0.1457							1.86			
	0.1008							1.86			

Volatilization in Soils

Volatilization Parameters		Soil Properties				
K_w	K_wK_D	OC (%)	Moisture Θ	Bulk Density	Temp. °C	Porosity

Compound Name: Dialifor

Compound Properties: Structure

M.W.
M.P.
B.P.
Sp. gr.
Vapor pressure

Water solubility
Log K_{ow}
Chemical class
Chemical reactivity

Adsorption in Soils

Adsorption Parameters			Soil Properties					
					Percent (%)			
K	N	K_{oc}	pH	CEC	OC	Sand	Silt	Clay

Degradation in Soils

Degradation Parameters				Soil Properties							
								Percent (%)			
$t_{1/2}$	K_s	n	Initial conc.	Temp. °C	Moisture Θ	pH	CEC	OC	Sand	Silt	Clay
140	-	-	1								
150	-	-	0.1								

Volatilization in Soils

Volatilization Parameters		Soil Properties				
K_w	K_wK_D	OC (%)	Moisture Θ	Bulk Density	Temp. °C	Porosity

Compound Name: 1,2,5,6-dibenzanthracene [dibenz[a,b]anthracene]

Compound Properties: Structure

M.W. 278.4
M.P. 262°C
B.P.
Sp. gr.
Vapor pressure 1×10^{-10} torr (20°C)

Water solubility 0.5 μg/l
Log K_{ow} 5.97
Chemical class
Chemical reactivity

Adsorption in Soils

Adsorption Parameters			Soil Properties					
					Percent (%)			
K	N	K_{oc}	pH	CEC	OC	Sand	Silt	Clay

Degradation in Soils

Degradation Parameters				Soil Properties							
								Percent (%)			
$t_{1/2}$	K_s	n	Initial conc.	Temp. °C	Moisture θ	pH	CEC	OC	Sand	Silt	Clay
21	0.033		97000	>25							
18	0.039		25000	>25							
130	0.005		57			6.1	0.75				
183	0.004		57			6.1	0.75				
99	0.007		93			6.1	0.75				
141	0.005		4.54			6.1	0.75				
119	0.007		147			6.1	0.75				
190	0.004		3.46			6.1	0.75				

Volatilization in Soils

Volatilization Parameters		Soil Properties				
K_w	K_wK_D	OC (%)	Moisture θ	Bulk Density	Temp. °C	Porosity

Compound Name: 1,2,5,6-dibenzanthracene

Compound Properties:

Structure

M.W. 278.3
M.P.
B.P.
Sp. gr.
Vapor pressure

Water solubility 249 g/l
Log K_{OW} 6.50
Chemical class
Chemical reactivity

Adsorption in Soils

Adsorption Parameters			Soil Properties					
					Percent (%)			
K	N	K_{OC}	pH	CEC	OC	Sand	Silt	Clay
20461		1690971						
34929		1687404						
18361		805292						
18882		2622453						
1759		1172847						
2506		2277875						
14497		3020262						
25302		2663317						
20192		3059425						
7345		565014						
55697		2962603						
39809		2383765						
19254		808991						
39840		2691870						

Degradation in Soils

Degradation Parameters				Soil Properties							
								Percent (%)			
$t_{1/2}$	K_s	n	Initial conc.	Temp. °C	Moisture Θ	pH	CEC	OC	Sand	Silt	Clay

Volatilization in Soils

Volatilization Parameters		Soil Properties				
K_w	K_wK_D	OC (%)	Moisture Θ	Bulk Density	Temp. °C	Porosity

Compound Name: Dibenzothiophene

Compound Properties: Structure

M.W. 184.27
M.P. 99-100°C
B.P.
Sp. gr.
Vapor pressure

Water solubility 1.47 ppm
Log K_{ow} 4.38
Chemical class
Chemical reactivity

Adsorption in Soils

Adsorption Parameters			Soil Properties					
					Percent (%)			
K	N	K_{oc}	pH	CEC	OC	Sand	Silt	Clay
117.5	-	9711	6.35	3.72	1.21			18.6
180.6	-	8725	7.79	23.72	2.07			55.2
167.1	-	7329	7.44	19.00	2.28			31.0
60.8	-	8444	7.83	33.01	0.72			68.6
9.4	-	6267	8.32	3.72	0.15			6.8
5.8	-	5273	8.34	12.40	0.11			17.4
49.7	-	10354	4.54	18.86	0.48			63.6
179.9	-	18937	7.79	11.30	0.95			35.7
65.1	-	9864	7.76	15.43	0.66			39.5
101.4	-	7800	5.50	8.50	1.30			28.6
276.0	-	14681	7.60	8.33	1.88			7.1
176.3	-	10557	7.55	8.53	1.67			21.2
338.6	-	16328	6.70	31.15	2.38			69.1
134.5	-	9088	7.75	20.86	1.48			42.9

Degradation in Soils

Degradation Parameters				Soil Properties							
								Percent (%)			
$t_{1/2}$	K_s	n	Initial conc.	Temp. °C	Moisture Θ	pH	CEC	OC	Sand	Silt	Clay
24	0.029		69			6.1		0.75			
24	0.029		0.57			6.1		0.75			

Volatilization in Soils

Volatilization Parameters		Soil Properties				
K_w	K_wK_D	OC (%)	Moisture Θ	Bulk Density	Temp. °C	Porosity

Compound Name: 1,2-bibromo, 3-chloropropane

Compound Properties:

Structure

M.W. 236.3
M.P.
B.P.
Sp. gr.
Vapor pressure

Water solubility
Log K_{ow}
Chemical class
Chemical reactivity

Adsorption in Soils

Adsorption Parameters			Soil Properties					
					Percent (%)			
K	N	K_{oc}	pH	CEC	OC	Sand	Silt	Clay

Degradation in Soils

Degradation Parameters				Soil Properties							
								Percent (%)			
$t_{1/2}$	K_s	n	Initial conc.	Temp. °C	Moisture Θ	pH	CEC	OC	Sand	Silt	Clay

Volatilization in Soils

Volatilization Parameters		Soil Properties				
K_w	K_wK_D	OC (%)	Moisture Θ	Bulk Density	Temp. °C	Porosity

Compound Name: Dibromomethane [methylene bromide]

Compound Properties: Structure

M.W.	173.9	Water solubility	3218 mg/l (20°C)
M.P.	-52.8°C		5033 mg/l (30°C)
B.P.		Log K_{ow}	
Sp. gr.		Chemical class	
Vapor pressure	340 mm (20°C)	Chemical reactivity	

Adsorption in Soils

Adsorption Parameters			Soil Properties					
					Percent (%)			
K	N	K_{oc}	pH	CEC	OC	Sand	Silt	Clay

Degradation in Soils

Degradation Parameters				Soil Properties							
$t_{1/2}$	K_s	n	Initial conc.	Temp. °C	Moisture Θ	pH	CEC	Percent (%) OC	Sand	Silt	Clay

Volatilization in Soils

Volatilization Parameters		Soil Properties				
K_w	K_wK_D	OC (%)	Moisture Θ	Bulk Density	Temp. °C	Porosity

Compound Name: Dicamba [3,6-dichloro-2-methoxybenzoic acid]

Compound Properties: Structure

M.W.	221.0	Water solubility	4470 mg/l
M.P.		Log K_{ow}	-1.69
B.P.		Chemical class	
Sp. gr.		Chemical reactivity	
Vapor pressure			

Adsorption in Soils

Adsorption Parameters			Soil Properties					
					Percent (%)			
K	N	K_{oc}	pH	CEC	OC	Sand	Silt	Clay
0.11	0.72		5.9		6.08	47.5	33.2	20.3
0.00	-		6.5		3.75	53.3	27.5	19.2
0.00	-		7.7		2.41	5.3	25.3	69.5
0.00	-		7.8		2.36	69.3	12.3	18.5
0.00	-		7.5		1.03	81.6	10.4	8.0

Degradation in Soils

Degradation Parameters				Soil Properties							
								Percent (%)			
t1/2	Ks	n	Initial conc.	Temp. °C	Moisture Θ	pH	CEC	OC	Sand	Silt	Clay
	0.197				5%	7.5		1.86			
	0.2140				10%	7.5		1.86			
	0.2140				-	5.2		6.79			
	0.0486				25%	7.7		2.43			
	0.0982				35%	7.7		2.43			
	0.0217				-			1.91			
	0.0407							1.62			
	0.0267							2.20			

Volatilization in Soils

Volatilization Parameters		Soil Properties				
K_W	K_WK_D	OC (%)	Moisture Θ	Bulk Density	Temp. °C	Porosity

Compound Name: Dicapthon [O-(2-chloro-4-nitrophenyl) O,O-dimethyl-phosphorothioate]

Compound Properties: Structure

M.W.	297.7	Water solubility	
M.P.	53°C	Log K_{OW}	
B.P.		Chemical class	
Sp. gr.		Chemical reactivity	
Vapor pressure			

Adsorption in Soils

Adsorption Parameters			Soil Properties					
					Percent (%)			
K	N	K_{oc}	pH	CEC	OC	Sand	Silt	Clay

Degradation in Soils

Degradation Parameters				Soil Properties							
			Initial	Temp.	Moisture			Percent (%)			
$t_{1/2}$	K_s	n	conc.	°C	Θ	pH	CEC	OC	Sand	Silt	Clay
43			0.1								

Volatilization in Soils

Volatilization Parameters		Soil Properties				
K_W	K_WK_D	OC (%)	Moisture Θ	Bulk Density	Temp. °C	Porosity

Compound Name: 1,2-dichlorobenzene

Compound Properties: Structure

M.W.	147.0	Water solubility	150 mg/l
M.P.		Log K_{OW}	3.38
B.P.	180.5°C	Chemical class	
Sp. gr.		Chemical reactivity	
Vapor pressure			

Adsorption in Soils

Adsorption Parameters			Soil Properties					
					Percent (%)			
K	N	K_{OC}	pH	CEC	OC	Sand	Silt	Clay
3.54	-	321.6		14	1.10	9	68	21

Degradation in Soils

Degradation Parameters				Soil Properties							
								Percent (%)			
$t_{1/2}$	K_S	n	Initial conc.	Temp. °C	Moisture Θ	pH	CEC	OC	Sand	Silt	Clay

Volatilization in Soils

Volatilization Parameters		Soil Properties				
K_W	K_WK_D	OC (%)	Moisture Θ	Bulk Density	Temp. °C	Porosity

Compound Name: 1,3-dichlorobenzene

Compound Properties: Structure

M.W.	147.0	Water solubility	134 mg/l
M.P.		Log K_{ow}	3.38
B.P.	173°C	Chemical class	
Sp. gr.		Chemical reactivity	
Vapor pressure			

Adsorption in Soils

Adsorption Parameters			Soil Properties					
					Percent (%)			
K	N	K_{oc}	pH	CEC	OC	Sand	Silt	Clay
3.23	-	293.3		14	1.1	9	68	21

Degradation in Soils

Degradation Parameters				Soil Properties							
								Percent (%)			
$t_{1/2}$	K_s	n	Initial conc.	Temp. °C	Moisture Θ	pH	CEC	OC	Sand	Silt	Clay

Volatilization in Soils

Volatilization Parameters		Soil Properties				
K_w	K_wK_D	OC (%)	Moisture Θ	Bulk Density	Temp. °C	Porosity

Compound Name: 1,4-dichlorobenzene

Compound Properties: Structure

M.W. 147.0
M.P. 53.5°C
B.P. 174.1°C
Sp. gr.
Vapor pressure

Water solubility 137 mg/l
Log K_{ow} 3.39
Chemical class
Chemical reactivity

Adsorption in Soils

Adsorption Parameters			Soil Properties					
					Percent (%)			
K	N	K_{oc}	pH	CEC	OC	Sand	Silt	Clay
3.01	-	273.8		14	1.1	9	68	21

Degradation in Soils

Degradation Parameters				Soil Properties							
								Percent (%)			
$t_{1/2}$	K_s	n	Initial conc.	Temp. °C	Moisture Θ	pH	CEC	OC	Sand	Silt	Clay

Volatilization in Soils

Volatilization Parameters		Soil Properties				
K_w	K_wK_D	OC (%)	Moisture Θ	Bulk Density	Temp. °C	Porosity

Compound Name: 1,4-Dichloro-2-butene

Compound Properties: Structure

M.W.		Water solubility	
M.P.	3.5°C	Log K_{ow}	
B.P.		Chemical class	
Sp. gr.		Chemical reactivity	
Vapor pressure			

Adsorption in Soils

Adsorption Parameters			Soil Properties					
					Percent (%)			
K	N	K_{oc}	pH	CEC	OC	Sand	Silt	Clay

Degradation in Soils

Degradation Parameters				Soil Properties							
								Percent (%)			
$t_{1/2}$	K_s	n	Initial conc.	Temp. °C	Moisture Θ	pH	CEC	OC	Sand	Silt	Clay

Volatilization in Soils

Volatilization Parameters		Soil Properties				
K_w	K_wK_D	OC (%)	Moisture Θ	Bulk Density	Temp. °C	Porosity

Compound Name: Dichlorodifluoromethane (gas)

Compound Properties: Structure

M.W.	120.9	Water solubility	280 mg/l (25°C)
M.P.	-111°C	Log K_{OW}	2.16
B.P.		Chemical class	
Sp. gr.		Chemical reactivity	
Vapor pressure			

Adsorption in Soils

Adsorption Parameters			Soil Properties					
					Percent (%)			
K	N	K_{OC}	pH	CEC	OC	Sand	Silt	Clay

Degradation in Soils

Degradation Parameters				Soil Properties							
								Percent (%)			
$t_{1/2}$	K_S	n	Initial conc.	Temp. °C	Moisture Θ	pH	CEC	OC	Sand	Silt	Clay

Volatilization in Soils

Volatilization Parameters		Soil Properties				
K_W	K_WK_D	OC (%)	Moisture Θ	Bulk Density	Temp. °C	Porosity

Compound Name: 1-1-Dichloroethane

Compound Properties: Structure

M.W.	o9.0	Water solubility	5500 mg/l (20°C)
M.P.	-97.4°C	Log K_{ow}	1.79
B.P.		Chemical class	
Sp. gr.		Chemical reactivity	
Vapor pressure	70 mm (0°C)		
	180 mm (20°C)		
	234 mm (25°C)		
	270 mm (30°C)		

Adsorption in Soils

Adsorption Parameters			Soil Properties					
					Percent (%)			
K	N	K_{oc}	pH	CEC	OC	Sand	Silt	Clay

Degradation in Soils

Degradation Parameters				Soil Properties							
								Percent (%)			
$t_{1/2}$	K_s	n	Initial conc.	Temp. °C	Moisture Θ	pH	CEC	OC	Sand	Silt	Clay

Volatilization in Soils

Volatilization Parameters		Soil Properties				
K_w	K_wK_D	OC (%)	Moisture Θ	Bulk Density	Temp. °C	Porosity

Compound Name: 1-2-Dichloroethane

Compound Properties:

Structure

M.W.	99	Water solubility	8690 mg/l (20°C)
M.P.	-35.4°C		9200 mg/l (0°C)
B.P.		Log K_{ow}	1.48
Sp. gr.		Chemical class	
Vapor pressure	40 mm (10°C)	Chemical reactivity	
	61 mm (20°C)		
	105 mm (30°C)		

Adsorption in Soils

Adsorption Parameters			Soil Properties					
					Percent (%)			
K	N	K_{oc}	pH	CEC	OC	Sand	Silt	Clay

Degradation in Soils

Degradation Parameters				Soil Properties							
								Percent (%)			
$t_{1/2}$	K_s	n	Initial conc.	Temp. °C	Moisture Θ	pH	CEC	OC	Sand	Silt	Clay

Volatilization in Soils

Volatilization Parameters		Soil Properties				
K_w	K_wK_D	OC (%)	Moisture Θ	Bulk Density	Temp. °C	Porosity

Compound Name: 1,2 Dichloropropane

Compound Properties: Structure

M.W. 113
M.P. -100/-80°C
B.P.
Sp. gr.
Vapor pressure 42 mm (20°C)
50 mm (25°C)
66 mm (30°C)

Water solubility 2700 mg/l (20°C)
Log K_{ow}
Chemical class
Chemical reactivity

Adsorption in Soils

Adsorption Parameters			Soil Properties		Percent (%)			
K	N	K_{oc}	pH	CEC	OC	Sand	Silt	Clay

Degradation in Soils

Degradation Parameters				Soil Properties				Percent (%)			
$t_{1/2}$	K_s	n	Initial conc.	Temp. °C	Moisture Θ	pH	CEC	OC	Sand	Silt	Clay

Volatilization in Soils

Volatilization Parameters		Soil Properties				
K_w	K_wK_D	OC (%)	Moisture Θ	Bulk Density	Temp. °C	Porosity

Compound Name: 1,3 Dichloropropane

Compound Properties: Structure

M.W. 113
M.P.
B.P.
Sp. gr.
Vapor pressure

Water solubility
Log K_{ow} 2.00
Chemical class
Chemical reactivity

Adsorption in Soils

Adsorption Parameters			Soil Properties					
					Percent (%)			
K	N	K_{oc}	pH	CEC	OC	Sand	Silt	Clay

Degradation in Soils

Degradation Parameters				Soil Properties							
								Percent (%)			
$t_{1/2}$	K_s	n	Initial conc.	Temp. °C	Moisture Θ	pH	CEC	OC	Sand	Silt	Clay

Volatilization in Soils

Volatilization Parameters		Soil Properties				
K_w	K_wK_D	OC (%)	Moisture Θ	Bulk Density	Temp. °C	Porosity

Compound Name: Dieldrin [1,2,3,4,10,10-hexachloro-b, 7-epoxy-1,4,4a,5,6,7,8, 8a-octahydro-exo-1,4-endo-5,8-dimethanonaphthalene]

Compound Properties: Structure

M.W.	380.9	Water solubility	
M.P.	176-177°C	Log K_{ow}	2.00
B.P.		Chemical class	
Sp. gr.		Chemical reactivity	
Vapor pressure	1.78×10^{-7} torr 25°C		

Adsorption in Soils

Adsorption Parameters			Soil Properties					
					Percent (%)			
K	N	Koc	pH	CEC	OC	Sand	Silt	Clay
39		9750		1.7	0.4			
147		12250		7.5	1.2			
265		9815		34.4	2.7			
198		6387		24.8	3.1			
260		7429		27.7	3.5			
297		3960		55.5	7.5			
1507		8970		72.4	16.8			
13937	1.08	31892	6.1		43.7	52	34	14
117	0.91	7222	6.6		1.62	71	22	7
66	0.89	4552	6.8		1.45	56	30	14
17.8	0.88	4341	7.0		0.41	91.5	1.5	7

Degradation in Soils

Degradation Parameters				Soil Properties							
								Percent (%)			
$t_{1/2}$	K_s	n	Initial conc.	Temp. °C	Moisture θ	pH	CEC	OC	Sand	Silt	Clay
	0.0002					7.8		0.58			
	0.0001					7.8		0.29			

Volatilization in Soils

Volatilization Parameters		Soil Properties				
K_W	$K_W K_D$	OC (%)	Moisture θ	Bulk Density	Temp. °C	Porosity

Compound Name: 7,12-dimethylbenz(a)anthracene

Compound Properties: Structure

M.W.	256.3	Water solubility	24.4 μg/l
M.P.	122-123°C	Log K_{ow}	5.98
B.P.		Chemical class	
Sp. gr.		Chemical reactivity	
Vapor pressure			

Adsorption in Soils

Adsorption Parameters			Soil Properties					
					Percent (%)			
K	N	K_{oc}	pH	CEC	OC	Sand	Silt	Clay
2371		195998						
2646		127812						
5210		228499						
1346		186986						
611		407496						
1028		934225						
562		117161						
3742		393907						
1895		287196						
1617		124347						
5576		296580						
2679		160391						
6777		284743						
3740		252735						

Degradation in Soils

Degradation Parameters				Soil Properties							
								Percent (%)			
$t_{1/2}$	K_s	n	Initial conc.	Temp. °C	Moisture Θ	pH	CEC	OC	Sand	Silt	Clay

Volatilization in Soils

Volatilization Parameters		Soil Properties				
K_W	K_WK_D	OC (%)	Moisture Θ	Bulk Density	Temp. °C	Porosity

Compound Name: 2,6-dinitro-N(3-pentyl)-α,α,α-trifluoro-p-toluidine

Compound Properties: Structure

M.W.	391.4	Water solubility	
M.P.		Log K_{ow}	
B.P.		Chemical class	
Sp. gr.		Chemical reactivity	
Vapor pressure			

Adsorption in Soils

Adsorption Parameters			Soil Properties					
					Percent (%)			
K	N	K_{oc}	pH	CEC	OC	Sand	Silt	Clay

Degradation in Soils

Degradation Parameters				Soil Properties							
								Percent (%)			
$t_{1/2}$	K_s	n	Initial conc.	Temp. °C	Moisture Θ	pH	CEC	OC	Sand	Silt	Clay
168[a]				15		7.6	11.4	0.93[b]	64	20	14
60[a]				30		7.6	11.4	0.93[b]	64	20	14
108[a]				15		6.6	19.4	1.28[b]	35	45	20
42[a]				30		6.6	19.4	1.28[b]	35	45	20

Volatilization in Soils

Volatilization Parameters		Soil Properties				
K_W	K_WK_D	OC (%)	Moisture Θ	Bulk Density	Temp. °C	Porosity

[a]Reported in months.
[b]Reported as OM.

Compound Name: Diuron [3-(3,4-dichlorophenyl)-1, 1-dimethylurea]

Compound Properties: Structure

M.W. 233.1
M.P. 158-159°C
B.P.
Sp. gr.
Vapor pressure

Water solubility 42 (25°C)
Log K_{ow}
Chemical class
Chemical reactivity

Adsorption in Soils

Adsorption Parameters			Soil Properties					
					Percent (%)			
K	N	K_{oc}	pH	CEC	OC	Sand	Silt	Clay
14.3	0.77	1144	6.7	20.7	1.25[a]	11	63	26
6.5	0.74	867	6.7	8.2	0.75[a]	90	5	5

Degradation in Soils

Degradation Parameters				Soil Properties							
								Percent (%)			
$t_{1/2}$	K_s	n	Initial conc.	Temp. °C	Moisture Θ	pH	CEC	OC	Sand	Silt	Clay
	0.0064					6.8					
	0.0072					6.4					

Volatilization in Soils

Volatilization Parameters		Soil Properties				
K_w	K_wK_D	OC (%)	Moisture Θ	Bulk Density	Temp. °C	Porosity

Compound Name: Ethylbenzene

Compound Properties: Structure

M.W.	106.2	Water solubility	140 mg/l (15°C)
M.P.	-94.97°C		152 mg/l (20°C)
B.P.	136°C		206 mg/l (30°C)
Sp. gr.		Log K_{ow}	3.15
Vapor pressure	7 mm (20°C)	Chemical class	
	12 mm (30°C)	Chemical reactivity	

Adsorption in Soils

Adsorption Parameters			Soil Properties					
					Percent (%)			
K	N	K_{oc}	pH	CEC	OC	Sand	Silt	Clay
1.8		165		14	1.10	9	68	21

Degradation in Soils

Degradation Parameters				Soil Properties							
								Percent (%)			
$t_{1/2}$	K_s	n	Initial conc.	Temp. °C	Moisture Θ	pH	CEC	OC	Sand	Silt	Clay

Volatilization in Soils

Volatilization Parameters		Soil Properties				
K_w	K_wK_D	OC (%)	Moisture Θ	Bulk Density	Temp. °C	Porosity

Compound Name: Fluometuron [1,1-Dimethyl-3-α,α,α-trifluoro-m-tolyl)urea]

Compound Properties: Structure

M.W.	232.2	Water solubility	79.4 mg/l
M.P.		Log K_{ow}	3.44
B.P.		Chemical class	
Sp. gr.		Chemical reactivity	
Vapor pressure			

Adsorption in Soils

Adsorption Parameters			Soil Properties					
					Percent (%)			
K	N	K_{oc}	pH	CEC	OC	Sand	Silt	Clay
0.90	1.08	120	7.1		0.75	25	43	32
0.80	-[a]	107	7.1		0.75	25	43	32
0.4	0.98	40.4	6.6	9.2	0.99	46	37.6	16.4

Degradation in Soils

Degradation Parameters				Soil Properties							
								Percent (%)			
$t_{1/2}$	K_s	n	Initial conc.	Temp. °C	Moisture Θ	pH	CEC	OC	Sand	Silt	Clay

Volatilization in Soils

Volatilization Parameters		Soil Properties				
K_W	K_WK_D	OC (%)	Moisture Θ	Bulk Density	Temp. °C	Porosity

[a]N is assumed to be one.

Compound Name: Fluoranthene

Compound Properties: Structure

M.W. 202.2
M.P. 111
B.P.
Sp. gr.
Vapor pressure 6.0×10^{-6} (20°C)

Water solubility 0.26 mg/l
Log K_{ow} 5.33
Chemical class
Chemical reactivity

Adsorption in Soils

Adsorption Parameters			Soil Properties					
					Percent (%)			
K	N	K_{oc}	pH	CEC	OC	Sand	Silt	Clay

Degradation in Soils

Degradation Parameters				Soil Properties							
			Initial	Temp.	Mois-ture			Percent (%)			
$t_{1/2}$	K_s	n	conc.	°C	Θ	pH	CEC	OC	Sand	Silt	Clay
44	0.016		3.9	15-25							
182	0.004		18.8	15-25							
105	0.007		23.0	15-25							
143	0.005		16.5	15-25							
109	0.006		20.9	15-25							
175	0.004		44.5	15-25							
133	0.005		72.8	15-25							

Volatilization in Soils

Volatilization Parameters		Soil Properties				
			Moisture	Bulk	Temp.	
K_w	K_wK_D	OC (%)	Θ	Density	°C	Porosity

Compound Name: Fonofos [O-Ethyl S-phenyl ethylphosphonodithioate]

Compound Properties:

Structure

M.W.	230.3	Water solubility	
M.P.		Log K_{ow}	
B.P.		Chemical class	
Sp. gr.		Chemical reactivity	
Vapor pressure			

Adsorption in Soils

Adsorption Parameters			Soil Properties		Percent (%)			
K	N	K_{oc}	pH	CEC	OC	Sand	Silt	Clay

Degradation in Soils

Degradation Parameters						Soil Properties		Percent (%)			
$t_{1/2}$	K_s	n	Initial conc.	Temp. °C	Moisture Θ	pH	CEC	OC	Sand	Silt	Clay
82						7.1		1.5	20	50	30
46						8.2		2.0	30	20	50

Volatilization in Soils

Volatilization Parameters		Soil Properties				
K_w	K_wK_D	OC (%)	Moisture Θ	Bulk Density	Temp. °C	Porosity

Compound Name: HCB [Hexachlorobenzene]

Compound Properties: Structure

M.W.	248.8	Water solubility	
M.P.	230°C	Log K_{ow}	4.13
B.P.	323-326°C	Chemical class	
Sp. gr.		Chemical reactivity	
Vapor pressure			

Adsorption in Soils

Adsorption Parameters			Soil Properties						
					Percent (%)			
K	N	K_{oc}	pH	CEC	OC	Sand	Silt	Clay
4193	1.12	24943	7.2	72.4	16.81	15.7	50.8	33.5
276	0.99	5835	7.1	18.1	4.73	6.1	61.4	32.5
136	0.93	11525	4.5	13.1	1.18	1.1	64.3	34.6

Degradation in Soils

Degradation Parameters				Soil Properties							
								Percent (%)			
$t_{1/2}$	K_s	n	Initial conc.	Temp. °C	Moisture Θ	pH	CEC	OC	Sand	Silt	Clay

Volatilization in Soils

Volatilization Parameters		Soil Properties				
K_w	K_wK_D	OC (%)	Moisture Θ	Bulk Density	Temp. °C	Porosity

Compound Name: HCH [Hexachlorocyclohexane]

Compound Properties: Structure

M.W. 290.8
M.P. 157-158°C (α)
309°C (β)
112.5°C (γ)
B.P.
Sp. gr.
Vapor pressure
2.15x10^{-5} torr (α)
2.0x10^{-7} torr (β)
0.14 mm Hg (γ) (40°C)

Water solubility 1.63 ppm @ 25°C (α)
0.70 ppm @ 25°C (β)
Log K_{ow} 4.14 (Lindane)
3.81 (α)
3.81 (β)
Chemical class
Chemical reactivity

Adsorption in Soils

Adsorption Parameters			Soil Properties					
						Percent (%)		
K	N	K_{oc}	pH	CEC	OC	Sand	Silt	Clay
28.18	1.60	6405	6.20	18.6	0.44	79.6	4.8	15.6
52.48	1.26	3143	6.30	42.8	1.67	69.6	6.8	23.6
70.79	0.94	2212	5.20	19.2	3.20	45.6	7.8	45.6
501.20	1.16	3510	3.30	28.9	14.28	63.6	6.8	29.6
5.13	0.82	1166	6.20	18.6	0.44	79.6	4.8	15.6
39.81	1.20	2384	6.30	42.8	1.67	69.6	6.8	23.6
79.43	0.80	2482	5.20	19.2	3.20	45.6	7.8	45.6
158.50	0.80	1110	3.30	28.9	14.28	63.6	6.8	29.6
8.91	0.90	2025	6.20	18.6	0.44	79.6	4.8	15.6
25.12	0.80	1504	6.30	42.8	1.67	69.6	6.8	23.6
31.62	0.80	988	5.20	19.2	3.20	45.6	7.8	45.6
446.70	1.10	3128	3.30	28.9	14.28	63.6	6.8	29.6
295	0.92	1635		18				
7.91	0.80	2274		0.35				
45.1	0.84	1313		3.5				
665	0.98	1522	5.1		43.7	52	34	14
13.1	0.96	809	5.6		1.62	.71	22	7
10.2	0.97	703	6.8		1.45	56	30	14
6.9	0.99	1679	7.0		91.5	91.5	1.5	7

Compound Name: HCH [Hexachlorocyclohexane] (Continued)

Degradation in Soils

Degradation Parameters				Soil Properties							
								Percent (%)			
$t_{1/2}$	K_S	n	Initial conc.	Temp. °C	Moisture Θ	pH	CEC	OC	Sand	Silt	Clay
	0.0022					7.8		0.58			
	0.0026					7.8		0.29			
	0.0011										
	0.0014										
	0.0048										
	0.0147										
	0.0264										
	0.0074										
	0.0263										
	0.0264										
	0.0139										

Volatilization in Soils

Volatilization Parameters		Soil Properties				
K_W	K_WK_D	OC (%)	Moisture Θ	Bulk Density	Temp. °C	Porosity

Compound Name: Hexachlorocyclopentadiene

Compound Properties: Structure

M.W.	272.8	Water solubility	0.8 ppm
M.P.	9/10°C		1.8 ppm
B.P.			2.0 ppm
Sp. gr.		Log K_{ow}	
Vapor pressure	0.08 mm (25°C)	Chemical class	
	1 mm (78-79°C)	Chemical reactivity	

Adsorption in Soils

Adsorption Parameters			Soil Properties					
					Percent (%)			
K	N	K_{oc}	pH	CEC	OC	Sand	Silt	Clay

Degradation in Soils

Degradation Parameters				Soil Properties							
			Initial	Temp.	Moisture			Percent (%)			
$t_{1/2}$	Ks	n	conc.	°C	Θ	pH	CEC	OC	Sand	Silt	Clay

Volatilization in Soils

Volatilization Parameters		Soil Properties				
			Moisture	Bulk	Temp.	
K_w	K_wK_D	OC (%)	Θ	Density	°C	Porosity

Compound Name: Hexachloroethane

Compound Properties: Structure

M.W. 236.76
M.P. 187°C
B.P.
Sp. gr.
Vapor pressure 0.4 mm (20°C)
0.8 mm (30°C)

Water solubility 50 mg/l (22°C)
Log K_{ow}
Chemical class
Chemical reactivity

Adsorption in Soils

Adsorption Parameters			Soil Properties					
					Percent (%)			
K	N	K_{oc}	pH	CEC	OC	Sand	Silt	Clay

Degradation in Soils

Degradation Parameters				Soil Properties							
								Percent (%)			
$t_{1/2}$	K_s	n	Initial conc.	Temp. °C	Moisture Θ	pH	CEC	OC	Sand	Silt	Clay

Volatilization in Soils

Volatilization Parameters		Soil Properties				
K_w	K_wK_D	OC (%)	Moisture Θ	Bulk Density	Temp. °C	Porosity

Compound Name: Iodomethane

Compound Properties: Structure

M.W. 142
M.P. -66.1 °C
B.P.
Sp. gr.
Vapor pressure 400 mm (25°C)

Water solubility 14 g/l (20°C)
Log K_{ow} 1.69
Chemical class
Chemical reactivity

Adsorption in Soils

Adsorption Parameters			Soil Properties					
					Percent (%)			
K	N	K_{oc}	pH	CEC	OC	Sand	Silt	Clay

Degradation in Soils

Degradation Parameters				Soil Properties							
								Percent (%)			
$t_{1/2}$	K_s	n	Initial conc.	Temp. °C	Moisture Θ	pH	CEC	OC	Sand	Silt	Clay

Volatilization in Soils

Volatilization Parameters		Soil Properties				
K_w	K_wK_D	OC (%)	Moisture Θ	Bulk Density	Temp. °C	Porosity

Compound Name: Indeno(1,2,3-cd)-pyrene

Compound Properties: Structure

M.W.	289.3	Water solubility	0.062 mg/l
M.P.	163	Log K_{ow}	7.66
B.P.		Chemical class	
Sp. gr.		Chemical reactivity	
Vapor pressure	1.0x10-10 torr (20°C)		

Adsorption in Soils

Adsorption Parameters			Soil Properties					
					Percent (%)			
K	N	K_{oc}	pH	CEC	OC	Sand	Silt	Clay

Degradation in Soils

Degradation Parameters				Soil Properties							
								Percent (%)			
$t_{1/2}$	K_s	n	Initial conc.	Temp. °C	Moisture θ	pH	CEC	OC	Sand	Silt	Clay

Volatilization in Soils

Volatilization Parameters		Soil Properties				
K_w	K_wK_D	OC (%)	Moisture θ	Bulk Density	Temp. °C	Porosity

Compound Name: Linuron[3-(3-4-dichlorophenyl)-1-methoxy-1-methylurea]

Compound Properties: Structure

M.W.	249.1	Water solubility	75 µg/l
M.P.		Log K_{ow}	
B.P.		Chemical class	
Sp. gr.		Chemical reactivity	
Vapor pressure			

Adsorption in Soils

Adsorption Parameters			Soil Properties					
					Percent (%)			
K	N	K_{oc}	pH	Texture	OC	Sand	Silt	Clay
2.30	0.84	1916	8.5	S	0.06			4.8
4.80	0.85	470	8.5	S	0.58			15.0
4.90	0.79	345	7.9	S	0.81			13.0
9.50	0.73	655	7.6	S	0.87			6.8
7.93	0.94	492	7.7	S cl	0.93			31.5
18.20	0.82	913	5.1	S	1.10			10.6
12.80	0.77	1049	7.8	S	0.70			18.3
23.10	0.86	497	4.6	S	2.67			4.5

Degradation in Soils

Degradation Parameters				Soil Properties							
								Percent (%)			
$t_{1/2}$	K_s	n	Initial conc.	Temp. °C	Moisture θ	pH	CEC	OC	Sand	Silt	Clay
	0.0047										

Volatilization in Soils

Volatilization Parameters		Soil Properties				
K_w	K_wK_D	OC (%)	Moisture θ	Bulk Density	Temp. °C	Porosity

Compound Name: Malathion [0,0-dimethyl S-(1,2-di)(ethoxycarbonyl)ethy)-phosphorodithioate]

Compound Properties: Structure

M.W.	330.4	Wa:er solubility	145 mg/l
M.P.	2.9°C	Log K_{ow}	
B.P.	156-157°C	Chemical class	
Sp. gr.		Chemical reactivity	
Vapor pressure			

Adsorption in Soils

Adsorption Parameters			Soil Properties					
					Percent (%)			
K	N	K_{oc}	pH	Texture	OC	Sand	Silt	Clay

Degradation in Soils

Degradation Parameters				Soil Properties							
$t_{1/2}$	K_S	n	Initial conc.	Temp. °C	Moisture Θ	pH	CEC	Percent (%) OC	Sand	Silt	Clay
	2.9173					7.2					
	2.4618					6.4					
	1.2681					3.8					
	0.4152					5.3	0.64				
	1.9832					7.4	1.80				
	1.9026					7.2	2.73				

Volatilization in Soils

Volatilization Parameters		Soil Properties				
K_W	K_WK_D	OC (%)	Moisture Θ	Bulk Density	Temp. °C	Porosity

Compound Name: MCPA [4-chloro-2-methylphenoxyacetic acid]

Compound Properties:

Structure

M.W.	200.6	Water solubility	890 mg/l
M.P.	120°C	Log K_{ow}	-1.41
B.P.		Chemical class	
Sp. gr.		Chemical reactivity	
Vapor pressure			

Adsorption in Soils

Adsorption Parameters			Soil Properties					
					Percent (%)			
K	N	K_{oc}	pH	Texture	OC	Sand	Silt	Clay

Degradation in Soils

Degradation Parameters				Soil Properties							
								Percent (%)			
$t_{1/2}$	K_s	n	Initial conc.	Temp. °C	Moisture Θ	pH	CEC	OC	Sand	Silt	Clay
41				23±2°		5.2		3.65	50.2	32.5	18.2

Volatilization in Soils

Volatilization Parameters		Soil Properties				
K_w	K_wK_D	OC (%)	Moisture Θ	Bulk Density	Temp. °C	Porosity

Compound Name: Methoxychlor [1,1,1-Trichloro-2, 2-bis(p-methoxyphenyl)-ethane]

Compound Properties: Structure

M.W.	345.65	Water solubility	0.12 mg/l
M.P.	89°C	Log K_{ow}	4.30
B.P.		Chemical class	
Sp. gr.		Chemical reactivity	
Vapor pressure			

Adsorption in Soils

Adsorption Parameters			Soil Properties					
					Percent (%)			
K	N	K_{oc}	pH	Texture	OC	Sand	Silt	Clay
53		41000			0.13			
2600		80000			3.27			
1800		91000			1.98			
1400		100000			1.34			
1100		92000			1.20			
8.3		9700			0.086			
2200		80000			2.78			
1700		73000			2.34			
2300		80000			2.84			
2400		73000			3.29			
95		17000			0.57			
2500		86000			2.92			
2000		100000			1.92			
2100		93000			1.99			

Degradation in Soils

Degradation Parameters				Soil Properties							
								Percent (%)			
$t_{1/2}$	K_s	n	Initial conc.	Temp. °C	Moisture Θ	pH	CEC	OC	Sand	Silt	Clay
	0.0046					4.8		0.58			
	0.0033					6.5		1.16			

Volatilization in Soils

Volatilization Parameters		Soil Properties				
K_w	K_wK_D	OC (%)	Moisture Θ	Bulk Density	Temp. °C	Porosity

Compound Name: 3-Methylcholanthrene

Compound Properties: Structure

M.W.	268.3	Water solubility	3.23 μg/l
M.P.	179-180°C	Log K_{ow}	6.42
B.P.	280°C	Chemical class	
Sp. gr.		Chemical reactivity	
Vapor pressure			

Adsorption in Soils

Adsorption Parameters			Soil Properties					
					Percent (%)			
K	N	K_{oc}	pH	CEC	OC	Sand	Silt	Clay
15140		1251210						
30085		1453404						
8273		362845						
15820		2197250						
2257		1504538						
2694		2449190						
30627		6380703						
23080		2429456						
20642		3127521						
16231		1248534						
24506		1303532						
20972		1255821						
17127		719633						
37364		2524581						

Degradation in Soils

Degradation Parameters				Soil Properties							
								Percent (%)			
$t_{1/2}$	K_s	n	Initial conc.	Temp. °C	Moisture Θ	pH	CEC	OC	Sand	Silt	Clay

Volatilization in Soils

Volatilization Parameters		Soil Properties				
K_w	K_wK_D	OC (%)	Moisture Θ	Bulk Density	Temp. °C	Porosity

Compound Name: Methyl parathion [0-0-dimethyl-0-p-nitrophenyl phosphorothioate]

Compound Properties: Structure

M.W.	263.2	Water solubility	55-60 mg/l (25°C)
M.P.	35-36°C	Log K_{ow}	1.91
B.P.		Chemical class	
Sp. gr.		Chemical reactivity	
Vapor pressure	9.5×10^{-6} mm Hg (20°C)		

Adsorption in Soils

Adsorption Parameters			Soil Properties					
					Percent (%)			
K	N	K_{oc}	pH	CEC	OC	Sand	Silt	Clay
13.39	0.75	346.0	7.3	54.7	3.87	18.4	45.3	38.3
3.95	0.85	438.6	5.6	6.8	0.90	65.8	19.5	14.7
2.72	0.86	486.4	5.6	5.2	0.56	93.8	3.0	3.2
3.57	0.61	714.5	7.4	35.8	0.50	50.7	16.4	22.9

Degradation in Soils

Degradation Parameters				Soil Properties							
								Percent (%)			
$t_{1/2}$	K_s	n	Initial conc.	Temp. °C	Moisture Θ	pH	CEC	OC	Sand	Silt	Clay
	0.2207										

Volatilization in Soils

Volatilization Parameters		Soil Properties				
K_w	K_wK_D	OC (%)	Moisture Θ	Bulk Density	Temp. °C	Porosity

Compound Name: Metobromuron [3-(p-bromophenyl)-1-methoxy-1-methylurea]

Compound Properties: Structure

M.W.	259.1	Water solubility	350 μg/l
M.P.		Log K_{ow}	
B.P.		Chemical class	
Sp. gr.		Chemical reactivity	
Vapor pressure			

Adsorption in Soils

Adsorption Parameters			Soil Properties					
					Percent (%)			
K	N	K_{oc}	pH	Texture	OC	Sand	Silt	Clay
2.52	0.58	2100	8.5	S	0.06			9.8
2.09	0.78	204	8.5	S	0.58			15.0
2.20	0.82	154	7.9	S	0.81			13.0
1.95	0.82	134	7.6	S	0.87			6.8
4.54	0.85	281	7.7	S cl	0.93			31.5
6.18	0.80	330	5.1	S	1.10			10.6
6.12	0.82	501	7.8	S	0.70			18.3
10.20	0.81	219	4.6	S	2.67			4.5

Degradation in Soils

Degradation Parameters				Soil Properties							
								Percent (%)			
$t_{1/2}$	K_s	n	Initial conc.	Temp. °C	Moisture Θ	pH	CEC	OC	Sand	Silt	Clay

Volatilization in Soils

Volatilization Parameters		Soil Properties				
K_w	K_wK_D	OC (%)	Moisture Θ	Bulk Density	Temp. °C	Porosity

Compound Name: Monolinuron [3-(p-chlorophenyl)-1-methoxy-1-methylurea]

Compound Properties: Structure

M.W.	214.6	Water solubility	580 µg/l
M.P.		Log K_{ow}	
B.P.		Chemical class	
Sp. gr.		Chemical reactivity	
Vapor pressure			

Adsorption in Soils

Adsorption Parameters			Soil Properties					
					Percent (%)			
K	N	K_{oc}	pH	Texture	OC	Sand	Silt	Clay
2.43	0.50	2025	8.5	S	0.06			9.8
3.28	0.86	321	8.5	S	0.58			15.0
3.20	0.60	225	7.9	S	0.81			13.0
4.90	0.81	334	7.6	S	0.87			6.8
3.41	0.91	211	7.7	S cl	0.93			31.5
4.19	0.82	224	5.1	S	1.10			10.6
6.35	0.87	520	7.8	S	0.70			18.3
12.70	0.90	273	4.6	S	2.67			4.5

Degradation in Soils

Degradation Parameters				Soil Properties							
								Percent (%)			
$t_{1/2}$	Ks	n	Initial conc.	Temp. °C	Moisture Θ	pH	CEC	OC	Sand	Silt	Clay

Volatilization in Soils

Volatilization Parameters		Soil Properties				
K_w	K_wK_D	OC (%)	Moisture Θ	Bulk Density	Temp. °C	Porosity

Compound Name: Naphthalene

Compound Properties:

Structure

M.W.	128.2	Water solubility	30 mg/l
M.P.	80.2°C	Log K_{ow}	3.37
B.P.		Chemical class	
Sp. gr.		Chemical reactivity	
Vapor pressure	4.92×10^{-2} torr (20°C)		

Adsorption in Soils

Adsorption Parameters			Soil Properties					
					Percent (%)			
K	N	K_{oc}	pH	CEC	OC	Sand	Silt	Clay
		1300						

Degradation in Soils

Degradation Parameters				Soil Properties							
								Percent (%)			
$t_{1/2}$	Ks	n	Initial conc.	Temp. °C	Moisture Θ	pH	CEC	OC	Sand	Silt	Clay
0.12	5.78		7.0	15-25							
125	0.005		7.0	<15							
4	0.173		25000	>25							

Volatilization in Soils

Volatilization Parameters		Soil Properties				
K_W	$K_W K_D$	OC (%)	Moisture Θ	Bulk Density	Temp. °C	Porosity

Compound Name: Parathion [O,O-Diethyl O-(p-nitrophenyl) phosphorothioate]

Compound Properties: Structure

M.W.	291.3	Water solubility	6.54 ppm
M.P.	6°C		24 ppm (25°C)
B.P.	375°C	Log K_{ow}	3.40
Sp. gr.		Chemical class	
Vapor pressure	3.78×10^{-5} torr (20°C)	Chemical reactivity	

Adsorption in Soils

Adsorption Parameters			Soil Properties					
						Percent (%)		
K	N	K_{oc}	pH	CEC	OC	Sand	Silt	Clay
4.61	0.83	904	7.30	5.71	0.51	77	15	8
8.98	0.83	839	6.83	6.10	1.07	83	9	8
29.16	0.88	1105	5.00	21.02	2.64	37	42	21
32.13	0.80	845	7.30	37.84	3.80	21	55	24
201.47	0.81	1097	6.98	77.34	18.36	42	39	19
7.67	1.04	1743	6.2	18.6	0.44	79.6	48	15.6
12.30	1.05	1309	6.25	26.6	0.94	75.9	3.4	20.7
38.02	1.11	2277	6.30	42.8	1.67	69.6	6.8	23.6
125.90	1.05	3934	5.20	19.2	3.20	45.6	7.8	45.6
213.80	1.03	4792	3.50	21.2	4.76	53.6	12.8	33.6
457.10	1.02	3200	3.30	28.9	14.28	63.6	6.8	29.6
349	0.95	799	6.1		43.7	52	34	14
18.9	0.99	1166	6.6		1.62	71	22	7
16.3	1.01	1124	6.8		1.45	56	30	14
5.2	0.98	1264	7.0		0.41	91.5	1.5	7

Degradation in Soils

Degradation Parameters				Soil Properties							
								Percent (%)			
$t_{1/2}$	K_s	n	Initial conc.	Temp. °C	Moisture Θ	pH	CEC	OC	Sand	Silt	Clay
180			1	20		6.2	20	1.74			51
110			0.1	20		6.2	20	1.74			51

Volatilization in Soils

Volatilization Parameters		Soil Properties				
K_w	K_wK_D	OC (%)	Moisture Θ	Bulk Density	Temp. °C	Porosity

Compound Name: PBBs [polybrominated biphenyls]

Compound Properties: Structure

M.W.	Water solubility
M.P.	Log K_{ow}
B.P.	Chemical class
Sp. gr.	Chemical reactivity
Vapor pressure	

Adsorption in Soils

Adsorption Parameters			Soil Properties					
					Percent (%)			
K	N	K_{oc}	pH	CEC	OC	Sand	Silt	Clay
361[a]	1.99	2142	7.2	72.4	16.81	15.7	50.8	33.5
144[a]	1.89	3044	7.1	18.1	4.73	6.1	61.4	32.5
88[a]	1.77	7458	4.50	13.1	1.18	1.1	64.3	34.6

Degradation in Soils

Degradation Parameters				Soil Properties							
								Percent (%)			
$t_{1/2}$	K_s	n	Initial conc.	Temp. °C	Moisture Θ	pH	CEC	OC	Sand	Silt	Clay

Volatilization in Soils

Volatilization Parameters		Soil Properties				
K_w	K_wK_D	OC (%)	Moisture Θ	Bulk Density	Temp. °C	Porosity

[a]Reported in

Compound Name: Pentachloroethane

Compound Properties: Structure

M.W.	202.3	Water solubility	
M.P.	-29°C	Log K_{ow}	
B.P.		Chemical class	
Sp. gr.		Chemical reactivity	
Vapor pressure	3.4 mm (20°C)		
	6 mm (30°C)		

Adsorption in Soils

Adsorption Parameters			Soil Properties		Percent (%)			
K	N	K_{oc}	pH	CEC	OC	Sand	Silt	Clay

Degradation in Soils

Degradation Parameters				Soil Properties				Percent (%)			
$t_{1/2}$	K_s	n	Initial conc.	Temp. °C	Moisture Θ	pH	CEC	OC	Sand	Silt	Clay

Volatilization in Soils

Volatilization Parameters		Soil Properties				
K_w	K_wK_D	OC (%)	Moisture Θ	Bulk Density	Temp. °C	Porosity

Compound Name: Phenanthrene

Compound Properties:

Structure

M.W. 178.2
M.P. 100°C
B.P. 340°C
Sp. gr.
Vapor pressure 6.8×10^{-4} torr (20°C)

Water solubility 1.6 mg/l (15°C)
Log K_{OW} 4.46
Chemical class
Chemical reactivity

Adsorption in Soils

Adsorption Parameters			Soil Properties					
					Percent (%)			
K	N	K_{OC}	pH	CEC	OC	Sand	Silt	Clay
		23000						

Degradation in Soils

Degradation Parameters				Soil Properties							
					Moisture			Percent (%)			
$t_{1/2}$	K_S	n	Initial conc.	Temp. °C	Θ	pH	CEC	OC	Sand	Silt	Clay
26	0.027		2.1	15-25							
35	0.198		25000	>25							

Volatilization in Soils

Volatilization Parameters		Soil Properties				
K_W	K_WK_D	OC (%)	Moisture Θ	Bulk Density	Temp. °C	Porosity

Compound Name: Phorate [O,O-Diethyl S-[(ethylthio)methyl phosphorodithioate; diethyl S-(ethylthiomethyl)phosphorothiolathionate]

Compound Properties: Structure

M.W.	260.4	Water solubility	20 ppm 50 ppm @ room temp.
M.P.	<-15°C		
B.P.	75-78°C	Log K_{ow}	3.33
Sp. gr.		Chemical class	
Vapor pressure	8.4×10^{-4} torr (20°C)	Chemical reactivity	

Adsorption in Soils

Adsorption Parameters			Soil Properties					
					Percent (%)			
K	N	K_{oc}	pH	CEC	OC	Sand	Silt	Clay
2.14	0.94	419.6	7.3	5.71	0.51	77	15	8
4.86	0.91	454.2	6.83	6.10	1.07	83	9	8
8.63	0.92	326.9	5.00	21.02	2.64	37	42	21
13.73	0.88	361.3	7.30	37.84	3.80	21	55	24
74.79	1.01	407.4	6.98	77.34	18.36	42	39	19

Degradation in Soils

Degradation Parameters				Soil Properties							
								Percent (%)			
$t_{1/2}$	K_s	n	Initial conc.	Temp. °C	Moisture Θ	pH	CEC	OC	Sand	Silt	Clay
18						7.1		1.5[a]	20	50	30
14						8.0		2.0[a]	30	20	50

Volatilization in Soils

Volatilization Parameters		Soil Properties				
K_w	K_wK_D	OC (%)	Moisture Θ	Bulk Density	Temp. °C	Porosity

[a]Reported as OM (OC = OM/1.724).

Compound Name: Phosmet

Compound Properties:

Structure

M.W.
M.P.
B.P.
Sp. gr.
Vapor pressure

Water solubility
Log K_{ow}
Chemical class
Chemical reactivity

Adsorption in Soils

Adsorption Parameters			Soil Properties					
					Percent (%)			
K	N	K_{oc}	pH	CEC	OC	Sand	Silt	Clay

Degradation in Soils

Degradation Parameters				Soil Properties							
			Initial	Temp.	Mois-ture			Percent (%)			
$t_{1/2}$	K_s	n	conc.	°C	Θ	pH	CEC	OC	Sand	Silt	Clay
60	-	-	1								
51	-	-	0.1								

Volatilization in Soils

Volatilization Parameters		Soil Properties				
			Moisture	Bulk	Temp.	
K_w	K_wK_D	OC (%)	Θ	Density	°C	Porosity

Compound Name: Picloram [4-Amino-3,5,6-trichloro-picolinic acid]

Compound Properties: Structure

M.W.	241.5	Water solubility	426 mg/l
M.P.		Log K_{ow}	3.47
B.P.		Chemical class	
Sp. gr.		Chemical reactivity	
Vapor pressure			

Adsorption in Soils

Adsorption Parameters			Soil Properties					
					Percent (%)			
K	N	K_{oc}	pH	CEC	OC	Sand	Silt	Clay
0.684[a]	0.841		5.60	20	2.44			18
0.664[a]	0.850		5.60	20	2.44			18
0.639[a]	0.861		5.60	20	2.44			18
0.614[a]	0.822		5.68	19	2.09			27
0.529[a]	0.840		5.68	19	2.09			27
0.499[a]	0.841		5.68	19	2.09			27
0.297[a]	0.886		7.40	41	1.39			33
0.314[a]	0.815		7.40	41	1.39			33
0.297[a]	0.835		7.40	41	1.39			33
0.04[a]	0.92		7.14	8	0.54			8
0.976[b]	0.849		5.60	20	2.44			18
0.533[b]	0.816		5.68	19	2.09			27
0.310[b]	0.829		6.97	14	1.39			83
0.409[b]	0.743		7.40	41	1.39			33
0.118[b]	0.838		6.40	12	0.93			9
0.070[b]	0.596		7.14	8	0.54			8
0.75	-		7.9		7.20			13.5
0.49	-		6.5		6.08			20.3
0.31	-		7.9		3.75			19.2
0.23	-		8.1		14.72			57.3
0.24	-		8.2		2.48			15.0
0.0	-		8.0		2.41			69.5
0.24	-		8.1		2.36			18.5
0.09	-		6.9		1.03			8.0

[a]These constants are determined at 1:5 soil/solution ratio.
[b]These constants are determined at 1:2 soil/solution ratio.

Compound Name: Picloram [4-Amino-3,5,6-trichloro-picolinic acid] (continued)

Degradation in Soils

Degradation Parameters				Soil Properties							
								Percent (%)			
$t_{1/2}$	K_S	n	Initial conc.	Temp. °C	Moisture Θ	pH	CEC	OC	Sand	Silt	Clay
	0.0025										
	0.0044										
	0.0050					4.8		1.68			
	0.0354					6.3		1.10			
	0.0258					5.5		0.99			
	0.0268					5.8		1.10			
	0.0269					5.8		1.10			
	0.0048							1.62			

Volatilization in Soils

Volatilization Parameters		Soil Properties				
K_W	K_WK_D	OC (%)	Moisture Θ	Bulk Density	Temp. °C	Porosity

Compound Name: Prometone [2-methoxy-4,6-bis(isopropylamino)-1,3,5-triazine]

Compound Properties: Structure

M.W.	225.3	Vapor pressure	
M.P.		Water solubility	750 ppm
PK_1 = 4.3		Log K_{OW}	1.94
B.P.		Chemical class	
Sp. gr.		Chemical reactivity	

Adsorption in Soils

Adsorption Parameters			Soil Properties					
						Percent (%)		
K	N	K_{OC}	pH	CEC	OC	Sand	Silt	Clay
1.4			7.0	25.4	0.1			42.8
0.4			6.6	10.2	0.4			8.2
0.5			6.2	10.6	0.5			15.8
1.1			4.9	11.0	0.9			13.9
0.9			5.7	9.2	0.9			20.3
1.3			5.9	12.4	1.2			28.5
1.5			6.1	14.0	1.3			19.3
1.4			5.8	11.0	1.5			17.9
0.7			6.7	30.1	1.5			27.8
1.6			6.2	13.9	1.5			33.5
1.4			5.8	10.3	1.0			21.0
1.5			6.9	20.2	1.6			22.0
1.8			6.4	21.5	1.7			35.2
1.3			6.7	22.3	1.8			37.1
2.1			6.6	24.8	2.7			61.6
2.3			6.5	25.9	3.1			63.4
1.6			6.9	29.2	3.4			31.9
2.7			6.3	26.9	3.8			25.9
2.7			5.7	27.4	4.6			30.2
2.7			7.6	25.7	6.5			28.1
2.4			7.2	42.0	7.8			53.8
5.4			6.8	44.6	9.4			20.1
7.9			5.3	51.3	9.9			29.5
6.7			7.1	37.6	12.6			28.4
7.4			7.1	11.2	13.6			8.1
9.8			6.9	50.4	14.0			17.6
9.5			6.5	61.3	14.1			31.2
11.3			6.5	82.1	15.4			28.7
11.9			6.5	84.3	15.8			19.4
13.1			5.9	92.3	16.6			31.4
14.4			6.8	94.2	18.9			11.2
18.5			6.9	85.1	20.0			12.8
17.2			6.8	106.7	22.9			7.3
25.6			5.6	120.9	23.6			32.1
21.7			6.7	131.1	25.9			24.9
16.7			6.9	123.1	27.1			18.8

Compound Name: Prometone [2-methoxy-4,6-bis(isopropylamino)-1,3,5-triazine] (continued)

Degradation in Soils

Degradation Parameters				Soil Properties							
								Percent (%)			
$t_{1/2}$	K_s	n	Initial conc.	Temp. °C	Moisture Θ	pH	CEC	OC	Sand	Silt	Clay

Volatilization in Soils

Volatilization Parameters		Soil Properties				
K_W	K_WK_D	OC (%)	Moisture Θ	Bulk Density	Temp. °C	Porosity

Compound Name: Prometryne [2-methylthio-4,6-bis(isopropylamino)-1,3,5-triazine]

Compound Properties: Structure

M.W. 241.3
M.P. 118-120°C
PK_1 = 4.05
B.P.
Sp. gr.

Vapor pressure
Water solubility 48 ppm
Log K_{ow}
Chemical class
Chemical reactivity

Adsorption in Soils

Adsorption Parameters			Soil Properties					
					Percent (%)			
K	N	K_{oc}	pH	CEC	OC	Sand	Silt	Clay
3.3			7.0	25.4	0.1			42.8
1.8			6.6	10.2	0.4			8.2
3.1			6.2	10.6	0.5			15.8
4.7			4.9	11.0	0.9			13.9
6.2			5.7	9.2	0.9			20.3
3.8			5.9	12.4	1.2			28.5
3.6			6.1	14.0	1.3			19.3
6.2			5.8	11.0	1.5			17.9
8.2			6.7	30.1	1.5			27.8
3.8			6.2	13.9	1.5			33.5
4.8			5.8	10.3	1.0			21.0
5.8			6.9	20.2	1.6			22.0
5.4			6.4	21.5	1.7			35.2
5.4			6.7	22.3	1.8			37.1
6.2			6.6	24.8	2.7			61.6
6.5			6.5	25.9	3.1			63.4
8.2			6.9	29.2	3.4			31.9
10.2			6.3	26.9	3.8			25.9
9.8			5.7	27.4	4.6			30.2
12.2			7.6	25.7	6.5			28.1
14.2			7.2	42.0	7.8			53.8
28.3			6.8	44.6	9.4			20.1
39.7			5.3	51.3	9.9			29.5
33.1			7.1	37.6	12.6			28.4
34.7			7.1	11.2	13.6			8.1
56.4			6.9	50.4	14.0			17.6
49.2			6.5	61.3	14.1			31.2
44.9			6.5	82.1	15.4			28.7
47.5			6.5	84.3	15.8			19.4
54.9			5.9	92.3	16.6			31.4
53.3			6.8	94.2	18.9			11.2
63.8			6.9	85.1	20.0			12.8
75.1			6.8	106.7	22.9			7.3
77.6			5.6	120.9	23.6			32.1
106.6			6.7	131.1	25.9			24.9
83.8			6.9	123.1	27.1			18.8

Compound Name: Prometryne [2-methylthio-4,6-bis(isopropylamino)-1,3,5-triazine] (continued)

Degradation in Soils

Degradation Parameters				Soil Properties							
								Percent (%)			
$t_{1/2}$	K_S	n	Initial conc.	Temp. °C	Moisture Θ	pH	CEC	OC	Sand	Silt	Clay
	0.0238	-				7.0		1.16			18

Volatilization in Soils

Volatilization Parameters		Soil Properties				
K_W	K_WK_D	OC (%)	Moisture Θ	Bulk Density	Temp. °C	Porosity

[a]Reported as OM (OC = OM/1.724).

Compound Name: Propazine [2-chloro-4,6-bis(isopropylamino)-1,3,5-triazine]

Compound Properties: Structure

M.W.	229.7	Water solubility	8.6 ppm (20°C)
M.P.		Log K_{ow}	
B.P.		Chemical class	
Sp. gr.		Chemical reactivity	
Vapor pressure			

Adsorption in Soils

Adsorption Parameters			Soil Properties					
					Percent (%)			
K	N	K_{oc}	pH	CEC	OC	Sand	Silt	Clay
1.1			7.0	25.4	0.1			
0.5			6.6	10.2	0.4			
0.9			6.2	10.6	0.5			
1.4			4.9	11.0	0.9			
1.4			5.7	9.2	0.9			20.3
2.3			5.9	12.4	1.2			28.5
2.4			6.1	14.0	1.3			19.3
1.5			5.8	11.0	1.5			17.9
2.7			6.7	30.1	1.5			27.8
1.2			6.2	13.9	1.5			33.5
2.0			5.8	10.3	1.6			21.0
2.1			6.9	20.2	1.6			22.0
2.6			6.4	21.5	1.7			35.2
3.0			6.7	22.3	1.8			37.1
3.9			6.6	24.8	2.7			61.6
3.9			6.5	25.9	3.1			63.4
3.5			6.9	29.2	3.4			31.9
4.3			6.3	26.9	3.8			25.9
4.7			5.7	27.4	4.6			30.2
5.7			7.6	25.7	6.5			28.1
4.9			7.2	42.0	7.8			53.8
12.9			6.8	44.6	9.4			20.1
16.2			5.3	51.3	9.9			29.5
16.5			7.1	37.6	12.6			28.4
19.0			7.1	11.2	13.6			8.1
20.5			6.9	50.4	14.0			17.6
18.0			6.5	61.3	14.1			31.2
24.8			6.5	82.1	15.4			28.7
21.4			6.5	84.3	15.8			19.4
27.4			5.9	92.3	16.6			31.4
34.0			6.8	94.2	18.9			11.2
31.6			6.9	83.1	20.0			12.8
42.8			6.8	106.7	22.9			7.3
45.7			5.6	126.9	23.6			32.1
42.5			6.7	131.1	25.9			24.9
40.0			6.9	123.1	27.1			18.8

Compound Name: Propazine [2-chloro-4,6-bis(isopropylamino)-1,3,5-triazine] (continued)

Degradation in Soils

Degradation Parameters				Soil Properties							
								Percent (%)			
$t_{1/2}$	K_S	n	Initial conc.	Temp. °C	Moisture Θ	pH	CEC	OC	Sand	Silt	Clay
	0.0108					4.8		0.58			
	0.0056					6.5		1.16			

Volatilization in Soils

Volatilization Parameters		Soil Properties				
K_W	K_WK_D	OC (%)	Moisture Θ	Bulk Density	Temp. °C	Porosity

Compound Name: Pyrene (benzo(def)phenanthrene)

Compound Properties: Structure

M.W.	202.24	Water solubility	0.14 mg/l
M.P.	156°C	Log K_{ow}	4.88
B.P.		Chemical class	
Sp. gr.		Chemical reactivity	
Vapor pressure	$6.85x10^{-7}$ torr @ 20°C		

Adsorption in Soils

Adsorption Parameters			Soil Properties					
					Percent (%)			
K	N	K_{oc}	pH	CEC	OC	Sand	Silt	Clay
42		32000			0.13			
3000		92000			3.27			
2500		130000			1.98			
1500		110000			1.34			
1400		120000			1.20			
994		11000			0.086			
2100		76000			2.78			
3000		130000			2.34			
3600		120000			2.89			
3800		120000			3.29			
68		12000			0.57			
3200		110000			2.92			
2300		120000			1.99			
2500		110000			2.26			
760		62860						
1065		51469						
1155		50650						
614		82256						
101		67467						
71		64706						
277		57763						
783		82421						
504		76316						
723		59646						
1119		59515						
806		48236						
1043		43807						
994		67189						

Compound Name: Pyrene (benzo(def)phenanthrene) (continued)

Degradation in Soils

Degradation Parameters				Soil Properties							
								Percent (%)			
$t_{1/2}$	K_S	n	Initial conc.	Temp. °C	Moisture Θ	pH	CEC	OC	Sand	Silt	Clay
35	0.020	-	3.1								
10.5	0.067	-	500								
3	0.231	-	5								

Volatilization in Soils

Volatilization Parameters		Soil Properties				
K_W	K_WK_D	OC (%)	Moisture Θ	Bulk Density	Temp. °C	Porosity

Compound Name: Silvex [2-(2,4,5-Trichlorophenoxy)propionic acid]

Compound Properties: Structure

M.W.	269.5	Water solubility	140 mg/l (25°C)
M.P.	181.6°C	Log K_{ow}	
B.P.		Chemical class	
Sp. gr.		Chemical reactivity	
Vapor pressure			

Adsorption in Soils

Adsorption Parameters			Soil Properties					
					Percent (%)			
K	N	K_{OC}	pH	CEC	OC	Sand	Silt	Clay
42.1	0.639	2786	6.09	39.6	1.51	4.0	79.6	16.4
34.2	0.987	4682	6.07	34.5	0.73	6.5	81.5	12.0
162	1.05	440	6.25	42.0	36.8	8.3	80.2	11.5

Degradation in Soils

Degradation Parameters				Soil Properties							
								Percent (%)			
$t_{1/2}$	K_S	n	Initial conc.	Temp. °C	Moisture Θ	pH	CEC	OC	Sand	Silt	Clay
	0.0330							1.91			
	0.0495							1.62			
	0.0462							2.20			

Volatilization in Soils

Volatilization Parameters		Soil Properties				
K_W	$K_W K_D$	OC (%)	Moisture Θ	Bulk Density	Temp. °C	Porosity

Compound Name: Tebuthiuron [1-(5-tert-butyl-1,3,4-thiodiazol-2-yl)-1,3-dimethylurea]

Compound Properties: Structure

M.W.	216.3	Water solubility	2500 ppm (25°C)
M.P.		Log K_{ow}	
B.P.		Chemical class	
Sp. gr.		Chemical reactivity	
Vapor pressure			

Adsorption in Soils

Adsorption Parameters			Soil Properties					
					Percent (%)			
K	N	K_{oc}	pH	CEC	OC	Sand	Silt	Clay
2.4	-[a]		6.3	9.2	2.55[b]	30	52.2	17.8
1.2	-		6.1	13.3	2.49[b]	57	30.5	12.5
0.5	-		6.1	6.7	0.81[b]	58	33.8	8.2
0.1	-		6.7	0.6	0.17[b]	95	3	2

Degradation in Soils

Degradation Parameters				Soil Properties							
								Percent (%)			
$t_{1/2}$	K_s	n	Initial conc.	Temp. °C	Moisture Θ	pH	CEC	OC	Sand	Silt	Clay

Volatilization in Soils

Volatilization Parameters		Soil Properties				
K_w	K_wK_D	OC (%)	Moisture Θ	Bulk Density	Temp. °C	Porosity

[a]Assumed to be 1.
[b]Reported as OM.

Compound Name: Terbacil [3-tert-butyl-5 chloro-6 methyluracil]

Compound Properties: Structure

M.W.	216.7	Water solubility	710 mg/l (25°C)
M.P.	175-177°C	Log K_{ow}	
B.P.		Chemical class	
Sp. gr.		Chemical reactivity	
Vapor pressure	4.8×10^{-7} mm Hg (29.5°C)		

Adsorption in Soils

Adsorption Parameters			Soil Properties					
					Percent (%)			
K	N	K_{oc}	pH	CEC	OC	Sand	Silt	Clay
2.46	0.88	63.6	7.3	54.7	3.87	18.4	45.3	38.3
0.38	0.99	42.2	5.6	6.8	0.90	65.8	19.5	14.7
0.12	0.88	21.4	5.6	5.2	0.56	93.8	3.0	3.2
0.38	0.93	76.0	7.4	35.8	0.50	50.7	16.4	22.9

Degradation in Soils

Degradation Parameters				Soil Properties							
								Percent (%)			
$t_{1/2}$	K_s	n	Initial conc.	Temp. °C	Moisture θ	pH	CEC	OC	Sand	Silt	Clay

Volatilization in Soils

Volatilization Parameters		Soil Properties				
K_w	K_wK_D	OC (%)	Moisture θ	Bulk Density	Temp. °C	Porosity

Compound Name: Terbufos [0,0-diethyl S-[(1,1-dimethylethylthio)methyl] phosphorodithioate]

Compound Properties: Structure

M.W.	262.4	Water solubility	5.07 ppm
M.P.		Log K_{ow}	3.68
B.P.		Chemical class	
Sp. gr.		Chemical reactivity	
Vapor pressure			

Adsorption in Soils

Adsorption Parameters			Soil Properties					
					Percent (%)			
K	N	K_{oc}	pH	CEC	OC	Sand	Silt	Clay
3.14	0.95	616	7.3	5.71	0.51	77	15	8
10.58	0.94	989	6.83	6.10	1.07	83	9	8
7.90	0.96	299	5.00	21.02	2.64	37	42	21
20.55	0.97	541	7.30	37.84	3.80	21	55	24
50.79	0.97	277	6.98	71.34	18.36	42	39	19

Degradation in Soils

Degradation Parameters				Soil Properties							
								Percent (%)			
$t_{1/2}$	K_s	n	Initial conc.	Temp. °C	Moisture Θ	pH	CEC	OC	Sand	Silt	Clay
12						7.1		1.5	20	50	30
11						8.0		2.0	30	20	50

Volatilization in Soils

Volatilization Parameters		Soil Properties				
K_w	K_wK_D	OC (%)	Moisture Θ	Bulk Density	Temp. °C	Porosity

Compound Name: 1,1,1,2-Tetrachloroethane

Compound Properties: Structure

M.W.	167.9	Water solubility	
M.P.		Log K_{ow}	
B.P.		Chemical class	
Sp. gr.		Chemical reactivity	
Vapor pressure			

Adsorption in Soils

Adsorption Parameters			Soil Properties					
					Percent (%)			
K	N	K_{oc}	pH	CEC	OC	Sand	Silt	Clay

Degradation in Soils

Degradation Parameters				Soil Properties							
								Percent (%)			
$t_{1/2}$	K_s	n	Initial conc.	Temp. °C	Moisture Θ	pH	CEC	OC	Sand	Silt	Clay

Volatilization in Soils

Volatilization Parameters		Soil Properties				
K_w	K_wK_D	OC (%)	Moisture Θ	Bulk Density	Temp. °C	Porosity

Compound Name: 1,1,2,2-Tetrachloroethane

Compound Properties:

Structure

M.W. 167.9
M.P. -42.5/43.8°C
B.P.
Sp. gr.
Vapor pressure 5 mm (20°C)
8.5 mm (30°C)

Water solubility 2900 mg/l (20°C)
Log K_{OW}
Chemical class
Chemical reactivity

Adsorption in Soils

Adsorption Parameters			Soil Properties					
					Percent (%)			
K	N	K_{OC}	pH	CEC	OC	Sand	Silt	Clay

Degradation in Soils

Degradation Parameters				Soil Properties							
								Percent (%)			
$t_{1/2}$	K_S	n	Initial conc.	Temp. °C	Moisture Θ	pH	CEC	OC	Sand	Silt	Clay

Volatilization in Soils

Volatilization Parameters		Soil Properties				
K_W	$K_W K_D$	OC (%)	Moisture Θ	Bulk Density	Temp. °C	Porosity

Compound Name: Tetrachloroethylene

Compound Properties: Structure

M.W. 165.8
M.P. -22.7°C
B.P.
Sp. gr.
Vapor pressure 14 mm (20°C)
24 mm (30°C)
45 mm (40°C)

Water solubility 150 mg/l (25°C)
Log K_{ow} 2.60
Chemical class
Chemical reactivity

Adsorption in Soils

Adsorption Parameters			Soil Properties					
					Percent (%)			
K	N	K_{oc}	pH	CEC	OC	Sand	Silt	Clay

Degradation in Soils

Degradation Parameters				Soil Properties							
								Percent (%)			
$t_{1/2}$	K_s	n	Initial conc.	Temp. °C	Moisture Θ	pH	CEC	OC	Sand	Silt	Clay

Volatilization in Soils

Volatilization Parameters		Soil Properties				
K_w	K_wK_D	OC (%)	Moisture Θ	Bulk Density	Temp. °C	Porosity

Compound Name: Toluene (methylbenzene)

Compound Properties: Structure

M.W. 92.1
M.P. -95.1°C
B.P.
Sp. gr.
Vapor pressure 10 mm (6.4°C)
22 mm (20°C)
40 mm (31.8°C)

Water solubility 470 mg/l (16°C)
515 mg/l (20°C)
Log K_{ow} 2.60
Chemical class
Chemical reactivity

Adsorption in Soils

Adsorption Parameters			Soil Properties					
					Percent (%)			
K	N	K_{oc}	pH	CEC	OC	Sand	Silt	Clay
3.52	1.008	37.4	5.4		9.40			
2.69	1.002	46.4	5.1		5.80			
0.90	0.996	155.2	4.3		0.58			

Degradation in Soils

Degradation Parameters				Soil Properties							
								Percent (%)			
$t_{1/2}$	K_s	n	Initial conc.	Temp. °C	Moisture Θ	pH	CEC	OC	Sand	Silt	Clay

Volatilization in Soils

Volatilization Parameters		Soil Properties				
K_w	K_wK_D	OC (%)	Moisture Θ	Bulk Density	Temp. °C	Porosity

Compound Name: 1,2,4-trichlorobenzene

Compound Properties:

Structure

M.W.	181.4	Water solubility	49 mg/l
M.P.	17°C	Log K_{ow}	4.02
B.P.	213°C	Chemical class	
Sp. gr.		Chemical reactivity	
Vapor pressure			

Adsorption in Soils

Adsorption Parameters			Soil Properties					
					Percent (%)			
K	N	K_{oc}	pH	CEC	OC	Sand	Silt	Clay
9.52	-	865.7		14	1.1	9	68	21

Degradation in Soils

Degradation Parameters				Soil Properties							
								Percent (%)			
$t_{1/2}$	K_s	n	Initial conc.	Temp. °C	Moisture Θ	pH	CEC	OC	Sand	Silt	Clay

Volatilization in Soils

Volatilization Parameters		Soil Properties				
K_w	K_wK_D	OC (%)	Moisture Θ	Bulk Density	Temp. °C	Porosity

Compound Name: 1,1,1-trichloroethane

Compound Properties: Structure

M.W.	133.4	Water solubility	4400 mg/l (20°C)
M.P.	-32°C	Log K_{OW}	2.49
B.P.		Chemical class	
Sp. gr.		Chemical reactivity	
Vapor pressure	100 mm (20°C) 155 mm (30°C)		

Adsorption in Soils

Adsorption Parameters			Soil Properties					
					Percent (%)			
K	N	K_{OC}	pH	CEC	OC	Sand	Silt	Clay

Degradation in Soils

Degradation Parameters				Soil Properties							
								Percent (%)			
$t_{1/2}$	K_s	n	Initial conc.	Temp. °C	Moisture Θ	pH	CEC	OC	Sand	Silt	Clay

Volatilization in Soils

Volatilization Parameters		Soil Properties				
K_W	$K_W K_D$	OC (%)	Moisture Θ	Bulk Density	Temp. °C	Porosity

Compound Name: 1,1,2-trichloroethane

Compound Properties: Structure

M.W.	133.4	Water solubility	4500 mg/l (20°C)
M.P.	-35/-36.7°C	Log K_{ow}	
B.P.		Chemical class	
Sp. gr.		Chemical reactivity	
Vapor pressure	19 mm (20°C) 32 mm (30°C) 40 mm (35°C)		

Adsorption in Soils

Adsorption Parameters			Soil Properties					
					Percent (%)			
K	N	K_{oc}	pH	CEC	OC	Sand	Silt	Clay

Degradation in Soils

Degradation Parameters				Soil Properties							
$t_{1/2}$	K_s	n	Initial conc.	Temp. °C	Moisture Θ	pH	CEC	Percent (%) OC	Sand	Silt	Clay

Volatilization in Soils

Volatilization Parameters		Soil Properties				
K_w	K_wK_D	OC (%)	Moisture Θ	Bulk Density	Temp. °C	Porosity

Compound Name: Trichloroethylene

Compound Properties: Structure

M.W.	131.5	Water solubility	1.10 mg/l (25°C)
M.P.	-87°C	Log K_{ow}	2.29
B.P.		Chemical class	
Sp. gr.		Chemical reactivity	
Vapor pressure	20 mm (0°C)		
	60 mm (20°C)		
	95 mm (30°C)		

Adsorption in Soils

Adsorption Parameters			Soil Properties					
					Percent (%)			
K	N	K_{oc}	pH	CEC	OC	Sand	Silt	Clay

Degradation in Soils

Degradation Parameters				Soil Properties							
								Percent (%)			
$t_{1/2}$	K_s	n	Initial conc.	Temp. °C	Moisture Θ	pH	CEC	OC	Sand	Silt	Clay

Volatilization in Soils

Volatilization Parameters		Soil Properties				
K_w	K_wK_D	OC (%)	Moisture Θ	Bulk Density	Temp. °C	Porosity

Compound Name: Trichloromethane

Compound Properties: Structure

M.W.	137.4	Water solubility	1100 mg/l (25°C)
M.P.	-111°C	Log K_{OW}	2.53
B.P.		Chemical class	
Sp. gr.		Chemical reactivity	
Vapor pressure	0.904 atm (20°C) 1.29 atm (30°C)		

Adsorption in Soils

Adsorption Parameters			Soil Properties					
					Percent (%)			
K	N	K_{OC}	pH	CEC	OC	Sand	Silt	Clay

Degradation in Soils

Degradation Parameters				Soil Properties							
								Percent (%)			
$t_{1/2}$	K_S	n	Initial conc.	Temp. °C	Moisture Θ	pH	CEC	OC	Sand	Silt	Clay

Volatilization in Soils

Volatilization Parameters		Soil Properties				
K_W	K_WK_D	OC (%)	Moisture Θ	Bulk Density	Temp. °C	Porosity

Compound Name: Trifluraline [α,α,α-trifluoro-2,6-dinitro-N, N-dipropyl-p-toluidine]

Compound Properties: Structure

M.W. 335.3
M.P. 48.5-49°C
B.P. 139-140°C
Sp. gr.
Vapor pressure 1.99×10^{-4} mm Hg (29.5°C)

Water solubility 1 mg/l (27°C)
Log K_{ow}
Chemical class
Chemical reactivity

Adsorption in Soils

Adsorption Parameters			Soil Properties					
					Percent (%)			
K	N	K_{oc}	pH	CEC	OC	Sand	Silt	Clay
2.73	1.15	75.7	7.3	54.7	3.87	18.4	45.3	38.3
0.46	1.05	50.7	5.6	0.90	0.90	65.8	19.5	14.7
0.24	1.06	43.2	5.6	0.56	0.56	93.8	3.0	3.2
1.60	1.18	177.8	7.4	0.50	0.50	50.7	16.4	22.9

Degradation in Soils

Degradation Parameters				Soil Properties							
								Percent (%)			
$t_{1/2}$	K_s	n	Initial conc.	Temp. °C	Moisture Θ	pH	CEC	OC	Sand	Silt	Clay
375[a]				15		7.6	11.4	0.93[b]	64	20	14
90[a]				30		7.6	11.4	0.93[b]	64	20	14
273[a]				15		6.6	19.4	1.28[b]	35	45	20
75[a]				30		6.6	19.4	1.28[b]	35	45	20

Volatilization in Soils

Volatilization Parameters		Soil Properties				
K_W	$K_W K_D$	OC (%)	Moisture Θ	Bulk Density	Temp. °C	Porosity

[a]Reported in months.
[b]Reported as OM.

Compound Name: Carbaryl [N-methyl-α-naphthylurethene]

Compound Properties: Structure

M.W. 201.2
M.P.
B.P.
Sp. gr.
Vapor pressure

Water solubility
Log K_{ow}
Chemical class
Chemical reactivity

Adsorption in Soils

Adsorption Parameters			Soil Properties					
					Percent (%)			
K	N	K_{oc}	pH	CEC	OC	Sand	Silt	Clay
4.87×10^{-6}	0.438	2.85×10^{-4}	8.1	32.0	1.71	51.5	25.3	23.2
2.41×10^{-6}	0.432	1.41×10^{-4}	8.1	32.0	1.71	51.5	25.3	23.2
8.98×10^{-4}	0.933	0.37	7.95	11.6	0.24	51.6	21.6	27.0
8.42×10^{-4}	0.893	0.35	7.95	11.6	0.24	51.6	21.6	27.0
109	0.97	249	6.1		43.7	52	34	14
4.3	0.96	265	6.6		1.62	71	22	7
2.9	0.98	200	6.8		1.45	56	30	14
3.4	1.08	829	7.0		0.41	91.5	1.5	7

Degradation in Soils

Degradation Parameters				Soil Properties							
								Percent (%)			
$t_{1/2}$	K_s	n	Initial conc.	Temp. °C	Moisture Θ	pH	CEC	OC	Sand	Silt	Clay

Volatilization in Soils

Volatilization Parameters		Soil Properties				
K_W	$K_W K_D$	OC (%)	Moisture Θ	Bulk Density	Temp. °C	Porosity

Compound Name: Chlordane

Compound Properties: Structure

M.W.	409.8	Water solubility	0.056-1.85 ppm
M.P.	103-108.8°C	Log K_{OW}	2.78
B.P.		Chemical class	
Sp. gr.		Chemical reactivity	
Vapor pressure	$1x10^{-5}$ torr		

Adsorption in Soils

Adsorption Parameters			Soil Properties					
					Percent (%)			
K	N	K_{OC}	pH	CEC	OC	Sand	Silt	Clay

Degradation in Soils

Degradation Parameters				Soil Properties							
								Percent (%)			
$t_{1/2}$	K_S	n	Initial conc.	Temp. °C	Moisture Θ	pH	CEC	OC	Sand	Silt	Clay
	0.00072										
	0.0020										

Volatilization in Soils

Volatilization Parameters		Soil Properties				
K_W	K_WK_D	OC (%)	Moisture Θ	Bulk Density	Temp. °C	Porosity

Compound Name: Diazinon [0,0-diethyl 0-(2-isopropyl-4-methyl-6-pyrimidyl phosphorothioate]

Compound Properties: Structure

M.W.	304.4	Water solubility	40 ppm
M.P.		Log K_{ow}	
B.P.	83-84°C	Chemical class	
Sp. gr.		Chemical reactivity	
Vapor pressure	4.1×10^{-4} (20°C) 1.1×10^{-3} (40°C)		

Adsorption in Soils

Adsorption Parameters			Soil Properties					
					Percent (%)			
K	N	K_{oc}	pH	CEC	OC	Sand	Silt	Clay
325	1.00	744	6.1		43.7	52	34	14
20.1	1.07	1240	6.6		1.62	71	22	7
5.2	0.99	359	6.8		1.45	56	30	14
6.7	1.08	1630	7.0		0.41	91.5	15	7

Degradation in Soils

Degradation Parameters				Soil Properties							
								Percent (%)			
$t_{1/2}$	K_s	n	Initial conc.	Temp. °C	Moisture Θ	pH	CEC	OC	Sand	Silt	Clay
	0.0151			25		6.7		1.80			
	0.0067			15		6.7		1.80			
	0.0242					4.3		1.80			
	0.0239					4.8		0.58			
	0.0239					6.5		1.16			
	0.0248					6.5		1.16			
	0.0189					4.0		1.22			
	0.0260					5.4		1.74			
	0.0166					5.6		4.18			
	0.0171					5.4		23.2			

Volatilization in Soils

Volatilization Parameters		Soil Properties				
K_W	$K_W K_D$	OC (%)	Moisture Θ	Bulk Density	Temp. °C	Porosity

Compound Name: Endrin [1,2,3,4,10,10-hexachloro-6,7-epoxy-1,4,4a,5,6,7,8,8a-octahydroexo-5,8-dimethanonaphthalene]

Compound Properties: Structure

M.W. 376.9
M.P. 200°C°C
B.P. 184°C
Sp. gr.
Vapor pressure $2x10^{-7}$ torr @ 25°C

Water solubility 200 ppb °25°C
Log K_{ow} 5.6 (calculated)
Chemical class
Chemical reactivity

Adsorption in Soils

Adsorption Parameters			Soil Properties					
					Percent (%)			
K	N	K_{oc}	pH	CEC	OC	Sand	Silt	Clay
11115	1.08	25435	6.1		43.7	52	34	14
222.5	0.99	13735	6.6		1.62	71	22	7
660.5	1.12	45572	6.8		1.45	56	30	14
90.4	1.03	22049	7.0		0.41	91.5	1.5	7

Degradation in Soils

Degradation Parameters				Soil Properties							
								Percent (%)			
$t_{1/2}$	K_s	n	Initial conc.	Temp. °C	Moisture Θ	pH	CEC	OC	Sand	Silt	Clay

Volatilization in Soils

Volatilization Parameters		Soil Properties				
K_w	K_wK_D	OC (%)	Moisture Θ	Bulk Density	Temp. °C	Porosity

Compound Name: Mevinphos

Compound Properties: Structure

M.W. 208.1
M.P.
B.P.
Sp. gr.
Vapor pressure

Water solubility
Log K_{ow}
Chemical class
Chemical reactivity

Adsorption in Soils

K	N	K_{oc}	pH	CEC	OC	Sand	Silt	Clay
Adsorption Parameters			Soil Properties		Percent (%)			
8.8	0.95	20.1	6.1		43.7	52	34	14

Degradation in Soils

$t_{1/2}$	K_s	n	Initial conc.	Temp. °C	Moisture Θ	pH	CEC	OC	Sand	Silt	Clay
Degradation Parameters				Soil Properties				Percent (%)			
	0.2936					5.4		0.23			

Volatilization in Soils

K_w	K_wK_D	OC (%)	Moisture Θ	Bulk Density	Temp. °C	Porosity
Volatilization Parameters		Soil Properties				

Compound Name: α-naphthol (1-hydroxynaphthalene, 1-naphthol)

Compound Properties: Structure

M.W.	144.2	Water solubility	866 mg/l
M.P.	96.1°C	Log K_{ow}	2.85
B.P.	184°C	Chemical class	
Sp. gr.		Chemical reactivity	
Vapor pressure	1 mm (94°C)		
	10 mm (142°C)		
	100 mm (206°C)		

Adsorption in Soils

Adsorption Parameters			Soil Properties					
					Percent (%)			
K	N	K_{OC}	pH	CEC	OC	Sand	Silt	Clay
257.4	0.441	12435	7.79	24	2.07			55.2
76.5	0.550	3355	2.74	19	2.28			31.0
933.3	0.310	129625	7.83	33	0.72			68.6
18.2	0.609	12133	8.32	4	0.15			6.8
821.5	0.222	746818	8.34	12	0.11			17.4
243.4	0.357	8007	6.90	12	3.04			52.6
24.7	0.563	5146[b]	4.54	19	0.48			63.6
761.1	0.318	80116	7.79	11	0.95			35.7
489.7	0.283	74197	7.76	15	0.66			39.5
650.6	0.313	50046[b]	5.50	9	1.30			28.6
100.4	0.501	5340	7.60	8	1.88			7.1
52.1	0.642	3120	7.55	9	1.67			21.2
295	0.387	12395	6.70	31	2.38			69.1
163.9	0.442	11074	7.75	21	1.48			42.9
76.7	0.549	8522b	6.40	3	0.90			22.5
243.4	0.357	20116	6.35	4	1.21			18.6

Degradation in Soils

Degradation Parameters				Soil Properties							
								Percent (%)			
$t_{1/2}$	K_s	n	Initial conc.	Temp. °C	Moisture Θ	pH	CEC	OC	Sand	Silt	Clay

Volatilization in Soils

Volatilization Parameters		Soil Properties				
K_W	K_WK_D	OC (%)	Moisture Θ	Bulk Density	Temp. °C	Porosity

[a]Originally reported in μmole instead of μg.
[b]Soils and the others are sediments.

Compound Name: PCBs [polychlorinated biphenyl]

Compound Properties: Structure

M.W.
M.P.
B.P.
Sp. gr.
Vapor pressure

Water solubility
Log K_{ow} 3.47 3.54
Chemical class
Chemical reactivity

Adsorption in Soils

Adsorption Parameters			Soil Properties					
					Percent (%)			
K	N	K_{oc}	pH	CEC	OC	Sand	Silt	Clay
1300	-	54167	6.3		2.4	55.0	45	<1.0
1290	-	92143	6.1		1.4	56.0	44	<1.0
1370	-	171250	6.4		0.8	-	-	-
620	-	155000	6.5		0.4	93.0	6.0	2.0
1250	-	42083	6.3		2.4	55.0	45	<1.0
1090	-	77857	6.1		1.4	56.0	44	<1.0
1210	-	151250	6.4		0.8	-	-	-
540	-	135000	6.5		0.4	93.0	6.0	2.0
1180	-	49167	6.3		2.4	55.0	45	<1.0
990	-	70714	6.1		1.4	56.0	44	<1.0
580	-	72500	6.4		0.8	-	-	-
420	-	105000	6.5		0.4	93.0	6.0	2.0
1270	-	52917	6.3		2.4	55.0	45	<1.0
1080	-	77143	6.1		1.4	56.0	44	<1.0
650	-	81250	6.4		0.8	-	-	-
510	-	127500	6.5		0.4	93.0	6.0	2.0

Degradation in Soils

Degradation Parameters				Soil Properties							
								Percent (%)			
$t_{1/2}$	K_s	n	Initial conc.	Temp. °C	Moisture Θ	pH	CEC	OC	Sand	Silt	Clay

Volatilization in Soils

Volatilization Parameters		Soil Properties				
K_w	K_wK_D	OC (%)	Moisture Θ	Bulk Density	Temp. °C	Porosity

Compound Name: 2,4,5-T [2,4,5 tricholorophenoxy acetic acid]

Compound Properties: Structure

M.W.	225.5	Water solubility	228 ppm (25°C)
M.P.	153°C	Log K_{OW}	0.60
B.P.		Chemical class	
Sp. gr.		Chemical reactivity	
Vapor pressure			

Adsorption in Soils

Adsorption Parameters			Soil Properties					
					Percent (%)			
K	N	K_{oc}	pH	CEC	OC	Sand	Silt	Clay
5.02[a]	0.81	167	5.9		3.0		61	27
0.74[a]	0.85	92.5	7.7		0.8		31	30

Degradation in Soils

Degradation Parameters				Soil Properties							
								Percent (%)			
$t_{1/2}$	K_s	n	Initial conc.	Temp. °C	Moisture Θ	pH	CEC	OC	Sand	Silt	Clay
	0.0289							1.91			
	0.0330							1.64			
	0.0330							2.20			
	0.0495					5.5		0.99			
	0.0414					5.8		1.10			

Volatilization in Soils

Volatilization Parameters		Soil Properties				
K_W	K_WK_D	OC (%)	Moisture Θ	Bulk Density	Temp. °C	Porosity

[a]Reported in terms μmoles.

Appendix B: Glossary

Aggregated media	Media with two sets of pores, macropore and micropores.
Aliphatic compounds	Organic compounds in which the characteristic groups are linked in a straight or branched carbon chain.
Alkali earth metals	The elements of group II of the periodic table, including beryllium, magnesium, and calcium.
Alkali metals	The elements of group I of the periodic table including sodium and potassium.
Anions	Atoms or molecules that are negatively charged.
Aromatic compounds	Organic compounds in which characteristic groups are linked to a particular six-member carbon ring which contains three double bonds.
Autochthonous microorganisms	That part of soil microbial community which is capable of utilizing the refractory humic substances. The characteristic of these organisms is slow and constant activity.
Autotrophic bacteria	Bacteria that fix the carbon they need for growth from carbon dioxide and (usually) obtain their energy from light (photosynthetic) or the oxidation of inorganic compounds (lithotrophic).
Breakthrough curves (BTC)	A curve of C/C_0 vs. V/V_0 where: C = concentration of column effluent C_0 = concentration of column influent V = cumulative outflow volume V_0 = total water volume in soil column
CA	Constituent attenuation achieved.
CAR	Constituent attenuation required at the site.

CASSACI	Constituent attenuation achieved based on assimilation capacity for degradation (SSAC) treatment for immobilization.
Cations	Atoms or molecules that are positively charged.
CEC	Cation exchange capacity, usually is expressed in milliequivalents of cations per 100 grams of soil.
Chelate	A ligand which contains two donor atoms so arranged that both coordinate simultaneously to the same central elements.
CLS	Constituent level in the contaminated soil.
Cometabolism (cooxidation)	Metabolism by microorganism of a compound that the cell is unable to use as a source of energy or an essential nutrient.
Complex ions	Designates all ions other than monoatomic ones.
Critical Depth	End of microbially active zone, depth of groundwater, or end of soil zone.
Desorption	The reverse process of sorption, i.e., solute concentration decrease at the soil water interface.
Direct photodegradation	The excitation of substrate molecules directly via absorption of light particles or quanta. Loss of absorbed energy by the substrate molecules is via photodegradation reaction of dissociation, dehalogenation, isomerization, oxidation or other deactivation routes.
EH	Excavation and hauling of contamined soils.
Equilibrium	A state in which a chemical reaction and its reverse reaction are taking place at equal rates, so that the concentration of reacting constituents is constant.
ESP	Exchangeable sodium ratio $ESP = \frac{[Na+](me/l)}{\text{concentration of total cations (me/l)}} \times 100$
ET	Exogeneous treatment for constituent attenuation.
Exchange (outer sphere complex)	Ions held loosely in the vicinity of a charged surface site by electrostatic forces.

Fick's Law	The concentrations tend to diffuse from high to low concentration regions. Mathematically Fick's Law is: $J = D(\partial c/\partial x)$ where J = mass flux D = constant (diffusion constant) x = space coordinate c = concentration
Fugacity	A compound's chemical potential energy or "escaping tendency" expressed in units of pressure. Fugacity concepts allow a steady-state equilibrium estimation of compound partitioning in multicomponent environments.
Heavy metals	Defined as metals that precipitated in acid solution by hydrogen sulfide (Cd, Cr, Cu, Hg, Mn, Ni, Pb, and Zn).
Heterocyclic compounds	Compounds that have ring structure in which one member is an element other than carbon.
Heterotrophic bacteria	Bacteria that derive their energy and carbon for survival and growth from decomposition of organic materials.
Hydrolysis	Reaction of an ion with water to form a weak acid or base.
Hydrophobic bonding	Partitioning between a polar solvent (e.g., water) and a nonpolar adsorbent surface (e.g., soil humus).
Hysteresis (nonsingularity)	Retention of a residual on soil at a given equilibrium concentration, when desorption occurs.
Immobilization	Irreversible sorption of pollutants thus preventing leaching.
Ion pairs	Short range interactions between closely adjacent ions usually in pairs (e.g., $CdCO_3^{\circ}$, $ZnHCO_3^{+}$, and $CaSO_4^{\circ}$).
Ionic strength	Total electrolyte content of a solution.

K_{oc} Normalized Freundlich adsorption constant with respect to organic carbon content of soil. K_{oc} is defined as:

$K_{oc} = (K/oc\%) \times 100$ or

$K_{oc} = (K_d/oc\%) \times 100$

where K is nonlinear Freundlich adsorption isotherm ($N \neq 1$), K_d is distribution coefficient for linear adsorption isotherm ($N=1$), and oc% is percent organic carbon content of soil. K_{oc} values for hydrophobic compounds are essentially independent of soil type. Therefore, K_{oc} values offer a simple method to compute adsorption constant in different soils by using organic carbon content of the particular soil. Adsorption constants are important in mobility assessment and leaching potential of hazardous constituents.

K_{om} Normalized adsorption constant with respect to organic matter content of the soil. K_{om} is defined as:

$K_{om} = (K/om\%) \times 100$ or

$K_{om} = (K_d/om\%) \times 100$

for definition of terms see K_{oc}. K_{om} is related to K_{oc} by the following equation

$K_{oc} = 1.724\ K_{om}$

K_{ow} Octanol/water partition coefficient which has been correlated with K_{oc} for some families of hydrophobic compounds. Therefore, K_{ow} may be used to estimate K_{oc} values for compouns with no K_{oc} data available.

K_w The water to air partition coefficient

K_w
= (concentration in water/concentration in air)
= [T(WS)]/16.04 P M = 1/H

where
T = temperature, °K
WS = water solubility (mg/l)
P = vapor pressure of pure chemical (mm Hg)
M = chemical molecular weight
H = Henry's law constant

K_w indicates the relative partitioning of a compound between the water and air phases.

K_wK_{oc}	Parameter used interchangeably with K_wK_d to described the relative volatility rate of a compound. A large K_wK_{oc} or K_wK_d value indicates a small volatility potential, a small K_wK_{oc} or K_wK_d value indicates a large volatility potential.
LC	Constituent concentration in the leachate.
LD	Depth of leachate or water applied (cm).
Ligands	A group of atoms, ions, or molecules bound to the central element (atom or ion).
LR	Constituent concentration in soil (kg/ha).
Macro cation	Major cation constituent of uncontaminated soils, including Ca, Mg, Na and K.
MEGs	Multimedia environmental goals.
Mole	Mass numerically equal to the molecular weight.
Mutualistic (symbiotic) conditions	Positive interaction conditions.
Oxidation	A process in which the oxidation state of a substance is increased due to loss of electrons.
Oxidizing agent	A substance that does the oxidation.
Permanent surface charge	Charge on surface that is independent of surrounding media pH.
Photodegradation	The process of utilizing incident solar radiation to carry out photochemical reactions. Photodegradation reactions may include dissociation, dehalogenation, isomerization, oxidation or other deactivation routes.
Polymerization	A chemical process in which a large number of relatively simple molecules combine to form a chain-like macromolecule.
Recalcitrant compounds	Compounds that resist biodegradation.
Redox	Reduction/oxidation.
Redox potential	The energy gained in transfer of 1 mole of electrons to an oxidant from H_2, expressed in volts.

Reducing agent	A substance that does the reduction.
Reduction	A process in which the oxidation state of a substance is decreased due to gain of electrons.
Retardation factor (R)	The ratio between average pore water velocity and average pollutant front velocity.
Sensitized photooxidation	The transfer of absorbed light energy from sensitizing compounds to molecular oxygen to form a highly reactive singlet oxygen species, followed by the oxidation of substrate molecule by singlet oxygen.
Sensitizing compounds	Organic compounds which absorb light in the visible region and transfer this absorbed energy to other molecules as they return to a low energy or ground state. Energy transfer to molecular oxygen yields reactive singlet oxygen. Examples of sensitizers include riboflavin and methylene blue.
Singlet oxygen	A highly reactive species of oxygen with a life time of 3 μ sec which is formed through the transfer of energy from a sensitizing compound to molecular oxygen.
SAR	Sodium adsorption ratio.
Solubility constant	The product of the concentration of the ions of a substance in a saturated solution of the substance.
Sorption	A physical/chemical process in which the increase of solute concentration evolves at the soil water interface. Sorption term is used without distinquishing whether the process is exchange, hydrophobic bonding, adsorption, etc.
Sorption isotherm	A functional relation between the amount adsorbed per unit weight of soil (S) and the equilibrium solution phase concentration (C). As an example, Freundlich isotherm which is expressed as: $S = K\ C^N$ where K and N are constants.
SSAC	Site/soil assimilative or attenuation capacity.

SSACD	Assimilative capacity for degradation rate (biodegradation and photodegradation) per unit soil weight.
SSWI	Site/soil/waste interaction.
Stoichiometric	Pertaining to weight relations in a chemical reaction.
$t_{1/2}$	Half-life time, the time needed for 50% of the initial concentration to degrade. Usually, the concept of $t_{1/2}$ is associated with first order degradation rate, in which case the $t_{1/2}$ is independent of the initial concentration. However, the $t_{1/2}$ can be generalized as for other rate orders, which is not independent at the initial concentration. $t_{1/2}$ is a simple tool for comparing degradation potential of different compounds.

Copyright Notice

Figure 3-4	From Fundamentals of Soil Science by H. D. Foth. Sixth Edition. Copyright 1978 by John Wiley and Sons, New York, NY. Used by permission of the publisher.
Table 3-11	From Fundamentals of Soil Science by H. D. Foth. Sixth Edition. Copyright 1978 by John Wiley and Sons, New York, NY. Used by permission of the publisher.
Figure 3-5	From The Nature and Properties of Soils by N. C. Brady. Eighth Edition. Copyright 1974 by Macmillan Publishing Co., Inc., New York, NY. Used by permission of the publisher.
Table 3-12	From Fundamentals of Soil Science by H. D. Foth. Sixth Edition. Copyright 1978 by John Wiley and Sons, New York, NY. Used by permission of the publisher.
Figure 3-6	From The Nature and Properties of Soils by N. C. Brady. Copyright 1974 by Macmillan Publishing Co., Inc., New York, NY. Used by permission of the publisher.
Figure 3-7	From Fundamentals of Soil Science by H. D. Foth. Sixth Edition. Copyright 1978 by John Wiley and Sons, New York, NY. Used by permission of the publisher.
Figure 3-8	From Fundamentals of Soil Science by H. D. Foth. Sixth Edition. Copyright 1978 by John Wiley and Sons, New York, NY. Used by permission of the publisher.
Figure 3-9	From "The Soil Environment" by J. L. Ahlrichs, p. 3-26 in Organic Chemicals in the Soil Environment edited by C.A.I. Goring and J. W. Hamaker. Copyright 1972 by Marcel Dekker, Inc., New York, NY. Used by permission of the publisher.
Table 3-13	From Concepts in Soil Science by M. G. Cook and G. M. Pace. Copyright 1978 by North Carolina State University, Raleigh, NC. Used by permission of the publisher.
Table 3-13a	From "Chemical, Spectroscopic, and Thermal Methods for the Classification and Characterization of Humic Substances" by M. Schnitzer, p. 293-310 in Humic Substances, Their Structure and Function in the Biosphere edited by D. Povoledo and H. L. Golterman. Copyright 1975 by Center for Agricultural

Publishing and Documentation, Wageningen. Used by permission of the publisher.

Figure 3-10 From The Nature and Properties of Soils by N. C. Brady. Eighth Edition. Copyright 1974 by Macmillan Publishing Co., Inc., New York, NY. Used by permission of the publisher.

Table 3-14 From "The Mechanism of Reduction in Waterlogged Paddy Soil" by Y. Takai and T. Kamura. Folia Microbiol. (Prague) 11: 135-145, 1966. Used by permission of Folia Microbiologica.

Table 3-15 From Soil Fertility and Fertilizers by S. L. Tisdale and W. L. Nelson. Copyright 1975 by Macmillan Publishing Co., Inc., New York, NY. Used by permission of the publisher.

Figure 3-11 From Physical Edaphology by S. A. Taylor and G. L. Ashcroft. Copyright 1972 by W. H. Freeman and Co., San Francisco, CA. Used by permission of the publisher.

Figure 3-12 From Fundamentals of Soil Science by H. D. Foth. Sixth Edition. Copyright 1978 by John Wiley and Sons, New York, NY. Used by permission of the publisher.

Table 3-17 From Soil and Water Conservation Engineering by G. O. Schwab, R. K. Frevert, T. W. Edminster, and K. K. Barnes. Second Edition. Copyright 1966 by John Wiley and Sons, Inc., New York, NY. Used by permission of the publisher.

Figure 3-14 From "Factors Affecting the Solubility of Trace Metals in Soils" by S. W. Mattigod, G. Sposito, and A. L. Page in Chemistry in the Soil Environment. ASA Special Publication 40, 1981. Used by permission of the American Society of Agronomy.

Figure 3-15 From Chemical Equilibria in Soils by W. L. Lindsay. Copyright 1979 by John Wiley and Sons, New York, NY. Used by permission of the publisher.

Figure 3-16 From "The Chemistry of Lead and Cadmium in Soil: Solid Phase Formation" by J. Santillan-Medrano and J. J. Jurinak. Soil Sci. Soc. Am. Proc. 29:851-856, 1975. Used by permission of the Soil Science Society of America.

Figure 3-17 From "The Chemistry of Lead and Cadmium in Soil: Solid Phase Formation" by J. Santillan-Medrano and J. J. Jurinak. Soil Sci. Soc. Am. Proc. 29:851-856, 1975. Used by permission of the Soil Science Society of America.

Table 3-29 From "Factors Affecting the Solubility of Trace Metals in Soils" by S. W. Mattigod, G. Sposito, and A. L. Page in Chemistry in the Soil Environment. ASA Special Publication 40, 1981. Used by permission of the American Society of Agronomy.

Table 3-33	From "Effect of Soil pH on Adsorption of Lead, Copper, Zinc and Nickel" by R. D. Harter. Soil Sci. Soc. Am. J. 47:47-51, 1983. Used by permission of the Soil Science Society of America.
Figure 3-24	From "Copper and Cadmium Adsorption Characteristics of Selected Acid and Calcareous Soils" by N. Cavallaro and M. B. McBride. Soil Sci. Soc. Am. J. 42:550-556, 1978. Used by permission of the Soil Science Society of America.
Figure 3-25	From "Copper and Cadmium Adsorption Characteristics of Selected Acid and Calcareous Soils" by N. Cavallaro and M. B. McBride. Soil Sci. Soc. Am. J. 42:550-556, 1978. Used by permission of the Soil Science Society of America.
Table 3-34	From "Copper and Cadmium Adsorption Characteristics of Selected Acid and Calcareous Soils" by N. Cavallaro and M. B. McBride. Soil Sci. Soc. Am. J. 42:550-556, 1978. Used by permission of the Soil Science Society of America.
Table 3-35	From "Soil Sorption of Nickel: Influence of Solution Composition" by R. S. Bowman, M. E. Essington, and G. A. O'Connor. Soil Sci. Soc. Am. J. 45:860-865, 1981. Used by permission of the Soil Science Society of America.
Table 3-36	From "Soil Sorption of Nickel: Influence of Solution Composition" by R. S. Bowman, M. E. Essington, and G. A. O'Connor. Soil Sci. Soc. Am. J. 45:860-865, 1981. Used by permission of the Soil Science Society of America.
Table 3-37	From "Influence of Solution Composition on Sorption of Zinc by Soil" by M. A. Elrashidi and G. A. O'Connor. Soil Sci. Soc. Am. J. 46:1153-1158, 1982. Used by permission of the Soil Science Society of America.
Figure 3-28	From "The Chemistry of Soil Processes" by D. J. Greenland and M. H. B. Hayes. John Wiley and Sons, Chichester, England, 1981. Used by permission of the publisher.
Figure 3-29	From "Behavior of Chromium in Soils: I. Trivalent Forms, II. Hexavalent Forms" by R. S. Bartlett and J. M. Kimble. J. Environ. Qual. 5:379-386, 1976. Used by permission of the Journal of Environmental Quality.
Figure 3-31	From "Behavior of Chromium in Soils: I. Trivalent Forms, II. Hexavalent Forms" by R. S. Bartlett and J. M. Kimble. J. Environ. Qual. 5:379-386, 1976. Used by permission of the Journal of Environmental Quality.
Table 3-38	From "Volatility of Mercury from Soils Amended with Various Mercury Compounds" by J. D. Rogers. Soil Sci. Soc. Am. J. 43:289-291, 1979. Used by permission of the Soil Science Society of America.

Figure 3-37	From "Solubility and Redox Criteria for the Possible Forms of Selenium in Soils" by H. R. Geering, E. E. Cory, L. H. P. Jones. and W. H. Alloway. Soil Sci. Soc. Am. Proc. 32:35-40, 1968. Used by permission of the Soil Science Society of America.
Table 3-39	From "Interaction of Organic Pesticides with Particular Matter in Aquatic and Soil Systems" by J. B. Weber in Fate of Organic Pesticides in the Aquatic Environment edited by R. F. Gould. Copyright 1972 by American Chemical Society, Washington, DC. Used by permission of the American Chemical Society. Used by permission of the author.
Figure 3-38	From "Interaction of Organic Pesticides with Particular Matter in Aquatic and Soil Systems" by J. B. Weber in Fate of Organic Pesticides in the Aquatic Environment edited by R. F. Gould. Copyright 1972 by American Chemical Society, Washington, DC. Used by permission of the American Chemical Society. Used by permission of the author.
Table 3-40	From "Interaction of Organic Pesticides with Particular Matter in Aquatic and Soil Systems" by J. B. Weber in Fate of Organic Pesticides in the Aquatic Environment edited by R. F. Gould. Copyright 1972 by American Chemical Society, Washington, DC. Used by permission of the American Chemical Society. Used by permission of the author.
Table 3-41	From "Interaction of Organic Pesticides with Particular Matter in Aquatic and Soil Systems" by J. B. Weber in Fate of Organic Pesticides in the Aquatic Environment edited by R. F. Gould. Copyright 1972 by American Chemical Society, Washington, DC. Used by permission of the American Chemical Society. Used by permission of the author.
Table 3-42	From "Interaction of Organic Pesticides with Particular Matter in Aquatic and Soil Systems" by J. B. Weber in Fate of Organic Pesticides in the Aquatic Environment edited by R. F. Gould. Copyright 1972 by American Chemical Society, Washington, DC. Used by permission of the American Chemical Society. Used by permission of the author.
Table 3-43	From "A Survey of Sorption Relationships for Reactive Solutes in Soils" by C. C. Travis and E. K. Etnier. J. Environ. Qual. 10:8-17, 1981. Used by permission of the Journal of Environmental Quality.
Figure 3-39	From "Studies in Adsorption. Part XI. A System of Classification of Solution Adsorption Isotherms and Its Use in Diagnosis of Adsorption Mechanisms and in Measurement of Specific Surface Areas of Solids" by H. Giles, T. H. MacEwan, S. N. Nakhaw, and D. Smith. J. Chem. Soc. 111:3973-3993, 1960. Used by permission of the Royal Society of Chemistry, London, England.

Figure 3-40	From "Interaction of Organic Pesticides with Particular Matter in Aquatic and Soil Systems" by J. B. Weber in Fate of Organic Pesticides in the Aquatic Environment edited by R. F. Gould. Copyright 1972 by American Chemical Society, Washington, DC. Used by permission of the American Chemical Society. Used by permission of the author.
Table 3-45	From "Factors Influencing the Adsorption, Desorption, and Movement of Pesticides in Soils" by G. W. Bailey and J. L. White. Residue Reviews 32:29-93. Copyright 1970 by Springer-Verlag New York, Inc. Used by permission of the publisher.
Figure 3-43	From "Estimation of Pesticide Retention and Transformation Parameters Required in Non-Point Source Pollution Models" by P. S. C. Rao and J. M. Davidson in Environmental Impact of Nonpoint Source Pollution edited by M. R. Overcash and J. M. Davidson. Copyright 1980 by Ann Arbor Science Pub., Inc., Ann Arbor, MI. Used by permission of the publisher.
Figure 3-44	From "Increased Cooxidative Biodegradation of Malathion in Soil via Cosubstrate Enrichment" by G. J. Merkel and J. J. Perry. J. Agr. Food Chem. 25:1011-1012, 1977. Used by permission of the American Chemical Society.
Figure 3-45	From "Increased Cooxidative Biodegradation of Malathion in Soil via Cosubstrate Enrichment" by G. J. Merkel and J. J. Perry. J. Agr. Food Chem. 25:1011-1012, 1977. Used by permission of the American Chemical Society.
Table 3-46	From "Biodegradation of Polynuclear Aromatic Hydrocarbon Pollutants by Soil and Water Microorganisms" by E. J. McKenna. Presented at the 70th Annual Meeting of the Am. Inst. Chem. Eng., New York, NY. Used by permission of the American Institute of Chemical Engineers.
Figure 3-46	From "Turnover of Pesticide Residues in Soil" by J. W. Hamaker and C. A. I. Goring in Bound and Conjugate Pesticide Residues edited by D. D. Kaufman et al. ACS Symposium Series 29. Copyright 1976 by American Chemical Society, Washington, DC. Used by permission of the American Chemical Society.
Figure 3-47	From "Turnover of Pesticide Residues in Soil" by J. W. Hamaker and C. A. I. Goring in Bound and Conjugate Pesticide Residues edited by D. D. Kaufman et al. ACS Symposium Series 29. Copyright 1976 by American Chemical Society, Washington, DC. Used by permission of the American Chemical Society.
Figure 3-49	From Design of Land Treatment Systems for Industrial Wastes - Theory and Practice by M. R. Overcash and D. Pal. Copyright 1979 by Ann Arbor Science Pub., Inc., Ann Arbor, MI. Used by permission of the publisher.